How
PRODUCTS
Are MADE

How
PRODUCTS
Are MADE

An Illustrated Guide to

Product Manufacturing

**Kristine M. Krapp and
Jacqueline L. Longe, Editors**

GALE

DETROIT · NEW YORK · TORONTO · LONDON

STAFF

Kristine M. Krapp and Jacqueline L. Longe, *Editors*
Nicole Beatty, *Associate Editor*

Maureen Richards, *Research Specialist*

Shanna Heilveil, *Production Assistant*

Cynthia D. Baldwin, *Art Director*
Bernadette M. Gornie, *Page Designer*
Tracey Rowens, *Cover Designer*

Pamela A. Reed, *Photography Coordinator*

Electronic illustrations provided by Electronic Illustrators Group of Fountain Hills, Arizona

ISBN 0-7876-1547-1
ISSN 1072-5091

Contents

Introduction

About the Series

Welcome to *How Products Are Made: An Illustrated Guide to Product Manufacturing*. This series provides information on the manufacture of a variety of items, from everyday household products to heavy machinery to sophisticated electronic equipment. You will find step-by-step descriptions of processes, simple explanations of technical terms and concepts, and clear, easy-to-follow illustrations.

Each volume of *How Products Are Made* covers a broad range of manufacturing areas: food, clothing, electronics, transportation, machinery, instruments, sporting goods, and more. Some are intermediate goods sold to manufacturers of other products, while others are retail goods sold directly to consumers. You will find items made from a variety of materials, including products such as precious metals and minerals that are not "made" so much as they are extracted and refined.

Organization

Every volume in this series is comprised of many individual entries, each covering a single product. Although each entry focuses on the product's manufacturing process, it also provides a wealth of other information: who invented the product or how it has developed, how it works, what materials are used, how it is designed, quality control procedures, byproducts generated during its manufacture, future applications, and books and periodical articles containing more information.

To make it easier for you to find what you're looking for, the entries are broken up into standard sections. Among the sections you will find are the following:

- Background
- History
- Raw Materials
- Design
- The Manufacturing Process
- Quality Control
- Byproducts/Waste
- The Future
- Where To Learn More

Every entry is accompanied by illustrations. Uncomplicated and easy to understand, these illustrations generally follow the step-by-step description of the manufacturing process found in the text.

Bold-faced items in the text refer to other entries in this volume.

A general subject index of important terms, processes, materials, and people is found at the end of the book. Bold-faced items in the index refer to main entries. Main entries from previous volumes are also included in the index. They are listed with the volume in which they appear and page number within that volume.

About this Volume

This volume contains essays on 100 products, arranged alphabetically, and 15 special boxed sections. Written by curators at the Henry Ford Museum & Greenfield Village in Dearborn, Michigan, these boxed sections describe interesting historical developments related to a product. Photographs are also included.

Contributors/Advisor

The entries in this volume were written by a skilled team of technical writers and engineers, often in cooperation with manufacturers and industry associations. The advisor for this volume was William S. Pretzer, a manufacturing historian and curator at the Henry Ford Museum & Greenfield Village in Dearborn, Michigan.

Suggestions

Your questions, comments, and suggestions for future products are welcome. Please send all such correspondence to:

The Editor
How Products Are Made
Gale Research
835 Penobscot Building
Detroit, MI 48226

Contributors

Donna R. Braden

Nancy EV Bryk

Chris Cavette

Michael Cavette

Loretta Hall

Susan Bard Hall

Gillian S. Holmes

Jennifer Swift Kramer

Leo Landis

Erik R. Manthey

Jeanine Head Miller

Mary F. McNulty

Kristin Palm

Annette Petrusso

Henry J. Prebys

William S. Pretzer

Cynthia Read-Miller

Perry Romanowski

Jason Rude

Randy Schueller

Rose Secrest

Laurel M. Sheppard

Angela Woodward

Acknowledgments

The editors would like to thank the following individuals, companies, and associations for providing assistance with Volume 3 of *How Products Are Made:*

Accordion: American Accordionists' Association, Mineola, NY; Closet Accordion Players of America, Fort Collins, CO. **Asphalt Paver:** Walter Meinert, Barber-Greene Co., Caterpillar Paving Products, Inc. **Ballpoint Pen:** Linda Kwong, Bic Corp.; Judy Morse, Pilot Corp. of America. **CAT Scanner:** American Society of Radiologic Technologists, Albuquerque, NM. **Concrete Block:** Besser Co., Alpena, MI; National Concrete Masonry Association, Herndon, VA. **Fiberboard:** Georgia Pacific Corp., Atlanta, GA; Niagara Fiberboard Inc., Lockport, NY; Composite Panel Association, Gaithersburg, MD. **Football:** The Sherry Group, Inc., Communications Specialists, Parsippany, NJ; Spaulding Sports Worldwide, Chicopee, MA; Wilson Sporting Goods Co., Chicago, IL. **Football Helmet:** The Sherry Group, Inc. Communications Specialists, Parsippany, NJ; Spaulding Sports Worldwide, Chicopee, MA; Larry Maddux, Parkview Manufacturing Corp., Salem, IL; Riddell, Chicago, IL. **Garbage Truck:** Joe Green, Dempster Equipment Co., Toccoa, GA. **Gas Mask:** Jim Taylor, Survivair Corp, Santa Anna, CA. **Golf Ball:** TopFlite, Spaulding Sports Worldwide, Chicopee, MA; MacGregor Golf Co., Albany, GA. **Gummy Candy:** Fay Romanowski, Farley Foods, Chicago, IL. **Harmonica:** Society for the Preservation and Advancement of Harmonicas, Troy, MI. **Iron-on Decal:** Beacon Graphics Systems, Somerville, NJ; Flexible Products, Inc., Kennesaw, GA. **Latex:** Textile Rubber & Chemical Co., Latex Div. **Shampoo:** Cosmetic, Toiletry, and Fragrance Association, Washington, DC; Society of Cosmetic Chemists, New York, NY.

The historical photographs for the entries on Ballpoint Pen, Camera, Concrete Block, Cultured Pearl, Flour, Graham Cracker, Heavy Duty Truck, Maple Syrup, Marshmallow, Piano, Sewing Machine, Sofa, Soy Sauce, Teddy Bear, and Television are from the collections of **Henry Ford Museum & Greenfield Village**, Dearborn, Michigan.

Electronic illustrations in this volume were created by **Electronic Illustrators Group** of Fountain Hills, Arizona.

Accordion

The accordion is a portable, freely vibrating reed instrument. It consists of a keyboard and bass casing that are connected by a collapsible bellows. Within the instrument are metal reeds, which create sound when air, generated by the movement of the bellows, flows around them and causes them to vibrate. The accordion is constructed from hundreds of pieces, and much of it is hand assembled. First constructed in the early nineteenth century, the accordion continues to evolve into an ever more versatile instrument.

History

Development of the accordion is generally thought to have been inspired by the Chinese *cheng*, the first known instrument to use a free vibrating reed to create sound. This instrument was invented approximately 5,000 years ago. It consists of a series of bamboo pipes, a resonator box, a wind chamber, and a mouthpiece. It has a shape that resembles a phoenix and was introduced to European musicians in 1777.

The first accordions were invented in the early nineteenth century. In Germany, Christian Buschmann introduced and patented an instrument called the "Handaeoline" in 1822. It had an expandable bellows, a portable keyboard, and a series of free vibrating reeds inside. Seven years later, Cyrillus Damian refined the instrument by adding four bass keys that produced chords. He was awarded a patent for this instrument, which he called an accordion.

Over the next several decades, various improvements were made to the accordion. One major modification was made in 1850, when the chromatic accordion was introduced. The early diatonic accordions produced different notes when the bellows were drawn opened and pressed closed. The chromatic versions produce the same note regardless of the action of the bellows. Steel reeds were incorporated into the instrument in 1857. As several early companies, such as Hohner, Soprani, and Dallape, began manufacturing the instrument in the 1860s, other changes were made. The addition of more bass keys was particularly important. By the early twentieth century, manufacturers had settled on a standard size and shape for the instruments, which eventually led to the modern accordion.

The incorporation of electronics into accordions began around World War II. At first, they were wired to allow a hookup through an electronic organ. Eventually, accordions were connected to electronic boxes of their own, allowing for sound generation, amplification, and speakers. A recent development is the inclusion of Musical Instrument Digital Interface (MIDI) systems with conventional accordions. Instruments which have MIDI contacts can be connected to any MIDI-compatible device, such as synthesizers, electronic pianos, and sound modules.

Background

The modern accordion has three primary sections, the expandable bellows and the two wooden end units called the treble and bass ends. The treble end of the accordion has a keyboard attached. The bass end contains finger buttons that play bass notes and chords. The reeds and electronic components are located on the inside of the bellows.

In Germany, Christian Buschmann introduced and patented an instrument called the "Handaeoline" in 1822. It had an expandable bellows, a portable keyboard, and a series of free vibrating reeds inside.

The accordion is called a free reed instrument because it uses free-standing reeds to produce sound, similar to the **harmonica.** The reeds are made up of metal strips that are riveted to either side of a rectangular metal plate. Below the reed is a slot which allows air to flow through the bellows. When air passes through this slot in the appropriate direction (first on the reed, then through the slot) the reed vibrates, producing the characteristic accordion sound. Air flowing in the opposite direction does not create sound because the reed only bends instead of vibrating. To conserve air, a plastic or leather flap is placed on the opposite side of the slot away from the reed, preventing air flow in this direction. Each reed is arranged on the treble or bass reed blocks and is associated with a key on the keyboard or various buttons on the bass keyboard. The length and thickness of the reed determines the pitch of the note it produces. For example, a long reed produces a lower note than a shorter reed. Depending on the type of accordion, there can be multiple treble and bass reed blocks.

The keyboard on the treble side of the accordion can have various configurations. A popular style is the piano-type keyboard. Each key is extended into the body of the accordion and has a device attached to it called a pallet, which covers the holes of the reed block. When the key is left undisturbed, the hole in the reed block is closed and air can not reach the reed below. Depressing the key causes the pallet to open, allowing air to flow to the reed and producing sound. The treble grill covers the action of the keys on the pallet. Another set of keys on this side of the accordion are the register keys. These keys operate slides that can bring in different sets of reeds, thus increasing the variation in tonal quality available.

Like the treble end, the bass end is also attached to the bellows by a wooden plate. It also has a keyboard and register buttons. The bass keyboard is much different than the treble keyboard, though. Instead of traditional piano-style keys, it is made up of buttons. These buttons are attached to a series of rods and levers that control the airflow through the bass end reed block. When a button is pushed, multiple notes, or chords, are sounded. The standard Stradella bass keyboard has as many as 120 buttons.

When a musician plays the accordion, the instrument is typically held in place by shoulder-straps as the player sits or stands. The bellows are pulled apart or pressed together as keys are depressed, and air is forced through the reeds. As keys on the treble keyboard are pressed by the right hand fingers, the reeds associated with those keys vibrate and produce specific notes. The left hand, which is primarily responsible for moving the bellows, also operates the bass notes, which provide accompanying sounds and major and minor chords.

While the chromatic accordions, such as the Piano accordion or the Continental chromatic accordion, are the standard instrument in the United States, other types are available. Diatonic accordions are still manufactured since they are often used in folk music. Common types include the melodeon, the continental club model, and the British chromatic. A recent invention are the electronic piano accordions. Two types are made, one which has a normal bellows and reeds, but also an electronic tone generator. Another is fully electronic, and the bellows only serves to control the instrument's volume.

Raw Materials

Literally hundreds of different parts are used to make an accordion. These can be made of a variety of materials, including wood, metal, plastic, and others. The larger parts of the instrument, such as the frame, pallets, and reed block are typically made of poplar wood. This wood is useful because it is sturdy and lightweight. The bellows are made of strong manilla cardboard which is folded and pleated. Leather gussets are put on each inner corner, and metal protectors are fashioned on the outer corners to strengthen and protect the bellows. The treble grill is a fretted metal cover. It is often decorated with the manufacturer's logo and is vented to allow greater sound production.

Metal is also used to make many of the smaller pieces. For example, the reeds are made of highly tempered, watch-spring steel. They are riveted to an aluminum alloy reed plate. To minimize the amount of air that goes through a slot, leather or plastic flaps are used to cover the side opposite the

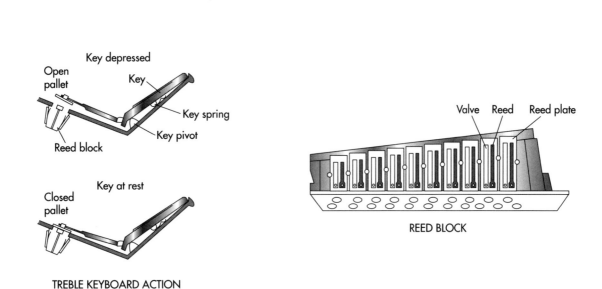

Key depressed

Open pallet

Key

Key spring

Key pivot

Reed block

Key at rest

Closed pallet

TREBLE KEYBOARD ACTION

Valve   Reed   Reed plate

REED BLOCK

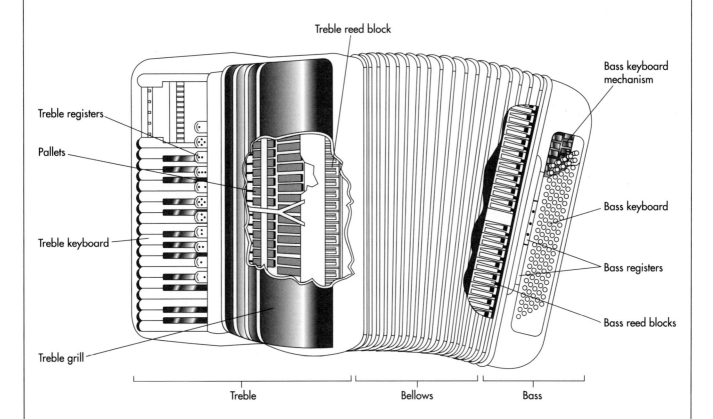

Treble reed block

Bass keyboard mechanism

Treble registers

Pallets

Treble keyboard

Bass keyboard

Bass registers

Treble grill

Bass reed blocks

Treble

Bellows

Bass

The modern accordion has three primary sections: the treble end, the bass end, and the expandable bellows section that connects them. As the player opens or closes the bellows and depresses keys on the treble keyboard, the depressed key opens a pallet, allowing air to flow into a slot in the reed block. The slot opens to a reed, a metal strip riveted to the reed plate. The air causes the reed to vibrate at a certain pitch. The bass keyboard is made up of buttons that are attached to a series of rods and levers that control the airflow through the reed block at the bass end. When a button is pushed, a bass note or chord is sounded.

reed. The rods which connect the bass buttons to the pallets and register slides that control the reed blocks on the inside of the accordion are also made of metal. The straps which allow the player to wear the accordion are made of strong leather and are usually padded. Leather or plastic washers are used throughout the instrument to keep it airtight. Additionally, wax is also used in some areas to prevent air leaks. Finally, the keys on the treble keyboard and the many buttons and switches are primarily made of plastic.

The Manufacturing Process

The manufacture of an accordion is not a completely automated process. In a sense, all accordions could be called handmade, since there is always some hand assembly of the small parts required. The general process involves making the individual parts, assembling the subsections, assembling the entire instrument, and final decorating and packaging.

Making the parts

1 Depending on the manufacturer, the parts of an accordion can be supplied by outside manufacturers or made in-house. The wooden parts are typically cut into the appropriate shapes by jigs and presses. This is an automated process in which the wood passes by these machines and is then cut. This system significantly simplifies the repeated making of identical components and ensures they are made with a high degree of precision.

2 The plastic components of the accordion such as the buttons and keys are usually produced by injection molding. In this approach, plastic is supplied as granules or powder and is fed into a large hopper. It is then heated, converting it into a liquid that can then be forcibly injected into a mold. After it cools, it solidifies and maintains its shape after the die is opened.

3 Various processes are used to construct the many metal parts of an accordion. These typically involve melting the metal to a liquid form, then placing it in a preformed mold. When the metal is cooled and hardened, the mold is opened and the part is

complete. In the case of the reeds, the metal is specially treated by a process called tempering. This reduces the hardness and brittleness of the metal, making it more ductile and tough.

Assembling the reeds, keyboards, and casings

4 After the individual parts are made, partial assembly begins. The reeds are riveted, or screwed, to an aluminum alloy reed plate. This plate has two slots, and the reeds are attached over each on opposite sides. On the open end of each slot, a leather or plastic valve is secured.

5 The reed plates are then arranged in a specific order and attached to a wooden reed block. Depending on the model, three or four of these blocks are put in the treble and bass side casings. The treble keyboard is attached to the reed block, and the bass side buttons and keyboards are also attached.

Final assembly

6 The bellows is typically supplied by outside manufacturers. It is made by folding and pleating strong cardboard and reinforcing it with leather and steel strips. The treble and bass casings are attached to it and sealed with wax to prevent air leakage.

Adding finishing touches

7 After the main parts have been assembled, various decorative finishing touches are put on the accordion. For example, the instrument is painted, the treble grill is attached, and the manufacturer's name is added. The accordion is then put in its case, packaged, and shipped to distributors for sale.

Quality Control

Quality control begins with the incoming raw materials and parts that are used to construct an accordion. If the manufacturer makes their own plastic parts, the starting resin is checked to ensure that it measures up to specifications related to physical appearance, melting point, and molecular weight, to name a few. Wood and steel are also checked similarly. For parts that are obtained by outside suppliers, the instrument manufacturer often relies on the supplier's quality control checks. During the production process, the quality of each accordion is verified by trained line inspectors and craftspersons. They perform visual inspections at each step and detect most flaws.

The Future

Improvements to the accordion have continued since its creation in the early nineteenth century. One recent invention is an accordion attachment that allows the musician to modify notes by "bending" the tone. This extra control over notes vastly improves on limitations of the current reed technology. Future instruments promise to utilize this type of technology and also to be more refined in the areas of tone and acoustic projection of sound, as well as in the areas of playability and handling. With the availability of increasingly lighter materials and the incorporation of computer technology, future accordions will certainly be much more versatile than their predecessors.

Where to Learn More

Books

Flynn, Ronald, Edwin Davison, and Edward Chavez. *The Golden Age of the Accordion*. Flynn Publications, 1992.

Liggett, Wallace. *The History of the Accordion in New Zealand*. New Zealand Accordion Association, 1992.

Periodicals

Spence, Scott. "The MIDI Polka." *Electronic Musician*, March 1, 1995, p.66.

Wallace, Len. "The Accordion—The People Instrument" *Canadian Folk Music Bulletin*, Fall 1992.

—*Perry Romanowski*

Acrylic Fingernail

Today acrylic chemistry is used to make a variety of nail enhancements, including nail tips, wraps, and sculpted nails.

Background

Acrylic nails are used to artificially enhance the appearance of natural fingernails. The term "acrylic nail" covers a range of product types, including press-on nails, nail tips, and sculpted nails. The first press-on acrylic nails were developed in the early 1970s; these were nail-shaped pieces of plastic that were glued on over natural nails. Early press-ons did not look natural and did nothing to strengthen real nails. Nonetheless, versions of this product could still be found on the market nearly 30 years later. Modern technology has advanced to allow development of more natural-looking nail enhancements which bond to the real nail. Early attempts at making these enhancements used the same plastic resin employed by the dental industry to make false teeth. This type of resin, known as an acrylic, is created by mixing a liquid and powder together to form a thick paste. The salon technician smooths the paste into place over the natural nail and allows it to dry. The resin then hardens to form a durable finish that is filed into the desired shape. Dental acrylic is no longer used because it caused allergic reactions in many people, but improvements in resin chemistry have essentially eliminated that problem. Today, acrylic chemistry is used to make a variety of nail enhancements, including nail tips, wraps, and sculpted nails. This article will focus on how sculpted acrylic nails are made.

Raw Materials

Monomer liquid

Artificial nail enhancements are made of acrylic plastic. Acrylic is the generic name given to the type of plastic made from a chemical called methacrylate. There are many types of acrylic resins based on different types of methacrylate molecules, but their chemistry is similar. The acrylic used in sculpted nails is formed by the reaction of a monomer liquid with a polymer powder. The monomers ("mono" meaning "one") contained in the liquid are microscopic chemical units which react together when mixed with chemicals in the powder. The monomers combine with one another in a head to tail fashion to form long fibers. These long chains of connected monomers are called polymers ("poly" meaning "many").

Polymer powder

A powdered polymer is then blended with the liquid to adjust the consistency of the plastic. The powdered polymer typically used in acrylics is polyethylmethylmethacrylate (PMMA). PMMA yields a very hard inflexible plastic, but it may be blended with softer polymers to improve its flexibility. When the polymer powder and monomer liquid are mixed, the polymer fibers react in a process known as crosslinking, forming a rigid netlike structure. The polymer strands will eventually dry to form a hard resin that can be made to resemble a fingernail.

Resin modifiers

Other ingredients are added to the monomer liquid and the polymer powder to control the properties of the resin. Crosslinking agents are used to hook the polymer chains together to make the plastic more rigid. The most common is ethylene glycol dimethacrylate. The polymer powder also carries an initiator, which starts the reaction that links the monomers together. A common initiator is benzoyl peroxide (BP), the same ingredient used in acne creams. When the liquid and powder are mixed together and applied to the client's fingers, the BP molecule is capable of exciting or energizing a monomer. Once energized, the monomers join together to form a polymer. Catalysts are also added to the formula to control the speed by which the initiator activates the reactions. A relatively small amount of catalyst is required to do the job, typically about only 1% of the monomer. Chemical inhibitors are added to the liquid monomer blend to prevent the monomers from reacting together prematurely, which turns the liquid into an unusable gel. Inhibitors help prolong the shelf life of the monomer solution. Plasticizers are used to improve resin performance. These liquids help lubricate the polymer chains so they are better able to resist breaking caused by stress.

Miscellaneous ingredients

A variety of ingredients are added to complete the resin. Dyes and pigments may be included to alter the resin's appearance. For example, titanium dioxide, a pigment commonly used in house paint, is used to whiten the nail and create a more natural appearance. It is also used to create special color effects like the white nail tips used in French manicures. Other colorants are added to give the polymer a pinkish or bluish color cast; these shades give a pleasing color to the nail bed. Flow agents are added to help control how polish spreads on the surface of the resin. Finally, color stabilizers are used to prevent yellowing. These materials absorb ultraviolet light that can cause discoloration of the resins.

Design

Every company that produces acrylic nail kits uses the same basic chemistry. However, each has designed its own formula with its distinct advantages and disadvantages. The real design work in creating acrylic nails is done by the nail technician. Each set of sculpted nails has its own idiosyncrasies which must be taken into account when designing the acrylic nails. In this sense, the technician designs the shape of the nail based on the requirements of the client.

The Manufacturing Process

Sculpted acrylic nails are not manufactured on an assembly line by a machine. Instead, as the name implies, they are "sculpted" by a nail technician. Each handcrafted nail is formed one at a time using a process which consists of the following steps: cleansing, priming, mixing, sculpting, and finishing.

Cleansing the nail

1 Before the new nail can be sculpted, the natural nail must be properly prepared. A nail bed cleanser is used to thoroughly clean the surface of the nail. These cleansers are typically solvents such as isopropyl alcohol, which dissolve oils and grease from the surface of the nail. They will also remove bacteria from the area to help reduce the chance of infection. Care must be used when applying these solvents because they may dry out the skin surrounding the cuticle. This occurs because the solvents also remove the skin's own natural moisturizing oils.

Priming the nail

2 After the nail bed has been cleansed, a primer is applied to the nail bed to make sure the acrylic will adhere properly. Primers are available in two types, non-etching and etching. The non-etching type works like double-sided tape; one side of the primer is very good at sticking to the natural nail, and the other end is equally attracted to the acrylic polymers used in the artificial nail. The etching type of primers are acids, such as methacrylic acid, which actually dissolve a thin layer of the nail itself. This etching process allows the acrylic to adhere to the nail better. The etching primers are more commonly used than non-etching. There is some debate regarding the proper use of etching primers; some chemists argue that the primer should dry thoroughly before applying the acrylic. Others believe that the acrylic should be applied while the primer is still wet to pull the acrylic deeper into the nail and anchor it more firmly.

Mixing the acrylic resin

3 The resin is made when the acrylic liquid is mixed with the acrylic powder. The

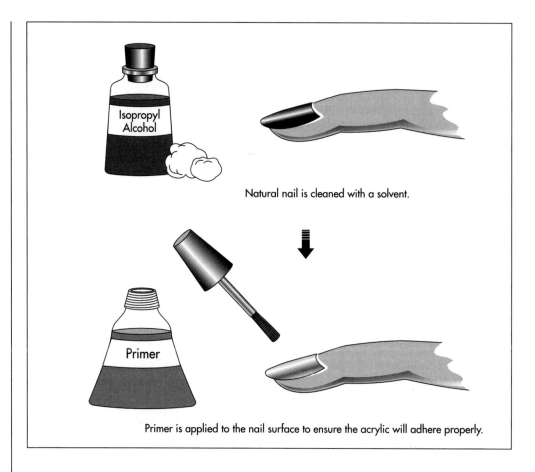

Natural nail is cleaned with a solvent.

Primer is applied to the nail surface to ensure the acrylic will adhere properly.

nail technician must work quickly with the resin once the liquid and powder are mixed. If not, the resin will harden before it can shaped into a nail and will not be useable.

Sculpting the nail

4 Before the resin is applied, a nail-shaped form is placed over each fingertip in order to hold the resin in place and ensure it takes the correct size and shape. These forms may be made of metalized foil or plastic. One common type consists of a thin metallic foil with an adhesive backing. The form is peeled off a roll (like a label) and carefully affixed to the fingers. The technician then applies the resin to the client's fingertips. The resin is sculpted to look as natural as possible before the resin hardens. The form is then removed.

Finishing the nail

5 After the acrylic dries, the new nail is filed and manicured to shape. Finally, coatings and polishes are used to complete the manicure. As the natural nail grows, further application of the liquid plastic is needed for the acrylic nail to maintain a regular contour. In some cases, an acetate tip is also applied to the end of the nail to provide a stronger base for the layers of acrylic resin.

Byproducts/Waste

Acrylic nail production creates chemical waste in the form of vapors, liquids, and solids. Any acrylic liquid or powder remaining at the end of day must be disposed of carefully in specially designated receptacles. Leftover liquids should never be returned to their bottles because they may have become contaminated during usage. Similar concerns apply to the powders used in nail production; care must be taken not to reuse powders that have been contaminated in any way. All the liquids used in nail production, including the monomer solutions, cleaning solvents, and primers, evaporate and give off vapors. Some of these vapors may be harmful if inhaled in sufficient quantities. Therefore it is crucial that salons have proper ventilation to ensure the technicians and their clients are not exposed to excessive concentrations of these vapors for

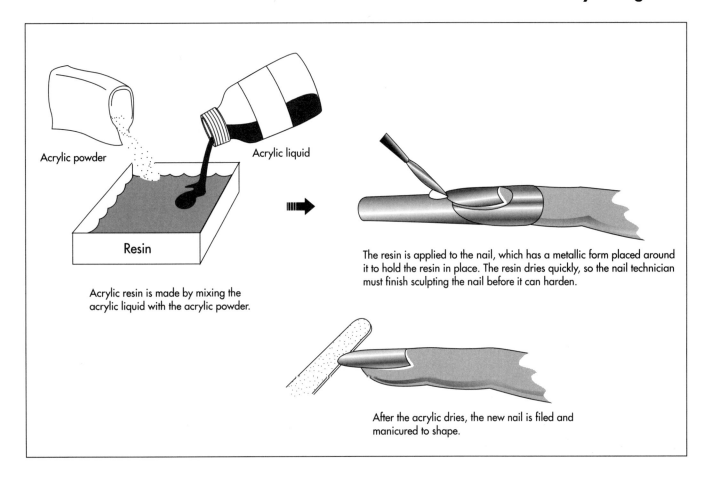

Acrylic powder

Acrylic liquid

Resin

Acrylic resin is made by mixing the acrylic liquid with the acrylic powder.

The resin is applied to the nail, which has a metallic form placed around it to hold the resin in place. The resin dries quickly, so the nail technician must finish sculpting the nail before it can harden.

After the acrylic dries, the new nail is filed and manicured to shape.

long periods of time. Likewise, technicians need to be protected against inhaling the dust that is created when they file acrylic nails. A simple dust mask is usually sufficient in this regard.

Quality Control

In the United States, the Food and Drug Administration deals with the safety of cosmetic chemicals. In the early 1970s, the FDA warned against the use of methyl methacrylate in nail care products because of consumers' allergic reactions to this monomer. The quality of the chemicals used in the nail-sculpting process should also be of key concern to nail technicians. The technicians must learn to recognize basic quality problems in the raw materials they use. For example, a common problem with the liquid monomer solution is caused by the early reaction of the monomers, creating a thickened gel rather than a thin liquid. This gelling essentially renders the monomers worthless. To prevent this problem, inhibitors are added to the monomer blend, ensuring the monomer solution will

maintain its quality for one or two years. Technicians must take care that all the raw materials are kept in usable condition; solvents should be tightly capped to prevent evaporation, powders must be kept clean and dry to prevent caking, and emulsion based products must be stored away from temperature extremes to avoid separation.

Another factor in assuring quality sculpted nails is to properly control the mixing and sculpting process. The liquid and powder must be added in the proper proportions, and they must be mixed to the correct consistency, or the strength of the nail will suffer. Typically, acrylic nails should contain 35-40% polymer. Too little polymer powder means less reinforcement and lower nail strength. Too much powder makes the nail too hard and brittle.

Quality problems can also arise if care is not taken in during nail sculpting. For example, ingredients in some of the adhesive chemicals can cause reactions with the nail and the nail bed. The nails can quickly become cracked, discolored, and misshapen; they may even be permanently disfigured.

Another problem is that tiny air pockets form between the real nails and the artificial ones. These spaces tend to become contaminated with fungi or bacteria. Proper mixing and molding techniques help prevent problems of this sort.

The Future

Future developments in nail enhancement production will be driven by several key factors. First, cosmetic chemists who formulate acrylic nail compounds are likely to continue to develop new formulations with improved properties, such as the ability to be molded more efficiently or to better resist chipping and breaking. Next, consumers and nail technicians may influence development of nail products by creating a demand for a particular style or for new types of nail enhancements. Finally, government regulation may impact the future of nail products. Various state legislatures have enacted laws regulating other aspects of the cosmetics industry. These laws limit a class of chemicals known as volatile organic compounds, which are used in hair sprays, antiperspirants, and other personal care products. If these same laws are expanded to include the nail industry, drastic changes in the way sculpted nails are made will be required.

Where to Learn More

Books

Chase, Deborah. *The New Medically Based No-Nonsense Beauty Book.* Henry Holt and Company, Inc., 1989.

Schoon, Douglas D. *Nail Structure and Product Chemistry.* Milady Publishing, 1996.

Periodicals

Anthony, Elizabeth. "ABC's of Acrylics," *NailPro Magazine,* October 1994.

Hamacker, Amy. "Dental Adhesives for Nails," *NailPro Magazine,* June 1994.

—*Randy Schueller*

Air Conditioner

Background

Residential and commercial space-cooling demands are increasing steadily throughout the world as what once was considered a luxury is now seemingly a necessity. Air-conditioning manufacturers have played a big part in making units more affordable by increasing their efficiency and improving components and technology. The competitiveness of the industry has increased with demand, and there are many companies providing air conditioning units and systems.

Air conditioning systems vary considerably in size and derive their energy from many different sources. Popularity of residential air conditioners has increased dramatically with the advent of central air, a strategy that utilizes the ducting in a home for both heating and cooling. Commercial air conditioners, almost mandatory in new construction, have changed a lot in the past few years as energy costs rise and power sources change and improve. The use of natural gas-powered industrial chillers has grown considerably, and they are used for commercial air conditioning in many applications.

Raw Materials

Air conditioners are made of different types of metal. Frequently, plastic and other non-traditional materials are used to reduce weight and cost. Copper or aluminum tubing, critical ingredients in many air conditioner components, provide superior thermal properties and a positive influence on system efficiency. Various components in an air conditioner will differ with the application, but usually they are comprised of stainless steel and other corrosion-resistant metals.

Self-contained units that house the refrigeration system will usually be encased in sheet metal that is protected from environmental conditions by a paint or powder coating.

The working fluid, the fluid that circulates through the air-conditioning system, is typically a liquid with strong thermodynamic characteristics like freon, hydrocarbons, ammonia, or water.

Design

All air conditioners have four basic components: a pump, an evaporator, a condenser, and an expansion valve. All have a working fluid and an opposing fluid medium as well.

Two air conditioners may look entirely dissimilar in both size, shape, and configuration, yet both function in basically the same way. This is due to the wide variety of applications and energy sources available. Most air conditioners derive their power from an electrically-driven motor and pump combination to circulate the refrigerant fluid. Some natural gas-driven chillers couple the pump with a gas engine in order to give off significantly more torque.

As the working fluid or refrigerant circulates through the air-conditioning system at high pressure via the pump, it will enter an evaporator where it changes into a gas state, taking heat from the opposing fluid medium and operating just like a heat exchanger. The working fluid then moves to the condenser, where it gives off heat to the atmosphere by condensing back into a liquid. After passing through an expansion valve, the working fluid returns to a low pressure

All air conditioners have four basic components: a pump, an evaporator, a condenser, and an expansion valve.

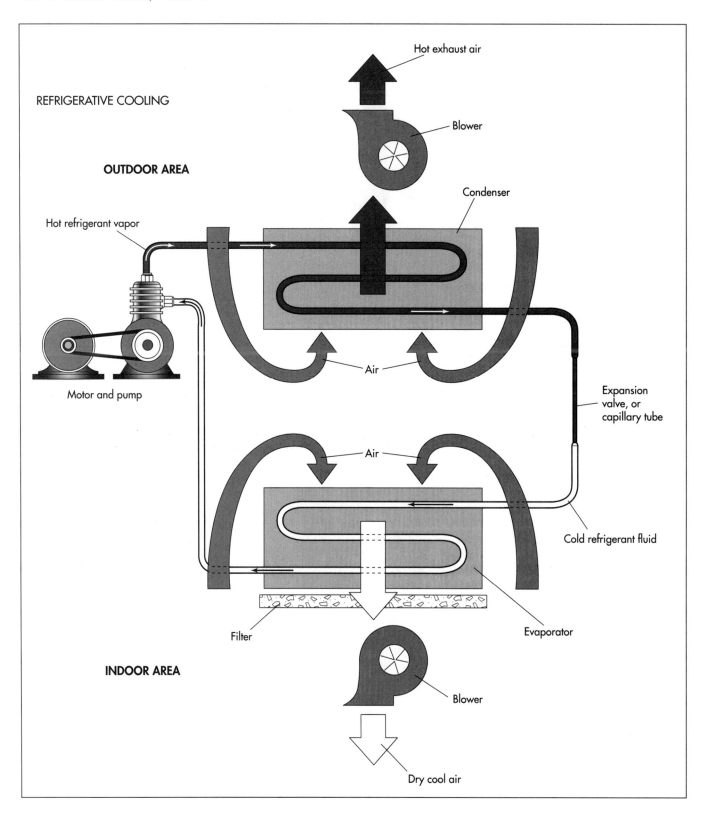

REFRIGERATIVE COOLING

OUTDOOR AREA

Hot exhaust air

Blower

Condenser

Hot refrigerant vapor

Motor and pump

Air

Expansion valve, or capillary tube

Air

Cold refrigerant fluid

Filter

Evaporator

INDOOR AREA

Blower

Dry cool air

state. When the cooling medium (either a fluid or air) passes near the evaporator, heat is drawn to the evaporator. This process effectively cools the opposing medium, providing localized cooling where needed in the building. Early air conditioners used freon as the working fluid, but because of the hazardous effects freon has on the environment, it has been phased out. Recent designs have met strict challenges to improve the efficiency of a unit, while using an inferior substitute for freon.

The Manufacturing Process

Creating encasement parts from galvanized sheet metal and structural steel

1 Most air conditioners start out as raw material, in the form of structural steel shapes and sheet steel. As the sheet metal is processed into fabrication cells or work cells, it is cut, formed, punched, drilled, sheared, and/or bent into a useful shape or form. The encasements or wrappers, the metal that envelopes most outdoor residential units, is made of galvanized sheet metal that uses a zinc coating to provide protection against corrosion. Galvanized sheet metal is also used to form the bottom pan, face plates, and various support brackets throughout an air conditioner. This sheet metal is sheared on a shear press in a fabrication cell soon after arriving from storage or inventory. Structural steel shapes are cut and mitered on a band saw to form useful brackets and supports.

Punch pressing the sheet metal forms

2 From the shear press, the sheet metal is loaded on a CNC (Computer Numerical Control) punch press. The punch press has the option of receiving its computer program from a drafting CAD/CAM (Computer Aided Drafting/Computer Aided Manufacturing) program or from an independently written CNC program. The CAD/CAM program will transform a drafted or modeled part on the computer into a file that can be read by the punch press, telling it where to punch holes in the sheet metal. Dies and other punching instruments are stored in the machine and mechanically brought to the punching arm, where it can be used to drive through the sheet. The NC (Numerically Controlled) press brakes bend the sheet into its final form, using a computer file to program itself. Different bending dies are used for different shapes and configurations and may be changed for each component.

3 Some brackets, fins, and sheet components are outsourced to other facilities or companies to produce large quantities. They are brought to the assembly plant only when needed for assembly. Many of the brackets are produced on a hydraulic or mechanical press, where brackets of different shapes and configurations can be produced from a coiled sheet and unrolled continuously into the machine. High volumes of parts can be produced because the press can often produce a complex shape with one hit.

Cleaning the parts

4 All parts must be completely clean and free of dirt, oil, grease, and lubricants before they are powder coated. Various cleaning methods are used to accomplish this necessary task. Large solution tanks filled with a cleaning solvent agitate and knock off the oil when parts are submersed. Spray wash systems use pressurized cleaning solutions to knock off dirt and grease. Vapor degreasing, suspending the parts above a harsh cleansing vapor, uses an acid solution and will leave the parts free of petroleum products. Most outsourced parts that arrive from a vendor have already been degreased and cleaned. For additional corrosion protection, many parts will be primed in a phosphate primer bath before entering a drying oven to prepare them for the application of the powder coating.

Powder coating

5 Before brackets, pans, and wrappers are assembled together, they are fed through a powder coating operation. The powder coating system sprays a paint-like dry powder onto the parts as they are fed through a booth on an overhead conveyor. This can be done by robotic sprayers that are programmed where to spray as each part feeds through the booth on the conveyor. The parts are statically charged to attract the powder to adhere to deep crevices and bends within each part. The powder-coated parts are then fed through an oven, usually with the same conveyor system, where the powder is permanently baked onto the metal. The process takes less than 10 minutes.

Bending the tubing for the condenser and evaporator

6 The condenser and evaporator both act as a heat exchanger in air conditioning systems and are made of copper or aluminum tubing bent around in coil form to maximize the distance through which the

Opposite page:
All air conditioners have four basic components: pump, evaporator, condenser, and expansion valve. Hot refrigerant vapor is pumped at high pressure through the condenser, where it gives off heat to the atmosphere by condensing into a liquid. The cooled refrigerant then passes through the expansion valve, which lowers the pressure of the liquid. The liquid refrigerant now enters the evaporator, where it will take heat from the room and change into a gaseous state. This part of the cycle releases cool air into the air-conditioned building. The hot refrigerant vapor is then ready to repeat the cycle.

working fluid travels. The opposing fluid, or cooling fluid, passes around the tubes as the working fluid draws away its heat in the evaporator. This is accomplished by taking many small diameter copper tubes bent in the same shape and anchoring them with guide rods and aluminum plates. The working fluid or refrigerant flows through the copper tubes and the opposing fluid flows around them in between the aluminum plates. The tubes will often end up with hairpin bends performed by NC benders, using the same principle as the NC press brake. Each bend is identical to the next. The benders use previously straightened tubing to bend around a fixed die with a mandrel fed through the inner diameter to keep it from collapsing during the bend. The mandrel is raked back through the inside of the tube when the bend has been accomplished.

7 Tubing supplied to the manufacturer in a coil form goes through an uncoiler and straightener before being fed through the bender. Some tubing will be cut into desired lengths on an abrasive saw that will cut several small tubes in one stroke. The aluminum plates are punched out on a punch press and formed on a mechanical press to place divots or waves in the plate. These waves maximize the thermodynamic heat transfer between the working fluid and the opposing medium. When the copper tubes are finished in the bending cell, they are transported by automatic guided vehicle (AGV) to the assembly cell, where they are stacked on the guide rods and fed through the plates or fins.

Joining the copper tubing with the aluminum plates

8 A major part of the assembly is the joining of the copper tubing with the aluminum plates. This assembly becomes the evaporator and is accomplished by taking the stacked copper tubing in their hairpin configuration and mechanically fusing them to the aluminum plates. The fusing occurs by taking a bullet, or mandrel, and feeding it through the copper tubing to expand it and push it against the inner part of the hole of the plate. This provides a thrifty, yet useful bond between the tubing and plate, allowing for heat transfer.

9 The condenser is manufactured in a similar manner, except that the opposing medium is usually air, which cools off the copper or aluminum condenser coils without the plates. They are held by brackets which support the coiled tubing, and are connected to the evaporator with fittings or couplings. The condenser is usually just one tube that may be bent around in a number of hairpin bends. The expansion valve, a complete component, is purchased from a vendor and installed in the piping after the condenser. It allows the pressure of the working fluid to decrease and re-enter the pump.

Installing the pump

10 The pump is also purchased complete from an outside supplier. Designed to increase system pressure and circulate the working fluid, the pump is connected with fittings to the system and anchored in place by support brackets and a base. It is bolted together with the other structural members of the air conditioner and covered by the wrapper or sheet metal encasement. The encasement is either riveted or bolted together to provide adequate protection for the inner components.

Quality Control

Quality of the individual components is always checked at various stages of the manufacturing process. Outsourced parts must pass an incoming dimensional inspection from a quality assurance representative before being approved for use in the final product. Usually, each fabrication cell will have a quality control plan to verify dimensional integrity of each part. The unit will undergo a performance test when assembly is complete to assure the customer that each unit operates efficiently.

The Future

Air conditioner manufacturers face the challenge of improving efficiency and lowering costs. Because of the environmental concerns, working fluids now consist typically of ammonia or water. New research is under way to design new working fluids and better system components to keep up with rapidly expanding markets and applications. The competitiveness of the industry should remain strong, driving more innovations in manufacturing and design.

Where to Learn More

Other

"HVAC Online." 1997. http://www.hvaconline.com (July 9, 1997).

"Cold Point Manufacturing." 1997. http://www.coldpoint.com/index3.htm (July 9, 1997).

—Jason Rude

Airship

During the first half of the twentieth century, Goodyear manufactured over 300 blimps, more than any other airship manufacturer.

Background

An airship is a large lighter-than-air gas balloon that can be navigated by using engine-driven propellers. There are three types of airships: rigid (has an internal metal frame to maintain the envelope's shape); semi-rigid (rigid keels run the length of the envelope to maintain its shape); and non-rigid (internal pressure of the lifting gas, usually helium, maintains the envelope's shape). This essay focuses on non-rigid airships (commonly called blimps) because they are the primary type of airship in general use today.

History

The history of airships begins, like the history of hot air balloons, in France. After the invention of the **hot air balloon** in 1783, a French officer named Meusnier envisioned an airship that utilized the design of the hot air balloon, but was able to be navigated. In 1784, he designed an airship that had an elongated envelope, propellers, and a rudder, not unlike today's blimp. Although he documented his idea with extensive drawings, Meusnier's airship was never built.

In 1852, another Frenchman, an engineer named Henri Giffard, built the first practical airship. Filled with hydrogen gas, it was driven by a 3 hp steam engine weighing 350 lb (160 kg), and it flew at 6 mi/hr (9 km/hr). Even though Giffard's airship did achieve liftoff, it could not be completely controlled.

The first successfully navigated airship, *La France,* was built in 1884 by two more Frenchman, Renard and Krebs. Propelled by a 9 hp electrically-driven airscrew, *La France* was under its pilots' complete control. It flew at 15 mi/hr (24 km/hr).

Military airships

In 1895, the first distinctly rigid airship was built by German David Schwarz. His design led to the successful development of the zeppelin, a rigid airship built by Count zeppelin. The zeppelin utilized two 15 hp engines and flew at a speed of 25 mi/hr (42 km/hr). Their development and the subsequent manufacture of 20 such vessels gave Germany an initial military advantage at the start of World War I.

It was Germany's successful use of the zeppelin for military reconnaissance missions that spurred the British Royal Navy to create its own airships. Rather than duplicating the design of the German rigid airship, the British manufactured several small non-rigid balloons. These airships were used to successfully detect German submarines and were classified as "British Class B" airships. It is quite possible this is where the term blimp originates—"Class B" plus limp or non-rigid.

Passenger-carrying airships

During the 1920s and 1930s, Britain, Germany, and the United States focused on developing large, rigid, passenger-carrying airships. Unlike Britain and Germany, the United States primarily used helium to give their airships lift. Found in small quantities in natural gas deposits in the United States, helium is quite expensive to make; however, it is not flammable like hydrogen. Because of the cost involved in its manufacture, the United States banned the exportation of helium to other countries, forcing Germany and Britain to rely on the more volatile hydrogen gas. Many of the large passenger-carrying airships using hydrogen instead of helium met with disaster, and because of such large losses of life, the heyday of the large passenger-carrying airship came to an abrupt end.

The first passenger-carrying non-rigid airship was invented in 1898 by Alberto Santos Dumount, a citizen of Brazil living in Paris. Under a sausage-shaped balloon with a ballonet or collapsible air bag inside, Dumount attached a propeller to his motorcycle's engine. He used both air and hydrogen, not helium, to lift the blimp.

The non-rigid airship of the 1940s and 1950s

After the rigid airship disasters of the 1920s and '30s, the United States as well as other countries refocused their attention on the non-rigid airship as a scientific/military tool. Aerial surveillance became the most common and successful use of the blimp. In the 1940s and '50s, blimps were used as early warning radar stations for merchant fleets along the eastern seaboard of the United States. They were also used and are still used in scientific monitoring and experiments.

Although as a company it no longer makes airships, Goodyear is a name sononymous with the manufacture of blimps. During the first half of the twentieth century, Goodyear manufactured over 300 blimps, more than any other airship manufacturer. Goodyear blimps were primarily used by the U.S. Army and Navy for aerial surveillance.

Modern resurgence of the non-rigid airship

Today, non-rigid airships are known more for their marketing power than for their surveillance capabilities. Blimps have been used commercially in the United States since about 1965. Advertising blimps measure about 150,000 cu ft (4,200 cu m). Since blimps can hover over one space and can be viewed over a large expanse with very little noise disturbance, they are excellent mediums for advertising at large outdoor events.

The use of the night billboard on blimps has been quite an advertising fad. The sign is a matte of multicolor incandescent lamps permanently fixed to the sides of the airship envelope, and it can be programmed to spell out different messages. Originally, the signs were developed by electromechanical relay. Now they are stored on magnetic tape, developed by composing equipment on the ground, which are fed into an airborne reader. The taped information is played back through a computer to the lamp driver circuits. The displayed messages can be seen over long distances. In the late 1980s, the use of blimps in advertising exploded. Its popularity does not seem to have let up.

Raw Materials

The envelope is usually made of a combination of man-made materials: Dacron, polyester, Mylar, and/or Tedlar bonded with Hytrel. The high-tech, weather-resistant plastic film is laminated to a rip-stop polyester fabric. The envelope's fabric also protects against ultraviolet light. Usually the envelope is smaller than the bladder to ensure that the envelope takes the load when the blimp is fully inflated. The bladder is made of a thin leak-resistant polyurethane plastic film.

Ballonets are usually made of a fabric lighter than the envelope's because they only retain gas tightness and do not have to withstand normal main envelope pressures. Air scoops channel air to the ballonets.

Blimps obtain much of their lift from lighter-than-air gases, most commonly helium, inside the envelope.

Most of the metal used on the blimp is riveted aircraft aluminum.

Earlier cars were fabric-covered tubing framework. Today's gondolas are made of metal monocoque design.

The nose cone is made of metal, wood, or plastic battens, laced to the envelope.

Design

The main body of the blimp is made up of an inner layer, the bladder, and an outer layer, the envelope. The bladder holds the

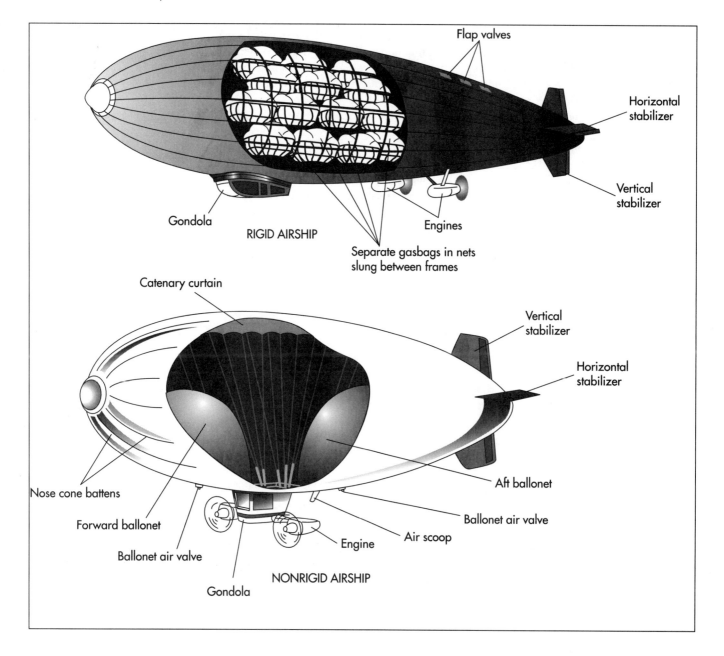

Flap valves

Horizontal stabilizer

Vertical stabilizer

Gondola

RIGID AIRSHIP

Engines

Separate gasbags in nets slung between frames

Catenary curtain

Vertical stabilizer

Horizontal stabilizer

Nose cone battens

Forward ballonet

Ballonet air valve

Gondola

Engine

Air scoop

Ballonet air valve

Aft ballonet

NONRIGID AIRSHIP

helium. Because the bladder is not resistant to punctures, it is protected by the envelope.

Inside the envelope are catenery curtains, which support the weight of the car by distributing the loads imposed by the airship into the fabric of the main envelope. Catenery curtains all consist of cable systems attached to the car, which terminate in the fabric curtains.

The envelope's shape is maintained by regulating internal pressure of helium gas inside. Within the bladder are one or more air cells/balloons called ballonets. These are filled with air (as opposed to the rest of the bladder which is filled with helium) and are attached to the sides or bottom of the blimp. The ballonets expand and contract to compensate for changes in helium volume due to varying temperature and altitude. The pilot has direct control of the ballonets via air valves.

The nose cone serves two purposes. It provides the point of attachment for mast mooring and adds rigidity to the nose (which encounters the greatest dynamic pressure loads in flight). On the ground, the inflated blimp is secured to a stationary pole called the mooring mast. The rigid nose dish is attached to the mooring mast. The secured blimp can move freely around the mast with wind changes. There are

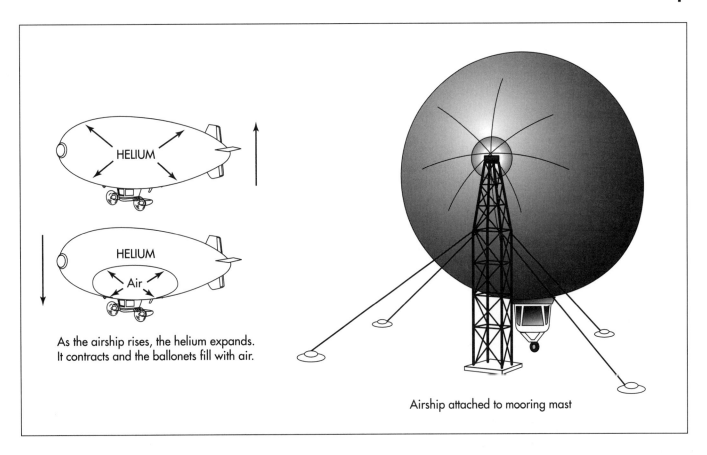

As the airship rises, the helium expands. It contracts and the ballonets fill with air.

Airship attached to mooring mast

nose lines attached to the nose dish used by the ground crew to maneuver blimp during takeoffs and landings.

Airship tail surfaces come in three configurations: the cruciform (+), the X, and the inverted Y. These tails are made up of a fixed main surface and a controllable smaller surface on the aft end. These surfaces weigh only 0.9 lb per sq ft (4.4 kg per sq m). Tail fins control flight direction. They are anchored at the rear of the ship and are supported by guide wires. The elevators and rudders also help to guide the blimp's movement and are mounted to the fin's edges with hinges.

The airship car, or gondola, is similar to conventional aircraft construction. The gondola contains a number of lead shot bags which are constantly adjusted based on the crew's analysis. The gondola is attached to the blimp by either an internal load curtain or externally, by being attached to envelope sides.

Inside the gondola, there a series of controls: the overhead control panel containing controls for communications, fuel, and electrical systems; throttles to regulate engine

speed and propeller pitch controls to regulate angles at which propeller blades "bite" the air; fuel mixture and heat controls to regulate the degree to which fuel is mixed with air in engine; temperature controls to prevent icing; envelope pressure controls to regulate helium and ballonet air pressure; communication equipment; main instrument panel; rudder pedals to control right/left direction of blimp; elevator wheels to control up/down direction of blimp; navigational instruments; and color weather radar.

The Manufacturing Process

Envelope

1 The envelope is made of patterns of fabric panels. Two or three plies of cloth are impregnated with an elastomer. One of these plies is placed in a bias direction with respect to others. The pieces of the envelope can be put together in a number of ways. They can be cemented and sewn together, or heat-welded (heat-sealed).

2 The outside of the envelope is coated with aluminized paint for protection

against sunlight. The envelope will have the required shape when filled with gas.

3 The catenery curtains are attached on to the main envelope proper in a similar manner.

4 The bladder is formed from strips that are welded together.

5 The tail construction consists mostly of lightweight metal structural beams covered with doped fabric. They are held to the envelope by cables which distribute the load into fabric patches cemented or heat-welded to the envelope proper. They are not directly attached to the envelope in the manufacturing process, but put on when the blimp is inflated.

Gondola

6 The frame of the gondola is made of material similar to the tail construction, covered with doped fabric.

Inflation

The erection of the blimp takes only a short amount of time. (The following is only one method of inflation. There are variations on this method.)

7 The envelope is spread out on the floor of the airship hangar, with a net placed over it. This net is held down by sandbags. Gas is fed into the enveloped from tank cars each containing 200,000 cu ft (5,700 cu m) of 99.9% pure helium compressed to 2100 psi (14.5 megapascals). The net is allowed to slowly rise, with the envelope underneath it.

8 Fins, nose cone, battens, air valves, and helium valves are attached while the envelope is still near the ground. After these parts are attached, the envelope is allowed to rise high enough to permit rolling the gondola underneath it. After the gondola is attached, the net is removed and the airship is rigged for flight.

Shipping

9 When transported, the uninflated fabric envelopes can be folded, shipped, and stored in a space that takes up less than 1% of its inflated volume. This feature makes

the non-rigid airship more practical than a rigid one.

Quality Control

A blimp requires a big crew, especially on the ground. Pilots must be certified in planes or helicopters and undergo special lighter-than-air pilot training. The FAA requires a separate license to command a blimp. As of 1995, there were only about 30 active blimp pilots in the world. Many blimps require 24-hour monitoring. The envelope and ballast are checked every hour to make sure the proper equilibrium is maintained.

The Future

Propulsive efficiency will be improved by using lightweight, two-stroke aviation diesel engines, gas turbines, or solar energy. New bow and stern thrusters will be developed to improve maneuverability. New lightweight plastics might change the hull design. More lightweight, high strength materials will probably be developed and inevitably improve the overall design and function of the airship. The Pentagon and the U.S. Navy have renewed interested in developing blimps for various defense, missile surveillance, radar-surveillance platforms, and reconnaissance purposes.

Where to Learn More

Books

Botting, Douglas, et al. *The Giant Airships.* Time-Life Books, 1980.

Ventry, Lord and Eugene M. Kolesnik. *Airship Saga: The history of airships seen through the eyes of the men who designed, built and flew them.* Blandford Press, 1982.

—*Annette Petrusso*

Animation

Background

Animation is a series of still drawings that, when viewed in rapid succession, gives the impression of a moving picture. The word animation derives from the Latin words *anima* meaning life, and *animare* meaning to breathe life into. Throughout history, people have employed various techniques to give the impression of moving pictures. Cave drawings depicted animals with their legs overlapping so that they appeared to be running. The properties of animation can be seen in Asian puppet shows, Greek bas-relief, Egyptian funeral paintings, medieval stained glass, and modern comic strips.

In 1640, a Jesuit monk named Althanasius Kircher invented a "magic lantern" that projected enlarged drawings on a wall. A fellow Jesuit, Gaspar Schott, developed this idea further by creating a straight strip of pictures, a sort of early filmstrip, that could be pulled across the lantern's lens. Schott further modified the lantern until it became a revolving disk. A century later, in 1736, a Dutch scientist named Pieter Van Musschenbroek created a series of drawings of windmill vanes that, when projected in rapid succession, gave the illusion of the windmill circling around and around.

The magic lantern became a popular form of entertainment. Traveling entertainers, visiting the villages and towns of Europe, included it in their shows. In London, the Swiss-born physician and scholar Peter Mark Roget, most famous for compiling the *Thesaurus of English Words and Phrases*, was fascinated by the scientific phenomenon at play and wrote an essay entitled "Persistence of Vision with Regard to Moving Objects" that was widely read and used as a basis for subsequent inventions. One of the first was the thaumatrope, developed in the 1820s by John Paris, also an English doctor. The thaumatrope was simply a small disk with a different image drawn on either side. Strings were knotted onto two edges so that the disk could be spun. As the disk twirled around, the two images appeared to blend. For example, a monkey on one side appeared to sit inside the cage on the opposite side.

The next major innovation was the phenakistoscope, created by Joseph Plateau, a Belgian physicist and doctor. Plateau's contribution was a flat disk perforated with evenly spaced slots. Figures were drawn around the edges, depicting successive movements. A stick attached to the back allowed the disk to be held at eye level in front of a mirror. The viewer then spun the disk and watched the reflection of the figures pass through the slits, once again giving the illusion of movement.

In Austria, Simon Ritter von Stampfer was toying with the same idea and called his invention a stroboscope. A number of other scopes followed, culminating in the zoetrope, created by William Horner. The zoetrope was a drum-shaped cylinder that was open at the top with slits placed at regularly spaced intervals. A paper strip with a series of drawings could be inserted inside the drum, so that when it was spun the images appeared to move.

By 1845, Baron Franz von Uchatius invented the first movie projector. Images painted on glass were passed in front of the projected light. Forty-three years later, George

Creating an animated short or full-length feature is a long, tedious process. Extremely labor-intensive, the average short cartoon has approximately 45,000 separate frames.

Eastman introduced celluloid film, a strip of cellulose acetate coated with a light-sensitive emulsion that retained and projected images better than those painted on glass. The first animated cartoon *Humorous Phases of Funny Faces* by J. Stuart Blackton, of the *New York Evening World,* was shown in the United States in 1906. Two years later, French animator Emile Cohl followed suit with *Phantasmagorie.* Winsor McCay introduced *Gertie the Dinosaur* in 1911. Other cartoonists who brought their characters to the screen included George McManus (Maggie and Jiggs) and Max Fleischer (Betty Boop and Popeye). By 1923, Walt Disney, the world's most famous animator, began turning children's stories into animated cartoons. Mickey Mouse was introduced in *Steamboat Willie* in 1928. Disney's first animated full-length film, *Snow White and the Seven Dwarfs,* debuted in 1937.

Yellow Submarine, a 1968 animated film starring the Beatles, featured the process of pixilation, in which live people are photographed in stop-motion to give the illusion of humanly-impossible movements. In the film *The Lord of the Rings,* directed in 1978 by Ralph Bakshi using rotoscoping, live action was filmed first. Then each frame was traced and colored to create a series of animation cels. By the late twentieth century, many in the industry were experimenting with computer technology to create animation. In 1995, John Lassiter directed *Toy Story,* the first feature film created entirely with computer animation.

Raw Materials

Although the most important raw material in creating animation is the imagination of the animator, a number of supplies are necessary to bring that imagination to life. Sometimes these items are purchased; sometimes they are constructed by the animator.

The animator works at an animation stand, a structure that holds a baseboard on which the drawings are attached by register pegs. The animation stand also supports a camera, lights, a work surface, and a platen (clear sheet of glass or plexiglass that holds the drawings in place).

The drawings are executed on cels, drawing paper, or on film. The majority of professional animation is drawn on cels, transparent acetate sheets five millimeters thick. Each cel measures approximately 10 in by 12 in (25.4 cm by 30.5 cm). Holes are punched along the top edge of the cels, paper, or film, corresponding to the register pegs on the animation stand and baseboard. The pegs keep the drawing surface rigid.

Opaque inks and paints, and transparent dyes are the most common media for drawing the story. Felt markers, crayons, and litho pencils can also be used.

Professional animation is photographed with 35mm **cameras.** However, it is possible to use Super 8 or 16mm models. A variety of camera lenses are employed, including standard, zoom, telephoto, wide angle, and fish-eye lenses.

The Manufacturing Process

Creating an animated short or full-length feature is a long, tedious process. Extremely labor-intensive, the average short cartoon has approximately 45,000 separate frames. To make a character say "Hello, Simon," can require 12 drawings to depict each movement of the character's lips.

The story is written

1 Sometimes the animator is also the writer. The animator makes a storyboard, a series of one-panel sketches pinned on a board. Dialogue and/or action summaries are written under each sketch. The sketches may be rearranged several times as a result of discussions between the writer, the animator, and the director.

The dialogue, music, and sound effects are recorded

2 Actors record the voices of each character. Background music and sound effects, such as doors slamming, footsteps, and weather sounds, are recorded. These recordings are generally preserved on magnetic tape. The music is timed for beats and accents; this information is recorded on a bar sheet so that the animation can be fitted around the music. Because Walt Disney was one of the first animators to fit the action to the music, this process is called

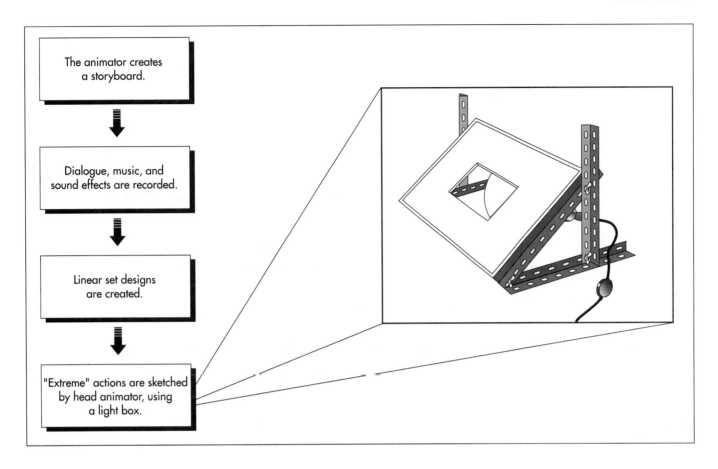

The animator creates a storyboard.

↓

Dialogue, music, and sound effects are recorded.

↓

Linear set designs are created.

↓

"Extreme" actions are sketched by head animator, using a light box.

"Mickey Mousing." Many professional studios now use an optical sound track on which voices, music, and sound effects are represented by varying lines. An electronic sound reader and synchronizer gives an accurate count of the number of frames required for each sound.

Dialogue measurements are entered on an exposure sheet

3 A technician known as a track reader measures each vowel and consonant in the dialogue. Words are recorded on exposure sheets (also called x-sheets or dope sheets), each of which represents a single film frame. This allows the animators to synchronize each movement of the character's lips with the dialogue. Footage, the time needed between lines of dialogue for the action to take place, is also charted on the exposure sheet. Slugs, or sections of film without sound, are inserted where the action occurs.

Model character sheets are created

4 A model is created for each character in order to keep their appearances uniform throughout the film. The models can be detailed descriptions or sketches of the characters in various positions with various facial expressions.

Artists create the layout or set design

5 A layout artist creates linear drawings that animators use as a guide for action and that the background artists use to paint the backgrounds.

Characters' actions are sketched

6 Using the model sheets, the head animator sketches the primary, or "extreme," action. For example, if the character is running, the head animator will draw the foot leaving the floor, the foot in the air, and the foot returning to the floor. Or if the story calls for the character to blink, the head animator will sketch the eyes going through the motions. Animation assistants then fill in the details.

The drawing is done on a transparent drawing board that is lighted from below. After one drawing is completed, a second sheet of

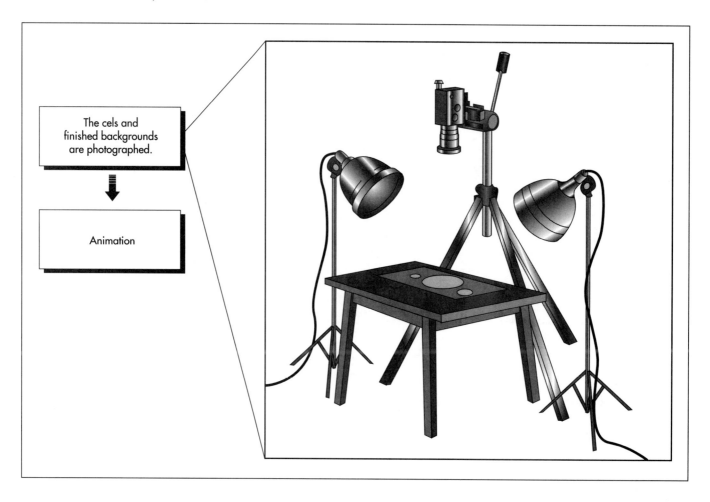

The cels and
finished backgrounds
are photographed.

↓

Animation

paper is laid on top of the first and the second drawing is varied slightly to signify movement.

Drawings are cleaned up and checked for accuracy

7 Artists check the characters against the model sheets. Drawings are enhanced but not altered. Scenes are checked to ensure that all action called for on the exposure sheet is included. All figures are checked for proper line-up with the background.

A video test is conducted

8 A computerized videotape is made of the sketches to check for smoothness of motion and proper facial expressions. Adjustments are made until the desired effect is achieved.

Artists create backgrounds

9 Artists create color background paintings, including landscapes, scenery, buildings and interiors, from the pencil lay-

outs. The color is filled in by computer. As the computer scans the layout, artists click on colors from a template.

Sketches are inked in and painted

10 If the animation drawings have been executed on paper, they are now transferred to cels using xerography, a process similar to photocopying. In a few studios, the inking is still done by hand, tracing the pencil sketches onto the cels.

Colors are applied to the reverse side of the cel, usually by computer, in the same manner that background colors are applied. All inked and painted materials are checked several times for accuracy.

The action is filmed

11 The cels and backgrounds are photographed according to the instructions on the exposure sheets. One scene of action can take several hours to photograph. The cels are laid on top of the backgrounds and photographed with a multiplane camera that is suspended high above. When more than one character appears in a frame, the number of cels stacked on top of the background increases. Each level is lit and staggered, creating the illusion of three-dimensional action. The film is sent to the photo lab where a print and a negative are made.

The sound is dubbed

12 Dialogue, music, and sound effects are re-recorded from 10 or more separate tracks onto one balanced track. Another set of two tracks, one with dialogue and the other with music and sound effects, is often made to facilitate translation when the film is sent to foreign markets.

The dubbing track and print are combined

13 The final dubbing track is combined with the print to make a married print. If the animated film is for television viewing, the negative and the tracks are often sent to a video post-production house to be put on videotape.

The Future

In the last decade of the twentieth century, computer-created animation began to make great strides. Although purists decry this development, it is unlikely that computer animation will disappear. What remains to be seen is whether or not traditional cel animation survives.

Anime, a cartoon form from Japan, is also changing the nature of animation. Story lines and characters are more detailed and reality-based. Varied camera angles bring the viewer further into the action.

Where to Learn More

Books

Cawley, John and Jim Korkis. *How to Create Animation*. Pioneer Books, 1990.

Locke, Lafe. *Film Animation Techniques*. Betterway Publications, 1992.

Periodicals

Harmon, Amy. "Making a Face." *Los Angeles Times*, March 25, 1996, p. D-1.

Considine, J.D. "Toon in Tomorrow." *The Baltimore Sun*, April 14, 1996, p. 1H.

—*Mary F. McNulty*

Antishoplifting Tag

The next generation of security tags will contain "smart" chips. Using radio waves, various people throughout the wholesale and retail supply chains will be able to read and write to integrated circuits within the tags.

Background

Ronald Assas' frustration with shoplifters came to a head the day he watched a man slip two bottles of wine under his shirt and run out of an Akron, Ohio, supermarket. Assas, the store manager, sprinted out the door in pursuit of the thief. Unable to catch the man and unsure what he would have done if he had caught him, Assas returned to the store. He commented that anyone who could figure out a way to deter such thieves would make a fortune. One of those who heard his remark was his cousin, Jack Welch. Welch was already working on electronic tagging of products, and he took up Assas' challenge. Several weeks later, Welch returned to the store with a 2 ft (61 cm) square of cardboard with a large foil tag taped to it, along with some bulky boxes filled with electronic components he had assembled in his garage. He showed Assas how an alarm would sound if someone tried to carry the tag out the door between the boxes. A couple of years later, Assas founded Sensormatic Electronics Corporation, which still holds a 65% share of the worldwide electronic security market.

Since they were first marketed in 1966, antishoplifting tags have become so popular that a billion dollars worth of them were manufactured last year to combat thefts that cost retailers 10 billion dollars a year. Using the tags is one of the most effective deterrents available to store owners. Some tags are hard tags or buttons that are attached to merchandise with pins that can be removed only with a special tool; these tags can be reused repeatedly by the merchant. Other tags look like thick, plastic labels; these are not removed from the merchandise during purchase, but they are electronically deactivated so the product can be taken from the store without activating the alarm. Tags of this type are disposable, although they can be reactivated if the purchased item is returned to the store for exchange or refund.

Within the retail industry, the devices are generally known as security tags or Electronic Article Surveillance (EAS) tags. The technology favored for modern tags involves a set of gates that transmits pulses of a low-range radio frequency. Inside each security tag is a resonator, a device that picks up the transmitted signal and repeats it. The set of gates also contains a receiver that is programmed to recognize whether it is detecting the target signal during the time gaps between the pulses being broadcast by the gates. Sensing a signal during these intervals indicates the presence of a signal being resonated (rebroadcast) by a security tag in the detection zone. When this occurs, the gates sound an alarm; in some systems, the alarm sound is accompanied by a flashing light.

For the first 20 years of their history, security tags used swept radio frequency (swept-RF) technology, which relied on a semiconductor diode to retransmit a high-frequency radio signal from the detection gates. Although the tags worked reasonably well, they had certain limitations. For instance, the older devices could be defeated by placing tagged merchandise in foil-lined pouches that could block the microwave signals, and they were not very reliable when used to tag metal or foil-wrapped products. Furthermore, widely spaced antenna gates (more than 4.5 ft[1.4 m]) were not effective, and false alarms could result when the deactivation process failed.

In the mid-1980s, acousto-magnetic technology was developed to overcome certain limitations of the swept-RF devices. These systems operate with low frequency radio waves that are not blocked by metal foil wrappings. Tags contain coils of an appropriate magnetic metal that resonates in response to the interrogation signal. Although these types of systems are somewhat more costly than those using the older technology, they work more reliably over wider detection zones.

Hard tags that are commonly attached to clothing items are difficult to remove without damaging the product. Several innovations have been introduced over the years to make security tags more effective. For example, ink tags, which were developed in the early 1980s, contain small vials of dye that break if the tag is forcibly removed from the garment. The resulting spillage not only spoils the tagged apparel, but it stains the thief's hands for easy identification. Another design causes a tag to sound a loud alarm if it is tampered with.

Disposable, label-style security tags are becoming increasingly popular, particularly when the tags are inserted inside the product or its packaging by the manufacturer. This "source tagging" makes the devices less accessible for tampering or premature removal, as well as eliminating the time spent by retail clerks to attach and remove tags.

Raw Materials

Hard tags are formed from durable plastic, and the pin used to attach the tag to the product is made of nickel-plated steel. Disposable tags are formed from more flexible plastic, such as polypropylene. Conductive and non-conductive components of the resonator units include such materials as copper, aluminum, cellulose acetate, acrylic, and polyester.

The Manufacturing Process

The following description applies generically to reusable hard tags; details may vary among manufacturers. Disposable security tags are manufactured in a similar manner, except that the resonator is sealed inside a flexible plastic envelope, which may be backed with adhesive.

The case

1 The plastic casing for the tag is vacuum formed or injection molded. In the former process, plastic that has been softened by heat is drawn into a mold by creation of a vacuum. In the latter, semi-molten plastic is squirted under pressure into a cooled mold, where it hardens quickly.

The resonator

2 There are several ways a resonator can be made. One technique involves laminating copper or aluminum coils onto a web of nonconductive material. This is done by passing the adhesive-coated base web between rollers that apply a spiral-shaped mask of non-sticky material, after which the web passes through a dryer to set the mask. A thin, flat strip of metal is then laminated to the uncoated (sticky) portion of the base web. The laminated strip subsequently passes between a backup roller and a cutting roller, which cuts through the metal but not the base web, disconnecting the individual metal coils from one another. This masking and laminating process is repeated, adding a layer of web with metal spirals atop the first layer so that the two layers of spirals are face to face, separated by a layer of dielectric (nonconductive) material. Finally, the laminated strip is cut into individual resonators that can be inserted into security tags.

Another type of resonator is made by winding insulated (encased in plastic) copper wire into a flat spiral of about a dozen loops, with the ends of the wire connected through a diode. One company makes a button-shaped tag that can operate with a very small-diameter coil because the wire is spiraled into a cone shape.

The lock

3 After the resonator is inserted into the security tag's plastic casing, the locking mechanism is installed. This usually consists of a clutch that will accept and lock a metal pin that can be inserted through a

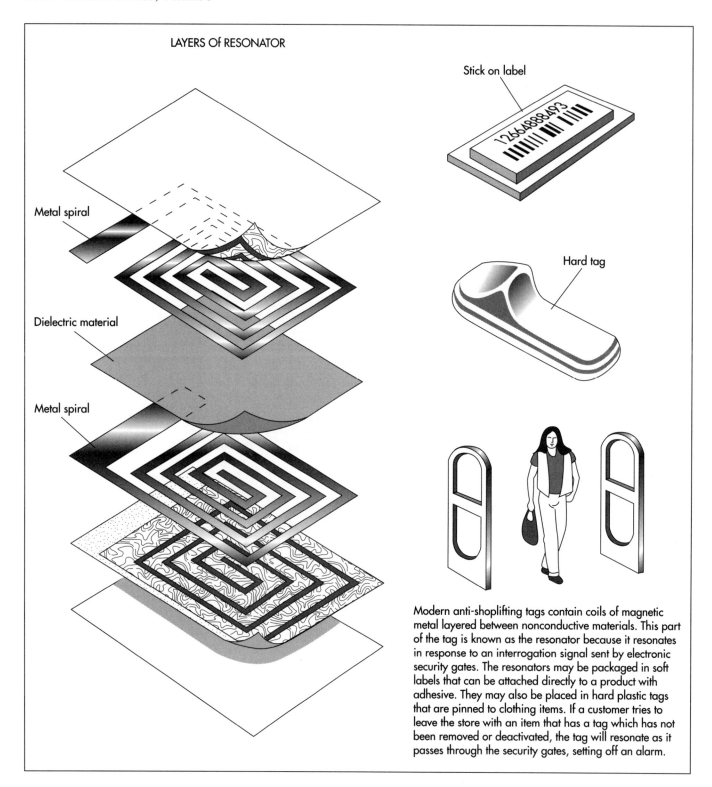

LAYERS Of RESONATOR

Metal spiral

Dielectric material

Metal spiral

Stick on label

1266.4888493

Hard tag

Modern anti-shoplifting tags contain coils of magnetic metal layered between nonconductive materials. This part of the tag is known as the resonator because it resonates in response to an interrogation signal sent by electronic security gates. The resonators may be packaged in soft labels that can be attached directly to a product with adhesive. They may also be placed in hard plastic tags that are pinned to clothing items. If a customer tries to leave the store with an item that has a tag which has not been removed or deactivated, the tag will resonate as it passes through the security gates, setting off an alarm.

product at the retail store. There are numerous designs of clutches, but one example is a metal plate with a small hole in the middle. The hole is too small for the pin's shaft to pass through unless the metal plate is flexed to enlarge the hole. Once the pin is inserted, the plate flattens, and the minimized hole fits around a grooved section in the shaft of the pin. To release this grip, the sales clerk inserts the tag into a magnetic device that flexes the clutch plate, allowing the pin to slide free. Another example of a clutch type is a ring of tiny balls that encircle the pin, with a spring mechanism pressing the balls into a groove in the pin's shaft; a magnetic deactivator retracts the balls from the groove, releasing the pin. Still other tag designs use a mechanical deactivator that inserts a probe into the tag to physically disengage a locking device.

Finishing

4 With the resonator and clutch assemblies in place, the upper and lower portions of the plastic tag casing are attached together. They are sealed by heat or welded by ultrasound. Finally, the completed tags are counted and boxed for shipment.

The Future

Concealing antishoplifting tags inside a product's package is becoming more prevalent, as some shoplifters manage to remove or disable visible tags. In fact, some label-style tags are so small they can be hidden within the seam of a garment while it is being manufactured. The next generation of security tags will contain "smart" chips. Using radio waves, various people throughout the wholesale and retail supply chains will be able to read and write to integrated circuits within the tags. Coded information about the dates and places of manufacture and purchase could remain with an article indefinitely for warranty or refund purposes.

The technology developed for antishoplifting tags has found other applications too. For example, some hospitals include tiny security tags in identification bracelets to alert the staff if a senile patient wanders out of his or her room.

Where to Learn More

Periodicals

"Let Shoplifters Beware: The Macbeth Solution." *Discover,* October 1986, p.14.

Ryan, Joseph, Jr. "Antishoplifting Labels." *Scientific American,* May 1997, p.120.

Schmuckler, Eric. "Attention, Shoplifters!" *Forbes,* November 14, 1988, pp. 258-59.

Sieder, Jill Jordan. "To Catch a Thief, Try This." *U.S. News & World Report,* September 23, 1996, p. 71.

Other

"Just the Facts." *Sensormatic.* http://www.sensormatic.com:80/html/news/execsum.htm (20 May 1997).

"United States Patent Number 5,494,550." *Patent   Server.* http://patent.womplex.ibm.com (20 May 1997).

"United States Patent Number 5,528,914." *Patent Server.* http://patent.womplex.ibm.com (20 May 1997).

—*Loretta Hall*

Artificial Eye

According to the Society for the Prevention of Blindness, between 10,000 and 12,000 people per year lose an eye. Though 50% or more of these eye losses are caused by an accident (in one survey more males lost their eyes to accidents compared to females), there are a number of inherited conditions that can cause eye loss or require an artificial eye.

Background

An artificial eye is a replacement for a natural eye lost because of injury or disease. Although the replacement cannot provide sight, it fills the cavity of the eye socket and serves as a cosmetic enhancement. Before the availability of artificial eyes, a person who lost an eye usually wore a patch. An artificial eye can be attached to muscles in the socket to provide eye movement.

Today, most artificial eyes are made of plastic, with an average life of about 10 years. Children require more frequent replacement of the prosthesis due to rapid growth changes. As many as four or five prostheses may be required from infancy to adulthood.

According to the Society for the Prevention of Blindness, between 10,000 and 12,000 people per year lose an eye. Though 50% or more of these eye losses are caused by an accident (in one survey more males lost their eyes to accidents compared to females), there are a number of inherited conditions that can cause eye loss or require an artificial eye. Microphthalmia is a birth defect where for some unknown reason the eye does not develop to its normal size. These eyes are totally blind, or at best might have some light perception.

Some people are also born without one or both eyeballs. Called anophthalmia, this presents one of the most difficult conditions for properly fitting an artificial eye. Sometimes the preparatory work can take a year or more. In some cases, surgical intervention is necessary.

Retinoblastoma is a congenital (existing at birth) cancer or tumor, which is usually inherited. If a person has this condition in just one eye, the chances of passing it on are one in four, or 25%. When the tumors are in both eyes, the chances are 50%. Other congenital conditions that cause eye loss include cataracts and glaucoma. One survey showed that 63% of eye loss due to disease occurs before 50 years of age.

There are two key steps in replacing a damaged or diseased eye. First, an ophthalmologist or eye surgeon must remove the natural eye. There are two types of operations. The enucleation removes the eyeball by severing the muscles, which are connected to the sclera (white of eyeball). The surgeon then cuts the optic nerve and removes the eye from the socket. An implant is then placed into the socket to restore lost volume and to give the artificial eye some movement, and the wound is then closed.

With evisceration, the contents of the eyeball are removed. In this operation, the surgeon makes an incision around the iris and then removes the contents of the eyeball. A ball made of some inert material such as plastic, glass, or silicone is then placed inside the eyeball, and the wound is closed.

At the conclusion of the surgery, the surgeon will place a conformer (a plastic disc) into the socket. The conformer prevents shrinking of the socket and retains adequate pockets for the prosthesis. Conformers are made out of silicone or hard plastic. After the surgery, it takes the patient from four to six weeks to heal. The artificial eye is then made and fitted by a professional ocularist.

History

Early artificial eye makers may not have been creating prostheses at all, but rather decorations for religious and aesthetic purposes. In the millennia B.C., the people of Babylon, Jericho, Egypt, China, and the Aegean area all had highly developed arts and a belief in the afterlife. Radiographs of mummies and tombs have revealed numerous artificial eyes made of silver, gold, rock crystal, lapis lazuli, shell, marble, enamel, or glass. The Aztec and Inca also used artificial eyes for similar reasons. The skill of the Egyptian artists was so great that they were probably asked to create artificial eyes for human use, especially if the afflicted were royalty.

In 1579, the Venetians invented the first prosthesis to be worn behind the eyelids. These artificial eyes were very thin shells of glass, and therefore, did not restore the lost volume of an atrophied or missing eyeball. Because the edges were sharp and uncomfortable, the wearers had to remove the eyes at night in order to get relief from discomfort and to avoid breakage.

After the invention of this glass shell prosthesis, there were no significant advances in artificial eyes until the nineteenth century. In the early 1800s, a German glassblower by the name of Ludwig Muller-Uri, who made life-like eyes for dolls, developed a glass eye for his son. Though it took 20 years to perfect his design, his success forced him to switch occupations to making artificial eyes full-time.

In 1880, Dutch eye surgeon Hermann Snellen developed the Reform eye design. This design was a thicker, hollow glass prosthesis with rounded edges. The increase in thickness restored most of the lost volume of the eye and the rounded edges gave the patient much more comfort. Germany became the center for manufacturing glass artificial eyes.

Several years later in 1884, a glass sphere was implanted for the first time in the scleral cavity (the hollowed out interior of the white of the eyeball) after evisceration. An English doctor, Phillip Henry Mules, used the implant to restore lost volume and to give the prosthesis some movement. The sphere implant was subsequently adapted for the enucleated socket as well.

Many materials such as bone, sponge, fat, and precious metals have been used for implants since then, but 100 years later, the Mules sphere is still used in the majority of cases. Eye sockets with spheres within the scleral cavity following evisceration continue to result in excellent cosmetic results. For the enucleated socket another solution had to be found.

During World War II, the glass eyes from Germany were unavailable, and therefore, the United States had to find an alternate material. In 1943, the U.S. Army dental technicians made the first plastic artificial eye. This material had the advantage of being unbreakable as well as malleable. Though these plastic prosthesis were impression-fitted, the back surface was not completely polished, leading to irritation of the eye socket due to a poor fit.

An alternative was introduced by German-American glass blowers who were learning to make artificial eyes out of plastic using the Reform design. Though this type of artificial eye was an improvement, there were still problems with a persistent discharge of mucus from the eye socket. The wearers could sleep with the prosthesis in place, but were required to remove it every morning for cleaning. Despite these limitations, demand outpaced what the ocularists could handle, and therefore, a few large optical companies began mass producing the 12 most commonly used glass eye shapes. Called stock eyes, they have the disadvantage of not being properly fitted to the individual's eye socket.

In the late 1960s the modified impression method was developed by American Lee Allen. This method included accurately duplicating the shape of the individual socket, as well as modifying the front surface of the prosthesis to correct eyelid problems. The back surface of the prosthesis must also be properly polished for an optimum fit. This method is widely used today.

Raw Materials

Plastic is the main material that makes up the artificial eye. Wax and plaster of paris are used to make the molds. A white powder called alginate is used in the molding process. Paints and other decorating materials are used to add life-like features to the prosthesis.

For a bioocular implant, the surgeon makes an incision around the iris and then removes the contents of the eyeball. A ball made of some inert material such as plastic, glass, or silicone is then placed inside the eyeball and the wound is closed.

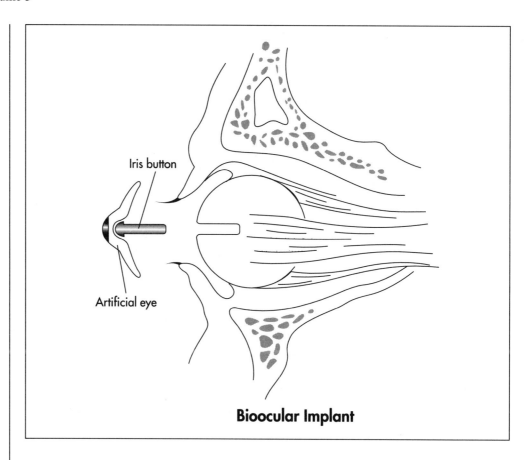

Iris button

Artificial eye

Bioocular Implant

The Manufacturing Process

The time to make an ocular prosthesis from start to finish varies with each ocularist and the individual patient. A typical time is about 3.5 hours. Ocularists continue to look at ways to reduce this time.

There are two types of prostheses. The very thin, shell type is fitted over a blind, disfigured eye or over an eye which has been just partially removed. The full modified impression type is made for those who have had eyeballs completely removed. The process described here is for the latter type.

1 The ocularist inspects the condition of the socket. The horizontal and vertical dimensions and the periphery of the socket are measured.

2 The ocularist paints the iris. An iris button (made from a plastic rod using a lathe) is selected to match the patient's own iris diameter. Typically, iris diameters range from 0.4-0.52 in (10-13 mm). The iris is painted on the back, flat side of the button and checked against the patient's iris by

simply reversing the buttons so that the color can be seen through the dome of plastic. When the color is finished, the ocularist removes the conformer, which prevents contraction of the eye socket.

3 Next, the ocularist hand carves a wax molding shell. This shell has an aluminum iris button imbedded in it that duplicates the painted iris button. The wax shell is fitted into the patient's socket so that it matches the irregular periphery of the socket. The shell may have to be reinserted several times until the aluminum iris button is aligned with the patient's remaining eye. Once properly fitted, two relief holes are made in the wax shell.

4 The impression is made using alginate, a white powder made from seaweed that is mixed with water to form a cream, which is also used by dentists to make impressions of gums. After mixing, the cream is placed on the back side of the molding shell and the shell is inserted into the socket. The alginate gels in about two minutes and precisely duplicates the individual eye socket. The wax shell is removed, with the alginate

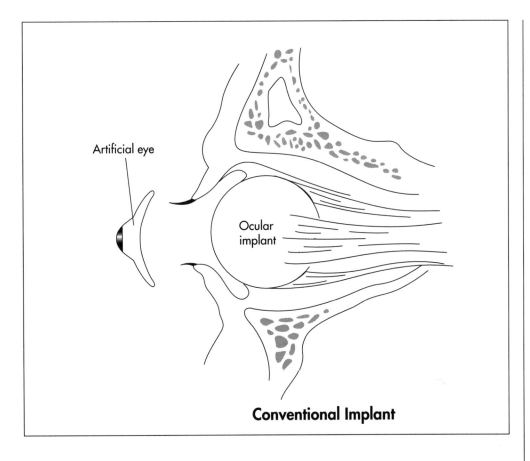

Artificial eye

Ocular implant

Conventional Implant

For a conventional implant, the surgeon removes the eyeball by severing the muscles, which are connected to the sclera (white of eyeball). The surgeon then cuts the optic nerve and removes the eye from the socket. An implant is then placed into the socket to restore lost volume and to give the artificial eye some movement, and the wound is closed.

impression of the eye socket attached to the back side of the wax shell.

5 The iris color is then rechecked and any necessary changes are made. The plastic conformer is reinserted so that the final steps can be completed.

6 A plaster-of-paris cast is made of the mold of the patient's eye socket. After the plaster has hardened (about seven minutes), the wax and alginate mold is removed and discarded. The aluminum iris button has left a hole in the plaster mold into which the painted iris button is placed. White plastic is then put into the cast, the two halves of the cast are put back together and then placed under pressure and plunged into boiling water. This reduces the water temperature and the plastic is thus cured under pressure for about 23 minutes. The cast is then removed from the water and cooled.

7 The plastic has hardened in the shape of the mold with the painted iris button imbedded in the proper place. About 0.5 mm of plastic is then removed from the anterior surface of the prosthesis. The white plastic, which overlaps the iris button, is ground down evenly around the edge of the button. This simulates how the sclera of the living eye slightly overlaps the iris. The sclera is colored using paints, chalk, pencils, colored thread, and a liquid plastic syrup to match the patient's remaining eye. Any necessary alterations to the iris color can also be made at this point.

8 The prosthesis is then returned to the cast. Clear plastic is placed in the anterior half of the cast and the two halves are again joined, placed under pressure, and returned to the hot water. The final processing time is about 30 minutes. The cast is then removed and cooled, and the finished prosthesis is removed. Grinding and polishing the prosthesis to a high luster is the final step. This final polishing is crucial to the ultimate comfort of the patient. The prosthesis is finally ready for fitting.

Quality Control

In 1957, the American Society of Ocularists (ASO) was established to raise standards and provide education for the ocularist. In 1971, the ASO began to certify ocularists. Those who already had well established practices were automatically certified. Others had to complete a five year apprenticeship under the direct supervision of a previously certified ocularist and complete 750 credit hours of related instruction approved by ASO.

In 1980, the National Commission of Health Certifying Agencies (NCHCA) created an independent testing organization for ocularists called the National Examining Board for Ocularists (NEBO). In November of 1981, NEBO administered the first National Boards certifying exam. Board certified ocularists must be recertified every three years. To achieve Fellowship in ASO, a board certified ocularist must accumulate 375 additional credit hours of related instruction and have demonstrated outstanding ability in their practice.

The Future

Improvements will continue in the ocular prosthesis, which will benefit both patient and ocularist. Several developments have already occurred in recent years. A prosthesis with two different size pupils which can be changed back and forth by the wearer was invented in the early 1980s. In the same period, a soft contact lens with a large black pupil was developed that simply lays on the cornea of the artificial eye.

In 1989, a patented implant called the Bioeye was released by the United States Food and Drug Administration. Today, over 25,000 people worldwide have benefited from this development, which is made from hydroxyapatite, a material converted from ocean coral and has both the porous structure and chemical structure of bone. In addition to natural eye movement, this type of implant has reduced migration and extrusion, and prevents drooping of the lower lid by lending support to the artificial eye via a peg connection.

With advancements in computer, electronics, and biomedical engineering technology, it may someday be possible to have an artificial eye that can provide sight as well. Work is already in progress to achieve this goal, based on advanced microelectronics and sophisticated image recognition techniques.

Though it may take several more years before a prosthesis will both look and see just like a natural eye, a Canadian company is developing an artificial eye that will be connected either to the optical nerve or directly to the visual cortex. This eye consists of a rubbery lens that can change focus, a high-precision color processing system, and microscopic photo-receptors that sense the presence of objects and pick up motion.

Researchers at MIT and Harvard University are also developing what will be the first artificial retina. This is based on a biochip that is glued to the ganglion cells, which act as the eye's data concentrators. The chip is composed of a tiny array of etched-metal electrodes on the retina side and a single sensor with integrated logic on the pupil side. The sensor responds to a small infrared laser that shines onto it from a pair of glasses that would be worn by the artificial-retinal recipient.

Where to Learn More

Books

Tillman, Walter. *An Eye for an Eye, A Guide for the Artificial Eye Wearer.* F.A.S.O., 1987.

Periodicals

Johnson, R. Colin. "Joint 'biochip' project eyes artificial retina," *Electronic Engineering Times,* September 18, 1995.

Munro, Margaret. "Building a better eyeball," *Montreal Gazette,* April 19, 1995.

Other

American Academy of Ophthalmology, 655 Beach St., San Francisco, CA 94109, 415-561-8500. http://www.eyenet.org/aao_ index. html.

Digital Journal of Ophthalmology. http://netope.harvard.edu:80/meei/.

"Integrated Orbital Implants." Movements On-Line. http://www.ioi.com.

Mie University School of Medicine Department of Ophthalmology. http://www.medic. mie-u.ac.jp/ophthalmology/index.html.

Ocular Surgery News. http://www.slack-inc.com/eye/osn/osnhome.htm.

—Laurel M. Sheppard

Artificial Skin

Background

Skin, the human body's largest organ, protects the body from disease and physical damage, and helps to regulate body temperature. It is composed of two major layers, the epidermis and the dermis. The epidermis, or outer, layer is composed primarily of cells: keratinocytes, melanocytes, and langerhans. The dermis, composed primarily of connective tissue fibers such as collagen, supplies nourishment to the epidermis.

When the skin has been seriously damaged through disease or burns, the body cannot act fast enough to manufacture the necessary replacement cells. Wounds, such as skin ulcers suffered by diabetics, may not heal and limbs must be amputated. Burn victims may die from infection and the loss of plasma. Skin grafts were developed as a way to prevent such consequences as well as to correct deformities. As early as the sixth century B.C., Hindu surgeons were involved in nose reconstruction, grafting skin flaps from the patient's nose. Gaspare Tagliacozzo, an Italian physician, brought the technique to Western medicine in the sixteenth century.

Until the late twentieth century, skin grafts were constructed from the patient's own skin (autografts) or cadaver skin (allografts). Infection or, in the case of cadaver skin, rejection were primary concerns. While skin grafted from one part of a patient's body to another is immune to rejection, skin grafts from a donor to a recipient are rejected more aggressively than any other tissue graft or transplant. Although cadaver skin can provide protection from infection and loss of fluids during a burn victim's initial healing period, a subsequent graft of the patient's own skin is often required. The physician is restricted to what skin the patient has available, a decided disadvantage in the case of severe burn victims.

In the mid-1980s, medical researchers and chemical engineers, working in the fields of cell biology and plastics manufacturing, joined forces to develop tissue engineering to reduce the incidences of infection and rejection. One of the catalysts for tissue engineering was the growing shortage of organs available for transplantation. In 1984, a Harvard Medical School surgeon, Joseph Vacanti, shared his frustration over the lack of available livers with his colleague Robert Langer, a chemical engineer at the Massachusetts Institute of Technology. Together, they pondered whether new organs could be grown in the laboratory. The first step was to duplicate the body's production of tissue. Langer came up with the idea of constructing a biodegradable scaffolding on which skin cells could be grown using fibroblasts, cells extracted from donated neonatal foreskins removed during circumcision.

In a variation of this technique developed by other researchers, the extracted fibroblasts are added to collagen, a fibrous protein found in connective tissue. When the compound is heated, the collagen gels and traps the fibroblasts, which in turn arrange themselves around the collagen, becoming compact, dense, and fibrous. After several weeks, keratinocytes, also extracted from the donated foreskins, are seeded onto the new dermal tissue, where they create an epidermal layer.

An artificial skin graft offers several advantages over those derived from the patient and cadavers. It eliminates the need for tis-

One foreskin can yield enough cells to make four acres of grafting material.

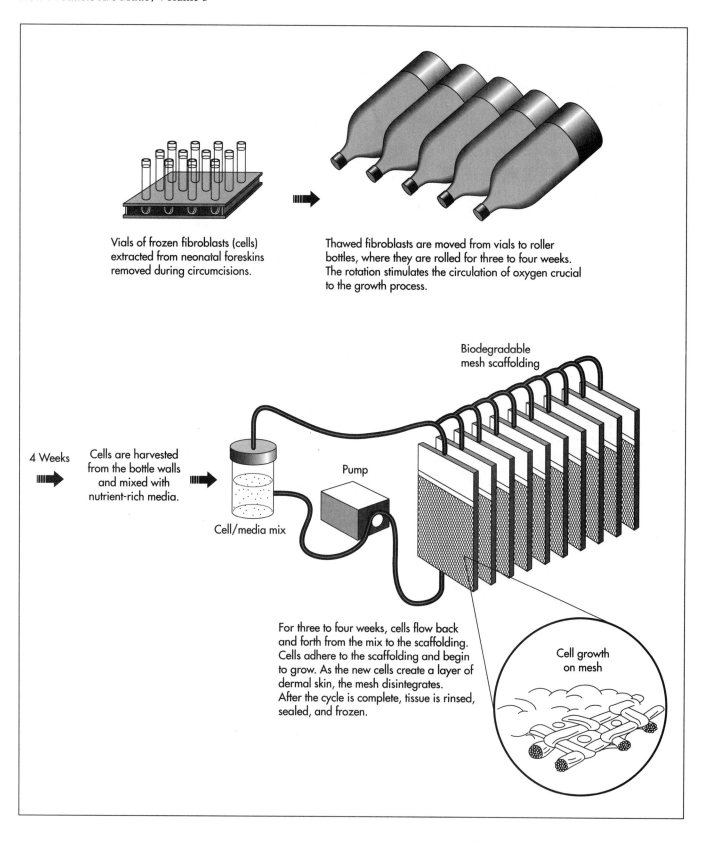

Vials of frozen fibroblasts (cells) extracted from neonatal foreskins removed during circumcisions.

Thawed fibroblasts are moved from vials to roller bottles, where they are rolled for three to four weeks. The rotation stimulates the circulation of oxygen crucial to the growth process.

Biodegradable mesh scaffolding

4 Weeks

Cells are harvested from the bottle walls and mixed with nutrient-rich media.

Cell/media mix

Pump

For three to four weeks, cells flow back and forth from the mix to the scaffolding. Cells adhere to the scaffolding and begin to grow. As the new cells create a layer of dermal skin, the mesh disintegrates. After the cycle is complete, tissue is rinsed, sealed, and frozen.

Cell growth on mesh

sue typing. Artificial skin can be made in large quantities and frozen for storage and shipping, making it available as needed. Each culture is screened for pathogens, severely curtailing the chance of infection.

Because artificial skin does not contain immunogenic cells such as dendritic cells and capillary endothelial cells, it is not rejected by the body. Finally, rehabilitation time is significantly reduced.

Raw Materials

The raw materials needed for the production of artificial skin fall into two categories, the biological components and the necessary laboratory equipment. Most of the donated skin tissue comes from neonatal foreskins removed during circumcision. One foreskin can yield enough cells to make four acres of grafting material. Fibroblasts are separated from the dermal layer of the donated tissue. The fibroblasts are quarantined while they are tested for viruses and other infectious pathogens such as HIV, hepatitis B and C, and mycoplasma. The mother's medical history is recorded. The fibroblasts are stored in glass vials and frozen in liquid nitrogen at -94°F (-70°C). Vials are kept frozen until the fibroblasts are needed to grow cultures. In the collagen method, keratinocytes are also extracted from the foreskin, tested, and frozen.

If the fibroblasts are to be grown on mesh scaffolding, a polymer is created by combining molecules of lactic acid and glycolic acid, the same elements used to make dissolving sutures. The compound undergoes a chemical reaction resulting in a larger molecule that consists of repeating structural units.

In the collagen method, a small amount of bovine collagen is extracted from the extensor tendon of young calves. The collagen is mixed with an acidic nutrient, and stored in a refrigerator at 39.2°F (4°C).

Laboratory equipment includes glass vials, tubing, roller bottles, grafting cartridges, molds, and freezers.

The Manufacturing Process

The manufacturing process is deceptively simple. Its main function is to trick the extracted fibroblasts into believing that they are in the human body so that they communicate with each other in the natural way to create new skin.

Mesh scaffolding method

1 *Fibroblasts are thawed and expanded.* The fibroblasts are transferred from the vials into roller bottles, which resemble liter soda bottles. The bottles are rotated on their sides for three to four weeks. The rolling action allows the circulation of oxygen, essential to the growth process.

2 *Cells are transferred to a culture system.* The cells are removed from the roller bottles, combined with a nutrient-rich media, flowed through tubes into thin, cassette-like bioreactors housing the biodegradable mesh scaffolding, and sterilized with e-beam radiation. As the cells flow into the cassettes, they adhere to the mesh and begin to grow. The cells are flowed back and forth for three to four weeks. Each day, leftover cell suspension is removed and fresh nutrient is added. Oxygen, pH, nutrient flow, and temperature are controlled by the culture system. As the new cells create a layer of dermal skin, the polymer disintegrates.

3 *Growth cycle completed.* When cell growth on the mesh is completed, the tissue is rinsed with more nutrient-rich media. A cryoprotectant is added. Cassettes are stored individually, labeled, and frozen.

Collagen method

4 *Cells are transferred to a culture system.* A small amount of the cold collagen and nutrient media, approximately 1-2% of the combined solution, is added to the fibroblasts. The mixture is dispensed into molds and allowed to come to room temperature. As the collagen warms, it gels, trapping the fibroblasts and generating the growth of new skin cells.

5 *Keratinocytes added.* Two weeks after the collagen is added to the fibroblasts, the extracted keratinocytes are thawed and seeded onto the new dermal skin. They are allowed to grow for several days and then exposed to air, inducing the keratinocytes to form epidermal layers.

6 *Growth cycle completed.* The new skin is stored in sterile containers until needed.

The Future

The medical profession is using artificial skin technology to pioneer organ reconstruction. It is hoped that this so-called engineered structural tissue will, for example, someday replace plastic and metal prostheses currently used to replace damaged joints and bones. Ears and noses will be reconstructed by seeding cartilage cells on polymer mesh. The regeneration of breast and urethral tissues is currently under study in the laboratory. Through this technology, it is possible that one day, livers, kidneys, and even hearts, will be grown from human tissues.

Where to Learn More

Periodicals

Langer, Robert and Joseph P. Vacanti. "Artificial Organs," *Scientific American*, September 1995, pp. 130-133.

Langer, Robert and Joseph P. Vacanti. "Tissue Engineering," *Science*, May 14, 1993, pp. 920-921.

McCarthy, Michael. "Bio-engineered tissues move towards the clinic," *The Lancet,* August 17, 1996, p. 466.

Rundle, Rhonda L. "Cells 'Tricked' To Make Skin For Burn Cases," *The Wall Street Journal*, March 17, 1994.

—*Mary F. McNulty*

Aspartame

Background

Aspartame is an artificial sweetener used in reduced calorie foods. It is derived primarily from two naturally occurring amino acids chemically combined and designated by the chemical name N-L-αaspartyl-L-phenylalanine-1-methyl ester (APM). Discovered inadvertently in 1965, it was later patented and is currently the most utilized artificial sweetener in the United States.

Aspartame is a white, odorless, crystalline powder. It is about 200 times sweeter than sugar and is readily dissolvable in water. It has a sweet taste without the bitter chemical or metallic aftertaste reported in other artificial sweeteners. These properties make it a good ingredient to use as a sugar replacement in many food recipes. However, aspartame does tend to interact with other food flavors, so it cannot perfectly replace sugar. Recipes for baked goods, candies, and other products must be modified if aspartame is utilized. Although aspartame can be used in microwave recipes, it is sensitive to extensive heating, which makes it unsuitable for baking.

The fact that aspartame provides sweetness and flavor without imparting other physical characteristics such as bulk or calories like other sweeteners makes it unique. Another useful trait is that it has a synergistic effect with other sweeteners, making it possible to use less total sweetener. In addition to sweetening foods, aspartame is used to reduce calories, and intensify and extend fruit flavors.

History

Humans have desired foods with a sweet taste for thousands of years. Ancient cave paintings at Arana in Spain show a neolithic man taking honey from a wild bee's nest. It has been suggested that early humans might have used the sweet taste of foods to tell them which ones would be safe to eat. It is even thought that the desire for sweet taste might be an innate human trait. Unfortunately, many of the foods that are naturally sweet contain relatively large amounts of calories and carbohydrates.

Alternative sweeteners were developed to provide the sweet taste without the unnecessary calories. They also provide the additional benefits of enhancing the palatability of pharmaceuticals, aiding in the management of diabetes, and providing a cost-effective source where sugar is not available. The first one, saccharin, was discovered in 1879 and has been used in products such as toothpaste, mouthwash, and sugarless gum.

The sugarlike taste of aspartame was discovered accidentally by James Schlatter, an American drug researcher at G.D. Searle and Co. in 1965. While working on an anti-ulcer drug, he inadvertently spilled some APM on his hand. Figuring that the material was not toxic, he went about his work without washing it off. He discovered APM's sweet taste when he licked his finger to pick up a piece of weighing paper. This initial breakthrough then led the company to screen hundreds of modified versions of APM. However, none of these materials offered all of the advantages found in the original compound, including economical manufacturing, excellent taste quality and potency, natural metabolic pathways for digestion, excellent stability, and very low toxicity. Consequently, the company pursued and was granted United States patent 3,492,131 and various international patents, and the initial discovery was com-

Alternative sweeteners were developed to provide the sweet taste without the unnecessary calories.

39

mercialized. The U.S. patent expired in 1992, and the technology is now available to any company who wants to use it.

After many years of toxicity testing, the FDA initially approved aspartame's use as a sweetener in 1980. However, a hallmark of synthetic chemicals used in food products is that their safety is under constant scrutiny. Aspartame is no exception and has been surrounded by some controversy concerning its safety since its introduction. Most of these concerns were put to rest in late 1984, when after investigating various aspartame-related complaints, the FDA and the Centers for Disease Control concluded that the substance is safe and does not represent a widespread health risk. This conclusion was further supported by the American Medical Association in 1985, and aspartame has been gaining market share ever since. In addition to its use in the United States, aspartame has also been approved for use in over 93 foreign countries.

Aspartame has been marketed since 1983 by Searle under the brand names NutraSweet® and Equal®. Currently, NutraSweet® is a very popular ingredient and is used in more than 4,000 products, including chewing gum, yogurt, diet soft drinks, fruit-juices, puddings, cereals, and powdered beverage mixes. In the U.S. alone, NutraSweet®'s sales topped $705 million in 1993, according to the company.

Raw Materials

Aspartame is primarily derived from compounds called amino acids. These are chemicals which are used by plants and animals to create proteins that are essential for life. Of the 20 naturally occurring amino acids, two of them, aspartic acid and phenylalanine, are used in the manufacture of aspartame.

All amino acids molecules have some common characteristics. They are composed of an amino group, a carboxyl group, and a side chain. The chemical nature of the side chain is what differentiates the various amino acids. Another characteristic of amino acids is the ability to form different molecular configurations known as isomers. These isomers are designated by the letters L and D. Aspartame is composed of only L,L isomers; none of the other isomer combinations taste sweet. The sweet taste of aspartame could not have been predicted by looking at the two amino acids that it is derived from. L-aspartic acid has a flat taste and L-phenylalanine tastes bitter. However, when the two compounds are chemically combined and the L-phenylalanine is slightly modified, a sweet taste is achieved.

Aspartic acid is one of five amino acids that have a "charged" side group. The charged side group on aspartic acid is ($-CH_2$-COOH). When put in water, this material ionizes and becomes negatively charged. Phenylalanine has a nonpolar, hydrophobic side group which is not compatible with water. It is made up of a six carbon ring and is attached to the main amino acid backbone via a methyl ($-CH_2$) group. Prior to synthesis into aspartame, it is reacted with methanol. This adds a methyl group which is linked to the molecule by an oxygen, and the compound is converted to a methyl ester. The methanol required for the synthesis of aspartame has the chemical structure (CH_3-OH). This is a very common material and is used extensively by organic chemists for various chemical syntheses.

The Manufacturing Process

Although its components—aspartic acid, phenylalanine, and methanol—occur naturally in foods, aspartame itself does not and must be manufactured. NutraSweet® (aspartame) is made through fermentation and synthesis processes.

Fermentation

Direct fermentation produces the starting amino acids needed for the manufacture of aspartame. In this process, specific types of bacteria which have the ability to produce certain amino acids are raised in large quantities. Over the course of about three days, the amino acids are harvested and the bacteria are destroyed.

1 To start the fermentation process, a sample from a pure culture of bacteria is put into a test tube containing the nutrients necessary for its growth. After this initial inoculation the bacteria begin to multiply. When their population is large enough, they are transferred to a seed tank. The bacterial

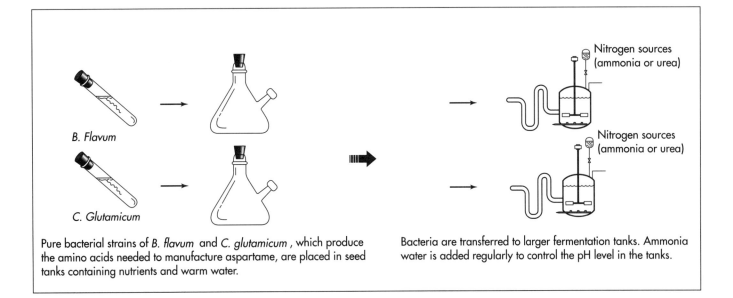

Pure bacterial strains of *B. flavum* and *C. glutamicum* , which produce the amino acids needed to manufacture aspartame, are placed in seed tanks containing nutrients and warm water.

Bacteria are transferred to larger fermentation tanks. Ammonia water is added regularly to control the pH level in the tanks.

strains used to make L-aspartic acid and L-phenylalanine are *B. flavum* and *C. glutamicum* respectively.

2 The seed tank provides an ideal environment for growing more bacteria. It is filled with the things bacteria need to thrive, including warm water and carbohydrate foods like cane molasses, glucose, or sucrose. It also has carbon sources like acetic acid, alcohols or hydrocarbons, and nitrogen sources such as liquid ammonia or urea. These are required for the bacteria to synthesize large quantities of the desired amino acid. Other growth factors such as vitamins, amino acids, and minor nutrients round out seed tank contents. The seed tank is equipped with a mixer, which keeps the growth medium moving, and a pump, which delivers filtered, compressed air. When enough bacterial growth is present, the contents from the seed tank are pumped to the fermentation tank.

3 The fermentation tank is essentially a larger version of the seed tank. It is filled with the same growth media found in the seed tank and also provides a perfect environment for bacterial growth. Here the bacteria are allowed to grow and produce large quantities of amino acids. Since pH control is vital for optimal growth, ammonia water is added to the tank as necessary.

4 When enough amino acid is present, the contents of the fermentation tank are transferred out so isolation can begin. This

process starts with a centrifugal separator, which isolates a large portion of the bacterial amino acids. The desired amino acid is further segregated and purified in an ion-exchange column. From this column, the amino acids are pumped to a crystallizing tank and then to a crystal separator. They are then dried and readied for the synthesis phase of aspartame production.

Synthesis

Aspartame can be made by various synthetic chemical pathways. In general, phenylalanine is modified by a reaction with methanol and then combined with a slightly modified aspartic acid which eventually forms aspartame.

5 The amino acids derived from the fermentation process are initially modified to produce aspartame. Phenylalanine is reacted with methanol resulting in a compound called L-phenylalanine methyl ester. Aspartic acid is also modified in such a way to shield various portions of the molecule from the effects of further reactions. One method is by reacting the aspartic acid with substances that result in added benzyl rings to protect these sites. This ensures that further chemical reactions will occur only on specific parts of the aspartic acid molecule.

6 After the amino acids are appropriately modified, they are pumped into a reactor tank, where they are allowed to mix at room temperature for 24 hours. The temper-

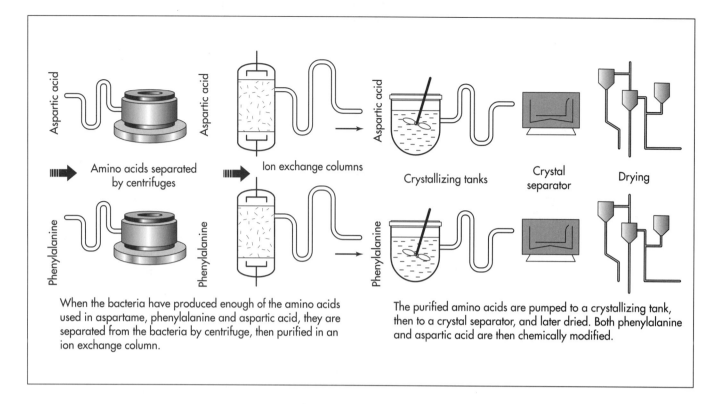

Aspartic acid

Aspartic acid

Aspartic acid

⟹ Amino acids separated by centrifuges ⟹ Ion exchange columns

Crystallizing tanks

Crystal separator

Drying

Phenylalanine

Phenylalanine

Phenylalanine

When the bacteria have produced enough of the amino acids used in aspartame, phenylalanine and aspartic acid, they are separated from the bacteria by centrifuge, then purified in an ion exchange column.

The purified amino acids are pumped to a crystallizing tank, then to a crystal separator, and later dried. Both phenylalanine and aspartic acid are then chemically modified.

ature is then increased to approximately 149°F (65°C) and maintained for another 24 hours. The reaction is then cooled to room temperature. It is diluted with an appropriate solvent and cooled to about 0°F (-18°C), causing crystallization. The crystals are then isolated by filtration and dried. These crystals are an intermediate of aspartame which must be further modified.

7 The intermediate is converted to aspartame by reacting it with acetic acid. This reaction is performed in a large tank filled with an aqueous acid solution, a palladium metal catalyst, and hydrogen. It is thoroughly mixed and allowed to react for about 12 hours.

Purification

8 The metal catalyst is removed by filtration, and the solvent is distilled, leaving a solid residue. This residue is purified by dissolving it in an aqueous ethanol solution and recrystallizing. These crystals are filtered and dried to provide the finished, powder aspartame.

Quality Control

The quality of the compounds is checked regularly during the manufacturing process. Of particular importance are frequent checks of the bacterial culture during fermentation. Also, various physical and chemical properties of the finished product are checked, such as pH level, melting point, and moisture content.

The Future

Currently, there are only three alternative sweeteners in the United States that can be used in food products. While aspartame is perhaps one of the best available, scientists are looking for new ways to make these sweeteners taste as much like sugar as possible. Their research has been focused in three areas, including finding new derivatives, blending sweeteners, and enhancing the efficiency of aspartame.

Most of the chemical derivative work has centered on finding compounds which will fit into the taste bud receptors better than traditional aspartame. Using aspartame as the model, researchers believe they will be able to improve various characteristics by making slight modifications. For example, they have found that when L-aspartic acid alone is modified in a certain way, it gives products that have a sweet taste. Future re-

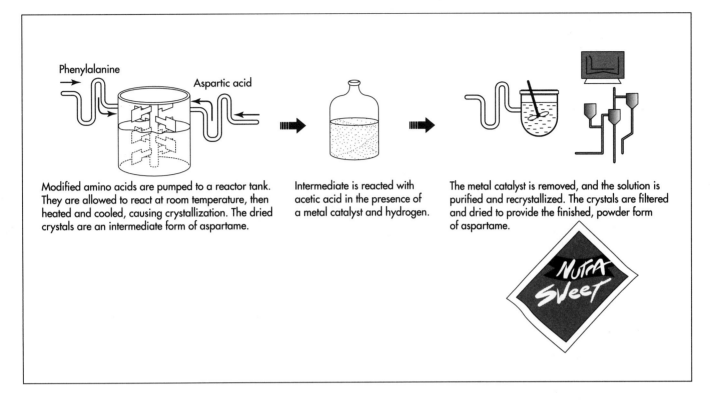

Phenylalanine

Aspartic acid

Modified amino acids are pumped to a reactor tank. They are allowed to react at room temperature, then heated and cooled, causing crystallization. The dried crystals are an intermediate form of aspartame.

Intermediate is reacted with acetic acid in the presence of a metal catalyst and hydrogen.

The metal catalyst is removed, and the solution is purified and recrystallized. The crystals are filtered and dried to provide the finished, powder form of aspartame.

search will likely focus on these kinds of derivatives.

Another area of research focuses on improving the heat stability of aspartame. Using encapsulation technology, aspartame has been developed which can be used in baked goods and baking mixes. Initial test results are positive, and FDA approval has been granted for bakery applications.

Since only three synthetic sugar substitutes are currently approved for use in food in the U.S., combining artificial sweeteners in products is becoming an important technological advance. Here, scientists combine two or three sweeteners in an effort to make the product taste more sugarlike.

Where to Learn More

Books

Nabors, Lyn, and Robert Gelardi. *Alternative Sweeteners*. Marcel Dekker, Inc., 1986.

Periodicals

Best, Daniel and Lisa Nelson. "Low-calorie foods and sweeteners." *Prepared Foods*, June 1993, p. 47.

Tomasula, Dean. "Sweet as sugar: artificial sweetener producers are blending products, in search of a market winning combination." *Chemical Marketing Reporter*, June 27, 1994, p. S22.

—*Perry Romanowski*

Asphalt Paver

The concept of asphalt as a paving material dates back to 1815, when Scottish road engineer John McAdam (or MacAdam) developed a road surface consisting of a compacted layer of small stones and sand sprayed with water.

Background

An asphalt paver is a machine used to distribute, shape, and partially compact a layer of asphalt on the surface of a roadway, parking lot, or other area. It is sometimes called an asphalt-paving machine. Some pavers are towed by the dump truck delivering the asphalt, but most are self-propelled. Self-propelled pavers consist of two major components: the tractor and the screed. The tractor provides the forward motion and distributes the asphalt. The tractor includes the engine, hydraulic drives and controls, drive wheels or tracks, receiving hopper, feeder conveyors, and distribution augers. The screed levels and shapes the layer of asphalt. The screed is towed by the tractor and includes the leveling arms, moldboard, end plates, burners, vibrators, and slope sensors and controls.

In operation, a dump truck filled with asphalt backs up to the front of the paver and slowly discharges its load into the paver's hopper. As the paver moves forward, the feeder conveyors move the asphalt to the rear of the paver, and the distribution augers push the asphalt outward to the desired width. The screed then levels the layer of asphalt and partially compacts it to the desired shape. A heavy, steel-wheeled roller follows the paver to further compact the asphalt to the desired thickness.

History

Asphalt as a paving material dates back to 1815, when Scottish road engineer John McAdam (or MacAdam) developed a road surface consisting of a compacted layer of small stones and sand sprayed with water.

The water dissolved the natural salts on the stones and helped cement the materials together. This type of road surface was named water macadam in his honor. Later, coal tar was used as a binding material instead of water, and the new pavement became known as tar macadam, from which we get the shortened term tarmac that is sometimes used to describe asphalt pavement.

Tar macadam pavement was used in the United States up through the beginning of the twentieth century. Modern mixed asphalt pavement, which provides a more durable road surface, was introduced in the 1920s. Unlike macadam, in which the stone and sand aggregates are laid on the road surface before being sprayed with the binding material, the aggregates in mixed asphalt are coated with the binding material before they are laid. At first, mixed asphalt was simply dumped on the roadway and raked or graded level before being rolled smooth. In 1931 Harry Barber, of Barber-Greene Company, developed the first mechanical asphalt paver in the United States. It traveled on a set of steel rails and included a combination loader and mixer to proportion and blend the components before spreading the asphalt evenly over the road surface. The rails were soon replaced by crawler tracks, and the first production paver came off the Barber-Greene line in 1934. This new machine quickly became popular with road builders because it allowed them to place asphalt more rapidly and with greater uniformity. Hydraulic drives replaced mechanical drives in pavers during the late 1950s to give the operator even smoother control. Today, almost all asphalt is placed using paving machines. When you consider that 98% of the roads in the United States are asphalt, you can understand the value of the asphalt paver.

Raw Materials

Most of the components of an asphalt paver are made of steel. The tractor mainframe is fabricated from heavy-gauge steel plate. The feeder conveyor is made of heavy-duty chain with forged steel sections, called flight bars. The distribution augers are made of cast Ni-Hard steel. The screed is fabricated from steel tubing, channel, and plate. The engine cover and access doors are formed from steel sheet.

Rubber-tired pavers have two large inflatable rear drive tires and four or more smaller solid rubber steering tires. Rubber-tracked pavers have a molded synthetic rubber track with several internal layers of flexible steel cable for reinforcement. The track is driven by a friction drive wheel on the rear, and the load is distributed among several intermediate rubber-coated steel bogie wheels. A hydraulic cylinder presses against the forward wheel to maintain tension in the track.

Purchased components on a paver include the engine, radiator, hydraulic components, batteries, electrical wiring, instruments, steering wheel, and operator's seat. Purchased fluids include hydraulic fluid, diesel fuel, engine oil, and antifreeze.

Design

Most manufacturers of asphalt pavers offer several sizes and models. Engine horsepower is usually in the 3-20 hp (2-15 kw) range for smaller, towed pavers, and may be in the 100-250 hp (75-188 kw) range for larger, self-propelled pavers. Most engines use diesel fuel because that is the fuel commonly used on other construction equipment.

Most larger, self-propelled pavers are about 19-23 ft (5.8-7.0 m) long, 10 ft (3.1 m) wide, and 10 ft (3.1 m) high. They weigh about 20,000-40,000 lb (9,090-18,180 kg) depending on the hopper capacity, engine size, and type of drive system. The typical rate of asphalt placement is 100-300 ft/min (31-92 m/min). The standard paving width is 8-12 ft (2.4-3.7 m) up to a maximum width of 40 ft (12.2 m) with the use of screed extensions on some machines. The maximum paving thickness on a single pass is 6-12 in (152-305 mm).

Options include lighting packages, manual and automatic screed extensions, and various sensors and controls to alter the grade (fore-aft dimensions) and slope (side-to-side dimensions) of the layer of asphalt.

The Manufacturing Process

Asphalt pavers are assembled from component parts. Some of these parts are fabricated in the assembly plant, while others are manufactured elsewhere and are shipped to the plant. All parts are given a primer coat of paint. The parts are stored in a warehouse and are brought to various work stations or areas as needed.

The tractor and the screed are assembled separately. The tractor assembly process starts as the mainframe is placed on an air-flotation pallet. As the assembly proceeds, the tractor is manually moved by attaching a compressed air line to the flotation pallet. This allows the heavy tractor to float on a thin cushion of air, and it can be easily pushed from one work station to another with the help of guide rails in the floor. The screed is assembled in a single area and does not move from one work station to another.

Here is a typical sequence of operation for the assembly of an asphalt paver:

Fabricating the tractor mainframe

1 The individual pieces of the mainframe are cut to size from steel plate with bandsaws or by flame cutting. The required holes are drilled or punched.

2 The pieces are held in position relative to each other using jigs and fixtures. They are then welded together with automatic wire-fed welders that are programmed to weld along the contour of the joints. When it is finished, the mainframe looks like the letter "H" with one long leg on each side to support the tires or tracks and a cross leg in the middle to support the engine, which is mounted sideways.

3 After the mainframe is welded together, it is shot blasted with a stream of high

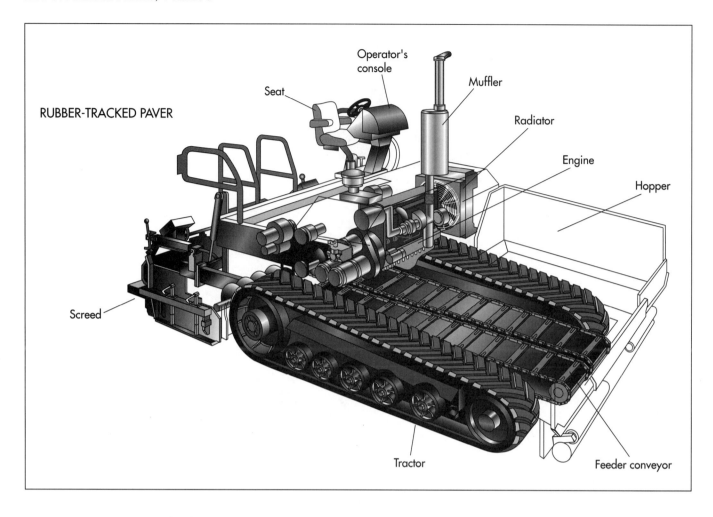

RUBBER-TRACKED PAVER

Operator's console

Seat

Muffler

Radiator

Engine

Hopper

Screed

Tractor

Feeder conveyor

velocity air, carrying small steel balls. This relieves any stresses in the metal caused by welding and removes any welding spatter. The mainframe is then painted with a primer and the paint is allowed to dry.

Assembling the tractor

4 The mainframe is placed on an air-flotation pallet and is moved to the first work station. The feeder conveyor chains and flights are installed first, followed by the hydraulic feeder drive motors and the feeder lubrication hoses. If the tractor is to have a tracked drive, the left and right drive hubs are installed. On some models, the fuel tank is also installed at this time.

5 While the mainframe is in the first work station, the engine is being prepared in a separate area. The engine is placed on a rolling support stand and the fan, oil filters, and various sensors are installed at this time. The disconnect clutch and pump drive gearbox are bolted to the rear of the engine. The gearbox is triangular-shaped and has

mounting locations for three sets of hydraulic pumps. The upper set of pumps provide power for the drive tires or tracks. The two lower sets of pumps provide power for the left and right conveyor feeders, distribution augers, and screed vibrators. Each set of pumps consists of two or more pumps sandwiched end to end and running off the same central shaft.

6 The mainframe is moved to the next work station. The engine is lifted from its support stand with an overhead hoist and is lowered into position crossways on the mainframe. It is bolted in place on several hard rubber mounts, which act to isolate the engine vibration. The radiator is bolted in place and coolant hoses are run between the engine and radiator.

7 The left and right distribution auger assemblies are bolted in place and the hydraulic auger drive motors and drive chains are installed. The rear hopper pieces are bolted in place, as are the hydraulic cylinders to raise and lower the screed leveling

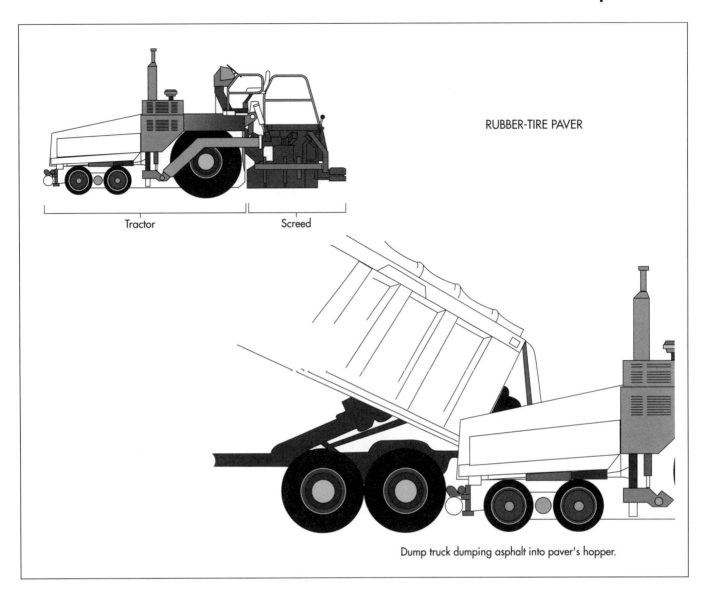

RUBBER-TIRE PAVER

Tractor   Screed

Dump truck dumping asphalt into paver's hopper.

arms. Various hydraulic hoses and electrical wiring are routed between components.

8 If the tractor is to have a tracked drive, the left and right variable-speed hydraulic drive motors and two-speed planetary gears are bolted to the drive hubs. If the tractor is to have a rubber-tired drive, the drive axle, two-speed gearbox, and two-speed hydraulic drive motor is installed.

9 At the next work station, the main electrical box is installed, the hydraulic tank and valves are installed and connected with hoses, and the wiring for the screed and tractor lights are routed.

10 As the tractor moves down the assembly line, the engine side covers and inlet air cleaner are installed, the rear plat-

form and open grate deck are put in place, and the operator's control console is mounted. Some pavers have two operator's consoles, one on each side, to give the operator a better view when paving close to curbing or other obstacles. Other pavers have a movable console that can slide to one side or the other. Any final electrical connections are made at this time.

11 The batteries and engine muffler are installed next and the various fluids are added as required. If the tractor has a tracked drive, the lower bogie wheels are installed at this point.

12 The tractor assembly is completed by attaching the screed leveling arms, hopper sides, engine access doors, lights, and other exterior components. The tires or

In operation, a dump truck filled with asphalt backs up to the front of the paver and slowly discharges its load into the paver's hopper. As the paver moves forward, the feeder conveyors move the asphalt to the rear of the paver, and the distribution augers push the asphalt outward to the desired width.

tracks are installed last. The engine is started and the finished tractor is given a preliminary check for proper operation.

Testing the tractor

13 The tractor is washed to remove any grease or oil that may have accumulated on the surfaces during assembly. A fluorescent dye is added to the hydraulic oil to help spot any leaks. The tractor is then hooked up to an automatic testing machine, which cycles it through various electrical and hydraulic functions. A computer records the results of these tests for future reference. An ultraviolet "black light" is used to detect leaks in the hydraulic system.

14 After the cycle test, the tractor is driven outside and given a short functional test to visually inspect its operation. If adjustments are required, they are made at this time. The tractor is then parked awaiting a customer's order.

Assembling the screed

15 The screed is assembled in a separate area from the tractor. The frame parts are fabricated and welded together. The burner assemblies and hydraulic vibrator motors are installed and plumbed with hoses. The burners provide heat along the length of the screed to keep the asphalt from sticking to it. The vibrators help provide partial compaction of the asphalt as it is being laid. Electrical wiring is routed to the various components. The hydraulic actuators to control the side-to-side slope of the screed are installed last.

Testing the screed

16 The finished screed is attached to a testing machine that duplicates the functions and controls of a tractor. The various screed functions — burner ignition, vibrator operation, slope control, and others — are then tested.

Finishing the paver

17 When a customer orders a paver, the customer may specify one of several models of tractors to be matched with one of several screed designs. The tractor, which has just a coat of primer paint, is now cleaned and given a final coat of paint. Any

warning labels, decorative striping, or name plates are then installed. The screed is usually painted black because it is in contact with the black, oily asphalt.

18 The screed is attached to the tractor. The electrical wiring, burner fuel lines, and hydraulic hoses are interconnected. The finished paver is then given a final functional test. The operator's seat is installed last.

Quality Control

All component suppliers are thoroughly checked and certified before they may begin shipping parts. Periodically, incoming parts are given a thorough dimensional and metallurgical inspection to ensure continued high quality. The air-operated wrenches used to tighten critical fasteners are checked and recalibrated to make sure they are delivering the proper torque. The tractor and screed are machine-tested separately in addition to several visual inspections by human operators, and then checked again once the tractor and screed are coupled together for delivery.

The Future

Many cities and states have placed an emphasis on reducing the surface variations, or waviness, of asphalt roadways. This is especially important when paving over an existing roadway, which may have significant surface variations from years of hard use. On some highway projects, a penalty is assessed against the road contractor for exceeding certain waviness limits. In order to meet these stringent requirements, contractors are asking asphalt paver manufacturers for more sophisticated slope and grade control systems. Future systems may include a laser-guided screed control, utilizing a computer-generated road profile as a reference.

Another area of future development for asphalt pavers involves a change in the formulation of the asphalt pavement itself. In the United States, the Strategic Highway Research Program, sponsored by the Federal Highway Administration, is developing a new asphalt pavement formulation known as Superpave. This new pavement is expected to produce smoother, more durable roads and is targeted for implementation in the year 2000. It will involve changes to both the asphalt binder material and the aggregates and may require different methods of placement.

Where to Learn More

Books

Barber-Greene. *Asphalt Construction Handbook.* Caterpillar, Inc., 1992.

Barber-Greene. *75 Years of Innovation: The Story of Barber-Greene.* Caterpillar, Inc., 1991.

Butler, John L. *First Highways of America.* Krause Publications, 1994.

Wallace, Hugh A. and J. Rogers Martin. *Asphalt Pavement Engineering.* McGraw-Hill, Inc., 1967.

Periodicals

Peterson, Eric. "Smooth Operators: A Start to Finish Look at the Highway Building Process," *Construction Monthly,* June, 1996, pp. 22-29.

Other

American Road and Transportation Builders Association. http://www.artba-hq.org.

Asphalt Contractor magazine. http://back40.global-image.com/group3/asphalt/site/index.html.

National Asphalt Pavement Association. http://www.hotmix.org.

"Paving products." Caterpillar Inc. http://www.cat.com/products/equip/ paving/paving.html.

—*Chris Cavette*

Automatic Drip Coffee Maker

The automatic drip coffee maker debuted in the United States in 1972, and as of 1996, some 73% of American households report owning an automatic drip coffee maker.

Background

Coffee was first cultivated in Ethiopia in the sixth century A.D. The coffee berries were consumed whole, or a wine was made out of the fermented fruits. Coffee as we know it, made from ground, roasted beans, dates to the thirteenth century, and by the fifteenth century, coffee was popular all across the Islamic world. The drink was introduced to Europe around 1615. The ancient method of preparing coffee was to boil the crushed roasted beans in water until the liquid reached the desired strength. The typical coffee pot was a long-handled brass pot with a narrow throat. This kind of pot is still used throughout the Arab world, and is known in the West as a Turkish coffee pot.

In England and America, boiling coffee in a sauce pan was for a long time the standard method. Sometimes the coffee was boiled for several hours; other classic recipes called for additions to the pot such as egg white, salt, and even mustard.

More sophisticated methods of brewing coffee evolved in France. The coffee bag, similar to the familiar tea bag, appeared in France in 1711. Ground coffee was placed in a cloth bag, the bag into a pot, and boiling water poured on top. Nearly a hundred years later, Jean Baptiste de Belloy, who was Archbishop of Paris, invented a three-part drip coffee pot. The top part of the pot held inside it a filter section made of perforated metal or china. Boiling water was poured through the filter section, and it slowly dripped down to fill the pot below. The percolator was invented in 1825. In a percolator, the pot full of water is placed directly on the stove burner. When the water boils, it condenses in the top of the pot, and then drips through a strainer basket filled with coffee. The Melitta filter—a plastic cone with several openings in the bottom, that holds a paper filter of finely ground coffee—appeared around 1910, as did the glass Silex, an hourglass-shaped filter pot.

The automatic drip coffee maker operates on the same principle as the Melitta and Silex, by dripping boiled water through finely ground coffee in a paper filter. This machine debuted in the United States in 1972 as the well-known Mr. Coffee™. Mr. Coffee™ was an immediate success, and popularized the automatic drip method. As of 1996, some 73% of American households report owning an automatic drip coffee maker. In an automatic drip coffee maker, a measured amount of cold water is poured into a reservoir. Inside the reservoir, a heating element heats the water to boiling. The steam rises through a tube and condenses. The condensed water is distributed over the ground coffee in the filter through a device like a shower head. The water flows through the filter, infusing with the coffee, and falls into a carafe. The carafe sits on a metal plate which has another heating element inside it. This keeps the coffee warm. Some models have timing features, so that they can be pre-filled at night to make coffee at dawn. Other units have a temporary shut-off function, so the carafe can be removed from the warmer plate while the coffee is filtering. Others pulse the water over the filter at intervals, for a slower drip and more concentrated brew.

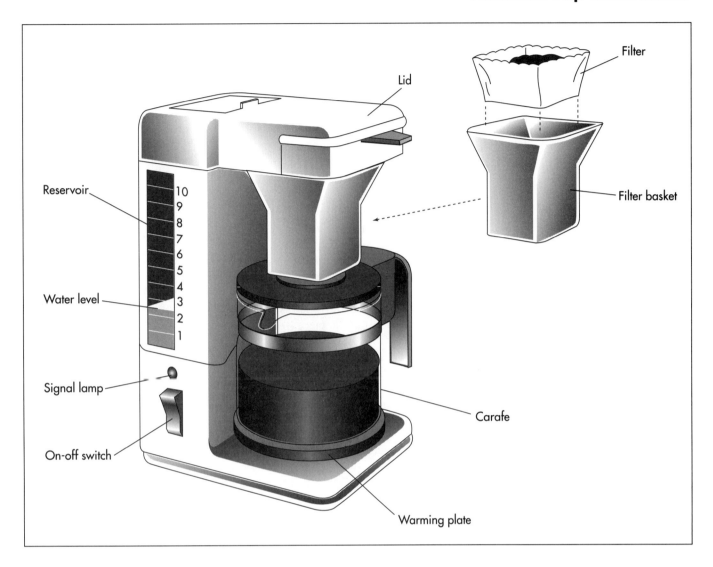

Filter

Lid

Filter basket

Reservoir

Water level

Signal lamp

On-off switch

Carafe

Warming plate

Raw Materials

Most automatic drip coffee maker parts are made out of plastic, including the body and the basket which holds the filter. The base plate, warmer plate, and heating unit are made out of various metals, usually steel or anodized aluminum. The carafe is made out of heat-proof glass. Other parts include timers, switches, and wiring.

The Manufacturing Process

The parts for the automatic drip coffee maker are typically made by specialized shops. The digital clocks, timers, and switches are all purchased from companies that produce those items. The plastic parts are made at a plastics company, and the metal parts at a metal stamping plant. The

manufacture of the actual coffee maker consists of putting all these parts together.

Injection molding

1 The plastic parts for automatic drip coffee makers are designed by the manufacturer and then outsourced to specialty plastics companies. The plastics company uses the manufacturer's design to make a mold. Then parts are produced by injection molding; heated plastic is forced into the mold under pressure, cooled, and released. These parts are then shipped to the manufacturer for assembly.

Stamping

2 The metal base plate is made at a specialized metal stamping plant. A sheet of metal is rolled out, and heavy machines punch out the specified shape. Then these are shipped to the manufacturer.

In an automatic drip coffee maker, a measured amount of cold water is poured into a reservoir. Inside the reservoir, a heating element heats the water to boiling. The steam rises through a tube and condenses. The condensed water is distributed over the ground coffee in the filter through a device like a shower head. The water flows through the filter, infusing with the coffee, and falls into a carafe.

Assembly

3 The parts of the coffee maker are put together on an assembly line. The electrical components are assembled first. These are designed so they can be simply snapped together. Workers standing at the assembly line each snap in a part as it comes to them, and the whole line may have 40 to 80 workers, each doing a specialized job. The timer device is snapped in, then the cord. The metal warmer plate is snapped on, and then the thermostat is wired. The heater for the warmer plate is assembled, then placed on a small pallet about the size of an index card. The pallets are placed on a conveyer belt that carries them through a sonic welding machine. This automatically welds the wiring for the heater. Once the internal wiring is complete, the rest of the pieces—the housing and the filter reservoir—are snapped together. Some pieces may be screwed in by workers using pneumatic screw drivers.

Packaging

4 After assembly, workers place the coffee makers in small cartons. Then workers place these cartons in larger packing cartons which might contain several units. These may be taped shut by hand, or they may be taped automatically by passing on a conveyor belt through a taping machine. Typically, another machine automatically imprints a bar code on the packing box, for tracking information. Then the boxes are stacked on large pallets and shipped or stored in a warehouse.

Quality Control

When the outsourced parts arrive at the coffee maker manufacturer, a receiving inspector checks them. Any defective parts are weeded out before they are taken to the assembly line. Then there may be several points along the assembly line where random pieces are removed and inspected. Typically, a hundred piece audit is done at the end of the assembly process. One hundred units are taken at random as they come off the assembly line, and these are thoroughly checked for internal and external defects.

The Future

European manufacturers are experimenting with coffee makers made out of a single plastic. The advantage of this is that the unit is recyclable. The single plastic can be melted down and re-used after the appliance is thrown away. There are distinct engineering problems to making a single-plastic coffee maker, since as many as six different plastics are used in some models to make up a single component. U.S. manufacturers do not seem as interested in single-plastic as European makers, but this may well become a global trend as recycling becomes more of an issue.

Where to Learn More

Books

Castle, Timothy James. *The Perfect Cup.* Addison-Wesley, 1991.

Roden, Claudia. *Coffee.* Faber & Faber, 1977.

Periodicals

Ellis, Beth R. "Mr. Coffee: the Man and His Machine." *Weekly Home Furnishing Newspaper*, June 15, 1987, pp. 1-3.

Huneve, Michelle. "Eat or Be Eaten." *The New York Times Magazine*, March 10, 1996, pp. 62-63.

—Angela Woodward

Ballpoint Pen

A ballpoint pen is a writing instrument which features a tip that is automatically refreshed with ink. It consists of a precisely formed metal ball seated in a socket below a reservoir of ink. As the pen is moved along a writing surface, ink is delivered. Even though ballpoint pens were first patented in the late nineteenth century, they only started to reach commercial significance in the early 1950s. Now, ballpoint pens dominate the writing instrument market, selling over one hundred million pens each year worldwide.

History

While the idea of a ballpoint pen had been around for many years, it took three different inventors and almost 60 years to develop this modern writing instrument. The first patent for this invention was issued on October 30, 1888, to a man named John J. Loud. His ballpoint pen consisted of a tiny rotating ball bearing that was constantly coated with ink by a reservoir above it. While this invention worked, it was not well suited for paper because it leaked and caused smearing. Two other inventors, Ladislas Biro and his brother Georg, improved on Loud's invention and patented their own version, which became the first commercially significant ballpoint pen. These pens still leaked, but not as badly. They became popular worldwide, reaching the height of sales in 1944. The next year another inventor, Baron Marcel Bich, finally solved the leakage problem and began manufacturing Bic pens in Paris. Over the years, many improvements have been made in the technology and quality of the various parts of the pen, such as the ink, the ball, the reservoir, and the body.

Background

The ballpoint pen was developed as a solution to the problems related to writing with a fountain pen. Fountain pens require the user to constantly refresh the pen by dipping its tip in ink. This is not necessary with a ballpoint pen because it is designed with its own ink reservoir, which uses capillary action to keep the ink from leaking out. At the tip of the pen is a freely rotating ball seated in a socket. Only part of this ball is exposed; the rest of it is on the inside of the pen and is constantly being bathed by ink from the reservoir. Pressing the tip of the pen on the writing surface causes the ball to roll. This rolling action then transfers ink from the inside of the pen to the writing surface.

While different designs of ballpoint pens are available, many of the components are the same. Common components include a ball, a point, ink, an ink reservoir or cartridge, and an outer housing. Some pens are topped with a cap to prevent it from leaking or having its point damaged. Other pens use a retractable point system for the same reason. Here a small spring is attached to the outside of the ink reservoir, and when a button is pushed, the point is either exposed or retracted. Still other varieties of ballpoint pens have multiple ink cartridges, making it possible to write in different colors using one pen. Other pens have refillable ink cartridges. One type of pen has a pressurized cartridge that enables the user to write underwater, over grease, and in space.

Raw Materials

A variety of raw materials are used for making the components of a ballpoint pen,

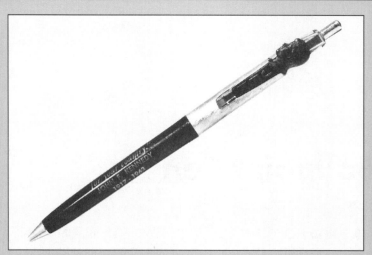

A 1963 plastic and metal ballpoint pen commemorating the assassination of President John F. Kennedy, Jr. (From the collections of Henry Ford Museum & Greenfield Village.)

Until the advent of the computer, humans have scrambled to find writing instruments to record story and song. The earliest scribbles were made with a burnt stick in sand. By the fourth century B.C., the Sumerians used wedge-shaped reed pens to cut pictorial shapes into clay tablets. Eygptians painted hieroglyphics with brushes made from marsh reeds and the ancient Chinese wrote with brushes of stiff hair. Ancient Greeks and Romans sharpened stiff reeds to a point, resulting in chirography that was taut and precise.

The quill pen, made from goose or swan feathers, was favored by writers for over 1,000 years. The soft quill was honed to a point, split at the tip to permit ink to flow freely, and constantly resharpened. A monumental improvement over the quill pen was Joseph Gillott's invention of the steel pen nib in the late nineteenth century, which required no sharpening and could be separated from the pen body and changed as needed. Still, the writer constantly dipped pen into ink, hoping to avoid drips.

Fountain pens store ink inside a reservoir within the pen, the nib thus supplied with a constant stream of ink. Alonzo Cross featured a "stylographic pen" with an ink-depositing needle point in the late 1860s, but blots and smears were still common. However, the ballpoint pen virtually eradicated messes. Ballpoint pens manufactured early in the century leaked, skipped, and dropped ink until 1950, when a new ink was developed that made the ballpoint reliable.

Nancy EV Bryk

including metals, plastics, and other chemicals. When ballpoint pens were first developed, an ordinary steel ball was used. That ball has since been replaced by a textured tungsten carbide ball. This material is superior because it is particularly resistant to deforming. The ball is designed to be a perfect sphere that can literally grip most any writing surface. Its surface is actually composed of over 50,000 polished surfaces and pits. The pits are connected by a series of channels that are continuous throughout the entire sphere. This design allows the ink to be present on both the surface and interior of the ball.

The points of most ballpoint pens are made out of brass, which is an alloy of copper and zinc. This material is used because of its strength, resistance to corrosion, appealing appearance, and ability to be easily formed. Other parts, like the ink cartridge, the body, or the spring can also be made with brass. Aluminum is also used in some cases to make the pen body, and stainless steel can be used to make pen components. Precious metals such as gold, silver, or platinum are plated onto more expensive pens.

The ink can be specially made by the pen manufacturer. To be useful in a ballpoint pen, the ink must be slightly thick, slow drying in the reservoir, and free of particles. These characteristics ensure that the ink continues to flow to the paper without clogging the ball. When the ink is on the paper, rapid drying occurs via penetration and some evaporation. In an ink formulation, various pigments and dyes are used to provide the color. Other materials, such as lubricants, surfactants, thickeners, and preservatives, are also incorporated. These ingredients are typically dispersed in materials such as oleic acid, castor oil, or a sulfonamide plasticizer.

Plastics have become an important raw material in ballpoint pen manufacture. They have the advantage of being easily formed, lightweight, corrosion resistant, and inexpensive. They are primarily used to form the body of the pen, but are also used to make the ink cartridge, the push button, the cap, and part of the tip. Different kinds of plastics are used, based on their physical characteristics. Thermosetting plastics, like phenolic resins, which remain permanently hard after being formed and cooled, are typically used in constructing the body, cap, and other pieces. Thermoplastic materials remain flexible. These include materials like high-density polyethylene (HDPE) and

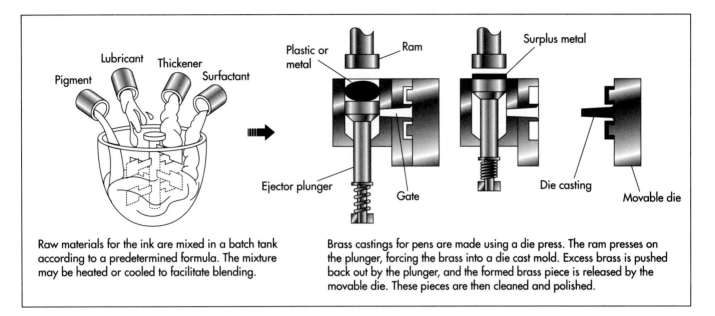

Raw materials for the ink are mixed in a batch tank according to a predetermined formula. The mixture may be heated or cooled to facilitate blending.

Brass castings for pens are made using a die press. The ram presses on the plunger, forcing the brass into a die cast mold. Excess brass is pushed back out by the plunger, and the formed brass piece is released by the movable die. These pieces are then cleaned and polished.

vinyl resins, which can be used to make most of the pen components.

The Manufacturing Process

Ballpoint pens are made to order in mass quantities. While each manufacturer makes them slightly differently, the basic steps include ink compounding, metal component formation, plastic component molding, piece assembly, packaging, labeling, and shipping. In advanced shops, pens can go from raw material to finished product in less than five minutes.

Making the ink

1 Large batches of ink are made in a designated area of the manufacturing plant. Here workers, known as compounders, follow formula instructions to make batches of ink. Raw materials are poured into the batch tank and thoroughly mixed. Depending on the formula, these batches can be heated and cooled as necessary to help the raw materials combine more quickly. Some of the larger quantity raw materials are pumped and metered directly into the batch tank. These materials are added simply by pressing a button on computerized controls. These controls also regulate the mixing speeds and the heating and cooling rates. Quality control checks are made during different points of ink batching.

Stamping and forming

2 While the ink is being made, the metal components of the pen are being constructed. The tungsten carbide balls are typically supplied by outside vendors. Other parts of the pen, such as the point and the body, are made using various molds. First, bands of brass are automatically inserted into stamping machines, which cut out thousands of small discs. The brass discs are next softened and poured into a compression chamber, which consists of a steel ram and a spring-backed ejector plunger. The steel ram presses on the metal, causing the plunger to retract and forcing the metal into a die cast mold. This compresses the metal and forms the various pen pieces. When the ram and plunger return to their original positions, the excess metal is then scraped off and recycled. The die is then opened, and the pen piece is ejected.

3 The formed pieces are then cleaned and cut. They are immersed in a bath to remove oils used in the molding process. After they emerge from the bath, the parts are then cut to the dimensions of the specific pen. The pen pieces are next polished by rotating brushes and cleaned again to remove any residual oils. The ball can then be inserted into the point cavity.

Molding the housing

4 The plastic components of the pen are constructed simultaneously with the

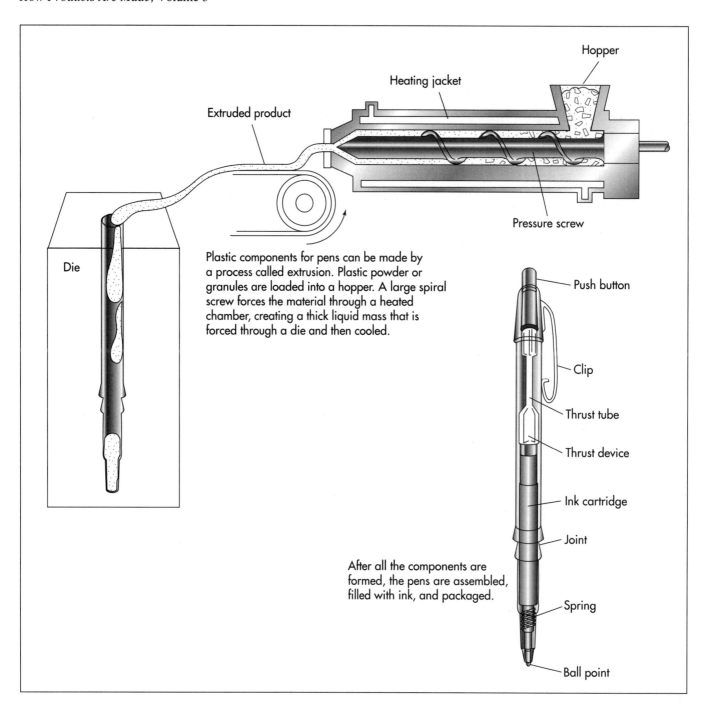

Plastic components for pens can be made by a process called extrusion. Plastic powder or granules are loaded into a hopper. A large spiral screw forces the material through a heated chamber, creating a thick liquid mass that is forced through a die and then cooled.

After all the components are formed, the pens are assembled, filled with ink, and packaged.

other pen pieces. They can be produced by either extrusion or injection molding. In each approach, the plastic is supplied as granules or powder and is fed into a large hopper. The extrusion process involves a large spiral screw, which forces the material through a heated chamber, making it a thick, flowing mass. It is then forced through a die, cooled, and cut. Pieces such as the pen body and ink reservoir are made by this method.

5 For pieces that have more complex shapes, like caps, ends, and mechanical components, injection molding is used. In this process the plastic is heated, converting it into a liquid that can then be forcibly injected into a mold. After it cools, it solidifies and maintains its shape after the die is opened.

Ink filling and assembly

6 After the components are formed, assembly can take place. Typically, the ballpoint is first attached to the ink reservoir. These pieces are then conveyored to

injectors, which fill the reservoir with the appropriately colored ink. If a spring is going to be present, it is then placed on the barrel of the reservoir.

Final assembly, packaging, and shipping

7 The point and reservoir are then placed inside the main body of the pen. At this stage, other components such as the cap and ends are incorporated. Other finishing steps, such as adding coatings or decorations or performing a final cleaning, are also done. The finished pens are then packaged according to how they will be sold. Single pens can be put into blister packages with cardboard backings. Groups of pens are packed into bags or boxes. These sales units are then put into boxes, stacked on pallets, and shipped to distributors.

Quality Control

The quality of pen components is checked during all manufacturing stages. Since thousands of parts are made each day, inspecting each one is impossible. Consequently, line inspectors take random samples of pen pieces at certain time intervals and check to ensure that they meet set specifications for size, shape, and consistency. The primary testing method is visual inspection, although more rigorous measurements are also made. Various types of measuring equipment are available. Length measurements are made with a vernier caliper, a micrometer, or a microscope. Each of these differ in accuracy and application. To test the condition of surface coatings, an optical flat or surface gauge may be used.

Like the solid pieces of the pens, quality tests are also performed on the liquid batches of ink. After all the ingredients are added to the batch, a sample is taken to the Quality Control (QC) laboratory for testing. Physical characteristics are checked to make sure the batch adheres to the specifications outlined in the formula instructions. The QC group runs tests such as pH deter-mination, viscosity checks, and appearance evaluations. If the batch is found to be "out of spec," adjustments can be made. For instance, colors can be adjusted by adding more dye.

In addition to these specific tests, line inspectors are also posted at each phase of manufacture. They visually inspect the components as they are made and check for things such as inadequately filled ink reservoirs, deformed pens, and incorrectly assembled parts. Random samples of the final product are also tested to ensure a batch of pens writes correctly.

The Future

Ballpoint pen technology has improved greatly since the time of Loud's first patented invention. Future research will focus on developing new inks and better designed pens that are more comfortable and longer lasting. Additionally, manufacturers will strive to produce higher quality products at the lowest possible cost. One trend that will continue will be the development of materials and processes which use metals and plastics that have undergone a minimum of processing from their normal state. This should minimize waste, increase production speed, and reduce the final cost of the pens.

Where to Learn More

Books

Carraher, Charles, and Raymond Seymour. *Polymer Chemistry*. Marcel Dekker, 1992.

Periodicals

Peeler, Tom. "The Ball-Point's Bad Beginnings." *Invention & Technology*, Winter 1996, p. 64.

Trebilcock, Bob. "The Leaky Legacy of John J. Loud." *Yankee*, March 1989, p. 141.

—*Perry Romanowski*

Black Box

Flight Data Recorders (FDRs) can now track such in-flight characteristics as speed, altitude, flap position, auto-pilot mode, even the status of onboard smoke alarms.

Background

Black box is a generic term used to describe the computerized flight data recorders carried by modern commercial aircraft. The Flight Data Recorder (FDR) is a miniaturized computer system which tracks a variety of data regarding the flight of the plane, such as airspeed, position, and altitude. This device is typically used in conjunction with a second black box known as the Cockpit Voice Recorder (CVR), which documents radio transmissions and sounds in the cockpit, such as the pilots' voices and engine noises. In the event of a mishap, the information stored in these black boxes can be used to help determine the cause of the accident.

Black boxes have been used since the earliest days of aviation. The Wright brothers carried the first flight recorder aloft on one of their initial flights. This crude device registered limited flight data such as duration, speed, and number of engine revolutions. Another early aviation pioneer, Charles Lindbergh, used a somewhat more sophisticated version consisting of a barograph, which marked ink on paper wrapped around a rotating drum. The entire device was contained in a small wooden box the size of an index card holder. Unfortunately, these early prototypes were not sturdily constructed and could not survive a crash.

In the 1940s, as commercial aviation grew by leaps and bounds, a series of crashes spurred the Civil Aeronautics Board to take the importance of flight data more seriously. They worked with a number of companies to develop a more reliable way of collecting data. Rising to the challenge, General Electric developed a system called the "selsyns," which consisted of a series of tiny electrodes attached directly to the plane's instruments. These sensors wired information to a recorder in the back of the plane. (Recorders are typically stored in the plane's tail section because it is the most crash-survivable area of the plane.) GE engineers overcame a number of technical challenges in the design of the selsyns. For example, they cleverly recognized that the high altitude conditions of low pressure and temperature would cause the ink typically used in recording devices to freeze or clog the pens. Their solution was a recording system that relied on a stylus to cut an image into black paper coated with white lacquer. However, despite their efforts, the unit was never used in an actual flight. Around the same time, another engineering company, Frederick Flader, developed an early magnetic tape recorder; however, this device was also never used.

Black box technology did not advance further until 1951, when professor James J. Ryan joined the mechanical division of General Mills. Ryan was an expert in instrumentation, vibration analysis, and machine design. Attacking the problem of FDRs, Ryan came up with his own VGA Flight Recorder. The "V" stands for Velocity (airspeed); "G" for G forces (vertical acceleration); and "A" is for altitude. The Ryan Recorder was a 10 lb (4.5-kg) device about the size of a bread box with two separate compartments. One section contained the measuring devices (the altimeter, the accelerometer, and the airspeed indicator) and the other contained the recording device, which connected to the three instruments.

Ryan's basic compartmentalized design is still used in flight recorders today, although it has undergone numerous improvements. The stylus and lacquer film recording device was replaced by one-quarter-inch (6.4-mm) magnetic tape, which was in turn replaced by digital memory chips. The number of variables that recorders can track has also dramatically increased, from three or four parameters to about 300. FDRs can now track such in-flight characteristics as speed, altitude, flap position, auto-pilot mode, even the status of onboard smoke alarms. In the early 1960s, the airline industry added voice recording capability with the Cockpit Voice Recorder (CVR). But perhaps the most significant advance in flight recorder manufacture has been the improvements made in its construction, allowing the units to better withstand the destructive force of a crash. Early models had to withstand only about 100 Gs (100 times the force of gravity), which is loosely equivalent to the force of being dropped from about 10 ft (3 m) off the ground onto a concrete surface. To better simulate actual crash conditions, in 1965 the requirements were increased to 1,000 Gs for five milliseconds and later to 3,400 Gs for 6.5 milliseconds.

Today, large commercial aircraft and some smaller commercial, corporate, and private aircraft are required by the FAA to be equipped with a Cockpit Voice Recorder and a Flight Data Recorder. In the event of a crash, the black boxes can be recovered and sent, still sealed, to the National Transportation Safety Board (NSTB) for analysis.

Components

The Flight Data Recorder and the Voice Data Recorder (or Cockpit Voice Recorder) are built from similar components. Both include a power supply, a memory unit, electronic controller board, input devices, and a signal beacon.

Power supply

Both FDRs and CVRs run off of a dual voltage power supply (115 VAC or 28 DC) which gives the units the flexibility to be used in a variety of aircraft. The batteries

are designed for 30-day continuous operation and have a six-year shelf life.

Crash Survivable Memory Unit (CSMU)

The CSMU is designed to retain 25 hours of digital flight information. The stored information is of very high quality because the unit's state of the art electronics allow it to hold data in an uncompressed form.

Integrated Controller and Circuitry Board (ICB)

This board contains the electronic circuitry that acts as switchboard for the incoming data.

Aircraft Interface

This port serves as the connection for the input devices from which black boxes obtain all their information about the plane. The FDR interface receives and processes signals from a variety of instruments on board the plane, such as the airspeed indicator, on-board warning alarms, altimeter, etc. The interface employed for the CVR receives and processes signals from a cockpit-area microphone, which is usually mounted somewhere on the overhead instrument panel between the two pilots. The microphone is intended to pick up sounds that may aid investigators in determining the cause of a crash, such as engine noise, stall warnings, landing gear extension and retraction, and other clicks and pops. These sounds can help determine the time at which certain crash-related events occurred. The microphone also relays communications with Air Traffic Control, automated radio weather briefings, and conversation between the pilots and ground or cabin crew.

Underwater Locater Beacon (ULB)

Each recorder may be equipped with an Underwater Locator Beacon (ULB) to assist in identifying its location in the event of an overwater accident. The device, informally known as a "pinger," is activated when the recorder is immersed in water. It transmits an acoustical signal on 37.5 KHz that can be detected with a special receiver. The

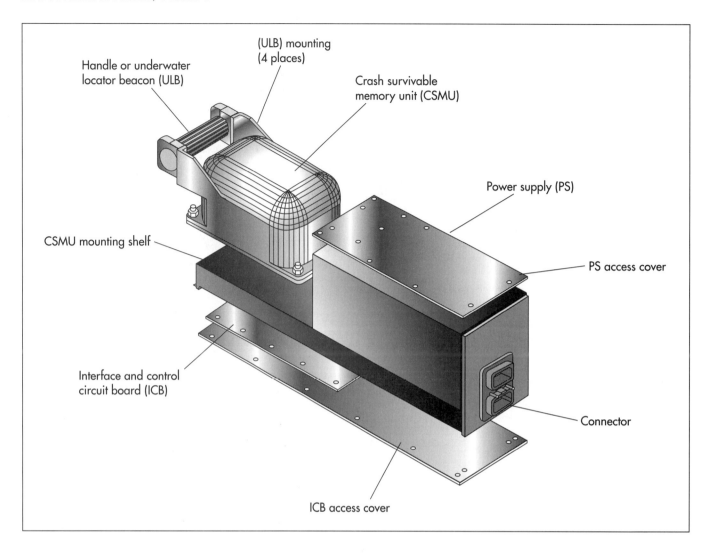

Handle or underwater locator beacon (ULB)

(ULB) mounting (4 places)

Crash survivable memory unit (CSMU)

Power supply (PS)

CSMU mounting shelf

PS access cover

Interface and control circuit board (ICB)

Connector

ICB access cover

The Flight Data Recorder (FDR) is a miniaturized computer system that tracks a variety of data regarding the flight of the plane, including its airspeed, position, and altitude. The system is housed in a heavy metal container that is built to withstand the stress of a crash.

beacon can transmit from depths to 14,000 ft (4,200 m).

The Manufacturing Process

The key to manufacturing a successful black box is to make it as indestructible as possible. This is done by sheathing the components inside a multi-layer protective shell. The different makers of recorders each have their own proprietary design, but in general the manufacturing process can be described as follows:

1 The key components (the power supply, the interface/controller board, and the memory circuits) are built as separate units and then assembled to form the completed black box. This modular approach allows the components to be easily replaced without disassembly of the entire device. Each of these components has its own special as-

sembly requirements, but primary attention is given to the protecting the memory unit, since it contains the data that will be of interest to investigators.

2 A multi-layered configuration is used to ensure the memory unit's integrated circuits are adequately protected. The outermost layer is the housing, which consists of steel armor plate.

3 Below that is a layer of insulation, followed by a thick slab of paraffin, which forms a thermal block. As the paraffin melts, it absorbs heat and therefore keeps the temperature of the memory core lower.

4 Beneath the paraffin lies the board containing the memory chips.

5 Underneath the memory board is another paraffin thermal block, followed by another layer of insulation. The entire as-

sembly is mounted on a steel plate that serves as an access cover.

6 The assembled Crash Survivable Memory Unit is then bolted onto the front of a heavy metal plate mounting shelf with four large retaining bolts. The power supply is attached just behind the CSMU.

7 The Interface and Control Circuit Board (ICB) is attached by screws to the underside of the mounting shelf. A metal access cover protects the board and provides easy access.

8 The Underwater Locator Beacon (ULB) is affixed to the two arms extending from the front of the memory unit. The ULB protrudes from the casing and has a cylindrical shape that allows it to be used as a handle for the entire device. If the recorder is to be sold without a ULB, a hollow metal handle tube is installed in its place.

9 The outer casing is painted bright orange or red to make it more visible in a crash.

Quality Control

After manufacture, the units are exposed to a series of grueling and somewhat bizarre torture test conditions. Black boxes are shot from cannons, stabbed by thin steel rods, attached to 500 lb (227 kg) weights and dropped from 10 ft (3 m) above the ground, crushed in a vice at 5,000 lb (2,270 kg) of pressure, cooked with a blow torch for an hour at 2,012°F (1,100°C), and submerged under the equivalent of 20,000 ft (6,000 m) of seawater for one month. After such testing, the onboard microprocessor allows a variety of diagnostics to be run to ensure the unit is operating correctly. The high speed interface allows the entire memory unit to be checked in under five minutes. This evaluation can be done at the factory to check that the unit is working perfectly, then again after installation to ensure it is still functioning properly. By regulation, flight recorders for newly manufactured aircraft must accurately monitor at least 28 critical factors, such as time, altitude, airspeed, heading, and aircraft attitude. The average time between failures for these de-

vices should be greater than 15,000 hours, and they are designed to be maintenance free. If the unit passes all of the tests described above, it meets the requirements established by the FAA (Federal Aviation Authority).

The Future

The future is already unfolding for manufacturers of black boxes. Smith Industries, a major supplier of flight recorders, has recently announced it is developing a single device which will replace separate FDR and CVR units. Their device is known as an Integrated Data Acquisition Recorder (IDAR), and it incorporates flight and voice data in a single box configuration, together with a data transfer system for maintenance data retrieval. The introduction of the IDAR allows a 25% reduction in critical system weight. Interestingly, this new direction in product development comes at the same time as new legislation that makes the recording of data linked to air traffic control messages mandatory. This new law would require black boxes to contain even more information. It is likely that the manufacturers of flight recording equipment will rise to the challenge and develop black boxes that can store more and more information in ever-shrinking packages.

Where to Learn More

Periodicals

AlliedSignal Aerospace Catalog. AlliedSignal, Inc.

Baldwin, Tom. "Black boxes Built to Survive Doom." *Journal of Commerce and Commercial,* July 29, 1996, p.1B.

Goyer, Robert. "The Secrets of Black Boxes." *Flying,* December 1996, p. 88.

Sendzimir, Vanda. "Black Box." *American Heritage of Invention & Technology,* Fall 1996.

—Randy Schueller

Bulldozer

Popularized in the 1920s and used heavily ever since, the bulldozer, commonly termed a dozer, is a clear offspring of the crawler tractor.

Background

Popularized in the 1920s and used heavily ever since, the bulldozer, commonly termed a dozer, is a clear offspring of the crawler tractor. Used in conjunction with other earthmoving vehicles, the bulldozer is a powerful and necessary tool utilized in almost every construction site in the world.

Primarily manufactured in the United States by Caterpillar, John Deere, and Case Tractor Company, the bulldozer provides for many industrial applications such as construction, waste management, and farming.

Raw Materials

Bulldozers and crawlers, characterized for their immense blade and versatile track, are comprised of many structural, hydraulic, and engine assemblies. The core body of the bulldozer, consisting of the mainframe and undercarriage, is primarily fabricated from low carbon structural steel plates and a giant casting. The cab contains many glass, rubber, and plastic components which enhance the ergonomic feel of the machine. Supplying the power for the dozer and its various systems, the engine contains many high strength steel parts, which endure high operating temperatures. The other necessary components, the blade, power train, and various systems components, are formed from structural and high carbon steel. The track, which is fashioned from many standard grade steel links, adds to the already tremendous weight of this mostly steel machine. Once the dozer is filled with fuel, hydraulic fluid, coolant, oil, and other types of fluids, its weight increases by several hundred pounds. Decorative trim, decals, and paint complete the dozer's aesthetics and add distinctive appeal.

Design

Two distinct features characterize the bulldozer, the long, vertical steel blade in the front of the vehicle and the rotating twin tracks, which facilitate the bulldozer movement. The blade, which can weigh up to 16,000 lb (7,264 kg), is useful for pushing material from one spot to another. Perpendicular to the ground, the curved blade is attached to the frame by a long lever arm that can tilt and move up and down under hydraulic power.

The familiar flexible track of a bulldozer is widely utilized in industrial machinery equipment and military tanks. In fact, some farming tractors are considered to be cousins to the bulldozer, since they also utilize the flexible track instead of standard wheels. Steel links, sometimes more than 2 ft (61 cm) in length, are connected with lubricated pins to provide for fluid motion and stability. Moreover, many bulldozers have incorporated an elevated sprocket design which suspends the power train, and thereby, improves its reactivity to the terrain. The diesel engine of the bulldozer can generate anywhere from 50-700 horsepower, so rough terrain and steep slopes are not a problem for this machine.

Mounted above the flexible track, the operator cabin contains the complex hydraulic mechanisms, which power the blade in a limited vertical range. The cabin design has seen many improvements in operator comfort and ergonomics and has provided for many improved automotive features, such as air conditioning, AM/FM radio, automatic seat adjustments, electronic controls, and systems-monitoring equipment. In these areas of dozer design, the engineering and research that precedes the manufacturing mimic the automotive industry in many ways.

The power train includes the transmission, differential, and gears that rotate the track. Coupled to the engine crankshaft, the power train will transmit power from the engine to the elevated sprocket gear. Many new bulldozers have independent steering, which allows each sprocket to rotate at full power even while one is rotating slower as the dozer is in a turn. Other innovations in recent years include differential steering, hydraulic power, and planetary gear transmissions.

The Manufacturing Process

The bulldozer, a seemingly endless network of bulky steel components, complex systems, and intricate assemblies, begins its manufacturing process on an assembly line. Prior to final assembly, much machining, fabrication, and sub-assembly must take place. Manufacturing begins with engineering prints and drawings taken from a computer-aided drafting (CAD) program that outlines the method of construction for each component part. Some of these programs can be used to set up machines for which most of the manufacturing will take place, that is, in fabrication cells, large machining centers, and sub-assembly lines. This is called computer-aided manufacturing (CAM) and is used to produce the components and assemblies that join together on the main line. Some of these components will then undergo heat treating, annealing, or painting after their respective fabrication cell, sub-assembly line, or machining center step. An overhead conveyor system will then transport the pieces through the rough paint or powder coating operation and lift them to the main assembly line, where they arrive in time to be assembled. These pieces may also be transported by lift truck, hand cart, or floor conveyor to arrive at the staging area before they are assembled to the bulldozer.

Mainframe core

1 The mainframe core, which forms the rigid inner body, is cut from steel plate and structural shaped, so that it easily resists high impact shock loads and torsional forces normally incurred by the dozer. The main structural skeleton, formed through the welding of steel plates to machined casting, is comprised of two boxed-in rail sections connected to the main casing. The fabrication is normally performed in a fabrication cell, where the burned plate arrives ready to be mounted into fixtures and manually or robotically welded to the stationary central casting. Far too massive to be lifted by hand, the frames are then transported by overhead crane to different stations, where steel mounting blocks and trunions, or cross members, are welded on as a support for the other components of the bulldozer. Once completed, the frame is rotary sanded on all plated surfaces and sent to the paint booth and the main assembly line.

Diesel engine and transmission

2 At the assembly line, the independently manufactured diesel engine and transmission join the mainframe. The engine is usually purchased completely assembled as it is a complex system with machined components that can be used in many different vehicular applications. In fact, the engine (which has been subjected to various performance tests) is certified to operate on arrival. The engine mounts in the front of the bulldozer; however, it is connected to the transmission, which sits in the back. The two are connected by a long shaft and supported by couplings and bearings. The transmission is then connected to a series of gears and differentials to comprise the rest of the power train. By mounting on pads previously welded to the frame, the engine/transmission assembly can be bolted directly to the base on the main assembly line.

Radiator and additional assemblies

3 On the front of the bulldozer, an engine casing is mounted to support the radiator and hydraulic lifting cylinders. The radiator, another finished assembly, will then sit between the engine casing and mount to the front drive shaft. Connections can then be made to attach water lines from the engine to the radiator. Additional assemblies for the hydraulic, lubrication, cooling, and fuel systems are also constructed at other locations and purchased as a finished as-

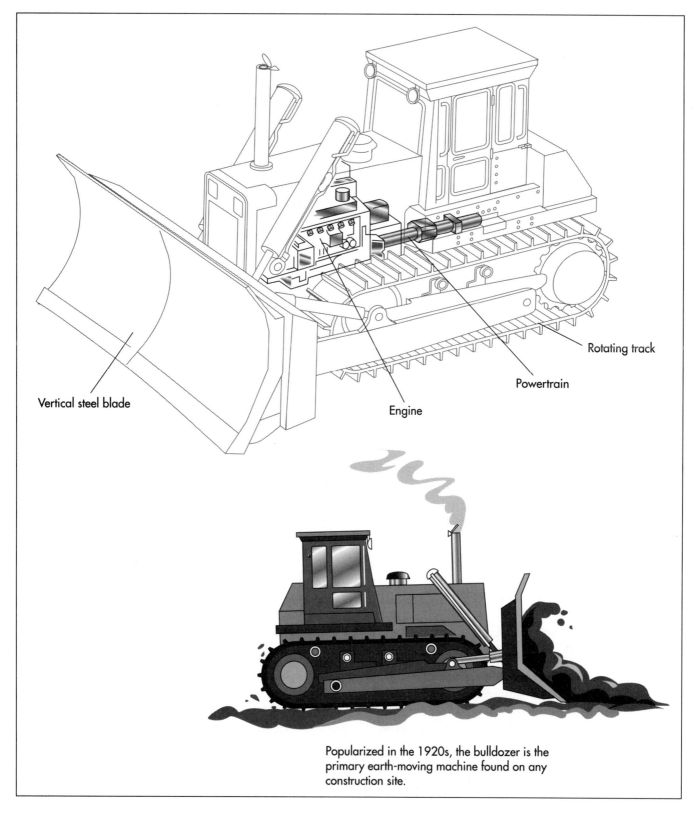

Vertical steel blade

Engine

Powertrain

Rotating track

Popularized in the 1920s, the bulldozer is the primary earth-moving machine found on any construction site.

sembly ready to be fastened directly to the engine or base. These include hydraulic lines composed of tubes, hoses, and fittings pre-assembled and mounted on the engine or frame and connected to pumps, valves, tanks, and cylinders, each of which can be brought to the main assembly line as a finished component. Fuel, exhaust, hydraulic, and coolant lines also arrive ready for assembly and mate to other finished components. Many of these components and subassemblies must be inspected and approved for dimensional compliance at an incoming inspection station prior to assembly.

Large component assembly

As the entire assembly of the frame, engine, transmission, and line groups move along the main production line, larger assemblies and components are brought in by overhead cranes, overhead conveyors, automatic guided vehicles (AGV), or lift trucks. These components include the cab, larger hydraulic cylinders, undercarriage components, and the front blade.

4 The cab, which can also be purchased as a finished assembly, is usually manufactured at a different facility and shipped for assembly. Usually complete in its array of electronics and controls, the cab will be mounted on steel blocks or pads located on the dozer frame. After mounting, connections will be made to the various controls, and power can supplied to the fully functioning cab.

5 Concurrent with the engine/transmission mounting, the undercarriage, composed of tubular roller frames, drive sprockets, and bogey independent suspension rollers, will be mounted on the frame and assembled to the drive train. The axle assembly will turn the outer sprockets that rotate the track, allowing the vehicle to maneuver. The sprockets, typically 2 ft (61 cm) in diameter, will fit into the track with case hardened teeth, which move the track as they rotate. In many manufacturing operations, the undercarriage can be machined, assembled, and painted in the same facility as the main assembly line, but various smaller components like bearings and lubrication bushings need to be outsourced to other facilities or outside contractors. The track, which is often pre-assembled from machined steel links, can be fitted around the drive sprockets, rollers, and front/back guide gears only after the engine/transmission and undercarriage components are in place. The exhaust stack, attached directly to the engine, is supported by brackets and flanges at its base.

6 After the cab controls are connected to the engine and hydraulic systems, prefabricated cowlings or body panels are mounted directly on the base frame to cover the engine, transmission, radiator, and fluid lines. The body panels are designed to fold back, making the inside of the dozer easily accessible for regular maintenance. They are assembled into hinges already fastened to structural supports. Tooling and storage compartments may also be built into the dozer once the lines have all been connected. Deck plates lie around the cabin and are welded to support brackets.

7 The front blade is attached to hydraulic cylinders, which can position the blade at different angles of tilt. The cylinders, each comprised of a hardened steel piston inside a honed cylinder, are attached at one end to engine casing in the front of the bulldozer to move the blade vertically. Initially in the assembly process, the cylinders are left unattached at the one end until the roll formed steel blade is assembled, and then hydraulic lines can be fitted and tightened. The lower end of the blade is attached at two joints with large steel pins which rotate and tilt the blade with two more cylinders. Arms extending from the undercarriage are attached to the blade and then are assembled along with the other undercarriage components.

Final assemblies

8 Once the dozer has been outfitted with its primary components, more hoses, electrical lines, and fluid lines are attached at fitted connections. Items such as the batteries, which are connected to the starter on the engine, lie underneath a cowling in a compartment located near the engine. Lights, one of the last items installed on the dozer, will be placed in a number of different areas and connected to their power source. In addition, hand or guard rails and foot pegs are bolted on the frame which complete main line assembly.

Paint

9 At Caterpillar's Track-Type Tractor (TTT) division located in Peoria, Illinois, Caterpillar bulldozers and crawlers use the same paint and final prep lines as many other tracked vehicles. Applied manually with spray guns, the final paint booth will deliver paint to any area not blocked off with paper or plastic wrapping. The paint dries quickly and the bulldozer will

Opposite page:
Two distinct features characterize the bulldozer, the long, vertical steel blade in the front of the vehicle and the rotating twin tracks, which facilitate the bulldozer movement. The blade, which can weigh up to 16,000 lb (7,264 kg), is useful for pushing material from one spot to another.

move to the next station where decals and trim are applied by hand templates.

Fluids

10 Various fluids are added, and the vehicle is then sent to a testing station where the operation of all systems is mechanically verified and recorded. The vehicle is transported from the manufacturing site to a staging area for customization and shipping. The completed bulldozer is shipped on a flat bed trailer and is ready for field operation upon arrival.

Byproducts/Waste

Waste produced by the manufacturing operations may include machining coolants, oils, parts-cleaning detergents, paint, and diesel fuel. The United States Environmental Protection Agency (EPA) places strict regulations on manufacturers to mandate that these potentially harmful liquids are disposed of in a proper manner. Companies contract a waste removal firm to recycle most of the liquid waste. Metal chips and shavings are recycled and sold to scrap dealers in an effort to reduce waste.

The Future

Bulldozers consistently undergo component design modernization efforts, and innovations appear inevitable. Improvements in cab comfort and diesel engine efficiency will probably be the driving force for many of these changes, while design and operational changes will be limited to individual components. In spite of the fact these enhancements in both the manufacturing process and streamlining of material flow will probably not change the face of bulldozers, costs may improve. Therefore, as a useful member of any earth-moving team, the bulldozer will continue to serve a unique purpose in building construction, waste management, and many industries.

Where to Learn More

Books

D7R Track-Type Tractor Specifications. Caterpillar, 1996.

D9R Track-Type Tractor Specifications. Caterpillar, 1995.

D11R Track-Type Tractor Specifications. *Caterpillar,* 1996.

—*Jason Rude*

Camera

Background

Photography has staked its claim as America's favorite hobby, and today, cameras are available in sizes and shapes to suit the needs of every kind of photographer and budget. Much like Henry Ford wanted a Model T in every driveway, George Eastman thought every consumer should be able to afford a camera. His developments in photographic film and portable, affordable cameras led to photo negatives from which prints can be made, color film, color positives or slides, pocket-sized cameras, and point-and-shoot cameras (including single-use or disposable cameras) known for their ease of operation. Photography has also branched into more complex directions with developments in the camera lens, the single-lens reflex (SLR) camera that allows the photographer to see through the viewfinder what the camera sees, state-of-the-art electronics, and an assortment of mechanical controls.

From the simplest amateur camera to the most complex, professional piece of equipment, all cameras have five common parts. The lens is made of glass or plastic (or groups of glass elements) and focuses light passing through it on the film to reproduce an image. The diaphragm is an opening or aperture that controls the amount of light entering the camera from the lens and so limits the film's exposure to light. The diaphragm ranges in complexity from a fixed lens, opening in a simple camera, to apertures that can be adjusted manually or automatically.

The three remaining parts common to all cameras are incorporated in the camera body (also called a chassis or housing). The shutter also limits the film's exposure to light by controlling the length of time the film is exposed. Shutter speed can be adjusted in many cameras to suit light conditions and the photographic subject matter; moving objects can be frozen on film with fast shutter speeds. The camera body encloses and protects the operating parts of the camera, including a light meter, the film transport system, built-in flash, the reflex viewing system, and electronic and mechanical components. The body must be lightproof, durable, and resistant to environmental changes. The viewfinder is a specialized lens the photographer uses to preview the photograph either through the lens, if the camera is a reflex-type, or in a separate view for simpler cameras.

History

The story of the camera may have begun thousands of years ago when people first noticed that a chink in a wall or hole in a tent let light into the room and made a colored, upside-down reflection. The word camera means room, and the first camera was a room (or tent, actually) called a camera obscura with an eye at the top of the tent much like a periscope that could be rotated. Artists used it by training the eye on an image, which was reflected down onto the artist's work table where it could be drawn. Euclid and Aristotle studied the principles of light, and Leonardo da Vinci described and diagrammed the camera obscura, although it was not his discovery.

The first portable cameras were boxes with lenses on the front over apertures and plates at the back. The plates were flat and covered with light-sensitive materials. By removing

Instead of a viewfinder or eyepiece, the digital still camera has a color LCD screen similar to the view-type screen on some video cameras, so photos can be viewed instantly. It can be connected by cables to a computer, television, or VCR, so pictures can be transferred to screen, tape, or digitized electronically.

THE KODAK CAMERA.

"You press the button, -

- - - we do the rest."

The only camera that anybody can use without instructions. Send for the Primer, free.

The Kodak is for sale by all Photo stock dealers.

The Eastman Dry Plate and Film Co.,

Price $25.00—Loaded for 100 Pictures. ROCHESTER, N. Y.

A full line Eastman's goods always in stock at LOEBER BROS., 111 Nassau Street, New York.

This Kodak Camera advertisement appeared in the first issue of The Photographic Herald and Amateur Sportsman, *November, 1889. The slogan "You press the button, we do the rest" summed up George Eastman's ground-breaking snapshot camera system. (From the collections of Henry Ford Museum & Greenfield Village.)*

George Eastman introduced his Kodak™ camera in 1888 and revolutionized popular photography. The Kodak camera was small, hand-held, inexpensive, and, for the first time, made especially to hold a roll of flexible film. Prior to this, light sensitive chemicals captured the black-and-white negative images on pieces of glass. Large cameras were used to hold the photographic plates and a tripod was needed for support. For ordinary Americans, photography consisted of posed portraits in a professional photographer's studio. The Kodak camera allowed the average person to take photographs of their families, their homes, and their surroundings. It inaugurated the snapshot era of do-it-yourself photography. Awarded a medal at the Photographers' Annual Convention as the photographic invention of 1888, thousands of $25.00 Kodak cameras sold during the first year.

By 1889, celluloid, a type of plastic, replaced the paper of the first flexible film base. Another unique feature for the time was that the amateur photographer returned the unopened camera to the Rochester, New York, factory. There the film negatives were processed and the 2.5 in (6.35 cm) circular images were printed on paper and mounted on cardboard. The camera was then re-loaded with an unexposed roll of flexible film and returned to the customer with the processed photographs and negatives. This cost $10.00 and produced 100 snapshots. This activity became so popular that the term *kodaking* soon meant a fun outing to take snapshots.

Cynthia Read-Miller

the only image was the one on the plate; photos, like those produced by Louis Daguerre and Joseph Niépce in France during the 1820s and 1830s, were unique artworks that were not reproducible. Plate-type photography continued to be refined, and, as plates were made more sensitive to light, the lens was improved to provide a variable aperture to control light exposure. The camera was also modified by adding a shutter, so exposure time could be limited to seconds or less. The shutter was made of several metal leaves that opened or closed completely. A rubber bulb was used to provide air pressure to operate the shutter.

The invention of roll film in 1889 by George Eastman made photography more portable because cameras (and their operators) did not need to carry cumbersome plates and chemicals. Eastman's invention and the cameras he also manufactured made photography a popular hobby. By 1896, the Eastman Kodak Company had sold 100,000 cameras. The camera was modified to include a film transport system with take-up spools, a winder, a lever for cocking the shutter, and shutter blinds. By the turn of the century, the major obstacles to taking photographs had been eliminated and, in the twentieth century, photographic history has branched from the basic concept and perfected each development. These developments are numerous, but include design and perfection of flash units including synchronized and high-speed flash; continued miniaturization of cameras; the Polaroid system of producing a finished print in the camera and without a negative; design of high quality equipment like Leica, Zeiss, and Hasselblad cameras and lenses; and advocacy of photography as an art form by photographers such as Matthew B. Brady, Alfred Stieglitz, Edward J. Steichen, and Ansel Adams.

Design

Camera design is an intricate and specialized field. All designs begin with conceptualizing a product and evaluating the potential market and the needs of the consumer for the proposed product. Designs begin at computer-aided design (CAD) work stations, where the product's configuration and workings are drawn. The designer se-

the cover over the lens, light entered the box and was focused by the lens on the rear plate. Early exposures took from several seconds to a number of minutes because the sensitivity of the plates was so poor. Also, lects the materials, mechanics, electronics, and other features of design and construction, including interfaces with lenses, flash units, and other accessories.

The computer design is also tested by computer simulation. Designs that pass the computer program's review are checked against the initial concept and marketing and performance goals. The camera may then be approved for production as a prototype. Manufacture of a prototype is needed to test actual performance and to prepare for mass production. The prototype is tested by a rigorous series of field and laboratory tests. Prototypes selected for manufacture are used by the engineers to prepare design details, specifications, and toolmaking and manufacturing processes. Many of these are adapted directly from the CAD designs by computer-aided manufacturing (CAM) systems. Additional design is needed for any systems or accessories that interface with the new product. Camera manufacturers can conceive a new product and have it ready for shipment in approximately a year by using CAD/CAM design methods.

The Manufacturing Process

Camera chassis and cover

1 The camera chassis or body and back cover are made of a polycarbonate compound, containing 10-20% glass fiber. This material is very durable, lightweight, and shock-resistant as well as tolerant to humidity and temperature changes. Its major disadvantage is that it is not resistant to chemicals. The polycarbonate is molded to very specific tolerances because the internal workings of the camera must fit precisely to work well and to use the strength of the chassis for protection against jarring and other shocks, to which mechanical and electronic parts are sensitive. After the chassis is molded and assembled, it becomes the frame to which other parts of the camera, like electrical connections in the battery housing and the auto focus module, are attached.

Shutter and film transport system

2 The shutter assembly and film transport system are manufactured on a separate assembly line. These parts are largely me-

chanical although the film transport system has electronics to read the speed of the film. DX film coding appears as silver bands on the roll of film, and these are detected by multiple contacts in the film chamber. More advanced cameras have microchips that see the data imprinted in the silver bands and adjust shutter speed, flash, and other camera actions. Again, all parts are precisely made; the film magazine size must be accurate to 60 thousandths of an inch.

3 The shutter functions like a curtain that opens and closes. It must operate exactly to expose the film for the correct length of time and to coordinate with other operations such as the flash. The shutter is made of different materials depending on the type of camera and manufacturer.

Viewfinder lens

4 The viewfinder lens is a specialized lens that is manufactured using the same methods as a camera lens. The viewfinder also is made of optical glass, plastic, or glass/plastic combinations. All but the simplest viewfinders contain reticles that illuminate a frame and other information on the eyelens to help the photographer frame the picture. An in-line mirror has specialized coatings for color splitting; as many as 17 coatings may be added to the mirror to correct and modify its reflective properties.

Single-lens reflex (SLR) cameras have through-the-lens viewing capabilities and are also called real image viewfinders because they let the photographer see as the lens sees. The SLR viewfinder uses a prism to bend the light from the lens to the photographer's eye, and the prism is made of optical glass to precise requirements to make the correct view possible.

LCD screen and electronics

5 Advanced cameras and most compact models include a liquid crystal display (LCD) screen that provides information to the photographer such as film speed, aperture, photographic mode (including landscape, portrait, close-up, and other modes), count of photos taken, operation of redeye and flash and other accessories, battery condition, and other data regarding the camera's workings. Integrated circuitry is constructed

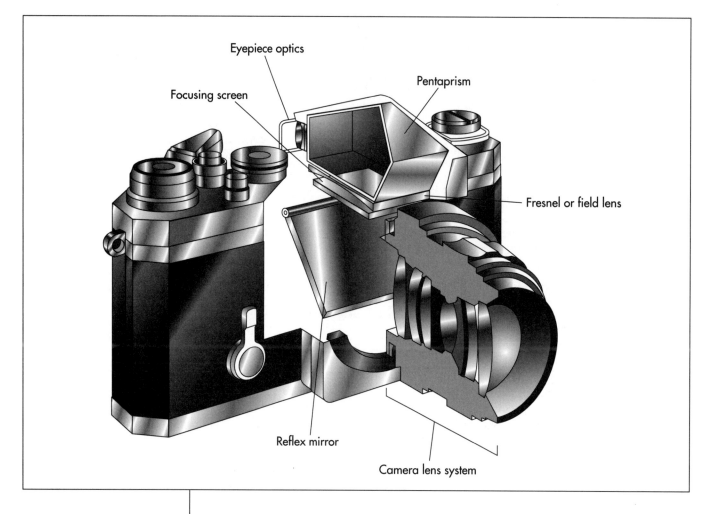

Eyepiece optics

Pentaprism

Focusing screen

Fresnel or field lens

Reflex mirror

Camera lens system

as subassemblies for the electronic brains of the camera and attached flash, if any.

Quality Control

Quality assurance and quality control practices are a matter of course among camera manufacturers. All departments from manufacturing to shipping have their own quality assurance procedures, and companywide quality assurance is also overseen by a separate division or department. The overseeing quality assurance divisions use statistical methods to monitor aspects of product quality such as camera function, performance, consistency, and precision. They also guide the flow of one assembly system into another and provide corrective measures if problems arise.

Byproducts/Waste

No byproducts result from camera manufacture, but a number of wastes are produced. The wastes include resins, oils such as cutting oil, solvents used for cleaning parts, and metals including iron, aluminum, and brass. The metals and resins are remainders or cuttings from manufactured parts and powder-fine cuttings and dust. The wastes are sorted by type and recovered; they are recycled or treated as industrial wastes by firms specializing in these activities. Camera manufacturers are well aware of the hazards associated with their processes and are careful to observe environmental regulations and sensitivities both in the country of manufacture and in receiving marketplaces. Japan's camera industry stopped using chlorofluorocarbons and trichloroethanes to clean printed circuit boards and camera lenses in 1993 on instruction of Japan's Ministry of International Trade and Industry (MITI), in response to import conditions of other countries, and in acknowledgment of industrywide respect of the environment.

The Future

For cameras like many other technical products, the future is electronic. The digital still camera introduced in 1995 stores approximately 100 pictures electronically. Instead of a viewfinder or eyepiece, the camera has a color LCD screen similar to the view-type screen on some video cameras, so photos can be viewed instantly. It can be connected by cables to a computer, television, or VCR, so pictures can be transferred to screen, tape, or digitized electronically. The digital camera has another advantage; after taking a photo and reviewing it, the photo can be erased if the photographer does not like the result. There is no wasted film or wasted space in the digital storage process. Also, the photograph can be edited, cropped, or enlarged as it is being taken. After photos have been taken, they remain in the camera as digital files rather than as negatives. To take more photos, these images have to be removed, and they can be stored on a computer disk. All the photos can be moved as a batch, or they can be stored on the computer one-by-one, or deleted from both the camera and computer storage. The transfer process requires software that also allows text to be attached to each picture to date it or write a caption. The camera or computer containing the photos can be hooked up to a video printer to print out copies on paper, or the photos can be transferred to videotape for viewing.

Where to Learn More

Books

Bailey, Adrian, and Adrian Holloway. *The Book Of Color Photography*. Alfred A. Knopf, 1979.

Collins, Douglas. *The Story of Kodak*. Harry N. Abrams, Inc., 1990.

Sussman, Aaron. *The Amateur Photographer's Handbook*. Thomas Y. Crowell Company, 1973.

Periodicals

Antonoff, Michael. "Digital Snapshots from my Vacation." *Popular Science*, June 1995, pp. 72-76.

From Glass Plates to Digital Images, Eastman Kodak Company, 1994.

—*Gillian S. Holmes*

CAT Scanner

First developed in the early 1970s, steady technological improvements have made this type of scanner an invaluable radiologic diagnostic device.

A computed tomography (CT) or computerized axial tomography (CAT) scanner is a medical imaging tool that provides clear pictures of the internal structures of the body. Utilizing a beam of x rays and a radiation detector, it supplies data to a computer, which then constructs a three-dimensional image. The CAT scanner is made up of various complex electronic components, which are produced by various subcontractors and assembled into a complete unit by the scanner manufacturers. First developed in the early 1970s, steady technological improvements have made this type of scanner an invaluable radiologic diagnostic device.

History

The invention of the CAT scanner was made possible by Wilhelm Roentgen, who discovered x rays in 1895. Around this time, various scientists were investigating the movement of electrons through a glass apparatus known as a Crookes tube. Roentgen wanted to visually capture the action of the electrons, so he wrapped his Crookes tube in black photographic paper. When he ran his experiment, he noticed that a plate coated with a fluorescent material, which just happened to be lying nearby the tube, fluoresced or glowed. This was unexpected because no visible light was being emitted from the wrapped tube. Upon further investigation, he found that indeed there was some kind of invisible light produced by this tube, and it could penetrate materials such as wood, aluminum, or human skin.

After this initial finding, Roentgen quickly realized the importance of his discovery to medicine. Using x rays, he determined that it was possible to create an image of structures beneath the skin. To this end he published the first x ray, an image of his wife's hand. He received the first Nobel prize in physics in 1901 for this discovery. The first documented use of x rays for an actual diagnosis in the United States occurred in 1896. Dr. Gilman Frost and his brother, who was a physicist, used them to determine the severity of injuries suffered by a young boy who had an ice skating accident. This x ray was taken in the physics laboratory of Dartmouth College.

As the field of radiography expanded, x-ray technology steadily improved. One of the major limitations of conventional x rays was that they lacked depth; therefore many internal structures were superimposed on each other. With the help of computers, scientists developed methods to solve this problem. One such method was computed tomography (CT), or computerized axial tomography (CAT). The first CAT scanner was demonstrated in 1970 by Godfrey Hounsfield and Allen Cormack. Over the next two decades, significant advances were made in scanner design, which have resulted in the high quality imaging scanners used today.

Background

CAT scanners, like all other x-ray machines, employ x rays to produce images of internal body structures. X rays are a type of ionizing radiation that is capable of penetrating solid materials to differing degrees, depending their density and thickness. In conventional radiology, an image is produced by placing a detector, such as a photographic film, behind the patient and then directing a beam of x rays toward it. The radiation passes through the patient's body and interacts with the film. Since x rays that strike the film produce dark areas after processing, body structures that are easily penetrated by x rays, such as skin, show up as dark regions. Other structures such as muscle, soft tissue, and organs allow different amounts of x rays through them and show up as gray areas. Bones, which do not allow x rays to pass through them, show up as bright white areas.

The images produced by conventional film x rays are often fuzzy because many of the internal structures are superimposed on each other. Tomography was developed to reduce this fuzziness and allow for the imaging of specific areas in the body. Early tomographic methods involved the simultaneous moving of the x-ray generator and the detecting film in opposite directions. As the two units move horizontally, only body structures that lie in a specific geometric plane will allow x rays to consistently pass through to the detector. In this way, these structures show up clearly on the film, while structures outside the plane are blurred. The image produced by this type of radiology is parallel to the long axis of the body.

Computerized axial tomography and computerized transaxial tomography represent a more complex and improved form of conventional tomography. The images are produced by rotating the x-ray generator and detectors around the patient in a circle. The amount of attenuated remnant radiation emitted from the body at various angles is measured and sent to a computer instead of being recorded directly on film. The computer then runs a series of complex algorithms to reconstruct the image, which can then be displayed on a monitor. Unlike conventional tomography, the image produced by computerized transaxial tomography is a cross section of the body and is called a transaxial image because it is perpendicular to the body's long axis.

X rays are called ionizing radiation because they are able to interact with and change certain types of matter, such as molecules in the body. While this is certainly a significant health risk to humans, the benefits of using x rays in medicine are overwhelming. However, care is taken by workers in the medical field to limit the amount of exposure to themselves and to patients.

Design

The CAT scanner is made up of three primary systems, including the gantry, the computer, and the operating console. Each of these are composed of various subcomponents. The gantry assembly is the largest of these systems. It is made up of all the equipment related to the patient, including the patient support, the positioning couch, the mechanical supports, and the scanner housing. It also contains the heart of the CAT scanner, the x-ray tube, as well as detectors that generate and detect x rays.

The x-ray tube is a special type of vacuum-sealed, electrical diode that is designed to emit x rays. It is made up of two electrodes, the cathode and anode. To produce x rays, a filament in the cathode is charged with electricity from a high voltage generator. This causes the filament to heat up and emit electrons. Using their natural attraction and a special focusing cup, the electrons travel directly toward the positively charged anode. X rays are emitted indiscriminately when the electrons strike the anode. The anode, which can be rotating or not, then conducts the electricity back to the high-voltage generator to complete the circuit. To focus the x rays into a beam, the x-ray tube is contained inside a protective housing. This housing is lined with lead except for a small window at the bottom. Useful x rays are able to escape out this window, while the lead prevents the escape of stray radiation in other directions.

Unlike other radiological devices, the detectors in a CAT scanner do not measure x rays directly. They measure radiation attenuated from the body structures due to their interaction with x rays. One type of detector is an ideal gas-filled detector. When radiation strikes one of these detectors, the gas is ionized and a radiation level can be determined.

The computer is specially designed to collect and analyze input from the detector. It is a large capacity computer capable of performing thousands of equations simultaneously. The reconstruction speed and image quality are all dependent on the computer's microprocessor and internal memory. A

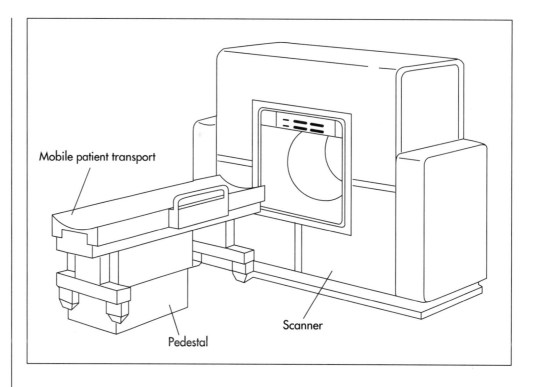

Mobile patient transport

Pedestal

Scanner

quick computer is particularly important because it greatly influences the speed and efficiency of the examination. Since the computer is so specialized, it requires a room with a strictly controlled environment. For example, the temperature is typically maintained below 68°F (20°C) and the humidity is below 30%.

The operating console is the master control center of the CAT scanner. It is used to input all of the factors related to taking a scan. Typically, this console is made up of a computer, a keyboard, and multiple monitors. Often there are two different control consoles, one used by the CAT scanner operator, and the other used by the physician. The operator's console controls such variables as the thickness of the imaged tissue slice, mechanical movement of the patient couch, and other radiographic technique factors. The physician's viewing console allows the doctor to view the image without interfering with the normal scanner operation. It also enables image manipulation, if this is required for diagnosis and image storage for later use. For this type of data storage, magnetic tapes or floppy disks are available.

The design of a CAT scanner improved incrementally over time. The original CAT scanners utilized a thin, pencil beam of x

rays and took 180 readings, one at each degree of rotation around a semicircle. The x-ray generator and detectors moved horizontally for each scan and then were rotated one degree to take the next scan. Two detectors were used, so that two different images could be generated from each scan. The drawback of this system was lengthy scanning times. A single scan could take up to five minutes. Designs improved as more detectors were added and the x-ray beam was fanned out using a special filter. This significantly reduced scanning time to about 20 seconds. The next major design improvement resulted in the elimination of the horizontal movement of the generator and detector, making it a rotate-only scanner. More detectors were added and grouped into a curvilinear detector array. The detector array eventually was designed to be stationary, and the resulting scan time was reduced to one second.

Raw Materials

A wide variety of materials, such as steel, glass and plastic, are used to construct the components of a CAT scanner. Some of the more specialized compounds can be found in the patient couch, detector array, and the x-ray tube. The patient couch is typically made from carbon fiber to prevent it from interfering with the x-ray beam transmis-

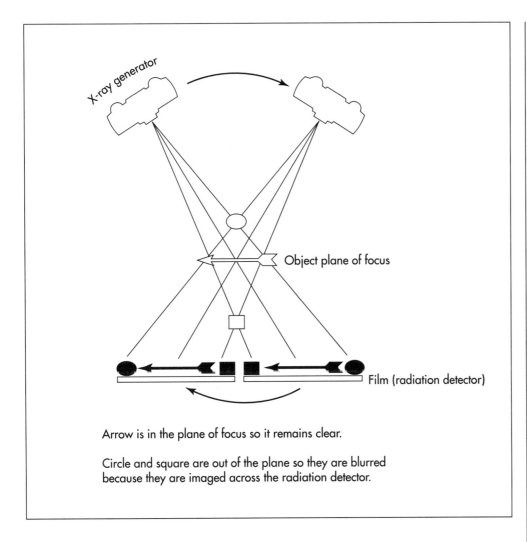

Arrow is in the plane of focus so it remains clear.

Circle and square are out of the plane so they are blurred because they are imaged across the radiation detector.

CAT scanners use X-ray technology to create three-dimensional images of the body's internal structures. Images are obtained by rotating the x-ray generator and detectors around the patient. This information is fed into a computer, which reconstructs images of the body structures within its plane of focus.

sion. The detector array of more modern scanners uses tungsten plates, a ceramic substrate, and xenon gas. Tungsten is also used to make the cathode and electron beam target of the x-ray tube. Other materials found in the tube are Pyrex™ glass, copper, and tungsten alloys. Throughout many parts of the CAT scanner system, lead can be found, which reduces the amount of excess radiation.

The Manufacturing Process

CAT scanner manufacture is typically an assembly of various components that are supplied by outside manufacturers. The following process discusses how the major components are produced.

Gantry assembly components

1 The x-ray tube is made much like other types of electrical diodes. The individual components, including the cathode and anode, are placed inside the tube envelope and vacuum sealed. The tube is then situated into the protective housing, which can then be attached to the rotating portion of the scanner frame.

2 Various detector arrays are available for CAT scanners. One type of detector array is the ideal gas-filled detector. This is made by placing strips of tungsten 0.04 inch (1 mm) apart around a large metallic frame. A ceramic substrate holds the strips in place. The entire assembly is hermetically sealed and pressure filled with an inert gas such as xenon. Each of the tiny chambers formed by the gaps between the tungsten plates are individual detectors. The finished detector is also attached to the scanner frame.

3 To create the large amount of voltage needed to produce x rays, an autotransformer is used. This power supply device is

made by winding wire around a core. Electric tap connections are made at various points along the coil and connected to the main power source. With this device, output voltage can be increased to approximately twice the input voltage.

Control consol and computer

4 The control consol and computer are specially designed and supplied by computer manufacturers. The primary model building computer is specifically programmed with the reconstruction algorithms needed to manipulate the x-ray data from the gantry assembly. The control consoles are also programmed with software to control the administration of the CAT scan.

Final assembly

5 The final assembly of the CAT scanner is a custom process which often takes place in the radiologic imaging facility. Rooms are specially designed to house each component and minimize the potential for excessive radiation exposure or electric shock. By following specific plans, equipment installation and wiring of the entire CAT scanner system is completed.

Quality Control

As with all electronic equipment, quality control tests are an important part of CAT scanner manufacturing. The scanner manufacturers typically rely on their suppliers to perform basic quality tests on the incoming components. When sections of the scanner are assembled, visual and electrical inspections are performed throughout the entire process to detect flaws. In addition to the quality specifications set by the manufacturers, the United States Food and Drug Administration (FDA) has regulations that require manufacturers to perform specific quality control tests. Examples of these tests

include calibration tests of the x-ray tube, mechanical tests of the patient table, and standardization tests of the visual output.

The Future

Research for future CAT scanners is focused on four basic goals, including the production of better quality images, reducing the amount of patient radiation exposure, optimizing computer reconstruction algorithms, and improving CAT scanner design. Various methods of achieving these aims have already been attempted. To improve image quality, some scanners incorporate unique movements of the x-ray tube, the detector, or both. Others change the position of the patient. Faster scanners are being developed to reduce patient exposure time. Different kinds of computer algorithms have been developed for a variety of examinations. Future CAT scanners will likely incorporate most of these new developments, along with a continuously rotating x-ray tube and detectors to provide the clearest and safest imaging procedure possible.

Where to Learn More

Books

Bushong, Stewart. *Radiologic Science for Technologist.* Mosby, 1993.

Curry, T.S. *Christenson's Physics of Diagnostic Radiology.* Lea and Febiger, 1990.

Tompson, Michael. *Principles of Imaging Science and Protection.* W.B. Saunders Co., 1994.

Periodicals

"Quality Assurance for Diagnostic Imaging Equipment." National Council on Radiation Protection and Measurements, 1988.

—*Perry Romanowski*

Cereal

Background

Breakfast cereal is a processed food manufactured from grain and intended to be eaten as a main course served with milk during the morning meal. Some breakfast cereals require brief cooking, but these hot cereals are less popular than cold, ready-to-eat cereals.

Prehistoric peoples ground whole grains and cooked them with water to form gruels and porridges similar to today's hot cereals. Cold cereals did not develop until the second half of the nineteenth century.

Ready-to-eat breakfast cereals were invented because of religious beliefs. The first step in this direction was taken by the American clergyman Sylvester Graham, who advocated a vegetarian diet. He used unsifted, coarsely ground flour to invent the **Graham cracker** in 1829. Influenced by Graham, Seventh-Day Adventists, who also believed in vegetarianism, founded the Western Health Reform Institute in Battle Creek, Michigan, in the 1860s. At this institute, later known as the Battle Creek Sanitarium, physician John Harvey Kellogg invented several grain-based meat substitutes.

In 1876 or 1877, Kellogg invented a food he called granola from wheat, oats, and corn that had been mixed, baked, and coarsely ground. In 1894, Kellogg and his brother W. K. Kellogg invented the first precooked flaked cereal. They cooked ground wheat into a dough, then flattened it between metal rollers and scraped it off with a knife. The resulting flakes were then cooked again and allowed to stand for several hours. This product was sold by mail order as Granose for 15 cents per 10-ounce (284 g) package.

Both W. K. Kellogg and C. W. Post, a patient at the sanitarium, founded businesses to sell such products as health foods. Their success led dozens of imitators to open factories in Battle Creek between 1900 and 1905. These businesses quickly failed, while Kellogg and Post still survive as thriving manufacturers of breakfast cereals.

Their success can be partially attributed to advertising campaigns, which transformed the image of their products from health foods to quick, convenient, and tasty breakfast foods. Another factor was the fact that Kellogg and Post both manufactured corn flakes, which turned out to be much more popular than wheat flakes. Breakfast cereals have continued to increase in popularity in the twentieth century. Ready-to-eat breakfast cereals are served in nine out of 10 American households.

Raw Materials

The most important raw material in any breakfast cereal is grain. The grains most commonly used are corn, wheat, oats, rice, and barley. Some hot cereals, such as plain oatmeal, and a few cold cereals, such as plain shredded wheat, contain no other ingredients. Most breakfast cereals contain other ingredients, such as salt, yeast, sweeteners, flavoring agents, coloring agents, vitamins, minerals, and preservatives.

The sweeteners used in breakfast cereals include malt (obtained from barley), white sugar, brown sugar, and corn syrup. Some natural cereals are sweetened with concentrated fruit juice. A wide variety of flavors may be added to breakfast cereals, including chocolate, cinnamon and other spices, and fruit flavors. Other ingredients added to

Ready-to-eat breakfast cereals are served in nine out of 10 American households.

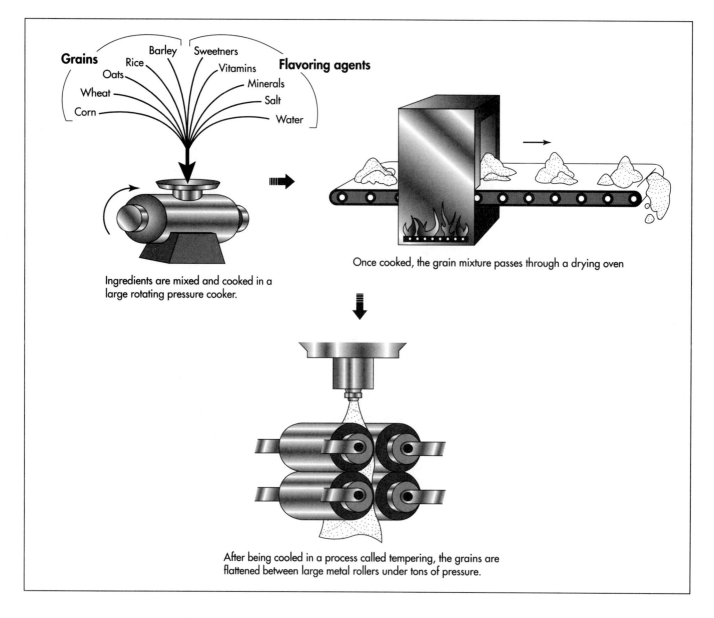

Grains Barley **Sweetners**
Rice Vitamins **Flavoring agents**
Oats Minerals
Wheat Salt
Corn Water

Ingredients are mixed and cooked in a
large rotating pressure cooker.

Once cooked, the grain mixture passes through a drying oven

After being cooled in a process called tempering, the grains are
flattened between large metal rollers under tons of pressure.

improve flavor include nuts, dried fruit, and
marshmallows.

Vitamins and minerals are often added to
breakfast cereals to replace those lost during cooking. The most important of these is
vitamin B-1, 90 % of which is destroyed by
heat. The antioxidants BHA and BHT are
the preservatives most often added to
breakfast cereals to prevent them from becoming stale and rancid.

The Manufacturing Process

Preparing the grain

1 Grain is received at the cereal factory,
inspected, and cleaned. It may be used in
the form of whole grains or it may require
further processing. Often the whole grain is
crushed between large metal rollers to remove the outer layer of bran. It may then be
ground more finely into flour.

2 Whole grains or partial grains (such as
corn grits) are mixed with flavoring
agents, vitamins, minerals, sweeteners, salt,
and water in a large rotating pressure cooker.
The time, temperature, and speed of rotation
vary with the type of grain being cooked.

3 The cooked grain is moved to a convey-
or belt, which passes through a drying
oven. Enough of the water remains in the
cooked grain to result in a soft, solid mass
which can be shaped as needed.

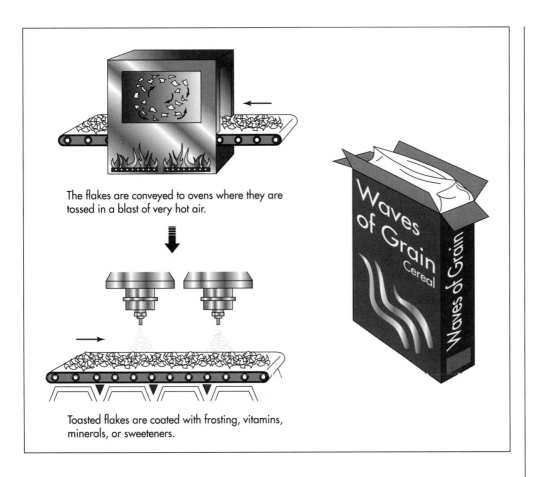

The flakes are conveyed to ovens where they are tossed in a blast of very hot air.

Toasted flakes are coated with frosting, vitamins, minerals, or sweeteners.

4 If flour is used instead of grains, it is cooked in a cooking extruder. This device consists of a long screw within a heated housing. The motion of the screw mixes the flour with water, flavorings, salt, sweeteners, vitamins, minerals, and sometimes food coloring. The screw moves this mixture through the extruder, cooking it as it moves along. At the end of the extruder, the cooked dough emerges as a ribbon. A rotating knife cuts the ribbon into pellets. These pellets are then processed in much the same way as cooked grains.

Making flaked cereals

5 The cooked grains are allowed to cool for several hours, stabilizing the moisture content of each grain. This process is known as tempering. The tempered grains are flattened between large metal rollers under tons of pressure. The resulting flakes are conveyed to ovens where they are tossed in a blast of very hot air to remove remaining moisture and to toast them to a desirable color and flavor. Instead of cooked grains, flakes may also be made from extruded pellets in a similar manner.

Making puffed cereals

6 Cereals may be puffed in ovens or in so-called "guns." Oven-puffed cereals are usually made from rice. The rice is cooked, cooled, and dried. It is then rolled between metal rollers like flaked cereals, but it is only partially flattened. This process is known as bumping. The bumped rice is dried again and placed in a very hot oven which causes it to swell.

7 Gun-puffed cereals may be made from rice or wheat. The rice grains require no pretreatment, but the wheat grains must be treated to partially remove the outer layer of bran. This may be done by abrading it off between grindstones, a process known as pearling. It may also be done by soaking the wheat grains in salt water. The salt water toughens the bran, which allows it to break off in large pieces during puffing. The grain is placed in the gun, a small vessel which can hold very hot steam and very high pressure. The gun is opened quickly to reduce the pressure suddenly, which puffs the grain. Extruded pellets can also be used to make gun-puffed cereals in the same way as grains.

Making shredded cereals

8 Shredded cereals are usually made from wheat. The wheat is cooked in boiling water to allow moisture to fully penetrate the grain. The cooked grain is cooled and allowed to temper. It is then rolled between two metal rollers. One roller is smooth and the other is grooved. A metal comb is positioned against the grooved roll with a tooth inside each groove. The cooked grain is shredded by the teeth of the comb and drops off the rollers in a continuous ribbon. A conveyor belt catches the ribbons from several pairs of rollers and piles them up in layers. The layers of shredded wheat are cut to the proper size, then baked to the desired color and dryness. Shredded cereals may also be made in a similar way from extruded pellets.

Making other cereals

9 Cereals can be made in a wide variety of special shapes (circles, letters of the alphabet, etc.) with a cooking extruder. A die is added to the end of the extruder which forms a ribbon of cooked dough with the desired cross-section shape. A rotating knife cuts the ribbon into small pieces with the proper shape. These shaped pieces of dough are processed in a manner similar to puffing. Instead of completely puffing, however, the pieces expand only partially in order to maintain the special shape.

10 Granolas and similar products are made by mixing grain (usually oats) and other ingredients (nuts, fruits, flavors, etc.) and cooking them on a conveyor belt which moves through an oven. The cooked mixture is then crumbled to the desired size. Hot cereals are made by processing the grain as necessary (rolling or cutting oats, cracking wheat, or milling corn into grits) and partly cooking it so the consumer can cook it quickly in hot water. Salt, sweeteners, flavors, and other ingredients may or may not be added to the partly cooked mixture.

Adding coatings

11 After shaping, the cereal may be coated with vitamins, minerals, sweeteners, flavors such as fruit juices, food colors, or preservatives. Frosting is applied by spraying a thick, hot syrup of sugar on the cereal in a rotating drum. As it cools the syrup dries into a white layer of frosting.

Packaging

12 Some cereals, such as shredded wheat, are fairly resistant to damage from moisture. They may be placed directly into cardboard boxes or in cardboard boxes lined with plastic. Most cereals must be packaged in airtight, waterproof plastic bags within cardboard boxes to protect them from spoiling.

13 An automated machine packages the cereal at a rate of about 40 boxes per minute. The box is assembled from a flat sheet of cardboard, which has been previously printed with the desired pattern for the outside of the box. The bottom and sides of the box are sealed with a strong glue. The bag is formed from moisture-proof plastic and inserted into the box. The cereal fills the bag and the bag is tightly sealed by heat. The top of the box is sealed with a weak glue which allows the consumer to open it easily. The completed boxes of cereal are packed into cartons which usually hold 12, 24, or 36 boxes and shipped to the retailer.

Quality Control

Every step in the manufacturing of breakfast cereal is carefully monitored for quality. Since cereal is a food intended for human consumption, sanitation is essential. The machines used are made from stainless steel, which can be thoroughly cleaned and sterilized with hot steam. Grain is inspected for any foreign matter when it arrives at the factory, when it is cooked, and when it is shaped.

To ensure proper cooking and shaping, the temperature and moisture content of the cereal is constantly monitored. The content of vitamins and minerals is measured to ensure accurate nutrition information. Filled packages are weighed to ensure that the contents of each box is consistent.

In order to label boxes with an accurate shelf life, the quality of stored cereal is tested over time. In order to be able to monitor freshness over a reasonable period of time, the cereals are subjected to higher than normal temperatures and humidities in order to speed up the spoiling process.

The Future

Breakfast cereal technology has advanced greatly since its origins in the late nineteenth century. The latest innovation in the industry is the twin-screw cooking extruder. The two rotating screws scrape each other clean as they rotate. This allows the dough to move more smoothly than in an extruder with only one screw. By using a twin-screw extruder, along with computers to precisely control temperature and pressure, cereals that usually require about 24 hours to make may be made in as little as 20 minutes.

Where to Learn More

Books

Bruce, Scott, and Bill Crawford. *Cerealizing America: The Unsweetened Story of American Breakfast Cereal*. Faber and Faber, 1995.

Fast, Robert B., and Elwood F. Caldwell, eds. *Breakfast Cereals and How They Are Made*. American Association of Cereal Chemists, 1990.

Periodicals

Dworetzky, Tom. "The Churn of the Screw." *Discover*, May 1988, pp. 28-29.

Fast, R. B. "Breakfast Cereals: Processed Grains for Human Consumption." *Cereal Foods World*, March 1987, pp. 241-244.

Other

Kellogg Company. "How Kellogg's® Cereal is Made." December 4, 1996. http://kelloggs.com/booth/cereal.html (July 9, 1997).

—*Rose Secrest*

Champagne

In the early days of champagne-making, 20-90% of the bottles exploded from the build up of carbonic acid in the bottles, giving rise to the practice of wearing iron face masks when walking through champagne cellars.

Background

Champagne is the ultimate celebratory drink. It is used to toast newlyweds, applaud achievements, and acknowledge milestones. A large part of its appeal is due to the bubbles that spill forth when the bottle is uncorked. These bubbles are caused by tiny drops of liquid disturbed by the escaping carbon dioxide or carbonic acid gas that is a natural by-product of the double fermentation process unique to champagne.

Today, fine champagne is considered a mark of sophistication. But this was not always so. Initially, wine connoisseurs were disdainful of the sparkling wine. Furthermore in 1688, Dom Perignon, the French monk whose name is synonymous with the best vintages, worked very hard to reduce the bubbles from the white wine he produced as Cellarer of the Benedictine Abbey of Haut-Villers in France's Champagne region. Ironically, his efforts were hampered by his preference for fermenting wine in bottles instead of casks, since bottling adds to the build-up of carbonic acid gas.

The Champagne province, which stretches from Flanders on the north to Burgundy in the south; from Lorraine in the east to Ile de France in the west, is one of the northernmost wine producing regions. For many years, the region competed with Burgundy to produce the best still red table wines. However, red grapes need an abundance of sun, something that the vineyards of Champagne do not receive on a regular basis. By the time Perignon took over the Abbey cellars in 1668, he was studying ways to perfect the harvesting of the Pinot Noir grape in order to produce a high-quality white wine.

Often called black grapes, the Pinot Noir actually bears a skin that is blue on the outside and red on the inside. The juice is white but care must be taken during harvesting so that the skin does not break and color the juice.

Climate is a major factor in winemaking and nowhere is this more apparent that in the case of champagne. The inconsistency and shortness of the Champagne region's summers lead inevitably to inconsistent harvests. Therefore, a supply of wine made during better years is saved so that it may be blended with the juice of grapes harvested during poorer seasons. When the wine is stored after the fall harvest, it begins to ferment but ceases when the cold winter months set in. In late spring or early summer, the wine begins to ferment again. Extra sugar is added to that which is left in the wine. The wine is then bottled and tightly corked. The carbonic acid that would normally escape into the air if the wine were stored in casks builds up in the bottle, ready to rush forth when the cork is released.

In the early days of champagne-making, this volatility was something of a problem. Twenty to 90% of the bottles exploded, giving rise to the practice of wearing iron face masks when walking through champagne cellars. By 1735, a royal ordinance established regulations governing the shape, size, and weight of champagne bottles. Corks were to be 1.5 in (3.75 cm) long and secured to the collar of the bottle with strong pack thread. Deep cellars with constant temperatures also keep the bottles from exploding. The chalky earth of the Champagne region make it ideal for these cellars.

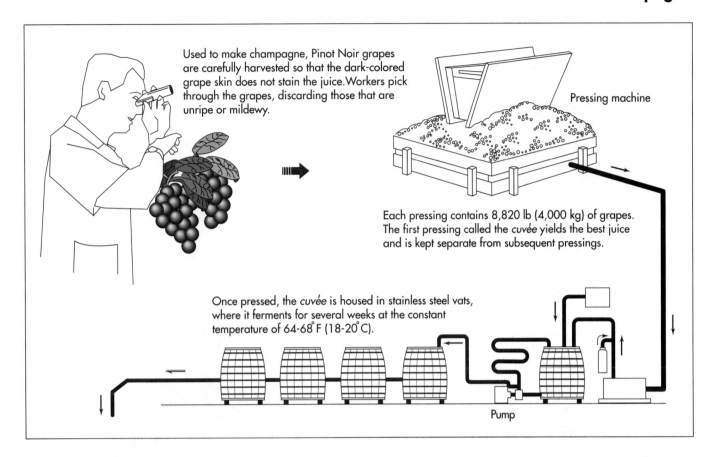

Used to make champagne, Pinot Noir grapes are carefully harvested so that the dark-colored grape skin does not stain the juice. Workers pick through the grapes, discarding those that are unripe or mildewy.

Pressing machine

Each pressing contains 8,820 lb (4,000 kg) of grapes. The first pressing called the *cuvée* yields the best juice and is kept separate from subsequent pressings.

Once pressed, the *cuvée* is housed in stainless steel vats, where it ferments for several weeks at the constant temperature of 64-68°F (18-20°C).

Pump

Three years after Perignon's death, Canon Godinot recorded the monk's specifications for the making of champagne:

- Use only Pinot Noir grapes.

- Prune the vine aggressively. Do not allow them to grow higher than three feet.

- Harvest the grapes carefully to keep the skins intact. Keep the grapes as cool as possible. Work the fields early in the morning or on showery days when the weather is very hot. Pick over the grapes while still in the fields. Reject all broken or bruised grapes.

- Set up the press as close to the fields as possible. If the grapes must be transported, use the slower pack animals such as mules or donkeys rather than horses to prevent the grapes from being jostled.

- Do not tread on the grapes or allow the skins into the juices.

Although modern champagne vintners have the use of technology to streamline certain parts of the champagne-making process, the steps have not changed significantly over the last three centuries.

Raw Materials

The main ingredient in champagne is the Pinot Noir grape. The grapes, left in bunches, are carefully picked so that the skin pigment does not stain the juice. Vineyard workers pick through the grapes, removing any that are unripe or mildewy. The grape bunches are weighed, generally 8,820 lb (4,000 kg) are used for a pressing. The grapes are taken directly to the press in a further effort to prevent the skin from coloring the juice.

During the double fermentation, several other natural ingredients are added to the wine. Yeast, usually saccharmonyces, is added during the first fermentation to help the grapes' natural sugar convert to alcohol. A *liquer de tirage*, cane sugar melted in still champagne wine, is added. In the second fermentation stage, a *liquer d'expédition* is added. This consists of cane sugar, still wine, and brandy. The amount of sugar added at this stage determines the type of champagne, from sweet to dry. Although each vintner has its own standards, the general guide is as follows: a 0.5% solution yields the driest champagne, known as brut;

Once the wine is blended and the *liquer de tirage* added, the mixture is bottled and sealed with crown caps.

Bottling   Crown caps added

$\frac{1}{8}$" Turn

Required by law to age for at least one year, the bottles are turned one-eighth of a turn each day by skilled workers. The turning prevents the sediment from settling at the bottom of the bottle.

After the sediment is removed in the process called *dégorgement* and the *liquor d' expedition* (mixture of reserve wines, sugar, and brandy) is added, the bottles are corked. The corks are secured by wire muzzles and the bottles are labelled and stored until shipment.

1% is added for extra sec; 3% for sec; and 5% for demi-sec, the sweetest type of champagne.

The Manufacturing Process

Pressing

1 The grapes are carefully loaded into the press, a square wooden floor surrounded by adjustable wooden rails and topped by a heavy oak lid. The lid is mechanically lowered and raised at intervals, causing the grapes to burst and the juice to pour out. The juices run through the rails into a sloped groove that carries the juice to stainless steel vats. The first pressing is called the *cuvée* and is the best juice from a batch of grapes. It is kept separate from subsequent pressings. The cuvée begins to ferment immediately. As scum rises to the top it is thrown off. Some of the scum falls to the bottom of

the vat; this sediment is called lees. The juice, called must, continues to ferment for 24-36 hours when it gradually returns to its normal temperature.

First fermentation

2 The cuveés are moved into temperature-controlled stainless steels vats and fermented for several weeks at 64-68°F (18-20°C). The amount of time varies depending on the house specifications. Some champagne producers also put the wine through a malolactic fermentation process at this point to reduce acidity.

Blending the wines

3 The head cellarer (*chef de caves*) and cellar assistants taste and blend wines from several different pressings to obtain the desired taste. The blended wines are churned in vats by sweeping mechanical arms.

Bottling and the second fermentation

4 The blended wine is drawn off into bottles. The liquer de tirage is added and the bottles are sealed with crown caps. Because the carbonic acid cannot escape through the glass, it builds up to a tremendous pressure, equal to that in a bus tire.

Aging

5 French law requires that non-vintage wines be aged for at least one year. Vintage wines must be aged for at least three years. Each wine house adds to this minimum requirement as desired. Non-vintage wines are those that result from a thin harvest and are combined with reserves from past good vintages. Non-vintage wine is not sold under a particular year. Vintage wines, on the other hand, are made from Champagne grapes harvested in the same year. Vintage wines are rare, produced only when the summer has been unusually hot and sunny. The year is printed on the cork and the label.

Racking (Remuerurage)

6 During the aging period, the bottles of champagne are turned daily to keep the sediment caused by dead yeast cells from settling on the bottom. Skilled workers, with quick hands, twist the bottles one-eighth of a turn each day. The bottles start out in the horizontal position; by the end of the aging period, the bottles are vertical with the necks pointed towards the floor so that the sediment has collected on the inside face of the cork.

Dégorgement

7 The bottleneck is plunged into freezing liquid, causing a pellet of frozen champagne to form in the neck. The crown cap is carefully removed and the ice expels the sediment.

Liquor d'expedition is added

8 The mixture of reserve wines, sugar, and brandy is added to the bottles of champagne to create the desired sweetness.

Corkage

9 A long, fat cork that has been branded with the house name is hand-driven halfway into the bottleneck. Then the exposed portion is squashed down into the neck and secured with a wire muzzle. The bottles are labelled and stored in the cellar until shipment at which time they are packed into crates or cartons.

Quality Control

Guided by government regulations, each champagne house sets its own standards for the aging of its wines. In France, where the finest champagne is produced, the Institute National des Appelations d'Origin also places strict standards on the quality of soil that may be used for the growing of Champagne grapes. However, every champagne producing country regulates the production and marketing of its wines to some extent. Furthermore, each step of the champagne-making process is presided over by veteran experts who are skilled in tasting and blending.

The Future

It is inevitable that the labor-intensive process of making champagne will be further mechanized in the twenty-first century. Already, agricultural advances have reduced the threat of rot in the vineyard, thus reducing the number of workers needed to pick over the grapes in the fields. Some of the larger champagne houses have replaced the traditional round wooden press with a horizontal model inside of which a rubber bag inflates and gently presses the grapes against the sides of the press. Experiments are underway to develop a mechanized method for rotating the bottles to replace the costly hand-turning method. To date, none have proved effective, but industry observers believe that the change is inescapable.

Where to Learn More

Books

Johnson, Hugh. *Vintage: The Story of Wine.* Simon and Schuster, 1989.

Simon, André. *Wines of the World.* 2nd ed. Ed. Serena Sutcliffe. McGraw-Hill, 1981.

Other

"How Champagne is Made." Moët & Chandon Homepage. http://moet.com/taste/made.html (January 21, 1997).

"Know-How." Jacquart Homepage. http://www.jacquart-champagne.fr/sf_eng.html (January 21, 1997).

—*Mary F. McNulty*

Cigar

Background

A cigar is a tobacco leaf wrapped around a tobacco leaf filling. Bigger than a cigarette, and taking longer to smoke, the cigar is considered by aficionados to be the finest way to enjoy tobacco.

Cigars come in several shapes and sizes. The standard shape is the round-headed cigar with parallel sides. Perfecto refers to a cigar with a pointed head and tapering sides; Panatella is a long, thin, straight cigar; Cheroot is an open-ended cigar, usually made in India or Asia. A special vocabulary denotes cigar sizes. From the smallest [3.5 in (8.9 cm)] to the largest [7.5 in (19 cm)] they are the Half Corona, Tres Petit Corona, Petit Corona Corona, Corona Grande, Lonsdale, and Double Corona. A set of initials usually stamped on the bottom or side of a box of cigars refers to the color of the tobacco leaf: C C C is Claro (light); C C means Colorado-Claro (medium); C means Colorado (dark); and C M stands for Colorado-Maduro (very dark). The darker leaf is generally the stronger tobacco.

History

The earliest cigars were probably those rolled by native Cubans. Columbus encountered Cubans smoking crude cigars, and subsequent Spanish and Portuguese expeditions to the New World brought back cigars to Europe. Many sailors smoked cigars, and brought the habit to port cities, but the habit did not become widely popular until the end of the eighteenth century. Cigar factories existed in Spain at this time, and in the 1780s factories were established in France and Germany as well. English officers who fought in Spain during the Napoleonic Wars brought cigars home to England, where they became a fad with the upper classes. Cigars were expensive, especially because of high import duties on them, and by the end of the nineteenth century, they had become a mark of luxury. Smoking cigars was for men only (even smoking in sight of a woman was considered vulgar), and special smoking clubs called divans sprang up where men could enjoy their habit.

In the twentieth century, cigars were associated with notable public figures, from presidents to gangsters to entertainers. Winston Churchill, Calvin Coolidge, Al Capone, and Groucho Marx, to name a few, were all avid cigar smokers. After World War II, the cigar increasingly became the old man's smoke. Instead of being considered suave, the cigar became something conspicuously inelegant. This perception of the cigar has reversed recently, as cigar smoking became newly fashionable in the 1990s. Special cigar clubs and cigar "smoke out" dinners in cities across the United States in the 1990s put forth a revamped image of the cigar as a luxurious vice for men and also women to enjoy. By the mid-1990s, there were an estimated eight million cigar smokers in the United States, and cigar manufacturers were hard pressed to meet booming demand.

Though the finest cigars still come from Cuba, cigars are manufactured all across the globe. As early as 1610, cigar tobacco was grown in Massachusetts, and other early centers of tobacco cultivation were the Philippines, Java, Ceylon (Sri Lanka), and Russia. American cigar tobacco was mostly exported to the West Indies, rolled there, and then imported as finished cigars, until

With the resurgence in popularity of cigar smoking in the 1990s and an estimated eight million cigar smokers in the United States, cigar manufacturers have been hard pressed to meet the booming demand.

the beginning of the nineteenth century. A domestic cigar industry developed after 1801, and by 1870 there were cigar factories all across the country. Tampa, Florida, was a center for cigar manufacturing, though Pennsylvania, Connecticut, and New York also had hundreds of cigar factories.

Cigars were made by hand until the beginning of the twentieth century. The industry mechanized rapidly between 1910 and 1929. The number of cigar factories in the United States fell dramatically—from almost 23,000 in 1910 to only around 6,000 in 1929—but the mechanized factories produced many more cigars than the old handwork ones. Today, the finest cigars are still made entirely by hand. But the majority are made either entirely or partially by machine.

Raw Materials

The principle raw material of the cigar is the leaf of the tobacco plant *(Nicotiana tabacum).*The tobacco plant grows in many climates, but the finest cigar tobacco is grown in Cuba, Jamaica, and the Dominican Republic. A cigar requires three kinds of tobacco leaf as its raw material. Small or broken tobacco leaves are used for the filler. Whole leaves are used for an inside wrapper, called the binder. The binder leaf can be of second quality or imperfect. Its appearance is not important. A large, finely textured leaf of uniform appearance is used for the outside wrapper. Some cigars are made with the leaves all from the same region. Others may be wrapped in a high-quality leaf (from Cuba for example) but filled with poorer quality leaf from another region. Secondary raw materials include a tasteless gum to stick the end of the wrapper together, flavoring agents that are sometimes sprayed on the filler leaves, and paper used for the band placed around each cigar.

Most machine-made cigars use homogenized tobacco leaf (HTL) for the binder, and often for the wrapper as well. HTL is made from tobacco leaf scraps that are pulverized, mixed with vegetable gum, and rolled into sheets. HTL is stronger and more uniform than whole tobacco leaf, and so is more suitable for use in cigar-making machines. When HTL is used for the wrapper, the manufacturer may add flavorings to it.

The Manufacturing Process

Cultivation of tobacco

1 Tobacco plants are seeded indoors, and transplanted into fields after six to 10 weeks. The plants are carefully pruned so the leaves grow to the necessary size. Plants that produce the outer wrappers of cigars are usually kept covered with cloth to protect them from the sun. The plants take several months to mature in the fields.

Curing

2 After harvesting, the tobacco leaves must be cured in order to develop their characteristic aroma. The leaves are cured when they have passed from bright green flexible fresh leaves to dried brown or yellowish leaves. Chemically, the naturally occurring chlorophyll in the leaf gradually breaks down and is replaced by carotene. To cure, the harvested plants are strung to narrow strips of wood called laths. The laths are hung from the ceiling of a well-ventilated curing barn. In dry weather, they may cure simply by hanging, a process called air curing. The leaves may also be flue-cured. In this method, the laths are hung in a small barn which is heated from 90-170°F (32.2-77°C). The temperature must be carefully monitored in order to prevent extreme rapid drying. Sawdust or hardwood may also be burned in the curing barn, to aid in drying the leaves and impart an aroma.

Fermenting

3 After the leaves are cured, they are sorted by color and size. Small or broken leaves are used for the cigar filler, large leaves for the inner wrapper or binder, and large, fine leaves, usually grown in shade or under cloth, are set aside for the outer wrapper. The leaves are tied into bundles called hands of 10 or 15 leaves each. The hands are packed in boxes or in large casks called hogsheads. The tobacco is kept in the hogshead for a period of from six months to five years. The leaves undergo chemical changes during this period referred to as fermentation. During fermentation, the aroma and taste of the leaf develops. Cigar tobacco is usually fermented longer than

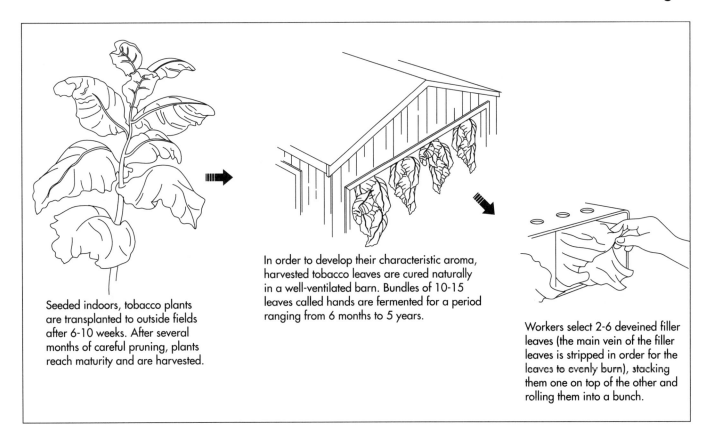

Seeded indoors, tobacco plants are transplanted to outside fields after 6-10 weeks. After several months of careful pruning, plants reach maturity and are harvested.

In order to develop their characteristic aroma, harvested tobacco leaves are cured naturally in a well-ventilated barn. Bundles of 10-15 leaves called hands are fermented for a period ranging from 6 months to 5 years.

Workers select 2-6 deveined filler leaves (the main vein of the filler leaves is stripped in order for the leaves to evenly burn), stacking them one on top of the other and rolling them into a bunch.

other tobacco. Fermentation for two to five years is typical for high quality cigars. After fermentation, the leaves are manually sorted again by highly trained workers.

Stripping

4 The filler leaves must have their main vein (or stem) removed, or else the cigar will not burn evenly. This can be done by hand or machine. Manually, a worker with a thimble knife fitted to his or her finger clips the vein near the tip and pulls it down. Then the worker stacks the stripped leaves in piles (called books or pads). Mechanically, a worker inserts the tobacco leaves into a machine under a grooved, circular knife. By depressing a foot treadle, the worker causes the knife to lower and cut out the vein. The worker can stop the machine with the foot treadle, and stack the stripped leaves.

The stripped leaves are wrapped in bales and stored for further fermentation. The bales may be shipped at this point, if final production resides elsewhere. Just before the leaves are ready for manufacture into cigars, they are steamed to restore lost humidity, and sorted again.

Hand rolling

5 Fine cigars are rolled by hand. Cigar rolling is skilled work: it may take a year for a roller to become proficient. The filler must be packed evenly for the cigar to burn smoothly, and the wrapper should be wound in an even spiral around the cigar. Hand cigar makers usually work in small factories. Each worker sits at a small table with a tray of sorted tobacco leaves on it and space to roll out the cigar. First the worker selects from two to six leaves for the filler. These are placed one on top of the other and rolled into a bunch. Then the worker places the bunch on the binder leaf and rolls the binder leaf cylindrically around the filler. The unfinished cigars are placed in an open wooden mold that holds them in shape until they can be wrapped.

6 Wrapping is the most difficult step. The worker takes the partially completed cigar out of the mold and places it on the wrapper leaf. With a special rounded knife called a chaveta, the worker trims off any irregularities from the filler. Then the worker rolls the wrapper leaf around the filler and binder three and a half times, and secures it at the end with a small amount of

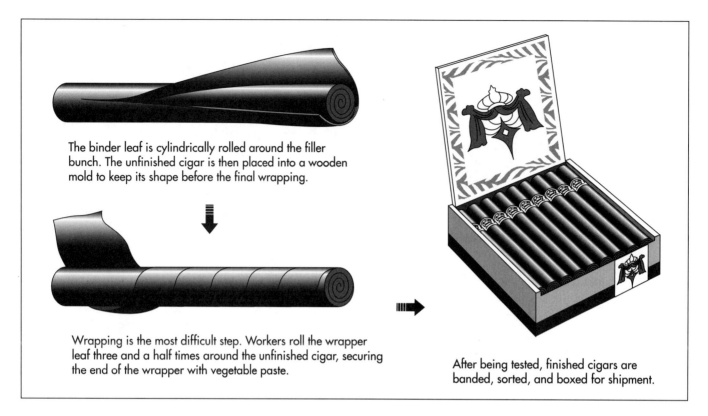

The binder leaf is cylindrically rolled around the filler bunch. The unfinished cigar is then placed into a wooden mold to keep its shape before the final wrapping.

Wrapping is the most difficult step. Workers roll the wrapper leaf three and a half times around the unfinished cigar, securing the end of the wrapper with vegetable paste.

After being tested, finished cigars are banded, sorted, and boxed for shipment.

Cigars come in several shapes and sizes. Perfecto refers to a cigar with a pointed head and tapering sides; Panatella is a long, thin, straight cigar; Cheroot is an open-ended cigar, usually made in India or Asia. From the smallest [3.5 in (8.9 cm)] to the largest [7.5 in (19 cm)] cigars are labeled the Half Corona, Tres Petit Corona, Petit Corona, Corona, Corona Grande, Lonsdale, and Double Corona.

vegetable paste. The worker cuts a small round piece out of a different wrapper leaf. This is sometimes done by tracing around a coin. This circle is then attached to the end of the cigar with paste. The worker has completed the cigar, though it still must be tested, sorted and packed.

Cigars may be made by hand in teams. Some workers may make the bunch and wrap it in the binder, and then the more delicate finishing work of rolling the wrapper is left to more skilled workers.

Machine rolling

7 The majority of cigars are made today by machine. A typical cigar machine may require several workers to tend to its different functions. One worker feeds tobacco leaves onto a feed belt between guide bars that are adjusted for the length of cigar desired. The machine bunches the leaves, forming the filler. A second worker places binder leaf (or HTL) onto the binder die. The leaf is held down by suction, and the machine cuts it to the proper size. The filler then drops onto the binder die. The machine rolls the binder around the filler. A third worker places the wrapper leaf (or HTL) on a wrapper die. The partially completed

cigar drops onto the wrapper die, and the machine rolls the wrapper around the cigar. A fourth worker inspects the completed cigars and places them in trays.

The finished cigars are passed to an examiner. The examiner inspects the cigars for imperfections and checks them for proper weight, size, shape, and condition of the wrapper. The examiner may correct imperfections by patching wrappers or re-shaping heads.

Finishing and packing

8 Cigars that pass inspection are placed on trays and passed to a banding and wrapping machine. A worker places the cigars in a hopper, and the machine places a band around them. The same machine may also wrap the cigars in cellophane. The ringed cigars may be also passed to workers expert in sorting by shade. They sort the finished cigars according to minute variations in wrapper color. Cigars with the same wrapper shade are then boxed together.

Quality Control

Cigars are checked for quality during each step of the manufacturing process. The quality of the tobacco leaves is very important, and leaves are sorted and inspected after curing, after fermentation, and before they are made into cigars. The finished cigars must be checked for consistent diameter, weight, size, draw (how well smoke can be sucked through them), and for any imperfections in the wrapper or in the shape. Cigar factories employ personnel to maintain the manufacturing machinery so that cigar measurements are consistent. In many smaller tobacco factories the final inspections are done by eye. A worker places cigars through a ring to check diameter and measures their length with a ruler. Appearance is critical to the individual cigar, and a box of cigars must also be inspected so that at least the top layer is consistent in color. The quality of the wrapping must be inspected for hand-rolled cigars. The veins of the wrapper should appear in a uniform spiral, and the leaf must be smooth and taut.

Where to Learn More

Books

Sherman, Joel and Nat Sherman. *A Passion for Cigars*. Andrews and McMeel, 1996.

Periodicals

DeGeorge, Gail, and Ivette Diaz. "I'm Rolling As Fast As I Can." *Business Week*, September 2, 1996, p. 46.

Flanagan, William G. and Toddi Gutner. "Cigar Power." *Forbes*, August 1, 1994, pp. 100-101.

Pruzan, Todd. "Stogies for Fogies? Puffing Now Upscale." *Advertising Age*, August 21, 1995, p. 1, 12.

—*Angela Woodward*

Clarinet

An instrument similar to the clarinet—a cylindrical cane tube played with a cane reed—was in use in Egypt as early as 3000 B.C.

Background

The clarinet is a woodwind instrument played with a single reed. Clarinets come in many different sizes, with different pitch ranges. Though there are more than a dozen different modern clarinet types, the most common ones used in orchestras and bands are the B flat and A clarinets. The bass clarinet, which is much bigger than the standard and has an upwardly curved bell, is also frequently used in modern bands and orchestras. The standard clarinet consists of five parts—the mouthpiece, the barrel or tuning socket, the upper (or lefthand) joint, lower (or righthand) joint, and the bell. A thin, flattened, specially shaped piece of cane called a reed must be inserted in the mouthpiece before the instrument can be played. Different notes are produced as the player moves his fingers over metal keys which open and close air holes in the clarinet's body.

History

An instrument similar to the clarinet—a cylindrical cane tube played with a cane reed—was in use in Egypt as early as 3000 B.C. Instruments of this type were used across the Near East into modern times, and other clarinet prototypes were played in Spain, parts of Eastern Europe, and in Sardinia. A folk instrument found in Wales through the eighteenth century, called the hornpipe or pibgorn, was very similar to Greek and Middle Eastern cane single reed instruments, but it was made of bone or of elder wood. Through the Middle Ages and up to the seventeenth century such single reed instruments were played across Europe, but they were almost exclusively peasant or folk instruments.

The modern clarinet seems to have been originated by a Nuremberg instrument maker, Johann Cristoph Denner, sometime around 1690. Denner was a celebrated manufacturer of recorders, flutes, oboes, and bassoons. His early clarinets (the word is a diminutive of the Italian word for trumpet, *clarino*) looked much like recorders, made in three parts and with the addition of two keys to close the holes. A clarinet with a flared bell, like the modern clarinet, may have been made by Denner's son. Parts scored for clarinet were soon found in the music of notable eighteenth century composers, including Handel, Glück, and Telemann. The early clarinets were usually made of boxwood or occasionally plum or pear wood. Rarely, they were made of ivory, and some used a mouthpiece of ebony.

The design of the clarinet was improved by the end of the eighteenth century. The two keys gave way to five or six, giving the instrument more pitch control. Composers and virtuoso performers began to exploit one of the signal characteristics of the clarinet, its versatile dynamic range, from whisper soft to loud and penetrating. Mozart composed a concerto for clarinet in 1791, showing that he realized its possibilities as a solo instrument. By 1800, most orchestras included clarinets. The clarinet developed further in the nineteenth century. Its intonation was improved by a rearrangement of the holes, more keys were added, and the instrument's range was extended. Virtuoso performers toured Europe and influenced composers such as Spohr and Weber to write clarinet concertos and chamber works. Instruments continued to be made out of boxwood, though makers experimented with silver and brass as well. Some clarinets were made out of cocuswood, a tropical wood found mostly in Jamaica. French makers began making clarinets out of ebony, a heavy, dark wood from Africa, in the mid-nineteenth century. But gradually the preferred material became African blackwood, which is similar to ebony but less heavy and brittle.

Clarinets made after 1850 are generally the same as modern clarinets in size and shape. Nineteenth century makers experimented widely with different key and fingering systems, and today there are two main key systems in use. The simple, or Albert, system is used principally in German-speaking countries. The Bohm system has more keys than the Albert and is standard in most other parts of the world.

Raw Materials

Most modern clarinet bodies are made out of African blackwood (*Dalbergia melanoxylon*). There are actually many different trees in the African blackwood genus, such as black cocus, Mozambique ebony, grenadilla, and East African ebony. It is this heavy, dark wood that gives clarinets their characteristic color. Inexpensive clarinets designed for students may be made out of artificial resins. Very occasionally, clarinets are manufactured out of silver or brass. The clarinet mouthpiece is made out of a kind of hard rubber called ebonite. The keys are usually made out of an alloy called German silver. This is made from copper, zinc, and nickel. It looks like pure silver, but does not tarnish. Some fine instruments may be made with pure silver keys, and expensive models are available with gold-plated keys. The key pads require cardboard and felt or leather. The reed is made from cane. Other materials used in the clarinet are cork and wax, for lining the joints, and a metal such as silver or a cheaper alloy for the ligature, the screw clip that holds the reed in place, and stainless steel for the spring mechanisms that work the keys.

The Manufacturing Process

Preparing the body

1 When wood is harvested for clarinet-making, logs are sawed to between 3-4 ft

(1-1.2 m) in length. The logs must be seasoned, to prevent later warping. They may be seasoned by being kept in the open air for several months, or they may be dried in a kiln. Then the logs are split and sawed to lengths approximating the finished lengths of the clarinet body pieces, (upper and lower joints, barrel and bell). The body pieces look like narrow rectangular blocks, and pieces for the barrel are carved in a rough pyramidal shape. These pieces are known as billets. The manufacturer buys the billets in lots, and begins the manufacturing process from these roughed-out shapes.

2 When the manufacturer receives the billets, workers inspect the lot. Then skilled workers place the billets on a borer, which drills a hole lengthwise through the center of each piece. The diameter and shape of this hole, called the bore of the clarinet, is crucial to determining the tone of the instrument. The bore may be drilled in a straight cylinder, or the cylinder may be slightly tapered. After the bore is drilled, the body pieces are turned on a lathe. The rectangular billets become smooth, round, hollow cylinders. These cylinders are then seasoned again.

After the rough pieces have been seasoned for the second time, they are reduced to finished size. The pieces are turned on a lathe and trimmed to exceedingly precise diameters. The joints where the body pieces fit into each other are turned after the exterior is completed. The bore may be reamed more precisely, and then it is polished on the inside. Then the joints are painted with a black dye.

Plastic models

3 Body parts for clarinets made of plastic are produced by injection molding. Plastic pellets are melted and forced under pressure into molds. The molds for clarinet body parts produce hollow cylinders. In some cases, the molds are so precise that these cylinders do not need any additional reaming. Or they may be reamed and polished, as are wooden clarinets.

The steps that follow apply to both wooden and plastic models.

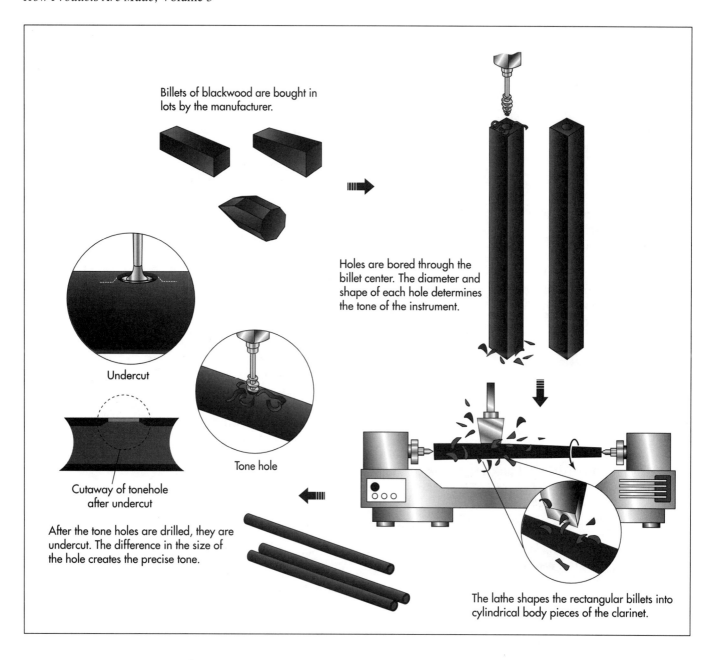

Billets of blackwood are bought in lots by the manufacturer.

Holes are bored through the billet center. The diameter and shape of each hole determines the tone of the instrument.

Undercut

Tone hole

Cutaway of tonehole after undercut

After the tone holes are drilled, they are undercut. The difference in the size of the hole creates the precise tone.

The lathe shapes the rectangular billets into cylindrical body pieces of the clarinet.

Boring the tone holes

4 Next, the maker bores the tone holes that the player's fingers cover to make the different notes. The most common method for mass-produced clarinets is to set the body pieces in a setting out machine. This is a table which holds the piece on a mount under a vertical drill. The holes are drilled at specified distances apart and with precise diameters. The exact dimension of the holes affects the tuning of the instrument, and the holes may be adjusted after the instrument is nearly complete. Not every hole is the same size, and the maker may have to insert a different drill bit for each hole. The holes are smaller on the out-side than on the inside, and to achieve their precise shape, after the holes are drilled they are undercut. The clarinet maker uses a small, flared tool placed in the tone hole to expand the underside of the hole. Next to the tone holes, tiny holes for holding the key mechanism are also drilled.

Construction of keys

5 Early clarinets were made with hand-forged keys. The modern method is usu-ally die-casting. Molten alloy (usually Ger-man silver) is forced under pressure into steel dies. A group of connected keys may be made in one piece in this method. Alter-nately, individual keys may be stamped out

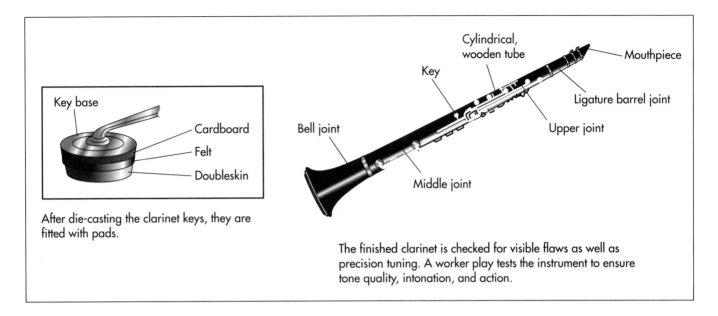

Key base — Cardboard — Felt — Doubleskin

After die-casting the clarinet keys, they are fitted with pads.

Bell joint · Middle joint · Cylindrical, wooden tube · Key · Ligature barrel joint · Upper joint · Mouthpiece

The finished clarinet is checked for visible flaws as well as precision tuning. A worker play tests the instrument to ensure tone quality, intonation, and action.

by a heavy stamping machine, and then trimmed. These individual keys are then soldered together with silver solder to make the connected group. Next the keys are polished. Keys for inexpensive models may be placed in a tumbling machine, where friction and agitation of pellets in a revolving drum polish the pieces. More expensive keys may be buffed individually by being held against the rotating wheel of a polishing machine. Some keys may be silver-plated, and then polished.

6 The keys are then fitted with pads. The pads are usually made of several layers—cardboard, felt, and skin or leather. The circular pads are stamped or cut, and then workers glue them by hand into the head of the key. This will muffle the sound of the tone hole closing when the instrument is played.

7 The keys are drilled, and then fitted with springs that will keep them either open or closed. These springs are made of fine steel wire.

Mounting the keys

8 The keys are mounted on small pillars called posts. The posts are first set in the holes previously drilled for them. In many models the posts are threaded, and they can be simply screwed in by hand. Using a very small drill bit, tiny holes are then drilled in the posts to hold the needle springs. Then the keys are screwed into the posts with

stainless steel hinge rods. The assembler uses a fine screwdriver, pliers, and a small leather mallet to fit the keys and adjust the spring action. The assembler also checks that the tone holes are covered completely by the key pad, inserting a tiny pick under the pad on each side. The pad may need to be adjusted or reset, or the assembler may clamp a key shut temporarily, to set the crease for a perfect, airtight closure.

Finishing

9 The joints of the body pieces are lined with cork and waxed, so that the pieces fit smoothly into each other. The ends of the body pieces are fitted with decorative metal rings, as is the bottom of the barrel. The barrel is usually embossed with the name of the maker. The mouthpiece, manufactured separately out of hard rubber, is fitted to the instrument. When a reed is inserted, the instrument can be played for the first time.

Quality Control

After the clarinet is fully assembled, a worker checks the instrument for visual flaws, checks the action of the keys, and then play tests it. By playing it, the worker can note the tone quality, intonation, and action of the new instrument.

The finished clarinet should be checked for precision tuning. The clarinet's sounding A natural should be at 440 cycles per second,

There are two main clarinet key systems in use. The simple, or Albert, system is used principally in German-speaking countries. The Bohm system has more keys than the Albert and is standard in most other parts of the world.

and the other notes in tune with this. If the instrument has been manufactured according to a standard model, with care to exact diameters of bore and tone holes, it should play in tune automatically. It may be tested with an electronic tuner, and the diameters of the tone holes made larger by more reaming, if necessary. If tone holes are too large (producing a flat note) they may be filled with a layer of shellac.

The wood of the clarinet body should not crack, and the action of the keys should be smooth and not too loud. Ideally, the instrument should last for decades without warping, cracking, or any serious defect.

The Future

Clarinet manufacturing itself is a fairly conservative industry, relying on highly skilled craftspeople who do much work by hand. Most of the innovations in clarinet design are now 100 years old. One area that is still in flux, however, is the manufacture of clarinet reeds. While the best reeds are said to come from a species of cane grown in France, some players and makers are experimenting with wild cane that grows in California. Synthetic reeds have also been developed recently, and more research is being done to improve them. As sources of natural cane diminish, and overall quality is not high, synthetic reeds may be what most clarinet players use in the future.

Where to Learn More

Books

Rendall, F. Geoffrey. *The Clarinet*. Norton, 1971.

Robinson, Trevor. *The Amateur Wind Instrument Maker*. University of Massachusetts Press, 1980.

Periodicals

Armato, Ben. "Raising 'Cain' with the Growers." *The Clarinet*. February/March 1994, pp. 32-33.

—Angela Woodward

Concrete Block

Background

A concrete block is primarily used as a building material in the construction of walls. It is sometimes called a concrete masonry unit (CMU). A concrete block is one of several precast concrete products used in construction. The term precast refers to the fact that the blocks are formed and hardened before they are brought to the job site. Most concrete blocks have one or more hollow cavities, and their sides may be cast smooth or with a design. In use, concrete blocks are stacked one at a time and held together with fresh concrete mortar to form the desired length and height of the wall.

Concrete mortar was used by the Romans as early as 200 B.C. to bind shaped stones together in the construction of buildings. During the reign of the Roman emperor Caligula, in 37-41 A.D., small blocks of precast concrete were used as a construction material in the region around present-day Naples, Italy. Much of the concrete technology developed by the Romans was lost after the fall of the Roman Empire in the fifth century. It was not until 1824 that the English stonemason Joseph Aspdin developed portland cement, which became one of the key components of modern concrete.

The first hollow concrete block was designed in 1890 by Harmon S. Palmer in the United States. After 10 years of experimenting, Palmer patented the design in 1900. Palmer's blocks were 8 in (20.3 cm) by 10 in (25.4 cm) by 30 in (76.2 cm), and they were so heavy they had to be lifted into place with a small crane. By 1905, an estimated 1,500 companies were manufacturing concrete blocks in the United States.

These early blocks were usually cast by hand, and the average output was about 10 blocks per person per hour. Today, concrete block manufacturing is a highly automated process that can produce up to 2,000 blocks per hour.

Raw Materials

The concrete commonly used to make concrete blocks is a mixture of powdered portland cement, water, sand, and gravel. This produces a light gray block with a fine surface texture and a high compressive strength. A typical concrete block weighs 38-43 lb (17.2-19.5 kg). In general, the concrete mixture used for blocks has a higher percentage of sand and a lower percentage of gravel and water than the concrete mixtures used for general construction purposes. This produces a very dry, stiff mixture that holds its shape when it is removed from the block mold.

If granulated coal or volcanic cinders are used instead of sand and gravel, the resulting block is commonly called a cinder block. This produces a dark gray block with a medium-to-coarse surface texture, good strength, good sound-deadening properties, and a higher thermal insulating value than a concrete block. A typical cinder block weighs 26-33 lb (11.8-15.0 kg).

Lightweight concrete blocks are made by replacing the sand and gravel with expanded clay, shale, or slate. Expanded clay, shale, and slate are produced by crushing the raw materials and heating them to about 2000°F (1093°C). At this temperature the material bloats, or puffs up, because of the rapid generation of gases caused by the

Some of the possible block designs for the future include the biaxial block, which has cavities running horizontally as well as vertically to allow access for plumbing and electrical conduits; the stacked siding block, which consists of three sections that form both interior and exterior walls; and the heatsoak block, which stores heat to cool the interior rooms in summer and heat them in winter.

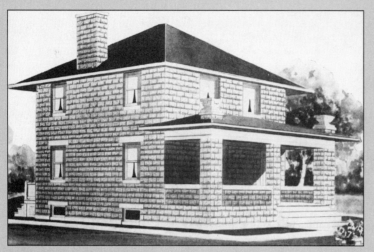

A Foursquare-style house design, appearing in the Radford Architectural Company's 1908 catalog Cement Houses and How to Build Them. *It was one of hundreds of concrete block house designs offered by the Radford company. They estimated that this design could be built for about $2,250.00, much less than traditional stone masonry houses of the time. (From the collections of Henry Ford Museum & Greenfield Village.)*

Concrete blocks were first used in the United States as a substitute for stone or wood in the building of homes. The earliest known example of a house built in this country entirely of concrete block was in 1837 on Staten Island, New York. The homes built of concrete blocks showed a creative use of common inexpensive materials made to look like the more expensive and traditional wood-framed stone masonry building. This new type of construction became a popular form of house building in the early 1900s through the 1920s. House styles, often referred to as "modern" at the time, ranged from Tudor to Foursquare, Colonial Revival to Bungalow. While many houses used the concrete blocks as the structure as well as the outer wall surface, other houses used stucco or other coatings over the block structure. Hundreds of thousands of these houses were built especially in the midwestern states, probably because the raw materials needed to make concrete blocks were in abundant supply in sand banks and gravel pits throughout this region. The concrete blocks were made with face designs to simulate stone textures: rock-faced, granite-faced, or rusticated. At first considered an experimental material, houses built of concrete blocks were advertised in many portland cement manufacturers' catalogs as "fireproof, vermin proof, and weatherproof" and as an inexpensive replacement for the ever-scarcer supply of wood. Many other types of buildings such as garages, silos, and post offices were built and continue to be built today using this construction method because of these qualities.

Cynthia Read-Miller

combustion of small quantities of organic material trapped inside. A typical lightweight block weighs 22-28 lb (10.0-12.7 kg) and is used to build non-load-bearing walls and partitions. Expanded blast furnace slag, as well as natural volcanic materials such as pumice and scoria, are also used to make lightweight blocks.

In addition to the basic components, the concrete mixture used to make blocks may also contain various chemicals, called admixtures, to alter curing time, increase compressive strength, or improve workability. The mixture may have pigments added to give the blocks a uniform color throughout, or the surface of the blocks may be coated with a baked-on glaze to give a decorative effect or to provide protection against chemical attack. The glazes are usually made with a thermosetting resinous binder, silica sand, and color pigments.

Design

The shapes and sizes of most common concrete blocks have been standardized to ensure uniform building construction. The most common block size in the United States is referred to as an 8-by-8-by-16 block, with the nominal measurements of 8 in (20.3 cm) high by 8 in (20.3 cm) deep by 16 in (40.6 cm) wide. This nominal measurement includes room for a bead of mortar, and the block itself actually measures 7.63 in (19.4 cm) high by 7.63 in (19.4 cm) deep by 15.63 in (38.8 cm) wide.

Many progressive block manufacturers offer variations on the basic block to achieve unique visual effects or to provide desirable structural features for specialized applications. For example, one manufacturer offers a block specifically designed to resist water leakage through exterior walls. The block incorporates a water repellent admixture to reduce the concrete's absorption and permeability, a beveled upper edge to shed water away from the horizontal mortar joint, and a series of internal grooves and channels to direct the flow of any crack-induced leakage away from the interior surface.

Another block design, called a split-faced block, includes a rough, stone-like texture on one face of the block instead of a smooth face. This gives the block the architectural appearance of a cut and dressed stone.

When manufacturers design a new block, they must consider not only the desired shape, but also the manufacturing process required to make that shape. Shapes that re-

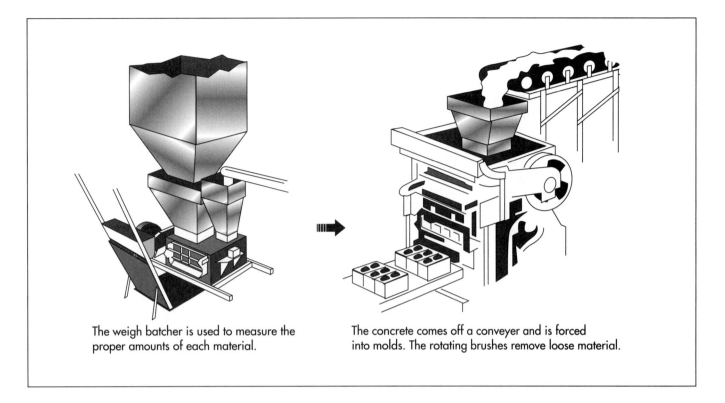

The weigh batcher is used to measure the proper amounts of each material.

The concrete comes off a conveyer and is forced into molds. The rotating brushes remove loose material.

quire complex molds or additional steps in the molding process may slow production and result in increased costs. In some cases, these increased costs may offset the benefits of the new design and make the block too expensive.

The Manufacturing Process

The production of concrete blocks consists of four basic processes: mixing, molding, curing, and cubing. Some manufacturing plants produce only concrete blocks, while others may produce a wide variety of precast concrete products including blocks, flat paver stones, and decorative landscaping pieces such as lawn edging. Some plants are capable of producing 2,000 or more blocks per hour.

The following steps are commonly used to manufacture concrete blocks.

Mixing

1 The sand and gravel are stored outside in piles and are transferred into storage bins in the plant by a conveyor belt as they are needed. The portland cement is stored outside in large vertical silos to protect it from moisture.

2 As a production run starts, the required amounts of sand, gravel, and cement are transferred by gravity or by mechanical means to a weigh batcher which measures the proper amounts of each material.

3 The dry materials then flow into a stationary mixer where they are blended together for several minutes. There are two types of mixers commonly used. One type, called a planetary or pan mixer, resembles a shallow pan with a lid. Mixing blades are attached to a vertical rotating shaft inside the mixer. The other type is called a horizontal drum mixer. It resembles a coffee can turned on its side and has mixing blades attached to a horizontal rotating shaft inside the mixer.

4 After the dry materials are blended, a small amount of water is added to the mixer. If the plant is located in a climate subject to temperature extremes, the water may first pass through a heater or chiller to regulate its temperature. Admixture chemicals and coloring pigments may also be added at this time. The concrete is then mixed for six to eight minutes.

Molding

5 Once the load of concrete is thoroughly mixed, it is dumped into an inclined

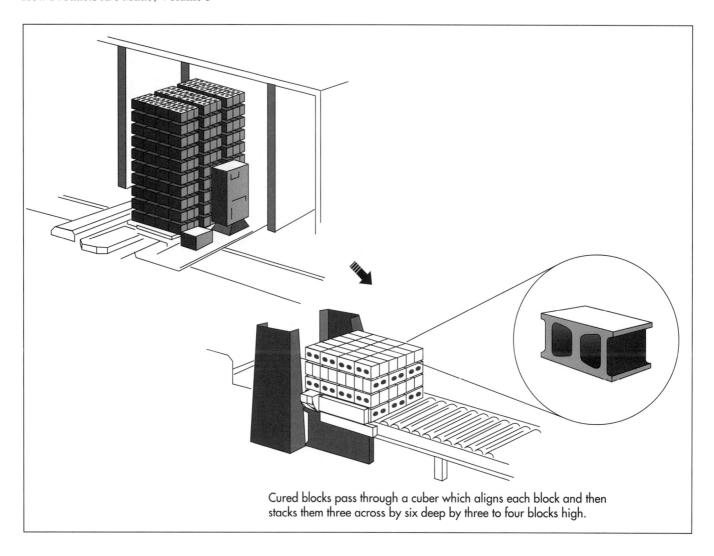

Cured blocks pass through a cuber which aligns each block and then stacks them three across by six deep by three to four blocks high.

bucket conveyor and transported to an elevated hopper. The mixing cycle begins again for the next load.

6 From the hopper the concrete is conveyed to another hopper on top of the block machine at a measured flow rate. In the block machine, the concrete is forced downward into molds. The molds consist of an outer mold box containing several mold liners. The liners determine the outer shape of the block and the inner shape of the block cavities. As many as 15 blocks may be molded at one time.

7 When the molds are full, the concrete is compacted by the weight of the upper mold head coming down on the mold cavities. This compaction may be supplemented by air or hydraulic pressure cylinders acting on the mold head. Most block machines also use a short burst of mechanical vibration to further aid compaction.

8 The compacted blocks are pushed down and out of the molds onto a flat steel pallet. The pallet and blocks are pushed out of the machine and onto a chain conveyor. In some operations the blocks then pass under a rotating brush which removes loose material from the top of the blocks.

Curing

9 The pallets of blocks are conveyed to an automated stacker or loader which places them in a curing rack. Each rack holds several hundred blocks. When a rack is full, it is rolled onto a set of rails and moved into a curing kiln.

10 The kiln is an enclosed room with the capacity to hold several racks of blocks at a time. There are two basic types of curing kilns. The most common type is a low-pressure steam kiln. In this type, the blocks are held in the kiln for one to three

hours at room temperature to allow them to harden slightly. Steam is then gradually introduced to raise the temperature at a controlled rate of not more than 60°F per hour (16°C per hour). Standard weight blocks are usually cured at a temperature of 150-165°F (66-74°C), while lightweight blocks are cured at 170-185°F (77-85°C). When the curing temperature has been reached, the steam is shut off, and the blocks are allowed to soak in the hot, moist air for 12-18 hours. After soaking, the blocks are dried by exhausting the moist air and further raising the temperature in the kiln. The whole curing cycle takes about 24 hours.

Another type of kiln is the high-pressure steam kiln, sometimes called an autoclave. In this type, the temperature is raised to 300-375°F (149-191°C), and the pressure is raised to 80-185 psi (5.5-12.8 bar). The blocks are allowed to soak for five to 10 hours. The pressure is then rapidly vented, which causes the blocks to quickly release their trapped moisture. The autoclave curing process requires more energy and a more expensive kiln, but it can produce blocks in less time.

Cubing

11 The racks of cured blocks are rolled out of the kiln, and the pallets of blocks are unstacked and placed on a chain conveyor. The blocks are pushed off the steel pallets, and the empty pallets are fed back into the block machine to receive a new set of molded blocks.

12 If the blocks are to be made into split-face blocks, they are first molded as two blocks joined together. Once these double blocks are cured, they pass through a splitter, which strikes them with a heavy blade along the section between the two halves. This causes the double block to fracture and form a rough, stone-like texture on one face of each piece.

13 The blocks pass through a cuber which aligns each block and then stacks them into a cube three blocks across by six blocks deep by three or four blocks high. These cubes are carried outside with a forklift and placed in storage.

Quality Control

The manufacture of concrete blocks requires constant monitoring to produce blocks that have the required properties. The raw materials are weighed electronically before they are placed in the mixer. The trapped water content in the sand and gravel may be measured with ultrasonic sensors, and the amount of water to be added to the mix is automatically adjusted to compensate. In areas with harsh temperature extremes, the water may pass through a chiller or heater before it is used.

As the blocks emerge from the block machine, their height may be checked with laser beam sensors. In the curing kiln, the temperatures, pressures, and cycle times are all controlled and recorded automatically to ensure that the blocks are cured properly, in order to achieve their required strength.

The Future

The simple concrete block will continue to evolve as architects and block manufacturers develop new shapes and sizes. These new blocks promise to make building construction faster and less expensive, as well as result in structures that are more durable and energy efficient. Some of the possible block designs for the future include the bi-axial block, which has cavities running horizontally as well as vertically to allow access for plumbing and electrical conduits; the stacked siding block, which consists of three sections that form both interior and exterior walls; and the heatsoak block, which stores heat to cool the interior rooms in summer and heat them in winter. These designs have been incorporated into a prototype house, called Lifestyle 2000, which is the result of a cooperative effort between the National Association of Home Builders and the National Concrete Masonry Association.

Where to Learn More

Books

Hornbostel, Caleb. *Construction Materials, 2nd Edition*. John Wiley and Sons, Inc., 1991.

Periodicals

Koski, John A. "How Concrete Block Are Made." *Masonry Construction,* October 1992, pp.374-377.

Schierhorn, Carolyn. "Producing Structural Lightweight Concrete Block." *Concrete Journal,* February 1996, pp. 92-94, 96, 98, 100-101.

Wardell, C. "Operation Foundation." *Popular Science,* December 1995, p. 31.

Yeaple, Judith Anne. "Building Blocks Grow Up." *Popular Science*, June 1991, pp. 80-82, 108.

—*Chris Cavette*

Cultured Pearl

Background

Thanks to its rarity and beauty, the pearl is as prized as a precious gem, but it is not formed by geologic processes like precious and semi-precious stones. Instead, the pearl is a product of some species of oysters and other shell-fish, formally called bivalve mollusks. It is formed when irritants become lodged in the soft tissue inside an oyster's shell, and, to protect itself, the oyster produces a coating for the irritant. This coating, called nacre, builds up in many thin layers and creates an iridescent cover over the irritant. The resulting product is a pearl.

Pearls have been treasured by rulers and the rich, and they have been described (and used as metaphors for desirable objects) in poetry and song. The lives of those able to find the elusive pearl have also been celebrated. Japanese women called *ama* who developed the extraordinary lung capacity to dive in deep waters for pearl-bearing oysters are featured in folk stories. French composer Georges Bizet wrote an opera called "The Pearl Fishers" about two romantic young men with the unusual occupation of diving or fishing for pearls.

Because of the romance associated with the pearl and its rarity, methods of producing pearls by extending nature's ability to do so have been identified and perfected in this century. The process of using natural methods to produce more pearls than nature can on her own is called culturing.

History

Pearls have been intertwined with history—and historical legend—since Cleopatra's time, when she supposedly dissolved a large pearl in vinegar and drank the potion to demonstrate her infinite wealth. Pearls have been found in the graves of women from Roman times. The largest known pearl weighs about 454 carats and is roughly the size of a chicken egg. The Indian pearl named "La Peregrina," a particularly beautiful specimen in shape and luster, weighs 28 carats, belonged to Mary Tudor for a time, and was housed in a museum in Moscow, Russia, until the 1960s when it was sold to the actor Richard Burton who presented it as a gift to his wife at the time, Elizabeth Taylor. In 1886, a remarkable natural creation named the "Great Southern Cross" was discovered in an Australian oyster; nine pearls had united during natural pearl formation to produce a perfect cross over 1 in (2.54 cm) long.

Pearl-fishing has long been practiced in oyster-bearing waters. Pearls themselves are rare; out of 30 to 40 pearl-forming shellfish, only one may carry a pearl. But the mother-of-pearl lining the mollusks' shell also has value and is another product of the pearl fisher. The Gulf of Manar on the northwest coast of Sri Lanka is the most important pearl fishery in the world. Other parts of the coast of Sri Lanka, the coast of India, the Persian Gulf near the islands of Bahrain, parts of the Red Sea near the Arabian coast, the island groups in the Indian Ocean, the Pacific Ocean near Japan and Hawaii, and the northwest coast of Western Australia are known for their pearl beds.

Monsieur Réaumur, a French naturalist who lived from 1683 to 1757, discovered that the outside layering on a pearl is identical to the inside layering of a mollusk's shell

The Chinese began pearl design in the twelfth century by cementing tiny Buddhas carved from wood, stone, or ivory or cast from metal inside the shells of freshwater mussels. The Buddhas became coated with nacre, or pearlized, and were a successful product.

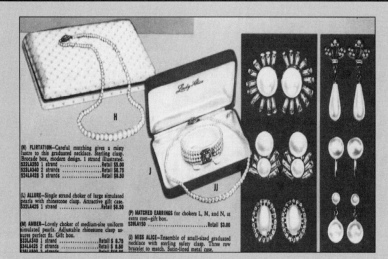

A costume pearl jewelry advertisement from Major's Appliance Store Catalog in Inkster, Michigan, circa 1958-59. (From the collections of Henry Ford Museum & Greenfield Village.)

For at least 1,300 years, men and women have searched for wild or natural pearls to adorn both clothing and jewelry, and their luster remains appealing today. As early as 650 A.D., metalworkers set delicate gold rings with naturally occurring pearls. Ancient Scottish jewelry often includes freshwater pearls, generally small and irregular pearls found in freshwater river mollusks. Those who happened on large, oddly-shaped natural pearls, called baroque pearls, often set them in gold and enameled them so that they resembled delicate, precious sculpture.

Because it was difficult to find matching, gem-quality pearls, imitation pearls were produced as early as 1300 A.D. by the French, who craved them for decorating luxurious clothing and accessories. They dipped hollow glass beads into an acid in order to produce an iridescence, and then filled the bead for solidity and weight. A few centuries later, these glass beads were coated with *essence d'orient*, a silvery substance concocted from the lining of fish scales suspended in a solvent. This essence d'orient was later painted or sprayed on hollow glass beads to simulate the pearl.

Once cultured pearls were developed by Kokichi Mikimoto and offered to the public in the 1920s, the "pearl craze" began—everyone wanted them, could afford them, and wore them. By the 1950s, actress Audrey Hepburn claimed her only earrings were a set of pearls, and every young woman yearned for a sweater clip or circle pin of cultured pearls.

Nancy EV Bryk

beads and bits of oyster mantle to stimulate pearl production. The champion of Japanese pearl "inventors" was Kokichi Mikimoto who, in 1893, began imbedding a variety of materials inside oysters to experiment with creating perfect saltwater pearls called *shinju*. By 1905, this son of a noodle vendor had succeeded, and, in 1908, he was awarded a Japanese patent for the process. Mikimoto's pearl farm had 12 million oysters at its peak and manufactured three-quarters of the world's pearls. Mikimoto lived near the city of Toba, Japan, next to Toba Bay. An island in the bay, now called Mikimoto Pearl Island, is home to monuments to this man who is considered a hero of Japanese industry and a pearl museum. By 1920, Mikimoto's technique dominated the world's pearl production, so that, by 1930, Japanese cultured pearls had completely supplanted the search for natural pearls. Mikimoto's pearl farm today is located in Ago Bay south of Toba.

Raw Materials

The materials for cultured pearls sound simple; they consist of an oyster or other mollusk, the shell nucleus that is to be implanted, a tiny bit of live tissue (from the mantle or lip) from another oyster, and water. Different types of oysters or mussels produce variety in pearls, and the akoya oyster from Japanese saltwater and the biwa mussel from that country's freshwater Lake Biwa may be the best known pearl-bearers. Producers claim freshwater pearls are more natural because nuclei are not used; instead only a piece of mantle is implanted to culture these pearls. All the materials are natural, although human intervention is required. Because the process occurs over several years, a perfect balance of conditions is required for the aquaculture, or growth in water, of pearls.

The Manufacturing Process

Implanting

1 The "irritant" that is surgically implanted in an oyster as the nucleus of the pearl is made from the shell of a mussel. As many as ten different mussel species, including the American pigtoe mussel and the wash-

called "mother-of-pearl." The pioneers of pearl culturing, however, were the Japanese. At the turn of the century, Tokichi Nishikawa and Tatsuhei Mise (a biologist and a carpenter, respectively) discovered independently the method of inserting board mussel, are used because their shells are as much as an inch thick near the hinge. Twenty of these beads or nuclei, called *kaku* in Japanese, can be carved from one mussel shell. In a cultured pearl, this nucleus is the major portion of the pearl, because the pearl coating is relatively thin. Many of the nuclei are manufactured in Tennessee where the Tennessee and Mississippi Rivers host the source mussels.

2 Baby oysters, called spat, are born in hatcheries and grown in tanks at the pearl farms. They are matured in baskets in ocean waters ("maricultured") after they are 60 days old, and after they have grown stronger after spending three years in the water, they are large enough to withstand removal and implanting. To implant the nuclei, harvested oysters are taken inside pearl farm "operating rooms" and held together on racks. When each is removed from the crowd, its shell opens slightly, a wedge is inserted so the shell stays open, and an operator opens the shell wider to insert the bead. A cut is made in the oyster's body, and the nucleus is inserted along with a tiny cutting of the mantle of another oyster that is sacrificed for the process. The mantle insert stimulates the production of nacre which is excreted by the mantle of the host animal. The oysters are then returned to the water by stringing them on plastic garlands or placing them in oyster baskets suspended from rafts.

Formation

3 While the oysters are in the water, the pearl exterior forms over time and fluctuates with water temperature and other conditions. A porous layer called the conchiolin forms around the nucleus and under the nacre. The nacre consists of micro-layers of a specialized form of the mineral calcium carbonate called aragonite. The layers are composed of microscopic plates that make cultured pearls feel rough when they are rubbed against the teeth—a test to distinguish simulated or imitation pearls from natural and cultured gems.

Harvesting

4 The oysters are harvested some time later (depending on the desired finished diameter of the pearl) in winter when cold water slows nacre production and also creates the best color, luster, and orient, which is the ability of the pearl to reflect light. The time of culturing typically ranges from one to three years, and the progress of pearl growth inside the oysters can be monitored with x rays. The pearls are carefully cut out of the oyster's flesh at the pearl farm, and productive oysters can be reseeded several times to produce larger pearls as the oysters continue to grow. The extracted pearls are processed for sale. Typically, the better specimens of pearls are sold in bulk at auctions regulated by governments. From auctions, the pearls move to dealers, then jewelers. The pearl farm may also drill and string its pearls for sale. Jewelry fasteners are usually added at another factory.

Design

Design of pearls seems an unlikely possibility, but, in fact, a wide variety of colors and shapes exist even through they are extensions of the natural process. The Chinese began pearl design in the twelfth century by cementing tiny Buddhas carved from wood, stone, or ivory or cast from metal inside the shells of freshwater mussels. The Buddhas became coated with nacre, or pearlized, and were a successful product. This practice still thrives in Chinese pearl culturing. The same general technique is used to make half-pearls called *mabes*. A molded or cut half shape is planted against the oyster's shell; after it becomes coated and is removed, the half-pearl can be mounted on a jewelry backing, like an earring.

In color, Japanese pearls range naturally from pink to blue to greenish yellow. Pearls are bleached to lighten these colors and eliminate any surface staining. Colored pearls are made by injecting dye into the porous conchiolin, and the pearl must be drilled to be dyed. The most exotic "designer" pearls are probably the large black pearls cultured in Australia and the South Seas. They are grown in the largest pearl oysters in the world, and a perfect specimen of black pearl can sell in the United States for $40,000. Black pearls also display a natural range of colors from silver to green pearls called "peacocks"—and white.

Shape is also designed not only by creating pearlized molds but through culturing of

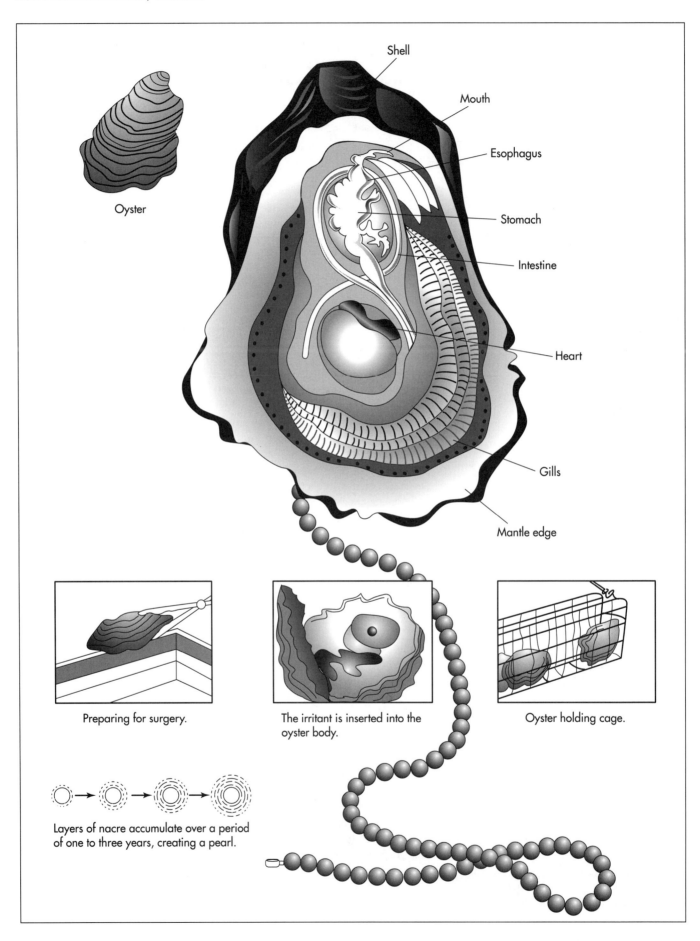

Oyster

Shell

Mouth

Esophagus

Stomach

Intestine

Heart

Gills

Mantle edge

Preparing for surgery.

The irritant is inserted into the oyster body.

Oyster holding cage.

Layers of nacre accumulate over a period of one to three years, creating a pearl.

freshwater pearls. These come in irregular shapes determined by the piece of mantle inserted to stimulate growth of a pearl. Human intervention is visible in these designs, because the implanter's skill influences the shape of the resulting pearl.

Quality Control

Raising baby oysters in tanks causes them to be less hardy, and pollution of oceans and freshwater sources has also caused oyster and pearl quality to decline. But pearl-bearing oysters are prized animals much like cattle or horses, and their health is carefully guarded by scrubbing them periodically to prevent disease. Sometimes, hurricanes or cyclones affect oysters and their pearl crop, but most often these disasters kill divers, not oysters. Pearl farmers ("pearlers") are also licensed by their governments, so the number is limited and controlled.

Some aspects of pearl culturing are closely guarded secrets. Producers often deny that pearls are dyed, enhanced, or otherwise treated, but authorities (and competitors) challenge these claims. Genetic engineering is also used to modify color variation. Culturing times are diminishing, sometimes to less than a year, resulting in a thinner coating of nacre. This is important to buyers because the nacre is soft and can be damaged by perfumes and body fluids. Some exporters and government pearl inspection agencies destroy unacceptable pearls, but the buyer should beware in the pearl market.

Byproducts/Waste

The mother-of-pearl lined shells from which pearls have been removed are also valuable products. Half shells are cleaned and sold as decorative dishes, and the shells can also be cleaned, cut into shapes, and the shapes polished and inset into all kinds of objects, particularly jewelry, buttons, and furniture. If an oyster's productive life is over or if it fails to produce pearls, the oyster meat is harvested and dried for sale as a delicacy.

Flawed pearls not acceptable as gems have other uses. They are ground into powder that is formed into tablets and sold for medicinal purposes. The calcium carbonate in the pearl nacre is valued in this form. Ground pearls are also used in cosmetics and toothpaste, particularly in Japan and China.

The Future

The cultured pearl has a future that is both iridescent, like the pearl itself, and murky. The pearl shows every promise of continuing value as an ornament and for jewelry. Like other gems and jewelry, it tends to go in and out of fashion. The future of the cultured pearl is compromised, however, by environmental concerns. Pearl-bearing animals tolerate only a limited range of ocean or fresh-water environments, and these have diminished with pollution. Commercial oyster beds are jeopardized by polluted water, as shown by decreased sizes of pearls produced, discoloration, and less translucent appearance.

Where to Learn More

Books

Bauer, Max. *Precious Stones*. Dover Publications, Inc., 1968.

Joyce, Kristin and Shellei Addison. *Pearls: Ornament & Obsession*. Simon & Schuster, 1993.

Kunz, George Frederick and Charles Hugh Stevenson. *The Book of the Pearl: The History, Art, Science, & Industry of the Queen of Gems*. Dover Publications, Inc., 1993.

Ward, Fred. *Pearls*. Gem Book Publishers, 1995.

Periodicals

Doubilet, David. "Australia's magnificent pearls." *National Geographic,* December 1991, pp. 230-240.

Doubilet, David. "Black Pearls of French Polynesia." *National Geographic,* June 1997, pp. 30-35.

Fankboner, Peter. "How baubles are born." *Bioscience,* May 1996, p. 384.

Fankboner, Peter. "Pearls and abalones." *Aquaculture,* November/December 1993, p. 28.

Fassler, C. R. "Farming jewels: The aquaculture of pearls." *Aquaculture,* September/October 1991, p. 34.

Fassler, C. R. "The American mussel crisis: Effects on the world pearl industry, part I." *Aquaculture,* July/August 1996, p. 42.

Fassler, C. R. "The return of the American pearl." *Aquaculture,* November/December 1991, p. 63.

Johnson, Julia Claiborne. "Pearl talk." *Working Woman,* January 1995, p. 68.

Ward, Fred. "The Pearl." *National Geographic,* August 1985, pp. 193-223.

—*Gillian S. Holmes*

Dental Drill

Background

The dental drill is a tool used by dentists to bore through tooth enamel as well as to clean and remove plaque from the tooth's surface. It is composed primarily of a handpiece, an air turbine, and a tungsten carbide drill bit. Since its development began in the mid 1700s, the dental drill has revolutionized the field of dentistry. The modern dental drill has enabled dentists to work more quickly and accurately than ever before, with less pain for the patient.

The teeth are composed of both living and nonliving tissue. The soft tissue inner layer, called the dentin, is similar in composition to skeletal bones. Enamel, the outer layer of teeth, which is highly calcified and harder than bone, cannot be regenerated by the body. Tooth decay, which damages to the enamel, is caused by various oral bacteria. One type of bacteria that resides in the mouth breaks down residual food particles that remain on teeth after eating. A byproduct of this bacteria's metabolism is plaque. Other bacteria attach themselves to this plaque and begin secreting an acid which causes small holes to form in the tooth enamel. This allows still other types of bacteria to enter these holes and crevices and erode the softer tissue below. The process weakens the tooth by creating a cavity. The breakdown of the soft tissue is responsible for the pain that is typically associated with cavities. Beyond the initial holes, the outer enamel is left primarily intact. Untreated, cavities can result in diseases such as dental caries and abscesses.

To prevent these diseases, dentists use a dental drill or other tools to remove the plaque from a cavity. As the tooth is drilled, the tiny diamond chips that cover its tip wear away the plaque and damaged enamel. Only by drilling into a tooth can dentists' ensure that all of the plaque is removed. With the plaque gone from the teeth, the enamel-damaging bacteria have nowhere to reside and cannot cause cavities. After the drilling is complete, the hole that is left is filled with a suitable material which strengthens the tooth and helps prevent further damage.

History

The earliest examples of dental drills were developed by the Mayans over 1,000 years ago. They used a stone tool made of jade, which was shaped as a long tube and sharpened on the end. By twirling it between the palms, a hole could be drilled into the teeth. They used this tool primarily in conjunction with a religious ritual for putting jewels in the teeth. Though this technology was ahead of its time, it was not known throughout the rest of the world. The early Greek, Roman, and Jewish civilizations also developed versions of a dental drill. While these early examples of tooth drilling are found, during the Middle Ages the technology was lost. In the mid 1600s, doctors discovered that temporary relief from dental diseases could be achieved by filling the natural holes in teeth with various substances. These early dentists even used a chisel to chip away bits of the damaged enamel. However, it was not until Pierre Fauchard came on the scene that dental drill technology was rediscovered.

Fauchard is said by some to be the father of modern dentistry. He first mentions the use of a bow drill on teeth for root canals in a book published in 1746. This device consisted of a long metal rod with a handle and

High speed drills were developed in 1911, but it was not until 1953 that the modern dental drill with its air turbine engine was introduced. These drills were over 100 times faster than their predecessors and significantly reduced the pain associated with tooth drilling.

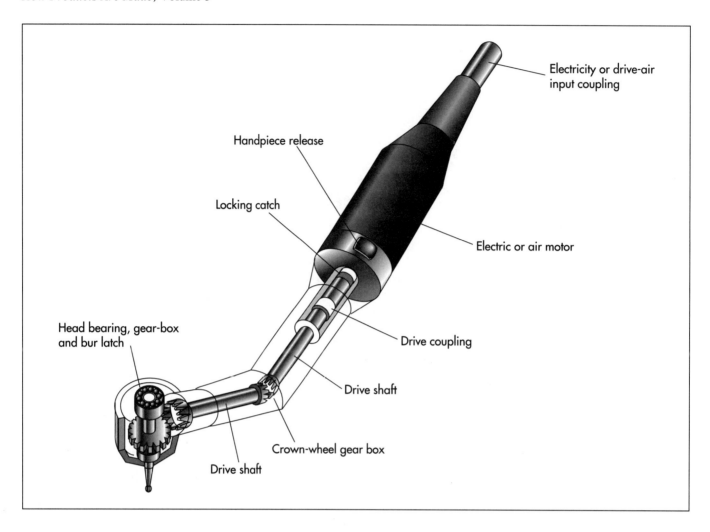

Electricity or drive-air input coupling

Handpiece release

Locking catch

Electric or air motor

Head bearing, gear-box and bur latch

Drive coupling

Drive shaft

Crown-wheel gear box

Drive shaft

Diagram of a dental drill. Although individual drills can vary in design, they all include a motor, handpiece, couplings, and a drill bit, or bur.

a bow that was used to power it. During this time, many innovations were developed. One of these was the 1778 introduction of a near-mechanical drill, which was powered by a hand crank that activated a rotating gear. Soon afterward, an inventor added a spinning wheel to power the drill head. The motion in this device was created by the dentist pushing a foot pedal to move a spinning wheel, which in turn moved the drill head. Other attempts at mechanical drills were made during the 1800s, but they were hard to handle and inefficient, so most dentists used simple, hand-operated steel drills.

Drill technology steadily improved over time, resulting in faster and more efficient drills. New types of foot-powered engines were attached to dental drills by 1870. Electrically powered drills soon followed, and the time it took to prepare a cavity was decreased from hours to less than 10 minutes. High speed drills were developed in 1911, but it was not until 1953 that the modern

dental drill with its air turbine engine was introduced. These drills were over 100 times faster than their predecessors and significantly reduced the pain associated with tooth drilling. To accommodate these faster speeds, tungsten carbide drill bits were introduced. Since then, manufacturers have made many modifications, such as adding fiber optic lights and cameras, incorporating sophisticated cooling systems, and making highly durable handpieces.

Design

There are various designs of dental drills available, however, each have the same basic features, including motors, a handpiece, couplings, and a drill bit. The high speed drilling is activated by an air turbine. These devices convert highly pressurized air into mechanical energy, enabling drill bits to rotate over 300,000 rpms. Slower speeds are also necessary for things such as polishing, finishing, and soft tissue drilling, so dental drills are typically equipped with secondary motors. Common types include electric motors and air-driven motors.

The handpiece is typically a slender, tube-shaped device which connects the drill bit with the driving motor. It is often light-weight and ergonomically designed. It also has an E-shaped attachment that ensures that the drill bit is properly angled for maximum system stability. These components of the dental drill were once quite delicate. How-ever, recent health concerns have forced de-signers to develop handpieces that can with-stand high-pressure steam sterilization. The couplings are used to connect the drill unit to the electric or air power sources and cool-ing water. They can either consist of two or four holes, depending on the type of fitting.

The drill bit, or bur, is the most important part of the dental drill. It is short and highly durable, able to withstand high speed rota-tion and the heat that is subsequently gener-ated. Many bur shapes are manufactured, each with varying cutting and drilling abili-ties. Some burs are even designed with dia-mond cutting flutes. Additional features may be added, such as coolant spray sys-tems or illumination devices. The most so-phisticated dental drill has an internal cool-ing system, an epicyclic speed-increasing gearbox, and fiberoptic illumination.

Raw Materials

Dental drills are constructed from a variety of raw materials, including metals and poly-mers. The handpiece, which houses the mo-tors, gears, and drive shaft, can be made from either lightweight, hard plastics or metal alloys such as brass. The most ad-vanced handpieces are made with titanium. The bur is made of tungsten carbide, one of the hardest substances known. Other mate-rials such as steel are used for the internal motors. The tubing that connects the drill to the main power sources is made of a flexi-ble material, such as polymeric silicone or polyvinyl chloride (PVC).

The Manufacturing Process

The production of a dental drill is an inte-grated process in which individual compo-nents are first made and then assembled to make the final product. While manufactur-ers could make each part individually, they usually depend on outside suppliers for many of the parts. A typical production method would include constructing the mo-tors and the drill bits, forming the hand-piece, final assembly, and packaging.

Handpiece

1 Although numerous designs and materi-als are used to make the handpiece, they are all typically made using a pre-formed mold. For plastic handpieces, this involves injection molding, a process in which the plastic is melted, injected into a mold, and released after it forms. Metal handpieces also use a similar molding process.

Drill bit

2 The drill bits are formed from tungsten carbide particles. They are made by first taking tungsten ore and chemically process-ing it to produce tungsten oxides. Hydrogen is then added to the system to remove the oxygen, resulting in a fine tungsten metal powder. This powder is then blended with carbon and heated, producing tungsten car-bide particles of varying sizes. These parti-cles are further processed to form the ap-propriately shaped drill bit.

Air turbine engine

3 The air turbine engine is constructed from small steel components. In one de-sign, the turbine is sandwiched between two sets of ball raced bearings and connected directly to the drill bit. The entire unit is en-cased in the drill head, with openings for in-coming air and exhaust air. Other types of turbine engines are farther up in the hand-piece and are connected to the drill bit by a series of driveshafts and gears.

Low-powered motors

4 The low-powered motors are put togeth-er much like the air turbine engines. The rotary vane air-powered motor consists of a core structure with sliding vanes protruding outward. It is placed in the handpiece and connected to the main driveshaft of the drill. It also has an opening for incoming and outgoing air. Electric motors are signif-

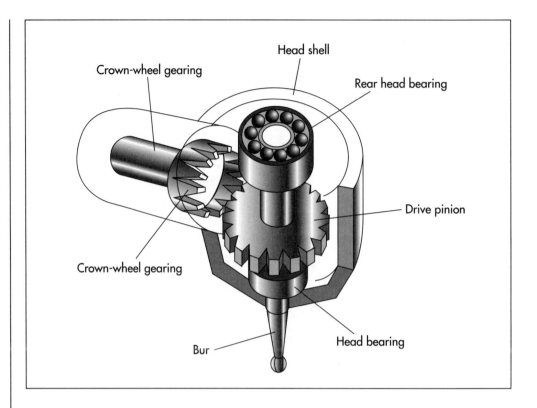

Crown-wheel gearing

Head shell

Rear head bearing

Drive pinion

Crown-wheel gearing

Head bearing

Bur

icantly more complex, consisting of a set of bearings, magnets, brushes, and armature coils.

Final assembly

5 After all the components are available, final assembly can begin. Depending on the design, the air turbine can be placed directly into casing of the handpiece or it can be attached along with the drill bit. The other parts of the drill are put into the handpiece, including air or electric motors, driveshaft, gears, and control switches. Other accessories are added, such as the cooling hoses and fiber optic lighting devices. The coupler is placed on one end of the handpiece, and the drill bit is attached to the other.

6 After an array of quality checks, the finished drills are placed in the appropriate packaging, along with accessories, manuals, and replacement parts, and are then shipped to distributors.

Quality Control

The quality of each drill component is checked during each stage of manufacturing. Since many parts are made each day, inspecting all of them is impossible. Therefore, line inspectors typically take random samples at certain time intervals and check to ensure that those samples meet set specifications for size, shape, and consistency. During this phase of quality control, the primary testing method is visual inspection, although more rigorous measurements can also be made.

The Future

During much of the developmental history of the dental drill, the focus of research had been on increasing the speed of the drill bits and correcting the problems related to these greater speeds. However, studies have shown that there is no benefit to increasing the drill bit speed any higher than it is today. Therefore, the focus of research has shifted to developing alternatives to conventional drills altogether. Two recent introductions are noteworthy and may be indicative of the direction dentistry is headed.

A new method of treating cavities is known as "air-abrasive" technology. Using this technique, a dentist blasts away parts of the tooth surface without using a drill. Small particles of alumina are forced by a stream of air, and the plaque is literally knocked off the tooth. Another technology that may replace the dental drill is the laser. The

FDA has recently approved the use of a laser drill for use on the soft tissue of teeth. However, approval for use on hard tissue is pending. This technology may allow for quicker and more accurate drilling. The result of both of these new technologies is optimal patient comfort as the pain and noise associated with conventional drilling are eliminated.

Where to Learn More

Books

Jedynakiewicz, Nicolas. *A Practical Guide to Technology in Dentistry*. Wolfe Publishing, 1992.

Glenner, Richard, Audrey Davis, and Stanley Burns. *The American Dentist*. Pictorial Histories Publishing Co., 1990.

Simonsen, Richard. *Dentistry in the 21st Century A Global Perspective*. Quintessence Publishing Co., 1989.

Periodicals

Ring, Malvin. "Behind the Dentist's Drill." *Invention & Technology,* Fall 1995, pp. 25-31.

—*Randy Schueller and Perry Romanowski*

Denture

Tooth replacement becomes necessary when the tooth and its roots have been irreparably damaged, and the tooth has been lost or must be removed.

Background

Dentures, or false teeth, are fixed or removable replacements for teeth. Tooth replacement becomes necessary when the tooth and its roots have been irreparably damaged, and the tooth has been lost or must be removed. Dentists have long known that a missing permanent tooth should always be replaced or else the teeth on either side of the space gradually tilt toward the gap, and the teeth in the opposite jaw begin to move toward the space.

There are several standard forms of tooth replacement in modern dentistry. A full denture is made to restore both the teeth and the underlying bone when all the teeth are missing in an arch. A smaller version is the fixed partial denture, also known as a fixed bridge, which can be used if generally healthy teeth are present adjacent to the space where the tooth or teeth have been lost. The partial is anchored to the surrounding teeth by attachment to crowns, or caps, that are affixed to the healthy teeth. A removable partial denture is used to replace multiple missing teeth when there are insufficient natural teeth to support a fixed bridge. This device rests on the soft tissues of the jaws, and is held in place with metal clasps or supports. Dental implants are the latest tooth-replacement technology. They allow prosthetic teeth to be implanted directly in the bones of the jaw.

History

Historically, a variety of materials have been used to replace lost teeth. Animal teeth and pieces of bone were among the earliest of these primitive replacement materials. Two such rudimentary false teeth (probably molars) were found wrapped in gold wire in the ancient Egyptian tomb of El Gigel. In the last few hundred years, artificial teeth have been fashioned from natural substances such as ivory, porcelain, and even platinum. These comparatively crude prototypes of earlier times were carved or forged by hand in an attempt to mimic the appearance and function of natural teeth. Such early denture workmanship is exemplified by George Washington's famous wooden teeth.

Modern technology has offered considerable advances in the materials used to make artificial teeth and improved techniques for affixing them in the mouth. Synthetic plastic resins and lightweight metal alloys have made teeth more durable and natural looking. Better design has resulted in dentures that provide more comfortable and efficient chewing. In the 1980s technology was developed to create the next generation of dentures, which are permanently anchored to the bones in the jaw. These new dentures, known as dental implants, are prepared by specialized dentists called denturists.

Raw Materials

Teeth

Most artificial teeth are made from high quality acrylic resins, which make them stronger and more attractive than was once possible. The acrylic resins are relatively wear-resistant, and teeth made from these materials are expected to last between five and eight years. Porcelain is also used as a tooth material because it looks more like natural tooth enamel. Porcelain is used par-

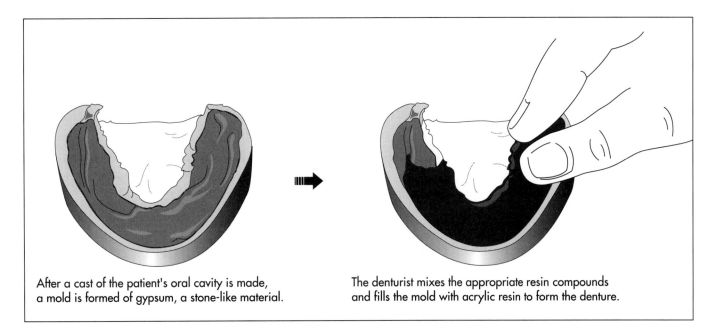

After a cast of the patient's oral cavity is made, a mold is formed of gypsum, a stone-like material.

The denturist mixes the appropriate resin compounds and fills the mold with acrylic resin to form the denture.

ticularly for upper front teeth, which are the most visible. However, the pressure of biting and chewing with porcelain teeth can wear away and damage natural teeth. Therefore, porcelain teeth should not be used in partial dentures where they will contact natural teeth during chewing.

Mounting frame

Artificial teeth are seated in a metal and plastic mount, which holds them in place in the mouth during chewing. The mount consists of a frame to provide its form and a saddle-shaped portion that is shaped to conform to the patient's gums and palate. This design allows for comfort and optimizes the dentures' appearance. Frames are typically constructed of metal alloys such as nobilium or chromium. The latest generations of plastic materials used in dentures are virtually indestructible and can be easily adjusted or repaired with a special kit at the dentist's office. These materials are also ultra lightweight and can eliminate problems in patients who are allergic to acrylic materials or who are bothered by the metallic taste left by a metal frame.

Design

Every individual's mouth is different, and each denture must be custom designed to fit perfectly and to look good. The latest methodology used in denture design, known as dentogenics, is based on research conducted in Switzerland in the early 1950s, which developed standards for designing teeth to fit specific smile lines, mouth shapes, and personalities. These standards are based on such factors as mouth size and shape, skull size, age, sex, skin color, and hair color. For example, through proper denture design, patients can be given a younger smile by simply making teeth longer than they normally would be at that patient's age. This rejuvenation effect is possible because a person's teeth wear down over time; slightly increasing the length of the front teeth can create a more youthful appearance.

The Manufacturing Process

1 The manufacturing process begins with a preliminary impression of the patient's mouth, which is usually done in wax. This impression is used to prepare a diagnostic cast. While making the impression, the dentist applies pressure to the soft tissues to simulate biting force and extends the borders of the mold to adjacent toothless areas to allow the dentures to better adapt to the gums.

2 Once an appropriate preliminary cast has been obtained, the final cast is cast from gypsum, a stone-like product. The final mold is inspected and approved before using it to manufacture the teeth.

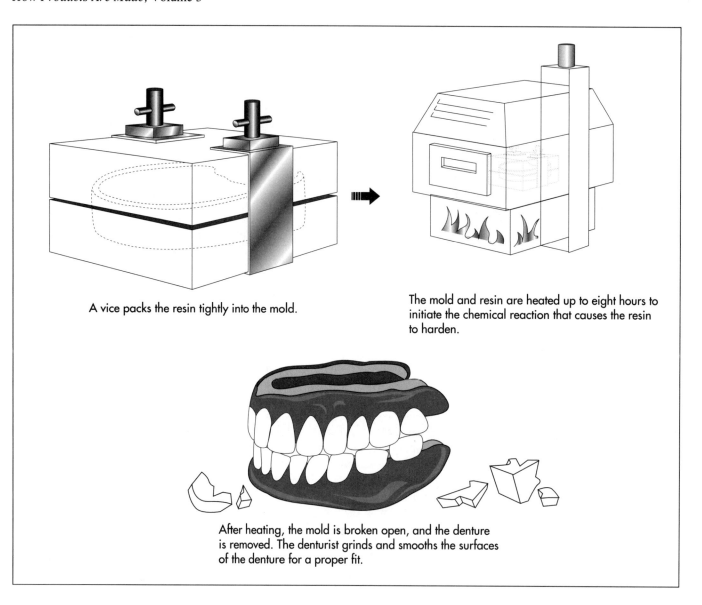

A vice packs the resin tightly into the mold.

The mold and resin are heated up to eight hours to initiate the chemical reaction that causes the resin to harden.

After heating, the mold is broken open, and the denture is removed. The denturist grinds and smooths the surfaces of the denture for a proper fit.

3 After the mold has been cast, it is filled with acrylic resin to form the denture. The mold is prepared with a release agent prior to adding the resin to ensure that the hardened acrylic can be easily removed once the process is completed. A sheet of separating film between the acrylic and the model is also helpful in this regard. The denturist then mixes the appropriate resin compounds in liquid form. Upon drying, the resin hardens to a durable finish.

4 This resin mixture is packed into the mold, while a vertical vise packs it tightly. At this point the model can be inspected to ensure it is filled properly, and if necessary additional resin can be added. Instead of vice packing, certain types of acrylic may be poured into the mold. This method is more prone to air bubbles than hand packing.

5 Once the mold is packed to the denturist's satisfaction, it is heated to initiate the chemical reaction which causes the resin to harden. This part of the process may take up to eight hours.

6 After the heating is done and the mold has cooled, the mold is broken apart so the denture may be removed.

7 The denture is then put in the model of the patient's mouth to ensure that it fits and that the bite is good. Because of the number of processing steps there may be a slight discrepancy in the fit. Usually just a minor grinding and smoothing of surfaces is all that is necessary to make the denture fit correctly.

At this point, if the denture is the removable type, it is ready for use. Implants require additional preparatory steps before they can be used. The denturist must drill the appropriate holes in the jaw bone and attach an anchor. After three to six months, when the hole has healed and the anchor is set in place, a small second surgical procedure is necessary to expose the implant and connect a metal rod to it that will be used to hold the crown or bridge. Finally, the replacement tooth is attached to the rod, where it is held firmly in place.

Quality Control

Good quality control is critical to ensure the denture fits and looks natural in the patient's mouth. No two dentures will be alike; even two sets of dentures made for the same person will not be exactly alike because they are manufactured in custom molds that must be broken in order to extract the denture. After the molding process is completed, the fine details of the denture are added by hand. This step is necessary to ensure the teeth look natural and fit properly.

The quality of the denture's fit can be controlled in two ways. Relining is a process by which the sides of the denture that contacts the gums are resurfaced. Such adjustments are necessary because the dental impressions used to make dentures cause the gums to move. As a result new dentures may not fit properly. Also, over time bone and gum tissues can shift, altering the fit of the denture. Rebasing is used to refit a denture by replacing or adding to the base material of the saddle. This process is required when the denture base degenerates or no longer extends into the proper gum areas. Most patients require relining or rebasing approximately five to eight years after initial placement of the dentures.

Byproducts/Waste

Denture manufacture generates little waste other than a minimal amount of the gypsum and plaster materials used in mold making. There is also little excess of the acrylic resins used in crafting the teeth and mounts. Large quantities of wasted materials are not generally produced since dentures are hand crafted and not mass produced on a production line.

The Future

Dentistry has shared in many of the successes experienced in other areas of science and medicine. For example, improved surgical techniques have led to the development of implants. Advances in polymer chemistry have resulted in improved resins that are more durable and better looking. However, other materials used in dental techniques still require significant improvement. For example, the adhesives used in bonding artificial constructs to natural teeth research must be improved because a high proportion of these bonding processes are not successful. Similarly, improved resins are necessary to make dentures even more comfortable and longer lasting. As breakthroughs are made in related fields of chemistry, they will be incorporated into denture manufacturing.

Where to Learn More

Books

Goldstein, Ronald. *Change Your Smile, 3rd ed.* Quintessence Publishing, 1997.

Woodforde, J. *The Strange Story of False Teeth.* Universe Books, 1968.

Periodicals

Weber, Hans-Peter. "A Tooth for a Tooth." *Harvard Health Letter,* April 1993, p. 6.

"Dental Implants: How Successful." *Healthfacts,* January 1996, p. 1.

—Randy Schueller

Disposable Diaper

Today's diapers are not only highly functional, they include advanced features such as special sizing and coloring for specific gender and age, color change indicators to show when the child is wet, and re-attachable Velcro™-type closures.

Background

A disposable diaper consists of an absorbent pad sandwiched between two sheets of nonwoven fabric. The pad is specially designed to absorb and retain body fluids, and the nonwoven fabric gives the diaper a comfortable shape and helps prevent leakage. These diapers are made by a multi-step process in which the absorbent pad is first vacuum-formed, then attached to a permeable top sheet and impermeable bottom sheet. The components are sealed together by application of heat or ultrasonic vibrations. Elastic fibers are attached to the sheets to gather the edges of the diaper into the proper shape so it fits snugly around a baby's legs and crotch. When properly fitted, the disposable diaper will retain body fluids which pass through the permeable top sheet and are absorbed into the pad.

Disposable diapers are a relatively recent invention. In fact, until the early 1970s mothers had no real alternative to classic cloth diapers. Cotton diapers have the advantage of being soft, comfortable, and made of natural materials. Their disadvantages include their relatively poor absorbency and the fact that they have to be laundered. Disposable diapers were developed to overcome these problems. The earliest disposables used wood pulp fluff, cellulose wadding, fluff cellulose, or cotton fibers as the absorbent material. These materials did not absorb very much moisture for their weight, however. Consequently, diapers made from these materials were extremely bulky. More efficient absorbent polymers were developed to address this issue.

Since the 1970s, disposable diaper technology has continued to evolve. In fact, nearly 1,000 patents related to diaper design and construction have been issued in the last 25 years. Today's diapers are not only highly functional, they include advanced features such as special sizing and coloring for specific gender and age, color change indicators to show when the child is wet, and re-attachable Velcro™-type closures. These innovations have enabled disposables to capture a large share of the diaper market. In 1996, disposable diaper sales exceeded $4 billion in the United States alone. Proctor and Gamble and Kimberly Clark are the two largest brand name manufacturers, and their sales account for nearly 80% of the market. Private label manufacturers that produce store brands and generic diapers account for most of the remaining 20%.

Raw Materials

Absorbent pad

The single most important property of a diaper, cloth or disposable, is its ability to absorb and retain moisture. Cotton material used in cloth diapers is reasonably absorbent, but synthetic polymers far exceed the capacity of natural fibers. Today's state-of-the-art disposable diaper will absorb 15 times its weight in water. This phenomenal absorption capacity is due to the absorbent pad found in the core of the diaper. This pad is composed of two essential elements, a hydrophilic, or water-loving, polymer and a fibrous material such as wood pulp. The polymer is made of fine particles of an acrylic acid derivative, such as sodium acrylate, potassium acrylate, or an alkyl acrylate. These polymeric particles act as tiny sponges that retain many times their weight in water. Microscopically these polymer molecules resemble long chains or ropes. Portions of these chemical "ropes" are designed to interact with water molecules. Other parts of the polymer have the ability to chemically link with different polymer molecules in a process known as cross linking. When a large number of these polymeric chains are cross linked, they form a gel network that is not water soluble but that can absorb vast amounts of water. Polymers with this ability are referred to as hydrogels, superabsorbents, or hydrocolloids. Depending on the degree of cross linking, the strength of the gel network can be varied. This is an important property because gel strength is related to the tendency of the polymer to deform or flow under stress. If the strength is too high the polymer will not retain enough water. If it too low the polymer will deform too easily, and the outermost particles in the pad will absorb water too quickly, forming a gel that blocks water from reaching the inner pad particles. This problem, known as gel blocking, can be overcome by dispersing wood pulp fibers throughout the polymer matrix. These wood fibers act as thousands of tiny straws which suck up water faster and disperse it through the matrix more efficiently to avoid gel blocking. Manufacturers have optimized the combinations of polymer and fibrous material to yield the most efficient absorbency possible.

Nonwoven fabric

The absorbent pad is at the core of the diaper. It is held in place by nonwoven fabric sheets that form the body of the diaper. Nonwoven fabrics are different from traditional fabrics because of the way they are made. Traditional fabrics are made by weaving together fibers of silk, cotton, polyester, wool, etc. to create an interlocking network of fiber loops. Nonwovens are typically made from plastic resins, such as nylon, polyester, polyethylene, or polypropylene, and are assembled by mechanically, chemically, or thermally interlocking the plastic fibers. There are two primary methods of assembling nonwovens, the wet laid process and the dry laid process. A dry laid process, such as the "meltblown" method, is typically used to make nonwoven diaper fabrics. In this method the plastic resin is melted and ex-

truded, or forced, through tiny holes by air pressure. As the air-blown stream of fibers cools, the fibers condense onto a sheet. Heated rollers are then used to flatten the fibers and bond them together. Polypropylene is typically the material used for the permeable top sheet, while polyethylene is the resin of choice for the non-permeable back sheet.

Other components

There are a variety of other ancillary components, such as elastic threads, hot melt adhesives, strips of tape or other closures, and inks used for printing decorations.

The Manufacturing Process

Formation of the absorbent pad

1 The absorbent pad is formed on a movable conveyer belt that passes through a long "forming chamber." At various points in the chamber, pressurized nozzles spray either polymer particles or fibrous material onto the conveyor surface. The bottom of the conveyor is perforated, and as the pad material is sprayed onto the belt, a vacuum is applied from below so that the fibers are pulled down to form a flat pad.

At least two methods have been employed to incorporate absorbent polymers into the pad. In one method the polymer is injected into the same feed stock that supplies the fibers. This method produces a pad that has absorbent polymer dispersed evenly throughout its entire length, width, and thickness. The problems associated with method are that loss of absorbent may occur because the fine particles are pulled through the perforations in the conveyor by the vacuum. It is therefore expensive and messy. This method also causes the pad to absorb unevenly since absorbent is lost from only one side and not the other.

A second method of applying polymer and fiber involves application of the absorbent material onto the top surface of the pad after it has been formed. This method produces a pad which has absorbent material concentrated on its top side and does not have much absorbency throughout the pad. Another disadvantage is that a pad made in

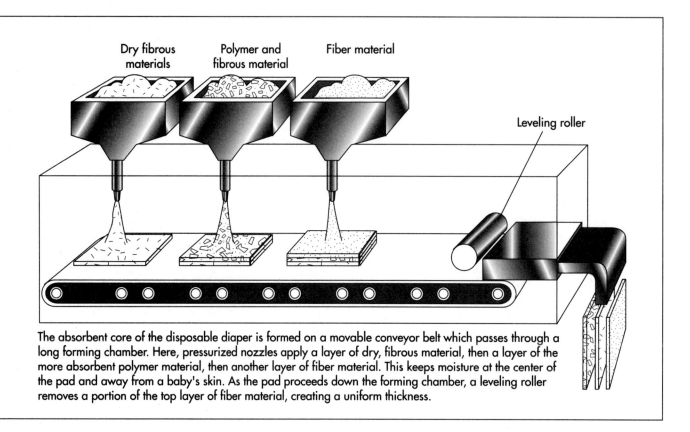

Dry fibrous materials

Polymer and fibrous material

Fiber material

Leveling roller

The absorbent core of the disposable diaper is formed on a movable conveyor belt which passes through a long forming chamber. Here, pressurized nozzles apply a layer of dry, fibrous material, then a layer of the more absorbent polymer material, then another layer of fiber material. This keeps moisture at the center of the pad and away from a baby's skin. As the pad proceeds down the forming chamber, a leveling roller removes a portion of the top layer of fiber material, creating a uniform thickness.

this way may lose some of the polymer applied to its surface. Furthermore, this approach tends to cause gel blocking, since all the absorbent is on the outside of the pad. The moisture gets trapped in this outer layer and does not have a chance to diffuse to the center. This blockage holds moisture against the skin and can lead to discomfort for the wearer.

These problems are solved by controlling the mixture polymer and fibrous material. Multiple spray dispensers are used to apply several layers of polymer and fiber. As the fiber is drawn into the chamber and the bottom of the pad is formed, a portion of the polymer is added to the mix to form a layer of combined polymer and fiber. Then more pure fiber is pulled on top to give a sandwich effect. This formation creates a pad with the absorbent polymer confined to its center, surrounded by fibrous material. Gel blockage is not a problem because the polymer is concentrated at core of pad. It also solves the problem of particle loss since all the absorbent is surrounded by fibrous material. Finally, this process is more cost effective because it distributes the polymer just where it is needed.

2 After the pad has received a full dose of fiber and polymer, it proceeds down the conveyor path to a leveling roller near the outlet of the forming chamber. This roller removes a portion of the fiber at the top of the pad to make it a uniform thickness. The pad then moves by the conveyor through the outlet for subsequent operations to form the competed diaper.

Preparation of the nonwoven

3 Sheets of nonwoven fabric are formed from plastic resin using the meltblown process as described above. These sheets are produced as a wide roll known as a "web," which is then cut to the appropriate width for use in diapers. There is a web for the top sheet and another for the bottom sheet. It should be noted that this step does not necessarily occur in sequence after pad formation because the nonwoven fabrics are often made in a separate location. When the manufacturer is ready to initiate diaper production these large bolts of fabric are connected to special roller equipment that feeds fabric to the assembly line.

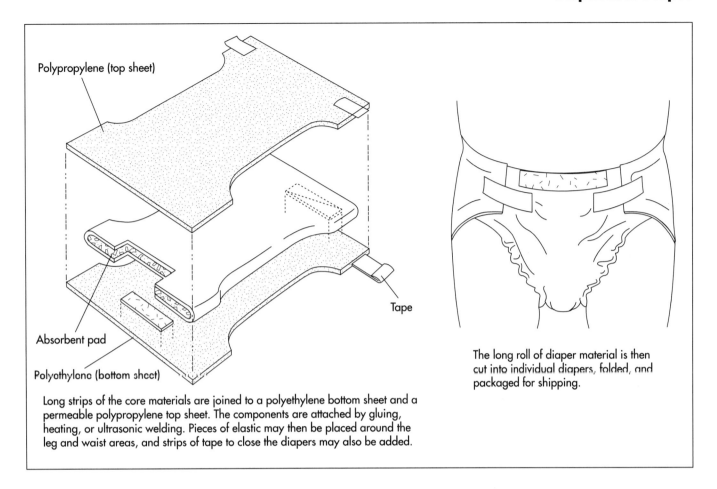

Polypropylene (top sheet)

Absorbent pad

Polyethylene (bottom sheet)

Tape

Long strips of the core materials are joined to a polyethylene bottom sheet and a permeable polypropylene top sheet. The components are attached by gluing, heating, or ultrasonic welding. Pieces of elastic may then be placed around the leg and waist areas, and strips of tape to close the diapers may also be added.

The long roll of diaper material is then cut into individual diapers, folded, and packaged for shipping.

4 At some point in the process, stretched elastic bands are attached to the backing sheet with adhesive. After the diaper is assembled, these elastic bands contract and gather the diaper together to ensure a snug fit and limit leakage.

Assembly of the components

5 At this point in the process there are still three separate components, the absorbent pad, the top sheet, and the backing sheet. These three components are in long strips and must be joined together and cut into diaper-sized units. This is accomplished by feeding the absorbent pad onto a conveyor with the polyethylene bottom sheet. The polypropylene top sheet is then fed into place, and the compiled sheets are joined by gluing, heating, or ultrasonic welding. The assembled diaper may have other attachments, such as strips of tape or Velcro™, which act as closures.

6 The long roll is then cut into individual diapers, folded, and packaged for shipping.

Byproducts/Waste

Diaper production does not produce significant byproducts; in fact the diaper industry uses the byproducts of other industries. The absorbent polymers used in diaper production are often left over from production lines of other chemical industries. The polymer particles are too small for other applications, but they are well suited for use in diapers. In diaper production, however, considerable amounts of both nonwoven material and polymer particles are wasted. To minimize this waste, the industry tries to optimize the number of diapers obtained from every square yard (meter) of material. Furthermore, every attempt is made to recover the excess fiber and polymer material used in the forming chamber. However, this is not always possible due to clogging of filters and other losses.

Quality Control

There are several methods used to control the quality of disposable diapers, and most of these relate to the product's absorbency.

One key is to make sure the polymer/fiber ratio in the absorbent pad is correct. Too much variation will impact the diaper's ability to soak up moisture. Industry trial and error has shown that for optimal performance and cost, the fiber to particle ratio should be about 75:25 to 90:10. Even more critical than this ratio are the size and distribution of these particles. It has been established that particles with mass median particle size greater than or equal to about 400 microns work very well with the fibers to enhance the rate at which the fluid is transported away from the body. If the particles vary much outside this range, gel blocking may occur.

There are several standard tests the industry uses to establish diaper absorbency. One is referred to as Demand Wettability or Gravimetric Absorbance. These tests evaluate what is are commonly referred to as Absorbance Under Load (AUL). AUL is defined as the amount of 0.9% saline solution absorbed by the polymers while being subjected to pressure equivalent to 21,000 dynes, or about 0.30 lb/sq in (0.021 kg/sq cm). This test simulates the effect of a baby sitting on a wet diaper. If the diaper has an absorbency of at least 24 ml/g after one hour, the quality is considered acceptable.

Other quality control factors besides absorbency are related to the diaper's fit and comfort. Particular attention must be paid to the melt characteristics of the nonwoven fabrics used to form the diaper's shell. If materials with different melting points are used, the material that melts the quickest may become too soft and stick to the assembly apparatus. When the fabric is pulled off it may be left with a rough surface that is uncomfortable to the user. Finally, the alignment of the components must be carefully checked or leakage may result.

The Future

Disposable diaper manufacture is a high technology field which has consistently shown innovation over the last few decades. Nonetheless, there are still a number of areas which require additional improvement. One such area is that of leakage reduction. It is likely that manufacturers will develop improved elastic bands to hold the waist more tightly without causing chafing or discomfort. It is also likely that current concern regarding the role of disposable diapers in landfills will impact manufacturing and formulation. This concern may to lead to the development of diapers which are less bulky and more biodegradable.

Where to Learn More

Periodicals

"Dueling diapers." *The Edell Health Letter*, August 1993, p. 6.

McAloney, Regina. "Thin is in." *Nonwovens Industry*, November 1994 p.52.

Lenzner, Robert, and Carrie Shooc. "The Battle of the Bottoms." *Forbes*, March 24, 1997, p. 98.

Ohmura, Kin. "Superabsorbent Polymers in Japan." *Nonwovens Industry*, January 1995, p. 32.

—*Randy Schueller*

EKG Machine

Background

An electrocardiogram (EKG or ECG) is a device which graphically records the electrical activity of the muscles of the heart. It is used to identify normal and abnormal heartbeats. First invented in the early 1900s, the EKG (derived from the German *electrokardiogramma*) has become an important medical diagnostic device.

The function of the EKG machine depends on the ability of the heart to produce electrical signals. The heart is composed of four chambers which make up two pumps. The right pump receives the blood returning from the body and pumps it to the lungs. The left pump gets blood from the lungs and pumps it out to the rest of the body. Each pump is made up of two chambers, an atrium and a ventricle. The atrium collects the incoming blood, and when it contracts, transfers the blood to the ventricle. When the ventricle contracts, the blood is pumped away from the heart.

The pumping action of the heart is regulated by the pacemaker region, or sinoatrial node, located in the right atrium. An electrical impulse is created in this region by the diffusion of calcium ions, sodium ions, and potassium ions across the membranes of cells. The impulse created by the motion of these ions is first transferred to the atria, causing them to contract and push blood into the ventricles. After about 150 milliseconds, the impulse moves to the ventricles, causing them to contract and pump blood away from the heart. As the impulse moves away from the chambers of the heart, these sections relax.

Using an EKG allows doctors to measure the relative voltage of these impulses at various positions in the heart. Electrocardiograms are possible because the body is a good conductor of electricity. When an electrical potential is generated in a section of the heart, an electrical current is conducted to the body surface in a specific area. Electrodes attached the body in these areas enable the measurement of these currents.

The electrical signals measured by the EKG have been characterized and represent various phases of a heartbeat. Each time the heart beats, it produces three distinct EKG waves. The first pulse that is seen is called the P wave. This measures the electrical signal generated by the pacemaker. The next pulse is the largest signal and is called the QRS complex. This segment of the graph represents the electrical signal created by the relaxing of the atria and the contraction of the ventricles. Completing the cycle is the T wave, which signifies the relaxing, or repolarization, of the ventricles. The characteristic sound of a heartbeat corresponds to the QRS complex and the T wave.

EKGs provide useful data and can help detect various problems related to heart function. One basic determination that can be made with an EKG is the heart rate, which can be determined by measuring the distance between peaks. Diagnosis of certain medical problems is also possible. For example, in patients with high blood pressure, the amplitude of the QRS complex is significantly increased. The balance of certain chemicals in the body can also be detected by an EKG, since the amplitude of the signals is related to the levels of chemicals in the body. Damage in the heart can also be

First invented in the early 1900s, the EKG (derived from the German electrokardiogramma) has become an important medical diagnostic device.

observed by a deformation in the Q wave. The most useful characteristic of the EKG is its ability to detect and describe arrhythmias, or abnormal heartbeats. EKG machines known as Holter monitors are for these detections. Finally, EKGs can be used to observe obstructions in the arteries. This is typically done by looking for a depressed segment between the S and T waves.

History

The development of the EKG began with the discovery of the electronic potential of living tissue. This electromotive effect was first investigated by Aloysio Luigi in 1787. Through his experiments, he demonstrated that living tissues, particularly muscles, are capable of generating electricity. Afterwards, other scientists studied this effect in electronic potential. The variation of the electronic potential of the beating heart was observed as early as 1856, but it was not until Willem Einthoven invented the string galvanometer that a practical, functioning EKG machine could be made.

The string galvanometer was a device composed of a coarse string that was suspended in a magnetic field. When the force of the heart current was applied to this device, the string moved, and these deflections were then recorded on photographic paper. The first EKG machine was introduced by Einthoven in 1903. It proved to be a popular device, and large-scale manufacturing soon began soon in various European countries. Early manufacturers include Edelmann and Sons of Munich and the Cambridge Scientific Instrument Company. The EKG was brought to the United States in 1909 and manufactured by the Hindle Instrument Company.

Improvements to the original EKG machine design began soon after its introduction. One important innovation was reducing the size of the electromagnet. This allowed the machine to be portable. Another improvement was the development of electrodes that could be attached directly to the skin. The original electrodes required the patient to submerge the arms and legs into glass electrode jars containing large volumes of a sodium chloride solution. Additional improvements included the incorporation of

amplifiers, which improved the electronic signal, and direct writing instruments, which made the EKG data immediately available. The modern EKG machine is similar to these early models, but microelectronics and computer interfaces have been incorporated, making them more useful and powerful. While these newer machines are more convenient to use, they are not more accurate than the original EKG built by Einthoven.

Raw Materials

The EKG machine consists of electrodes, connecting wires, an amplifier, and a storage and transmission device. The electrodes, or leads, used in an EKG machine can be divided into two types, bipolar and unipolar. The bipolar limb leads are used to record the voltage differential between the wrists and the legs. These electrodes are placed on the left leg, the right wrist, and the left wrist, forming a triangular movement of the electrical impulse in the heart that can then be recorded. Unlike bipolar leads, unipolar leads record the voltage difference between a reference electrode and the body surface to which they are attached. These electrodes are attached to the right and left arms and the right and left legs. Additionally, they are placed at specific areas on the chest and are used to view the changing pattern of the heart's electrical activity.

Various models of electrodes are made, including plate, suction, fluid column, and flexible, among others. Plate electrodes are metal disks which are constructed out of stainless steel, German silver, or nickel. They are held onto the skin with adhesive tape. Suction electrodes use a vacuum system to remain in place. They are designed out of nickel or silver and silver chloride and are attached to a compressor that creates the vacuum. Another type of electrode, the fluid column electrode, is less sensitive to patient movement because it is designed to avoid direct contact with the skin. The flexible electrode is most useful for taking EKG readings in infants. It is a mesh woven from fine stainless steel or silver wire with a flexible lead wire attached. The electrode attaches to the skin like a small bandage.

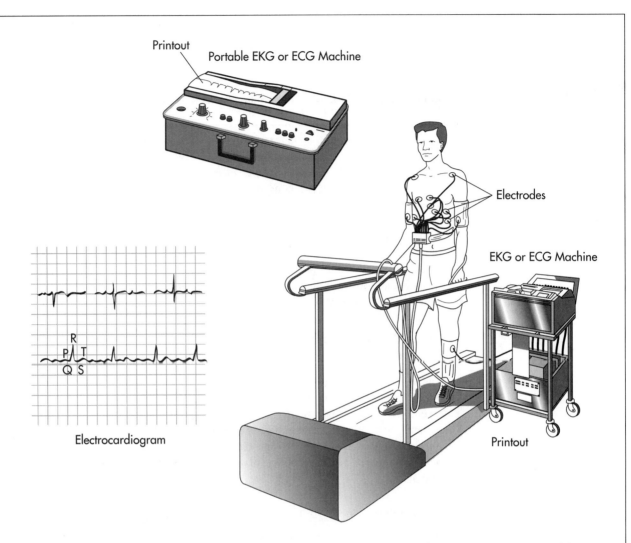

Printout

Portable EKG or ECG Machine

Electrodes

EKG or ECG Machine

R
P | T
Q S

Electrocardiogram

Printout

An electrocardiogram (EKG or ECG) is a graphical representation of the electrical activity of the heart muscle. To record these impulses, electrodes are attached to a patient at key points on the body. Standard leads are placed on the left and right arms and on the left leg. These record a triangular movement of the heart's electrical impulses. Additional electrodes are placed on the patient's chest to view the electrical patterns of the heart muscle. The resulting electrocardiogram shows a wave pattern that can be analyzed to diagnose an abnormal heartbeat. The P wave records the activity of the pacemaker region of the heart. The largest signal is the QRS complex, a representation of the relaxing of the atria and contraction of the ventricles. The T wave signifies the relaxing of the ventricles.

EKG amplifiers are needed to convert the weak electrical signal from the body into a more readable signal for the output device. A differential amplifier is useful when measuring relatively low level signals. During an EKG, the electrical signal from the body is transferred from the electrodes to the first section of the amplifier, the buffer amplifier. Here the signal is stabilized and amplified by a factor of five to 10. An electronic network follows, and the signal from the unipolar leads is translated. A differential pre-amplifier then filters and amplifies the signal by a factor of 10 to 100.

The sections of the amplifier which receive direct signals from the patient are separated from the main power circuitry of the rest of the EKG machine by optical isolators, preventing the possibility of accidental electric shock. The primary amplifier is found in the main power circuitry. In this powered amplifier, the signal is converted to a current suitable for output to the appropriate device.

The most common form of output for EKG machines is a paper-strip recorder. This device provide a hard copy of the EKG signal over time. Many other types of devices are also used, including computers, oscilloscopes, and magnetic tape units. Since the data collected is in analog form, it must be converted to digital form for use by most electronic output devices. For this reason the primary circuitry of the EKG typically has a built-in analog to digital converter section.

Various other parts are needed to complete the EKG unit. Since the signal is weakly transmitted through the skin to the electrodes, an electrolyte paste is usually used. This paste is applied directly to the skin. It is composed primarily of chloride ions which help form a conductive bridge between the skin and the electrode, allowing better signal transmission. Other components include mounting clips, various sensors, and thermal papers.

The Manufacturing Process

The components of an EKG machine are typically manufactured separately and then assembled prior to packaging. These components, including the electrodes, the amplifier, and the output device, can be supplied by outside manufacturers or made in-house.

Electrodes

1 The most common electrode used for an EKG machine is the silver and silver chloride electrode because the electrode potential of these electrodes is stable when exposed to biological tissue. The electrodes are received from outside suppliers and checked to see if they conform to set specifications. The type of electrode used depends on the EKG model. Often multiple types of electrodes will be supplied with one EKG machine. Each electrode contains a shielded cable that can be attached to the primary unit.

Internal electronics

2 The electronic components of an EKG device are quite sophisticated and use the latest in electronic processing technology. The amplifier and output device are assembled much like that of other electronic equipment. It begins with a board that has the electronic configuration mapped out. This board is then passed through a series of machines which place the appropriate chips, diodes, capacitors, and other electronic pieces in the appropriate places. When completed it is sent to the next step for soldering.

3 The electronic components are affixed to the board by a wave soldering machine. Boards that enter this machine are first washed to remove any contaminants. The board is then heated using infrared heat. The underside of the board is passed over a wave of molten solder which fills in the appropriate spots through capillary action. As the board is allowed to cool, the solder hardens, holding the pieces in place.

Display device

4 Depending on the type of display device, manufacture method differs. Certain devices such as magnetic tape recorders and paper printers can be supplied by outside manufacturers. Other components like computer microprocessors can be designed and made right along with the primary internal electronics.

Final assembly

5 The components of the EKG machine are assembled and placed into an appropriate metal frame. The finished devices are then put into final packaging along with accessories such as spare electrodes, printout paper, and manuals. They are then sent out to distributors and finally to customers.

Quality Control

To ensure the quality of each EKG device being manufactured, visual and electrical inspections occur throughout the entire production process, and most flaws are detected. The functional performance of each completed EKG device is tested to make sure it works. These tests are done under different environmental conditions such as excessive heat and humidity.

Most manufacturers set their own quality specifications for the EKG machines that they produce. However, standards and performance recommendations have been proposed by various medical organizations and governmental agencies. Some factors considered important are standardized input signal ranges, frequency response, accuracy of calibration signal, and recording duration.

The Future

In the future, more powerful and improved EKG machines will be developed. These machines will utilize the latest computer technology, making diagnosis quicker and more accurate. They will be more powerful and capable of measuring tiny electronic potentials such as fetal heart rates. They will also make it possible to construct three-dimensional models of the beating heart, providing doctors with more diagnostic data. New applications for EKG machines may also be found, such as the recent application of an EKG machine to determine the efficacy of drugs.

A recent innovation could mark a new direction in EKG development. One company has developed a portable EKG monitor which collects data that can be transmitted directly over the phone. The patients puts the electrodes under each arm and attaches a transmitter to the phone mouthpiece. The signal is sent to a monitoring center, where computers convert it to EKG readings. This information can then be transferred to a doctor, making it possible to detect heart problems in some patients much earlier.

Where to Learn More

Books

Lawrie, T.D. Veitch and Peter Macfarlane. *Comprehensive Electrocardiology. Theory and Practice in Health and Disease*. Pergamon Press, 1989.

Periodicals

Koyuncu, Baki. "Monitoring Heartbeat." *Electronics World & Wireless World,* July 1995, pp. 605-7.

Roberts, H. Edward. "Electrocardiograph I." *Radio-Electronics,* July 1991, pp. 31-40+.

Roberts, H. Edward. "Electrocardiograph II." *Radio-Electronics,* August 1991, pp. 44-49.

—*Perry Romanowski*

Escalator

The invention of the escalator is generally credited to Charles D. Seeberger who, as an employee of the Otis elevator company, produced the first step-type escalator manufactured for use by the general public. His creation was installed at the Paris Exhibition of 1900, where it won first prize.

Background

An escalator is a power-driven, continuous moving stairway designed to transport passengers up and down short vertical distances. Escalators are used around the world to move pedestrian traffic in places where elevators would be impractical. Principal areas of usage include shopping centers, airports, transit systems, trade centers, hotels, and public buildings. The benefits of escalators are many. They have the capacity to move large numbers of people, and they can be placed in the same physical space as stairs would be. They have no waiting interval, except during very heavy traffic; they can be used to guide people towards main exits or special exhibits; and they may be weather-proofed for outdoor use. It is estimated that there are over 30,000 escalators in the United States, and that there are 90 billion riders traveling on escalators each year. Escalators and their cousins, moving walkways, are powered by constant speed alternating current motors and move at approximately 1-2 ft (0.3-0.6 m) per second. The maximum angle of inclination of an escalator to the horizontal is 30 degrees with a standard rise up to about 60 ft (18 m).

The invention of the escalator is generally credited to Charles D. Seeberger who, as an employee of the Otis Elevator Company, produced the first step-type escalator manufactured for use by the general public. His creation was installed at the Paris Exhibition of 1900, where it won first prize. Seeberger also coined the term escalator by joining *scala*, which is Latin for steps, with a diminutive form of "elevator." In 1910 Seeberger sold the original patent rights for his invention to the Otis Elevator Company.

Although numerous improvements have been made, Seeberger's basic design remains in use today. It consists of top and bottom landing platforms connected by a metal truss. The truss contains two tracks, which pull a collapsible staircase through an endless loop. The truss also supports two handrails, which are coordinated to move at the same speed as the step treads.

Components

Top and bottom landing platforms

These two platforms house the curved sections of the tracks, as well as the gears and motors that drive the stairs. The top platform contains the motor assembly and the main drive gear, while the bottom holds the step return idler sprockets. These sections also anchor the ends of the escalator truss. In addition, the platforms contain a floor plate and a comb plate. The floor plate provides a place for the passengers to stand before they step onto the moving stairs. This plate is flush with the finished floor and is either hinged or removable to allow easy access to the machinery below. The comb plate is the piece between the stationary floor plate and the moving step. It is so named because its edge has a series of cleats that resemble the teeth of a comb. These teeth mesh with matching cleats on the edges of the steps. This design is necessary to minimize the gap between the stair and the landing, which helps prevent objects from getting caught in the gap.

The truss

The truss is a hollow metal structure that bridges the lower and upper landings. It is composed of two side sections joined together with cross braces across the bottom and just below the top. The ends of the truss are attached to the top and bottom landing platforms via steel or concrete supports. The truss carries all the straight track sections connecting the upper and lower sections.

The tracks

The track system is built into the truss to guide the step chain, which continuously pulls the steps from the bottom platform and back to the top in an endless loop. There are actually two tracks: one for the front wheels of the steps (called the step-wheel track) and one for the back wheels of the steps (called the trailer-wheel track). The relative positions of these tracks cause the steps to form a staircase as they move out from under the comb plate. Along the straight section of the truss the tracks are at their maximum distance apart. This configuration forces the back of one step to be at a 90-degree angle relative to the step behind it. This right angle bends the steps into a stair shape. At the top and bottom of the escalator, the two tracks converge so that the front and back wheels of the steps are almost in a straight line. This causes the stairs to lay in a flat sheet-like arrangement, one after another, so they can easily travel around the bend in the curved section of track. The tracks carry the steps down along the underside of the truss until they reach the bottom landing, where they pass through another curved section of track before exiting the bottom landing. At this point the tracks separate and the steps once again assume a stair case configuration. This cycle is repeated continually as the steps are pulled from bottom to top and back to the bottom again.

The steps

The steps themselves are solid, one-piece, die-cast aluminum. Rubber mats may be affixed to their surface to reduce slippage, and yellow demarcation lines may be added to clearly indicate their edges. The leading and trailing edges of each step are cleated with comb-like protrusions that mesh with the comb plates on the top and bottom platforms. The steps are linked by a continuous metal chain so they form a closed loop with each step able to bend in relation to its neighbors. The front and back edges of the steps are each connected to two wheels. The rear wheels are set further apart to fit into the back track and the front wheels have shorter axles to fit into the narrower front track. As described above, the position of the tracks controls the orientation of the steps.

The railing

The railing provides a convenient handhold for passengers while they are riding the escalator. It is constructed of four distinct sections. At the center of the railing is a "slider," also known as a "glider ply," which is a layer of a cotton or synthetic textile. The purpose of the slider layer is to allow the railing to move smoothly along its track. The next layer, known as the tension member, consists of either steel cable or flat steel tape. It provides the handrail with the necessary tensile strength and flexibility. On top of tension member are the inner construction components, which are made of chemically treated rubber designed to prevent the layers from separating. Finally, the outer layer, the only part that passengers actually see, is the rubber cover, which is a blend of synthetic polymers and rubber. This cover is designed to resist degradation from environmental conditions, mechanical wear and tear, and human vandalism. The railing is constructed by feeding rubber through a computer controlled extrusion machine to produce layers of the required size and type in order to match specific orders. The component layers of fabric, rubber, and steel are shaped by skilled workers before being fed into the presses, where they are fused together. When installed, the finished railing is pulled along its track by a chain that is connected to the main drive gear by a series of pulleys.

Design

A number of factors affect escalator design, including physical requirements, location, traffic patterns, safety considerations, and aesthetic preferences. Foremost, physical factors like the vertical and horizontal distance to be spanned must be considered. These factors will determine the pitch of the escalator and its actual length. The ability of the building infrastructure to support the heavy components is also a critical physical concern. Location is important because escalators should be situated where they can be easily seen by the general public. In department stores, customers should be able to view the merchandise easily. Furthermore, up and down escalator traffic should be physically separated and should not lead into confined spaces.

Traffic patterns must also be anticipated in escalator design. In some buildings the objective is simply to move people from one floor to another, but in others there may be a more specific requirement, such as funneling visitors towards a main exit or exhibit. The number of passengers is important because escalators are designed to carry a certain maximum number of people. For example, a single width escalator traveling at about 1.5 feet (0.45 m) per second can move an estimated 170 persons per five-minute period. Wider models traveling at up to 2 feet (0.6 m) per second can handle as many as 450 people in the same time period. The carrying capacity of an escalator must match the expected peak traffic demand. This is crucial for applications in which there are sudden increases in the number of passengers. For example, escalators used in train stations must be designed to cater for the peak traffic flow discharged from a train, without causing excessive bunching at the escalator entrance.

Of course, safety is also major concern in escalator design. Fire protection of an escalator floor-opening may be provided by adding automatic sprinklers or fireproof shutters to the opening, or by installing the escalator in an enclosed fire-protected hall. To limit the danger of overheating, adequate ventilation for the spaces that contain the motors and gears must be provided. It is preferred that a traditional staircase be located adjacent to the escalator if the escalator is the primary means of transport between floors. It may also be necessary to provide an elevator lift adjacent to an escalator for wheelchairs and disabled persons. Finally, consideration should be given to the aesthetics of the escalator. The architects and designers can choose from a wide range of styles and colors for the handrails and tinted side panels.

The Manufacturing Process

1 The first stage of escalator construction is to establish the design, as described above. The escalator manufacturer uses this information to construct the appropriately customized equipment. There are two types of companies that supply escalators, primary manufacturers who actually build the equipment, and secondary suppliers that design and install the equipment. In most cases, the secondary suppliers obtain the necessary equipment from the primary manufacturers and make necessary modifications for installation. Therefore, most escalators are actually assembled at the primary manufacturer. The tracks, step chains, stair assembly, and motorized gears and pulleys are all bolted into place on the truss before shipping.

2 Prior to installation, the landing areas must be prepared to connect to the escalator. For example, concrete fittings must be poured, and the steel framework that will hold the truss in place must be attached. After the escalator is delivered, the entire assembly is uncrated and jockeyed into position between the top and bottom landing holes. There are a variety of methods for lifting the truss assembly into place, one of which is a scissors lift apparatus mounted on a wheeled support platform. The scissors lift is outfitted with a locator assembly to aid in vertical and angular alignment of the escalator. With such a device, the upper end of the truss can be easily aligned with and then supported by a support wall associated with the upper landing. The lower end of the truss can be subsequently lowered into a pit associated with the floor of the lower landing. In some cases, the railings may be shipped separately from the rest of the equipment. In such a situation, they are carefully coiled and packed for shipping. They are then connected to the appropriate chains after the escalator is installed.

3 Make final connections for the power source and check to ensure all tracks and chains are properly aligned.

4 Verify all motorized elements are functioning properly, that the belts and chains

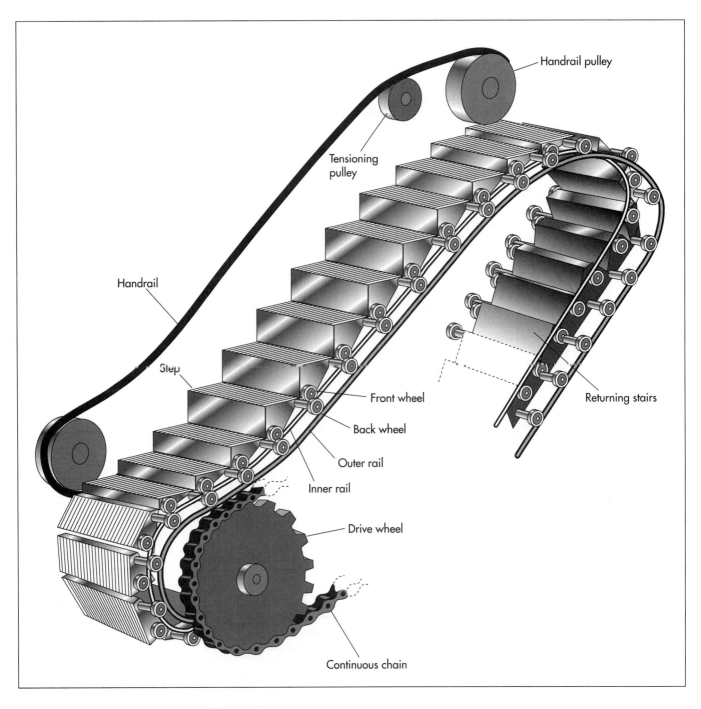

Handrail pulley

Tensioning pulley

Handrail

Step

Front wheel

Back wheel

Outer rail

Inner rail

Drive wheel

Continuous chain

Returning stairs

move smoothly and at the correct speed, and that the emergency braking system is activated. The step treads must be far enough apart that they do not pinch or rub against each other. However, they should be positioned such that no large gaps are present, which could increase the chance of injury.

Quality Control

The Code of Federal Regulation (CFR) contains guidelines for escalator quality control and establishes minimum inspection

standards. As stated in the code, "elevators and escalators shall be thoroughly inspected at intervals not exceeding one year. Additional monthly inspections for satisfactory operation shall be conducted by designated persons." Records of the annual inspections are to be posted near the escalator or be available at the terminal. In addition, the code specifies that the escalator's maximum load limits shall be posted and not exceeded. Additional safety standards can also be found in American Society of Mechanical Engineers Handbook.

An escalator is a continuously moving staircase. Each stair has a pair of wheels on each side, one at the front of the step and one at the rear. The wheels run on two rails. At the top and bottom of the escalator, the inner rail dips beneath the outer rail, so that the bottom of the stair flattens, making it easier for riders to get on and off.

The Future

Several innovations in escalator manufacture have been made in recent years. For example, one company recently developed a spiral staircase escalator. Another has developed an escalator suitable for transporting wheelchairs. Such advances are likely to continue as the industry expands to meet the changing needs of the marketplace. In addition, the industry is expecting a growth spurt as untapped markets such as China and Hungary begin to recognize the benefits of escalator technology.

Where to Learn More

Books

Barney, G.C., ed. *Elevator Technology.* Ellis Horwood, 1986.

Periodicals

Taninecz, George. "Schindler Elevator Corp." *Industry Week,* October 21, 1996, p. 54.

—*Randy Schueller*

Fake Fur

Fake fur is a type of textile fabric fashioned to simulate genuine animal fur. It is known as a pile fabric and is typically made from polymeric fibers that are processed, dyed, and cut to match a specific fur texture and color. First introduced in 1929, advances in polymer technology have tremendously improved fake fur quality. Today's fake furs can be nearly indistinguishable from the natural furs they imitate.

History

Fur is one of the oldest known forms of clothing, and has been worn by men and women for a variety of reasons throughout history. While quite desirable, real fur had the disadvantage of being expensive and in short supply. For this reason, fake furs were introduced on the market in 1929. These early attempts at imitation fur were made using hair from the alpaca, a South American mammal. From a fashion standpoint, they were of low quality, typically colored gray or tan, and could not compare to exquisite furs like mink or beaver. But the fabric was inexpensive and warm, so manufacturers continued to develop improved versions of the fake fur, trying to give it a denser look, better abrasion resistance, and more interesting colors.

In the 1940s, the quality of fake furs was vastly improved by advances in textile manufacture technology. However, the true modern fake furs were not developed until the mid 1950s, with the introduction of acrylic polymers as replacements for alpaca hair. These polymers were particularly important because they could provide the bulk required to imitate real fur without the weight associated with other fake fur fab-

rics. They were also easier to color and texture than alpaca fibers. Later in the decade, polymer producers found that acrylic polymers could be made even more fur-like and fire resistant by mixing them with other polymers. These new fabrics, called modacrylics, are now the primary polymer used in fake fur manufacture.

Background

Fake furs are known as pile fabrics, which are engineered to have the appearance and warmth of animal furs. They are attached to a backing using various techniques. Although they can never match the characteristics of natural furs, fake furs do have certain advantages over their natural counterparts. Unlike natural furs, fake furs can be colored almost any shade, allowing for more dramatic color combinations. Additionally, fake furs are more durable and resistant to environmental assaults. In fact, some are even labeled hand washable. With concerns over the environment and animal rights, more and more fashion designers are developing garments using fake fur. Lastly, fake furs are much less expensive than natural furs, making them an attractive option for many people.

Raw Materials

Fake furs are made with a variety of materials. The bulk fibers are typically composed of polymers, including acrylics, modacrylics, or appropriate blends of these polymers. Acrylic polymers are made from chemicals derived from coal, air, water, petroleum, and limestone. They are the result of a chemical reaction of an acrylonitrile monomer under conditions of elevated

First introduced in 1929, advances in polymer technology have tremendously improved fake fur quality. Today's fake furs can be nearly indistinguishable from the natural furs they imitate.

pressure and heat. For fake furs, secondary monomers are also added to improve the ability of the acrylic fibers to absorb dyes. Modacrylic polymers are copolymers made by the reaction of acrylonitrile and vinyl chloride monomers. These fibers are particularly useful for fake furs because they can be easily dyed with animal-like colors and have a natural fire retardance.

Modacrylic and acrylic polymers have other characteristics that make them useful in fake fur manufacture. They are lightweight and springy, imparting a fluffy quality to the garment. They are also highly resistant to heat, sunlight, soot, and smoke, are strong and resilient, and show good stability during laundering. Since they are thermoplastic polymers, they can be heatset. They resist mildew and are not susceptible to attack from insects. These polymers also have very low moisture absorbency and will dry quickly.

Other naturally occurring fabrics are also used to make fake furs and improve the look and feel of the overall garment. These include materials such as silk, wool, and mohair. Cotton or wool, along with polypropylene, are typically used to make the backings to which the fibers are attached. Rayon, a semisynthetic fiber made from cellulose and cotton linters, is used to supplement acrylic and modacrylic fibers on the garment, as are polyester and nylon. Materials such as silicones and various resins are used to improve the smoothness and luster of fake furs. To complete the look of a fake fur, dyes and colorants are used. If a true imitation is desired, designers match the color with natural fur. However, fashion designers have found that the fake fur fabric has merits of its own and have started using colors and styles that give it its own new, unique look.

The Manufacturing Process

The production of a fake fur can be a mostly automated process. The manufacturing steps involved include production of the synthetic fibers, construction of the garment, and modification of the garment.

Chemical synthesis of fibers

1 Making a fake fur begins with the production of the synthetic fibers. While different types of polymers are used, modacrylic polymers provide a good illustration of the fiber manufacturing process. First, the acrylonitrile and vinyl chloride monomers are mixed together in a large stainless steel container. They are forced into a chamber in which the pressure and temperature is increased. Mixing blades are constantly in motion and the polymerization process begins. A white powdery resin is produced, which is then converted into a thick liquid by dissolving it in acetone.

2 The liquid polymer mixture is then pumped through a filter to remove undissolved particles. From the filter, the material is pumped through spinnerets, which are submerged in a water bath. The spinnerets look similar to shower heads, and when the polymer is extruded through, it emerges as a group of continuous fibers called a tow.

3 The tow is then pulled along a conveyor belt and stretched through a series of pulleys. As the tow is stretched, it is also washed and dried. As it dries, the acetone is driven off, leaving only the polymer. The continuous fibers are then annealed, making them stronger, and are sent to a machine that cuts them to appropriate sizes.

4 After various quality control checks are performed on the fibers, they are moved to the next phase of processing. Here, the polymers are soaked in dye solutions and colored. While this is not the only phase of manufacture in which the fibers are colored, this is usually the point where solid background colors are obtained.

Producing the fur

5 While the fibers provide the primary texture and look for imitation fur, the backing provides most of the structure. Working off a specific garment design, the backing, which is made out of cotton or wool, is sent through a machine to be appropriately cut. It is then transferred to the next phase of production, in which the fibers will be attached.

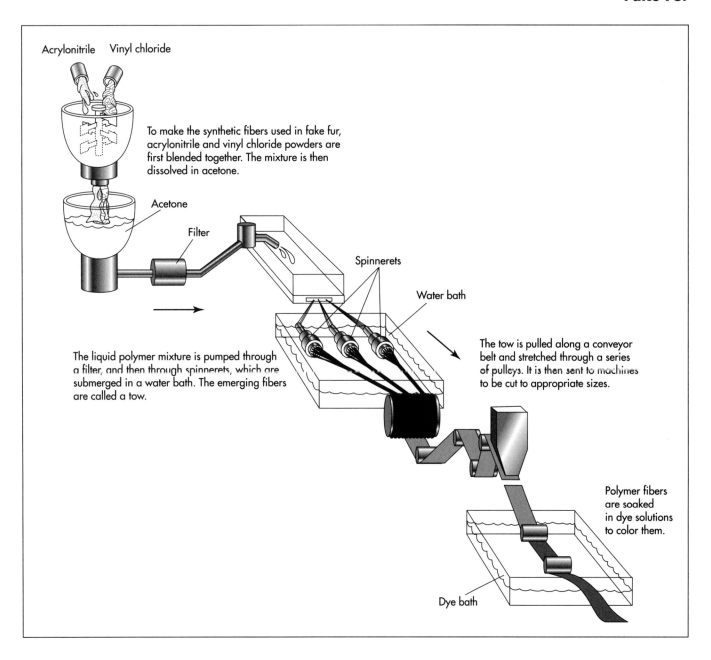

Acrylonitrile   Vinyl chloride

To make the synthetic fibers used in fake fur, acrylonitrile and vinyl chloride powders are first blended together. The mixture is then dissolved in acetone.

Acetone

Filter

Spinnerets

Water bath

The tow is pulled along a conveyor belt and stretched through a series of pulleys. It is then sent to machines to be cut to appropriate sizes.

The liquid polymer mixture is pumped through a filter, and then through spinnerets, which are submerged in a water bath. The emerging fibers are called a tow.

Polymer fibers are soaked in dye solutions to color them.

Dye bath

To convert the fibers into a garment, four different techniques can be employed. The most basic method is the weaving process. In this process, the fibers are looped through and interlaced with the backing fabric. While this technique is fairly slow, it can produce a large range of cloth shapes. Another method of fake fur production is called tufting. It is similar to weaving; however, it produces garments at a much higher rate of speed. Circular loop knitting and sliver knitting are other methods of fake fur garment production. Sliver knitting utilizes the same equipment used in jersey knitting. This makes it the fastest and most economical of all the fake fur garment production techniques, and it is also the one most used by manufacturers.

Finishing touches

To simulate a natural fur, the garments are treated in various ways. First, to ensure that the fake fur will remain unchanged after it is produced, the fabric is heated. This heat setting process preshrinks the fabric, giving it improved stability and expanded fiber diameters. Next, to remove loose fibers from the fabric, wire brushes are passed through the fabric. This process is known as tigering. Rough shearing of the fibers by cutting them with a set of helical knives gives

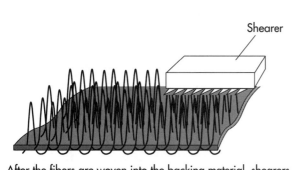

After the fibers are woven into the backing material, shearers cut the tops off the fibers, giving them a uniform length.

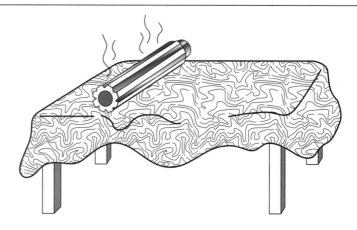

Fake fur can be polished with a technique known as electrofying. With this method, a grooved, heated cylinder is used to comb the fabric in both directions. The finished fabric is then ready to be made into garments.

them a uniform length. The luster of the fabric can be improved through a method known as electrofying. This is a polishing technique that involves combing the fabric with a heated, grooved cylinder in both directions. The next treatment is the application of chemicals such as resins and silicones, which improves the feel and look of the fiber. Coloring to simulate specific animals can also be enhanced at this staged. Another round of electrofying can be done, as well as a finishing shearing to remove any remaining loose fibers. Depending on the type of fake fur, embossing to simulate curls can also be done during this stage of manufacture.

8 After the fake fur has been produced, the government requires that they are labeled as imitation fur fabrics. These labels are typically sewn in the inside of the garment and must be legible throughout the life of the product. In the final steps of fake fur manufacturing, the garment is put in the appropriate packaging and shipped to distributors.

Quality Control

To ensure the quality of fake fur, manufacturers monitor the product during each phase of production. This process begins with an inspection of the incoming raw materials and continues with the finished fibers that are produced in the polymerization reactions. These fibers are subjected to a battery of physical and chemical tests to show that they meet the specifications pre-

viously developed. Some of the characteristics that are tested include pH, appearance, density, and melting point. Other things such as fiber elasticity, resilience, and absorbency can also be tested.

As the garments are being produced, line inspectors take random samples at certain time intervals and check to ensure that they meet set requirements for things such as appearance, sewing quality, fiber strength, size, and shape. The primary testing method is visual inspection, although more rigorous tests can also be performed. In addition to the manufacturer's own standards, the industry and government also set requirements. A set of governmental standards, known as L-22, has been voluntarily adopted by the industry. These tests outline minimum performance standards for things such as shrinkage, pilling, snagging, and wear.

The Future

The technology of producing fake furs has improved greatly since the early twentieth century. Future research will focus on developing new fibers and finishes. These polymeric fibers will have improved feel, look, and a lower cost. Additionally, quicker and more efficient methods of production are also being investigated. Special animal simulation techniques have recently been developed. One method simulates the long and short hair sections of mink or river otter fur by mixing shrinkable and non-shrinkable fibers. Another method simulates the feel of beaver fur by mixing certain fine and coarse fibers. Finally, manufacturers will strive to produce ever higher quality products at the lowest possible cost.

Where to Learn More

Books

Jerde, Judith. *Encyclopedia of Textiles.* Facts on File, 1992.

Keeler, Pat and Francis McCall. *Unraveling Fibers.* Atheneum Publishers, 1995.

Harris, Jennifer. *Textiles 5,000 Years: An International History and Survey.* Harry N. Abrams Publishing Co., 1993.

Seiler-Baldinger, Annemarie. *Textiles: A Classification of Techniques.* Smithsonian Institute Publications, 1995.

—Perry Romanowski

Fertilizer

The synthetic fertilizer industry experienced significant growth after the First World War, when facilities that had produced ammonia and synthetic nitrates for explosives were converted to the production of nitrogen-based fertilizers.

Background

Fertilizer is a substance added to soil to improve plants' growth and yield. First used by ancient farmers, fertilizer technology developed significantly as the chemical needs of growing plants were discovered. Modern synthetic fertilizers are composed mainly of nitrogen, phosphorous, and potassium compounds with secondary nutrients added. The use of synthetic fertilizers has significantly improved the quality and quantity of the food available today, although their long-term use is debated by environmentalists.

Like all living organisms, plants are made up of cells. Within these cells occur numerous metabolic chemical reactions that are responsible for growth and reproduction. Since plants do not eat food like animals, they depend on nutrients in the soil to provide the basic chemicals for these metabolic reactions. The supply of these components in soil is limited, however, and as plants are harvested, it dwindles, causing a reduction in the quality and yield of plants.

Fertilizers replace the chemical components that are taken from the soil by growing plants. However, they are also designed to improve the growing potential of soil, and fertilizers can create a better growing environment than natural soil. They can also be tailored to suit the type of crop that is being grown. Typically, fertilizers are composed of nitrogen, phosphorus, and potassium compounds. They also contain trace elements that improve the growth of plants.

The primary components in fertilizers are nutrients which are vital for plant growth. Plants use nitrogen in the synthesis of proteins, nucleic acids, and hormones. When plants are nitrogen deficient, they are marked by reduced growth and yellowing of leaves. Plants also need phosphorus, a component of nucleic acids, phospholipids, and several proteins. It is also necessary to provide the energy to drive metabolic chemical reactions. Without enough phosphorus, plant growth is reduced. Potassium is another major substance that plants get from the soil. It is used in protein synthesis and other key plant processes. Yellowing, spots of dead tissue, and weak stems and roots are all indicative of plants that lack enough potassium.

Calcium, magnesium, and sulfur are also important materials in plant growth. They are only included in fertilizers in small amounts, however, since most soils naturally contain enough of these components. Other materials are needed in relatively small amounts for plant growth. These micronutrients include iron, chlorine, copper, manganese, zinc, molybdenum, and boron, which primarily function as cofactors in enzymatic reactions. While they may be present in small amounts, these compounds are no less important to growth, and without them plants can die.

Many different substances are used to provide the essential nutrients needed for an effective fertilizer. These compounds can be mined or isolated from naturally occurring sources. Examples include sodium nitrate, seaweed, bones, guano, potash, and phosphate rock. Compounds can also be chemically synthesized from basic raw materials. These would include such things as ammonia, urea, nitric acid, and ammonium phosphate. Since these compounds exist in a number of physical states, fertilizers can be sold as solids, liquids, or slurries.

History

The process of adding substances to soil to improve its growing capacity was developed in the early days of agriculture. Ancient farmers knew that the first yields on a plot of land were much better than those of subsequent years. This caused them to move to new, uncultivated areas, which again showed the same pattern of reduced yields over time. Eventually it was discovered that plant growth on a plot of land could be improved by spreading animal manure throughout the soil.

Over time, fertilizer technology became more refined. New substances that improved the growth of plants were discovered. The Egyptians are known to have added ashes from burned weeds to soil. Ancient Greek and Roman writings indicate that various animal excrements were used, depending on the type of soil or plant grown. It was also known by this time that growing leguminous plants on plots prior to growing wheat was beneficial. Other types of materials added include sea-shells, clay, vegetable waste, waste from different manufacturing processes, and other assorted trash.

Organized research into fertilizer technology began in the early seventeenth century. Early scientists such as Francis Bacon and Johann Glauber describe the beneficial effects of the addition of saltpeter to soil. Glauber developed the first complete mineral fertilizer, which was a mixture of saltpeter, lime, phosphoric acid, nitrogen, and potash. As scientific chemical theories developed, the chemical needs of plants were discovered, which led to improved fertilizer compositions. Organic chemist Justus von Liebig demonstrated that plants need mineral elements such as nitrogen and phosphorous in order to grow. The chemical fertilizer industry could be said to have its beginnings with a patent issued to Sir John Lawes, which outlined a method for producing a form of phosphate that was an effective fertilizer. The synthetic fertilizer industry experienced significant growth after the First World War, when facilities that had produced ammonia and synthetic nitrates for explosives were converted to the production of nitrogen-based fertilizers.

Raw Materials

The fertilizers outlined here are compound fertilizers composed of primary fertilizers and secondary nutrients. These represent only one type of fertilizer, and other single nutrient types are also made. The raw materials, in solid form, can be supplied to fertilizer manufacturers in bulk quantities of thousands of tons, drum quantities, or in metal drums and bag containers.

Primary fertilizers include substances derived from nitrogen, phosphorus, and potassium. Various raw materials are used to produce these compounds. When ammonia is used as the nitrogen source in a fertilizer, one method of synthetic production requires the use of natural gas and air. The phosphorus component is made using sulfur, coal, and phosphate rock. The potassium source comes from potassium chloride, a primary component of potash.

Secondary nutrients are added to some fertilizers to help make them more effective. Calcium is obtained from limestone, which contains calcium carbonate, calcium sulphate, and calcium magnesium carbonate. The magnesium source in fertilizers is derived from dolomite. Sulfur is another material that is mined and added to fertilizers. Other mined materials include iron from ferrous sulfate, copper, and molybdenum from molybdenum oxide.

The Manufacturing Process

Fully integrated factories have been designed to produce compound fertilizers. Depending on the actual composition of the end product, the production process will differ from manufacturer to manufacturer.

Nitrogen fertilizer component

1 Ammonia is one nitrogen fertilizer component that can be synthesized from inexpensive raw materials. Since nitrogen makes up a significant portion of the earth's atmosphere, a process was developed to produce ammonia from air. In this process,

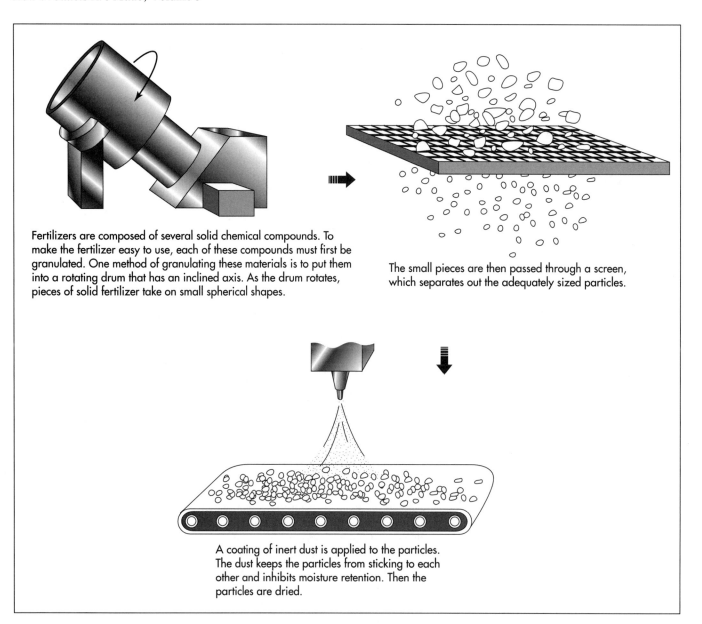

Fertilizers are composed of several solid chemical compounds. To make the fertilizer easy to use, each of these compounds must first be granulated. One method of granulating these materials is to put them into a rotating drum that has an inclined axis. As the drum rotates, pieces of solid fertilizer take on small spherical shapes.

The small pieces are then passed through a screen, which separates out the adequately sized particles.

A coating of inert dust is applied to the particles. The dust keeps the particles from sticking to each other and inhibits moisture retention. Then the particles are dried.

natural gas and steam are pumped into a large vessel. Next, air is pumped into the system, and oxygen is removed by the burning of natural gas and steam. This leaves primarily nitrogen, hydrogen, and carbon dioxide. The carbon dioxide is removed and ammonia is produced by introducing an electric current into the system. Catalysts such as magnetite (Fe_3O_4) have been used to improve the speed and efficiency of ammonia synthesis. Any impurities are removed from the ammonia, and it is stored in tanks until it is further processed.

2 While ammonia itself is sometimes used as a fertilizer, it is often converted to other substances for ease of handling. Nitric

acid is produced by first mixing ammonia and air in a tank. In the presence of a catalyst, a reaction occurs which converts the ammonia to nitric oxide. The nitric oxide is further reacted in the presence of water to produce nitric acid.

3 Nitric acid and ammonia are used to make ammonium nitrate. This material is a good fertilizer component because it has a high concentration of nitrogen. The two materials are mixed together in a tank and a neutralization reaction occurs, producing ammonium nitrate. This material can then be stored until it is ready to be granulated and blended with the other fertilizer components.

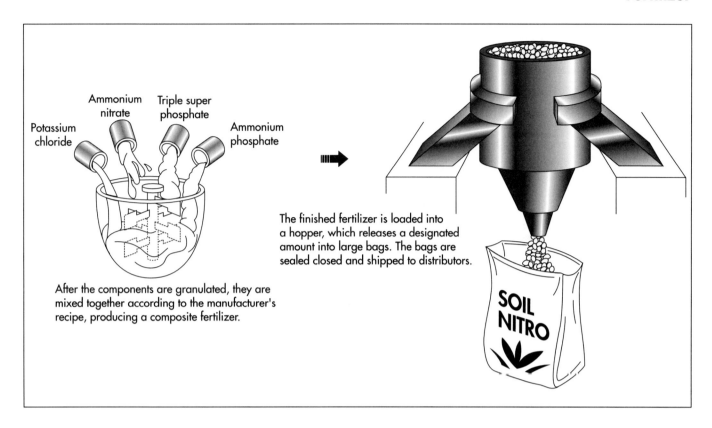

Potassium chloride

Ammonium nitrate

Triple super phosphate

Ammonium phosphate

The finished fertilizer is loaded into a hopper, which releases a designated amount into large bags. The bags are sealed closed and shipped to distributors.

After the components are granulated, they are mixed together according to the manufacturer's recipe, producing a composite fertilizer.

SOIL NITRO

Phosphorous fertilizer component

4 To isolate phosphorus from phosphate rock, it is treated with sulfuric acid, producing phosphoric acid. Some of this material is reacted further with sulfuric acid and nitric acid to produce a triple superphosphate, an excellent source of phosphorous in solid form.

5 Some of the phosphoric acid is also reacted with ammonia in a separate tank. This reaction results in ammonium phosphate, another good primary fertilizer.

Potassium fertilizer component

6 Potassium chloride is typically supplied to fertilizer manufacturers in bulk. The manufacturer converts it into a more usable form by granulating it. This makes it easier to mix with other fertilizer components in the next step.

Granulating and blending

7 To produce fertilizer in the most usable form, each of the different compounds, ammonium nitrate, potassium chloride, ammonium phosphate, and triple superphosphate are granulated and blended together.

One method of granulation involves putting the solid materials into a rotating drum which has an inclined axis. As the drum rotates, pieces of the solid fertilizer take on small spherical shapes. They are passed through a screen that separates out adequately sized particles. A coating of inert dust is then applied to the particles, keeping each one discrete and inhibiting moisture retention. Finally, the particles are dried, completing the granulation process.

8 The different types of particles are blended together in appropriate proportions to produce a composite fertilizer. The blending is done in a large mixing drum that rotates a specific number of turns to produce the best mixture possible. After mixing, the fertilizer is emptied onto a conveyor belt, which transports it to the bagging machine.

Bagging

9 Fertilizers are typically supplied to farmers in large bags. To fill these bags the fertilizer is first delivered into a large hopper. An appropriate amount is released from the hopper into a bag that is held open by a clamping device. The bag is on a vibrating surface, which allows better packing. When filling is complete, the bag is transported upright to a machine that seals it closed. The bag is then conveyored to a palletizer, which stacks multiple bags, readying them for shipment to distributors and eventually to farmers.

Quality Control

To ensure the quality of the fertilizer that is produced, manufacturers monitor the product at each stage of production. The raw materials and the finished products are all subjected to a battery of physical and chemical tests to show that they meet the specifications previously developed. Some of the characteristics that are tested include pH, appearance, density, and melting point. Since fertilizer production is governmentally regulated, composition analysis tests are run on samples to determine total nitrogen content, phosphate content, and other elements affecting the chemical composition. Various other tests are also performed, depending on the specific nature of the fertilizer composition.

Byproducts/Waste

A relatively small amount of the nitrogen contained in fertilizers applied to the soil is actually assimilated into the plants. Much is washed into surrounding bodies of water or filters into the groundwater. This has added significant amounts of nitrates to the water that is consumed by the public. Some medical studies have suggested that certain disorders of the urinary and kidney systems are a result of excessive nitrates in drinking water. It is also thought that this is particularly harmful for babies and could even be potentially carcinogenic.

The nitrates that are contained in fertilizers are not thought to be harmful in themselves. However, certain bacteria in the soil convert nitrates into nitrite ions. Research has shown that when nitrite ions are ingested, they can get into the bloodstream. There, they bond with hemoglobin, a protein that is responsible for storing oxygen. When a nitrite ion binds with hemoglobin, it loses its ability to store oxygen, resulting in serious health problems.

Nitrosamines are another potential byproduct of the nitrates in fertilizer. They are the result of a natural chemical reaction of nitrates. Nitrosamines have been shown to cause tumors in laboratory animals, feeding the fear that the same could happen in humans. There has, however, been no study that shows a link between fertilizer use and human tumors.

The Future

Fertilizer research is currently focusing on reducing the harmful environmental impacts of fertilizer use and finding new, less expensive sources of fertilizers. Such things that are being investigated to make fertilizers more environmentally friendly are improved methods of application, supplying fertilizer in a form which is less susceptible to runoff, and making more concentrated mixtures. New sources of fertilizers are also being investigated. It has been found that sewage sludge contains many of the nutrients that are needed for a good fertilizer. Unfortunately, it also contains certain substances such as lead, cadmium, and mercury in concentrations which would be harmful to plants. Efforts are underway to remove the unwanted elements, making this material a viable fertilizer. Another source that is being developed is manures. The first fertilizers were manures, however, they are not utilized on a large scale because their handling has proven too expensive. When technology improves and costs are reduced, this material will be a viable new fertilizer.

Where to Learn More

Books

Rao, N. S. *Biofertilizers in Agriculture & Forestry.* IBH, 1993.

Stocchi, E. *Industrial Chemistry.* Ellis Horwood, 1990.

Lowrison, George. *Fertilizer Technology.* John Wiley and Sons, 1989.

Periodicals

Kirschner, Elisabeth. "Fertilizer Makers Gear up to Grow." *Chemical & Engineering News,* March, 31 1997, p. 13-15.

—*Perry Romanowski*

Fiberboard

Background

Composite forest products, or engineered wood, refer to materials made of wood that are glued together. In the United States, roughly 21 million tons (21.3 million metric tons) of composite wood are produced annually. The more popular composites materials include plywood, blockboard, fiberboard, particleboard, and laminated veneer lumber. Most of these products are based on what were previously waste wood residues or little used or non-commercial species. Very little raw material is lost in composites manufacture.

Medium density fiberboard (MDF) is a generic term for a panel primarily composed of lignocellulosic fibers combined with a synthetic resin or other suitable bonding system and bonded together under heat and pressure. The panels are compressed to a density of 0.50 to 0.80 specific gravity (31-50 lb/ft.^{3}) Additives may be introduced during manufacturing to improve certain properties. Because fiberboard can be cut into a wide range of sizes and shapes, applications are many, including industrial packaging, displays, exhibits, toys and games, furniture and cabinets, wall paneling, molding, and door parts.

The surface of MDF is flat, smooth, uniform, dense, and free of knots and grain patterns, making finishing operations easier and consistent. The homogenous edge of MDF allows intricate and precise machining and finishing techniques. Trim waste is also significantly reduced when using MDF compared to other substrates. Improved stability and strength are important assets of MDF, with stability contributing to holding precise tolerances in accurately cut parts. It is an ex-cellent substitute for solid wood in many interior applications. Furniture manufacturers are also embossing the surface with three-dimensional designs, since MDF has such an even texture and consistent properties.

The MDF market has grown rapidly in the United States over the past 10 years. Shipments increased 62% and plant capacity grew 60%. Today, over a billion square feet (93 million sq m) of MDF is consumed in America every year. World MDF capacity increased 30% in 1996 to over 12 billion square feet (1.1 billion sq m), and there are now over 100 plants in operation.

History

MDF was first developed in the United States during the 1960s, with production starting in Deposti, New York. A similar product, hardboard (compressed fiberboard), was accidentally invented by William Mason in 1925, while he was trying to find a use for the huge quantities of wood chips that were being discarded by lumber mills. He was attempting to press wood fiber into insulation board but produced a durable thin sheet after forgetting to shut down his equipment. This equipment consisted of a blow torch, an eighteenth-century letter press, and an old automobile boiler.

Raw Materials

Wood chips, shavings, and sawdust typically make up the raw materials for fiberboard. However, with recycling and environmental issues becoming the norm, waste paper, corn silk, and even bagasse (fibers from sugarcane) are being used as well. Other materials are being recycled into MDF as well. One company is using dry waste ma-

In the United States, roughly 21 million tons (21.3 million metric tons) of composite wood are produced annually.

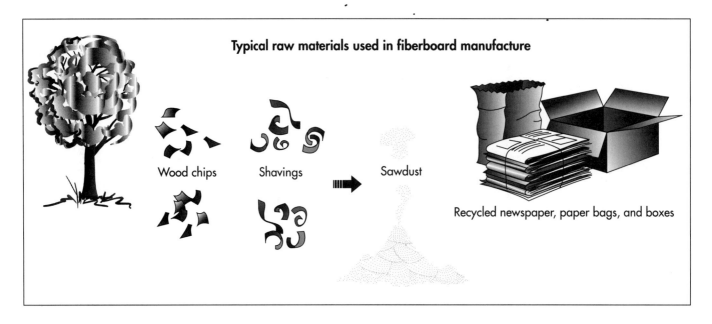

Typical raw materials used in fiberboard manufacture

Wood chips

Shavings

Sawdust

Recycled newspaper, paper bags, and boxes

Wood chips, shavings, and sawdust typically make up the raw materials for fiberboard. However, with recycling and environmental issues becoming the norm, waste paper, corn silk, bagasse (fibers from sugarcane), cardboard, cardboard drink containers containing plastics and metals, telephone directories, and old newspapers are being used.

terials at a rate of 100,000 tons a year. In addition to waste wood, cardboard, cardboard drink containers containing plastics and metals, telephone directories, and old newspapers are being used at this company. Synthetic resins are used to bond the fibers together and other additives may be used to improve certain properties.

The Manufacturing Process

Advanced technology and processing have improved the quality of fiberboard. These include innovations in wood preparation, resin recipes, press technology, and panel sanding techniques. Advanced press technology has shortened overall pressing cycles, while anti-static technology has also contributed to increased belt life during the sanding process.

Wood preparation

1 Producing quality fiberboard begins with the selection and refinement of the raw materials, most of which are recycled from shavings and chips reclaimed from sawmills and plywood plants. The raw material is first removed of any metal impurities using a magnet. Next, the material is separated into large chunks and small flakes. Flakes are separated into sawdust and wood chip piles.

2 The material is sent through a magnetic detector again, with the rejected materi-

al being separated for reuse as fuel. Good material is collected and sent into a presteaming bin. In the bin, steam is injected to heat and soften the material. The fibers are fed first into a side screw feeder and then into a plug screw feeder, which compresses the fibers and removes water. The compressed material is then fed into a refiner, which tears the material into usable fibers. Sometimes the fiber may undergo a second refining step in order to improve fiber purity. Larger motors on the refiners are sometimes used to sift out foreign objects from the process.

Curing and pressing

3 Resin is added before the refining step to control the formaldehyde tolerances in the mixture, and after refining, a catalyst is added. The fibers are then blown into a flash tube dryer, which is heated by either oil or gas. The ratio of solid resin to fiber is carefully controlled by weighing each ingredient. Next, the fiber is pushed through scalping rolls to produce a mat of uniform thickness. This mat goes through several pressing steps to produce a more usable size and then is trimmed to the desired width before the final pressing step. A continuous press equipped with a large drum compresses the mat at a uniform rate by monitoring the mat height. Presses are equipped with electronic controls to provide accurate density and strength. The resulting board is cut to the appropriate length using saws before cooling.

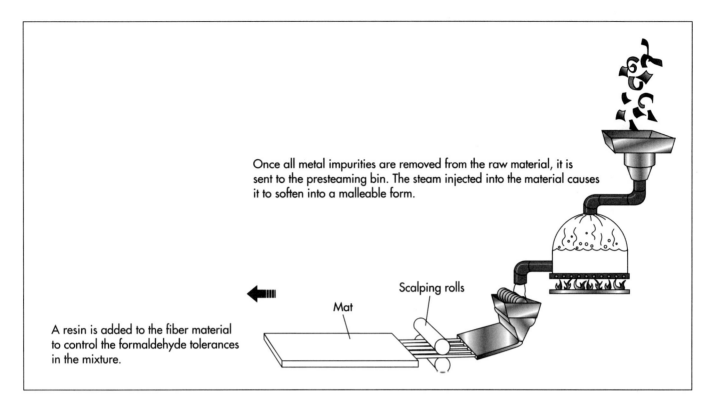

Once all metal impurities are removed from the raw material, it is sent to the presteaming bin. The steam injected into the material causes it to soften into a malleable form.

Scalping rolls

Mat

A resin is added to the fiber material to control the formaldehyde tolerances in the mixture.

Presses have counterbalanced, simultaneous closing systems that use hydraulic cylinders to effect platen leveling, which when operating in conjunction with a four-point position control gives greater individual panel thickness control. The hydraulics system can close the press at speeds and pressures that reduce board precure problems while shortening overall pressing cycles.

Panel sanding

4 To achieve a smooth finish, the panels are sanded using belts coated with abrasives. Silicon carbide has typically been used, but with the requirement for finer surfaces, other ceramic abrasives are utilized, including zirconia alumina and aluminum oxide. Eight-head sanding equipment and double-sided grading improves surface smoothness consistency. Anti-static technology is used to remove the static electricity that contributes to rapid loading and excessive sanding dust, thereby increasing belt life.

Finishing

5 Panels can undergo a variety of finishing steps depending on the final product. A wide variety of lacquer colors can be applied, as well as various wood-grain patterns. Guillotine cutting is used to cut the fiberboard into large sheets (for example 100 inches wide). For smaller sheet sizes such as 42 by 49 in (107 by 125 cm), die cutting is used. Specialty machines are used for cutting fiberboard into narrow strips of 1-24 in (2.5 -61 cm) widths.

6 Laminating machines are used to apply vinyl, foil, and other materials to the surface. This process involves unwinding a roll of fiberboard material, sending it between two rolls where the adhesive is applied, combining the adhesive-coated fiberboard with the laminating material between another set of rolls, and sending the combined materials into the laminator.

Quality Control

Most MDF plants use computerized process control to monitor each manufacturing step and to maintain product quality. In combination with continuous weight belts, basis weight gauges, density profile monitors, and thickness gauges, product consistency is maintained. In addition, the American National Standards Institute has established product specifications for each application, as well as formaldehyde emission limits. As environmental regulations and market conditions continue to change, these standards are revised.

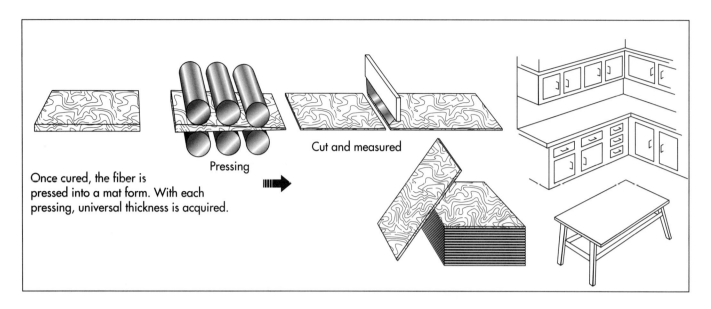

Once cured, the fiber is pressed into a mat form. With each pressing, universal thickness is acquired.

Pressing

Cut and measured

Panels can undergo a variety of finishing steps depending on the final product. A wide variety of lacquer colors can be applied, as well as various wood-grain patterns. For example, laminating machines are used to apply vinyl, foil, and other materials to the surface.

The most recent standard for MDF, ANSI Standard A208.2, is the third version of this industry standard. This standard classifies MDF by density and use (interior or exterior) and identifies four interior product grades. Specifications identified include physical and mechanical properties, dimensional tolerances, and formaldehyde emission limits. Specifications are presented in both metric and inch-pound limits.

Physical and mechanical properties of the finished product that are measured include density and specific gravity, hardness, modulus of rupture, abrasion resistance, impact strength, modulus of elasticity, and tensile strength. In addition, water absorption, thickness swelling, and internal bond strength are also measured. The American Society for Testing of Materials has developed a standard (D-1037) for testing these properties.

The Future

Though over 750 new plants were added in 1996, 1997 MDF consumption was expected to fall as much as 10% below the projected level. Usage rates have dropped for certain markets and exports have decreased. Despite this trend, some plants will continue to invest in high-tech equipment and environmental controls to produce a high-quality product.

Environmental regulations will continue to challenge the fiberboard industry. Though urea-formaldehyde resins are dominantly used in the MDF industry because of their low cost and fast curing characteristics, they have potential problems with formaldehyde emission. Phenol-formaldehyde resins are a possible solution, since they do not emit formaldehyde after cure. These resins are, however, more expensive, but preliminary research has shown that it can be used in far less quantities and achieve similar processing times as the urea resin.

Advances in manufacturing technology will also continue, including panel processing machinery and cutting tools. Pressing machinery will eventually be developed that eliminate precure and reduce individual panel thickness variation. MDF and other engineered wood products will become even more consistent in edge characteristics and surface smoothness, and have better physical properties and thickness consistency. These improvements will lead to more furniture and cabinet manufactures incorporating such products into their designs.

Where to Learn More

Periodicals

"Buyers and specifiers guide to particleboard and MDF." *Wood & Wood Products*, January 1996, pp. 67-75.

Koenig, Karen. "New MDF plant is high on technology and quality." *Wood & Wood Products*, April 1996, pp. 68-74.

"Lasani wood—the ideal wood replacement." *Economic Review*, April 1996, p. 48.

Margosian, Rich. "New standards for particleboard and MDF." *Wood & Wood Products*, January 1994, pp. 90-92.

Other

The Particle Board/Medium Density Fiberboard Institute. http://www.pbmdf.com (July 9, 1997).

—*Laurel M. Sheppard*

Flavored Coffee Bean

Flavored coffees in one form or another have been used for centuries, but the gourmet coffee boom of the 1990s resulted in an increased interest in exotic flavors of coffee. With current chemical technology, the beans can be produced with almost any flavor imaginable.

Background

Flavored coffee beans are coated with flavor compounds to supplement coffee beans' natural taste. In addition, these flavors help extend the shelf life of coffee by disguising changes in flavor due to decaffeination, oxidation, or aging processes. Flavored coffees in one form or another have been used for centuries, but the gourmet coffee boom of the 1990s resulted in an increased interest in exotic flavors of coffee. With current chemical technology, the beans can be produced with almost any flavor imaginable.

History

The origins of coffee, like those of so many other natural products long known to humans, are shrouded in legends. One entertaining story about the discovery of coffee involves an ancient Ethiopian goatherder, Kaldi, and his dancing goats. One day, so the story goes, Kaldi noticed his normally sluggish goats were dancing on their hind legs and bleating gleefully. The observant goatherder also noted they had been feeding on the red berries of a nearby shiny dark-green shrub. Tossing caution to the wind, he sampled the berries himself and experienced an immediate boost in his spirits and energies. Kaldi offered some of the berries to the head monk of the local monastery, who conducted a series of experiments on them, including parching them, crushing them in a mortar and pestle, and stirring the crushed berries in boiling water. The monk's efforts resulted in a fragrant beverage which he termed "heaven-sent," and henceforth gave it to all the monks in the evening to keep them from falling asleep during their prayers. News of this elixir quickly spread from the monastery to the nearby town and eventually throughout the world. The "magic" berries were actually coffee beans, and the heaven-sent beverage, of course, was coffee. Today coffee is harvested in nearly every tropical country within 1,000 miles (1,600 km) of the equator.

Although many people regard flavored coffee as a modern invention, its origins are nearly as old as the original beverage itself. History shows that a few hundred years ago in the Middle East, people enjoyed drinking coffee blended with nuts and spices. In modern times, innovative marketers have capitalized on coffee drinkers' desire for more flavors than nature can provide and have found new ways to introduce flavoring agents into coffee. First, flavored syrups were used to spike brewed coffee with a touch of a favored flavor. More recent improvements in food science have led to ways of introducing complex flavors directly onto the beans as part of a post-roasting process. When these flavored beans are used for brewing, the flavor is extracted into the resulting beverage. Today consumers can choose from a wide array of flavored coffee beans with names like "Chocolate Swiss Almond," "Hazelnut," "Amaretto Supreme," "Irish Creme," "French Vanilla," and "Georgia Pecan."

Raw Materials

Coffee beans

The type of bean used to make flavored coffee greatly impacts the taste of the finished product. It is estimated that coffee beans contain over 800 different compounds which contribute to their flavor, including sugars and other carbohydrates, mineral salts, organic acids, aromatic oils, and methylxanthines, a chemical class which includes caffeine. The bean's flavor is a function of where it was grown and how it was roasted. The name of the beans usually indicate their country of origin, along with additional information, such as the region within the country where the beans were grown, the grade of beans, or the type of roast. For instance, "Sumatra Lintong" denotes a specific growing region (Lintong) in Sumatra; "Kenya AA" designates AA beans, the highest grade of beans from Kenya; and "French Roast" is a blend of beans which are roasted very dark in the "French style." Some flavored coffees consist of only one kind of bean, like Kenya AA, which has distinctive regional taste characteristics.

In general *Coffea arabica* (or arabica) beans are used for flavored coffees due to their low levels of acidity and bitterness. Arabica was the earliest cultivated species of coffee and is still the most highly prized. These top quality beans are milder and more flavorful than the harsher *Coffea canefora* (or robusta) beans, which are used in many commercial and instant coffees. Some manufacturers create flavored coffees from a blend of beans from various regions. High quality beans are grown in Colombia, Mexico, Costa Rica, and Guatemala.

Flavoring oils

Flavoring oils are combinations of natural and synthetic flavor chemicals which are compounded by professional flavor chemists. Natural oils used in flavored coffees are extracted from a variety of sources, such as vanilla beans, cocoa beans, and various nuts and berries. Cinnamon, clove, and chicory are also used in a variety of coffee flavors. Synthetic flavor agents are chemicals which are manufactured on a commercial basis. For example, a nutty, woody, musty flavor can be produced with 2, 4-Dimethyl-5-acetylthiazole. Similarly, 2,5-Dimethylpyrazine is used to add an earthy, almost peanuty or potato-like flavor. Flavor chemists blend many such oils to achieve specific flavor combinations. While other food flavors may be composed of nine or 10 ingredients, coffee flavors may require up to 80 different compounds to achieve subtle

flavors. Virtually any taste can be reproduced. Marketers have found that consumers prefer coffee flavors with sweet creamy notes. The ideal flavor should mask some of the harsh notes of the coffee yet not interfere with its aromatic characteristics.

The pure flavor compounds described above are highly concentrated and must be diluted in a solvent to allow the blending of multiple oils and easy application to the beans. Common solvents include water, alcohol, propylene glycol, and fractionated vegetable oils. These solvents are generally volatile chemicals that are removed from the beans by drying. Older solvent system technology produced beans which dried up and lost flavor. Current technology uses more stable solvents which leave the beans with a glossy sheen and longer lasting flavor.

The flavor chemicals and the solvents used in flavors must not only be approved for use in foods, but they must also not adversely react with the packaging material and the processing equipment with which they come into contact. Furthermore, they must meet the desired cost constraints.

The Manufacturing Process

Processing the beans

1 Raw coffee beans are processed in two primary ways. The "dry method" allows the beans to dry on the plant or be dehydrated by the sun after harvesting. The beans are then separated from the rest of the plant debris by milling. In the "wet method," the beans are steeped and fermented up to 24 hours, then a water spray removes the pulp, and the beans are dried in the sun or in tumble dryers. A hulling machine then removes the protective membrane around the bean. In both cases the beans are cleaned, sorted and graded.

Roasting the beans

2 Roasting develops the beans' natural flavor by making the raw beans darker and bringing out the oils. Green, raw beans are roasted in ovens at a temperature between 380-480°F (193-249°C) for one to 17 minutes. The degree of roasting determines the depth of flavor—the darker the roast,

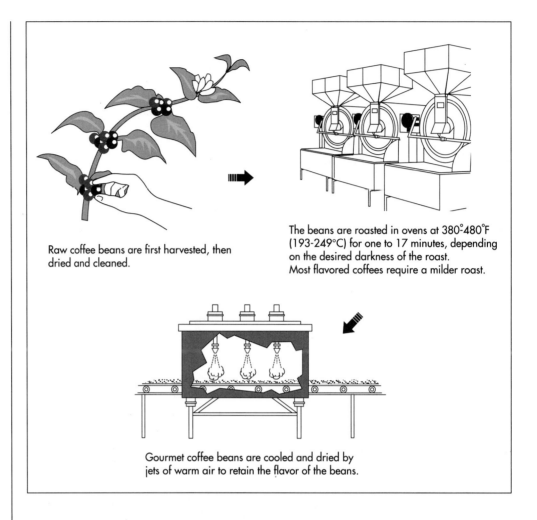

Raw coffee beans are first harvested, then dried and cleaned.

The beans are roasted in ovens at 380°-480°F (193-249°C) for one to 17 minutes, depending on the desired darkness of the roast. Most flavored coffees require a milder roast.

Gourmet coffee beans are cooled and dried by jets of warm air to retain the flavor of the beans.

the heavier the flavor. There are five commons roasts: American, Viennese, Italian, French Dark, and Espresso Black. The American, or Regular, roast has light to medium brown beans, with no oil on the bean. It makes mild to medium coffee with a definite acidic snap.The Viennese roast is slightly darker than American roast. The Italian, also known as Continental, roast features dark brown beans with an oily surface. It makes coffee that is dark flavored and bittersweet. French Dark roasting produces dark brown, almost black beans, with a shiny, oily surface. With its smoky, roasty flavors, it makes an authoritative coffee. Espresso Black is the highest roasting degree. This roast produces beans which are almost carbonized, and it yields the strongest brew.

If flavoring is added to beans which have too mild a roast, the coffee lacks significant flavor characteristics, and a flat-tasting beverage results. If the roast is too dark, the added flavor is overshadowed by the taste of the beans. For example, a French Vanilla flavor will be lost on a French Roast bean because the robust quality of the bean will overwhelm the sweet creamy tones of the flavor. The perfect roast color for flavored coffee is medium to brown.

After the beans are roasted, they must be quickly cooled before flavorings can be added. Flavoring the beans while they are still at high temperatures can destroy some of the flavor compounds. In large commercial operations, cooling is done by water quenching, which is a quick, economical process that has the undesirable effect of leaching out some of the natural flavor of the beans. Gourmet beans are dried more carefully, usually by jets of warm air.

Determining flavor usage

3 The appropriate amount of flavoring to be used must be determined before flavor oils can be added to the roasted beans. The rate of use typically varies between 2-

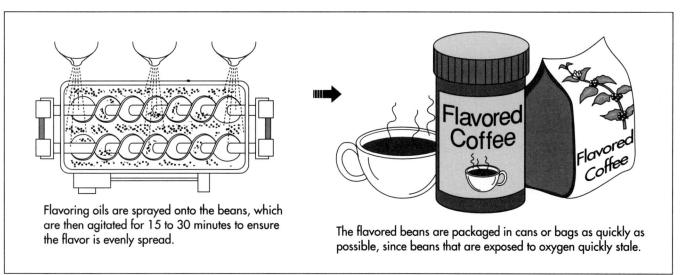

Flavoring oils are sprayed onto the beans, which are then agitated for 15 to 30 minutes to ensure the flavor is evenly spread.

The flavored beans are packaged in cans or bags as quickly as possible, since beans that are exposed to oxygen quickly stale.

3%, averaging 2.7% industrywide. A 3% usage rate means that three pounds of flavor oil are added to 100 pounds of roasted beans. The amount of flavoring required depends primarily on the type of flavor and its intensity, as well as the type of bean used and its roast level. Cost constraints also may play a role in determining how much flavoring to apply to the beans, because flavors are relatively expensive. The combination of flavors to be used and the quantity to be applied to the beans is established by experimental trial and error, in which test batches of beans are flavored with small quantities of oil until the desired characteristics are obtained. This formulation process is similar to the way one decides how much sugar to put in a cup of coffee or tea—add a small amount, taste it and, if necessary, add a little more. Once the precise amount is set, the dosage is held constant for that particular flavor oil and roasted bean combination. For different combinations of oils and beans, the usage level must be readjusted for optimal results.

Adding flavor oils

4 Flavors are typically added to roasted beans before they are ground. The beans are placed in a large mixer which is specially designed to gently tumble the beans without causing them damage. Examples of this type of mixer include ribbon blenders, drum rotators, and candy pan coaters. The flavors are usually introduced via a pressurized spray mechanism which breaks the oils into tiny droplets which allows for better mixing.

Oils must be added to the beans very gradually to guard against areas of highly concentrated flavor called hot spots. The beans are agitated for a set amount of time to ensure the flavor is evenly spread. This process may take 15-30 minutes, depending on the batch size and mixing characteristics of the oil. When the beans are properly coated, they take on a glossy finish that indicates a uniform distribution of oils.

It is also important to note that, instead of flavoring whole beans, flavors in dry form can be blended with ground coffee. In such cases, the flavors are encapsulated in starch or some other powdered matrix. There is enough moisture in the coffee to promote transfer of flavor and color from the encapsulated flavors to the coffee grounds in about 24 hours after mixing.

Packaging

5 The finished product is packed in bags or cans as quickly as possible and sealed to prevent contact with the atmosphere. Prior to packaging the container is flushed with nitrogen (an inert gas), a process that removes oxygen from the container. Oxygen can react with components of the flavor oils and the beans and cause deterioration. Coffee beans, once roasted, release their oils and begin to stale quickly when exposed to oxygen. Briefly flushing the container with nitrogen before filling pushes all the oxygen out and ensures freshness. Flavored beans should be stored in a cool, dark place if they are to be used within three or

four weeks. If longer storage is required, the beans may be frozen.

Quality Control

The quality of flavored coffees is assessed at various points throughout the manufacturing process. Before roasting, beans which do not meet standards for color or size are removed. This helps ensure a more even distribution of beans. After roasting, the color of the beans (which indicates the degree of roast) can be standardized by visual comparisons or with an analytical device known as a colorimeter, which measures the color of the beans. Beans which are over- or under-roasted are rejected. Similarly, the quality of the flavor oil is carefully checked. Flavorists use various analytical techniques, such as gas chromatography or spectrophotometry, to check flavor quality. These techniques can identify flavor compounds by analyzing their molecular structure. Individual natural and synthetic components are analyzed, as are the finished blended flavors, to ensure the consumer will taste the same quality of flavor from batch to batch. The quality of the final flavored product is checked with a sensory evaluation technique known as "cupping." This method involves placing 2.5 ounces (7.25 g) of ground coffee in a cup and adding 3.4 ounces (100 ml) boiling water. Both aroma and flavor can be evaluated in this manner. To communicate differences in flavor, the industry uses about 50 specialized terms to describe subjective flavor qualitites, such as earthy, nutty, spicy, and turpeny.

While there are no specific "coffee standards" the beans in particular must comply with, there are regulated Good Manufacturing Processes (GMPs) for food products. Relevant regulations are provided in the Code of Federal Regulations Title 21.

Byproducts/Waste

Production of flavored coffee beans does produce some waste in the form of beans that are rejected for one reason or another. There may be some degree of waste of the flavoring compounds due to batching or weighing errors. There is also waste in the form of solvent evaporation, which occurs during the curing process. These waste materials are not typically considered to be harmful, and therefore there are no special waste disposal requirements.

The Future

As advances in food technology are made, it is likely that improvements will be made in the manufacturing process for flavored coffee beans. Better mechanical methods of sorting and roasting beans will lead to more efficient production. More substantive heat resistant flavor compounds will be developed and, ideally, new technology will lead to flavors which cure onto the beans with no heat whatsoever. Of course, flavor chemists will continue to develop new exotic flavor compounds. It is also interesting to note other unconventional methods of flavoring coffees are gaining popularity. For example, instant flavored coffees have established a place in the mass market. These are made by entirely different processes, such as extracting the coffee flavor from the beans then spray drying, or by freeze-drying the coffee and blending it with flavor agents and other adjuncts. Also worthy of notice is an innovative new flavored coffee filter, which contains flavoring agents in the filter itself. It is touted as an economical way to serve flavored coffee and lets the consumer use his/her favorite coffee brand. Similar innovations will become common as the future of flavored coffees unfolds.

Where to Learn More

Periodicals

Kuntz, Lynn, "Coffee and Tea Beverages." *Food Product Design*, July 1996, pp. 78-100.

Mosciano, Gerard, et al. "Organoleptic Characteristics of Flavor Materials." *Perfumer and Flavorist,* November/December 1996, pp. 49-52.

Beck Flavor Brochure, Beck Flavor Company, 1996.

—*Randy Schueller*

Flour

Background

Flour is a finely ground powder prepared from grain or other starchy plant foods and used in baking. Although flour can be made from a wide variety of plants, the vast majority is made from wheat. Dough made from wheat flour is particularly well suited to baking bread because it contains a large amount of gluten, a substance composed of strong, elastic proteins. The gluten forms a network throughout the dough, trapping the gases which are formed by yeast, baking powder, or other leavening agents. This causes the dough to rise, resulting in light, soft bread.

Flour has been made since prehistoric times. The earliest methods used for producing flour all involved grinding grain between stones. These methods included the mortar and pestle (a stone club striking grain held in a stone bowl), the saddlestone (a cylindrical stone rolling against grain held in a stone bowl), and the quern (a horizontal, disk-shaped stone spinning on top of grain held on another horizontal stone). These devices were all operated by hand.

The millstone, a later development, consisted of one vertical, disk-shaped stone rolling on grain sitting on a horizontal, disk-shaped stone. Millstones were first operated by human or animal power. The ancient Romans used waterwheels to power millstones. Windmills were also used to power millstones in Europe by the twelfth century.

The first mill in the North American colonies appeared in Boston in 1632 and was powered by wind. Most later mills in the region used water. The availability of water power and water transportation made Philadelphia, Pennsylvania, the center of milling in the newly independent United States. The first fully automatic mill was built near Philadelphia by Oliver Evans in 1784. During the next century, the center of milling moved as railroads developed, eventually settling in Minneapolis, Minnesota. During the nineteenth century numerous improvements were made in mill technology. In 1865, Edmund La Croix introduced the first middlings purifier in Hastings, Minnesota. This device consisted of a vibrating screen through which air was blown to remove bran from ground wheat. The resulting product, known as middlings or farina, could be further ground into high-quality flour. In 1878, the first important roller mill was used in Minneapolis, Minnesota. This new type of mill used metal rollers, rather than millstones, to grind wheat. Roller mills were less expensive, more efficient, more uniform, and cleaner than millstones. Modern versions of middlings purifiers and roller mills are still used to make flour today.

Raw Materials

Although most flour is made from wheat, it can also be made from other starchy plant foods. These include barley, buckwheat, corn, lima beans, oats, peanuts, potatoes, soybeans, rice, and rye. Many varieties of wheat exist for use in making flour. In general, wheat is either hard (containing 11-18% protein) or soft (containing 8-11% protein). Flour intended to be used to bake bread is made from hard wheat. The high percentage of protein in hard wheat means the dough will have more gluten, allowing it to rise more than soft wheat flour. Flour intended to be used to bake cakes and pas-

A kernel of wheat consists of three parts, two of which can be considered byproducts of the milling process. The bran is the outer covering of the kernel and is high in fiber. The germ is the innermost portion of the kernel and is high in fat. The endosperm makes up the bulk of the kernel and is high in proteins and carbohydrates.

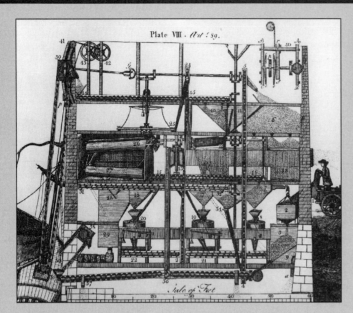

An illustration from The Young Millwright and Miller's Guide, depicting the processes of an automated grain mill. (From the collections of Henry Ford Museum & Greenfield Village.)

In 1795, an American engineer published a book called *The Young Millwright and Miller's Guide*. In the book, simple theories are transformed into a set of mechanical devices that form a flour mill. At the back of the book is a drawing, illustrating how these devices make a continuous production line in which the human hand is eliminated from the beginning of the process to the end of production. The author of this book was Oliver Evans, himself the son of a miller. He and his brothers ran their own mill, developed the systems, and perfected the operations that led to the automated grain mill.

Today, Evans is considered one of America's most ambitious mechanical innovators. He used his understanding of the way in which water turned a mill wheel and developed it into a viable grain-milling system.

Most important was the fact that his system contained the idea of the integrated and automated factory. When a machine substitutes human intervention, the problems of the fully automated assembly line are solved. This concept was not fully applied until the 1920s by Henry Ford, who was able to develop a successful, operational assembly line. Ford had the advantage of living at the end of the machine age, but Oliver Evans was the first to present the concept of automation before it was even possible.

Henry Prebys

try is made from soft wheat. All-purpose flour is made from a blend of soft and hard wheat. Durum wheat is a special variety of hard wheat, which is used to make a kind of flour called semolina. Semolina is most often used to make pasta.

Flour usually contains a small amount of additives. Bleaching agents such as benzoyl peroxide are added to make the flour more white. Oxidizing agents (also known as improvers) such as potassium bromate, chlorine dioxide, and azodicarbonamide are added to enhance the baking quality of the flour. These agents are added in a few parts per million. Self-rising flour contains salt and a leavening agent such as calcium phosphate. It is used to make baked goods without the need to add yeast or baking powder. Most states require flour to contain added vitamins and minerals to replace those lost during milling. The most important of these are iron and the B vitamins, especially thiamin, riboflavin, and niacin.

The Manufacturing Process

Grading the wheat

1 Wheat is received at the flour mill and inspected. Samples of wheat are taken for physical and chemical analysis. The wheat is graded based on several factors, the most important of which is the protein content. The wheat is stored in silos with wheat of the same grade until needed for milling.

Purifying the wheat

2 Before wheat can be ground into flour it must be free of foreign matter. This requires several different cleaning processes. At each step of purification the wheat is inspected and purified again if necessary.

3 The first device used to purify wheat is known as a separator. This machine passes the wheat over a series of metal screens. The wheat and other small particles pass through the screen while large objects such as sticks and rocks are removed.

4 The wheat next passes through an aspirator. This device works like a vacuum cleaner. The aspirator sucks up foreign matter which is lighter than the wheat and removes it.

5 Other foreign objects are removed in various ways. One device, known as a disk separator, moves the wheat over a series of disks with indentations that collect objects the size of a grain of wheat. Smaller or larger objects pass over the disks and are removed.

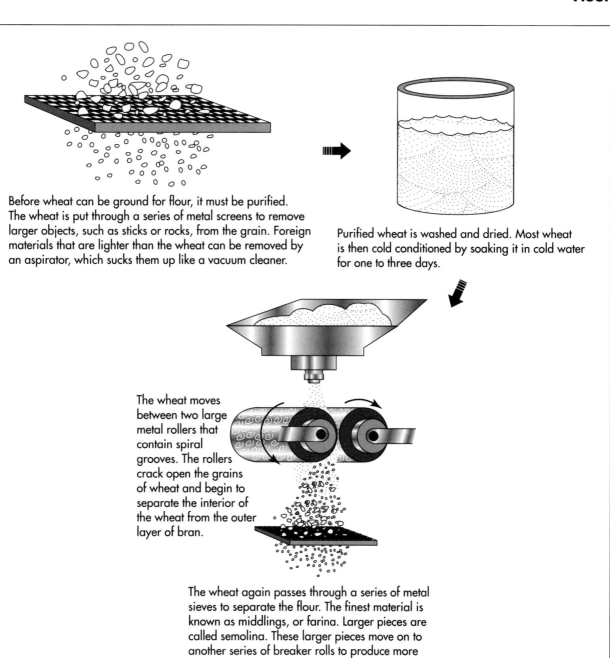

Before wheat can be ground for flour, it must be purified. The wheat is put through a series of metal screens to remove larger objects, such as sticks or rocks, from the grain. Foreign materials that are lighter than the wheat can be removed by an aspirator, which sucks them up like a vacuum cleaner.

Purified wheat is washed and dried. Most wheat is then cold conditioned by soaking it in cold water for one to three days.

The wheat moves between two large metal rollers that contain spiral grooves. The rollers crack open the grains of wheat and begin to separate the interior of the wheat from the outer layer of bran.

The wheat again passes through a series of metal sieves to separate the flour. The finest material is known as middlings, or farina. Larger pieces are called semolina. These larger pieces move on to another series of breaker rolls to produce more middlings.

6 Another device, known as a spiral seed separator, makes use of the fact that wheat grains are oval while most other plant seeds are round. The wheat moves down a rapidly spinning cylinder. The oval wheat grains tend to move toward the center of the cylinder while the round seeds tend to move to the sides of the cylinder, where they are removed.

7 Other methods used to purify wheat include magnets to remove small pieces of metal, scourers to scrape off dirt and hair, and electronic color sorting machines to remove material which is not the same color as wheat.

Preparing the wheat for grinding

8 The purified wheat is washed in warm water and placed in a centrifuge to be spun dry. During this process any remaining foreign matter is washed away.

9 The moisture content of the wheat must now be controlled to allow the outer

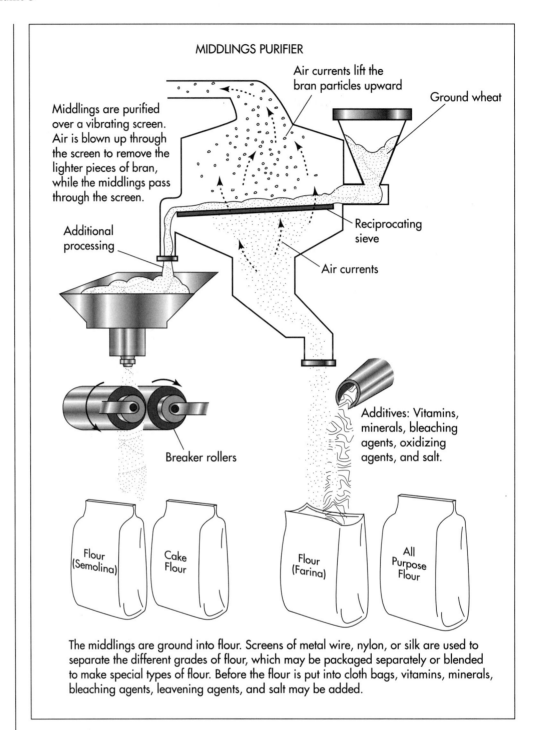

MIDDLINGS PURIFIER

Air currents lift the bran particles upward

Ground wheat

Middlings are purified over a vibrating screen. Air is blown up through the screen to remove the lighter pieces of bran, while the middlings pass through the screen.

Additional processing

Reciprocating sieve

Air currents

Breaker rollers

Additives: Vitamins, minerals, bleaching agents, oxidizing agents, and salt.

Flour (Semolina)

Cake Flour

Flour (Farina)

All Purpose Flour

The middlings are ground into flour. Screens of metal wire, nylon, or silk are used to separate the different grades of flour, which may be packaged separately or blended to make special types of flour. Before the flour is put into cloth bags, vitamins, minerals, bleaching agents, leavening agents, and salt may be added.

layer of bran to be removed efficiently during grinding. This process is known as conditioning or tempering. Several methods exist of controlling the amount of water present within each grain of wheat. Usually this involves adding, rather than removing, moisture.

10 Cold conditioning involves soaking the wheat in cold water for one to three days. Warm conditioning involves

soaking the wheat in water at a temperature of 115°F (46°C) for 60-90 minutes and letting it rest for one day. Hot conditioning involves soaking the wheat in water at a temperature of 140°F (60°C) for a short period of time. This method is difficult to control and is rarely used. Instead of water, wheat may also be conditioned with steam at various temperatures and pressures for various amounts of time. If conditioning results in too much moisture, or if the wheat happens to be too moist after purification, water can be removed by vacuum dryers.

Grinding the wheat

11 Wheat of different grades and moistures is blended together to obtain a batch of wheat with the characteristics necessary to make the kind of flour being manufactured. At this point, the wheat may be processed in an Entoleter, a trade name for a device with rapidly spinning disks which hurl the grains of wheat against small metal pins. Those grains which crack are considered to be unsuitable for grinding and are removed.

12 The wheat moves between two large metal rollers known as breaker rolls. These rollers are of two different sizes and move at different speeds. They also contain spiral grooves which crack open the grains of wheat and begin to separate the interior of the wheat from the outer layer of bran. The product of the breaker rolls passes through metal sieves to separate it into three categories. The finest material resembles a coarse flour and is known as middlings or farina. Larger pieces of the interior are known as semolina. The third category consists of pieces of the interior which are still attached to the bran. The middlings move to the middlings purifier and the other materials move to another pair of breaker rolls. About four or five pairs of breaker rolls are needed to produce the necessary amount of middlings.

13 The middlings purifier moves the middlings over a vibrating screen. Air is blown up through the screen to remove the lighter pieces of bran which are mixed with the middlings. The middlings pass through the screen to be more finely ground.

14 Middlings are ground into flour by pairs of large, smooth metal rollers. Each time the flour is ground it passes through sieves to separate it into flours of different fineness. These sieves are made of metal wire when the flour is coarse, but are made of nylon or silk when the flour is fine. By sifting, separating, and regrinding the flour, several different grades of flour are produced at the same time. These are combined as needed to produce the desired final products.

Processing the flour

15 Small amounts of bleaching agents and oxidizing agents are usually added to the flour after milling. Vitamins and minerals are added as required by law to produce enriched flour. Leavening agents and salt are added to produce self-rising flour. The flour is matured for one or two months.

16 The flour is packed into cloth bags which hold 2, 5, 10, 25, 50, or 100 lb (About 0.9, 2.3, 4.5, 11.3, 22.7, or 45.4 kg). For large-scale consumers, it may be packed in metal tote bins which hold 3000 lb (1361 kg), truck bins which hold 45,000 lb (20,412 kg), or railroad bins which hold 100,000 lb (45,360 kg).

Quality Control

The quality control of flour begins when the wheat is received at the flour mill. The wheat is tested for its protein content and for its ash content. The ash content is the portion which remains after burning and consists of various minerals.

During each step of the purification process, several samples are taken to ensure that no foreign matter ends up in the flour. Since flour is intended for human consumption, all the equipment used in milling is thoroughly cleaned and sterilized by hot steam and ultraviolet light. The equipment is also treated with antibacterial agents and antifungal agents to kill any microscopic organisms which might contaminate it. Hot water is used to remove any remaining traces of these agents.

The final product of milling is tested for baking in test kitchens to ensure that it is suitable for the uses for which it is intended. The vitamin and mineral content is measured in order to comply with government standards. The exact amount of additives present is measured to ensure accurate labeling.

Byproducts/Waste

A kernel of wheat consists of three parts, two of which can be considered byproducts of the milling process. The bran is the outer covering of the kernel and is high in fiber. The germ is the innermost portion of the kernel and is high in fat. The endosperm makes up the bulk of the kernel and is high in proteins and carbohydrates. Whole wheat flour uses all parts of the kernel, but white flour uses only the endosperm.

Bran removed during milling is often added to breakfast cereals and baked goods as a source of fiber. It is also widely used in animal feeds. Wheat germ removed during milling is often used as a food supplement or as a source of edible vegetable oil. Like bran, it is also used in animal feeds.

Where to Learn More

Books

Besant, Lloyd. *Grains: Production, Processing, Marketing.* Chicago Board of Trade, 1982.

Kent, N. L. *Technology of Cereals: With Special Reference to Wheat.* Pergamon Press, 1975.

Periodicals

Sokolov, Raymond. "Through a Mill, Coarsely." *Natural History*, February 1994, pp. 72-74.

Wrigley, Colin W. "Giant Proteins With Flour Power." *Nature*, June 27, 1996, pp. 738-739.

Other

"How Flour is Made." The Story of Wheat. University of Saskatchewan College of Agricultural Sciences. December 7, 1996. http://pine.usask.ca/cofa/displays/college/story/wheat.html.

—*Rose Secrest*

Football

Background

Although the game of football as we know it today supposedly dates back to the nineteenth century, there is some evidence to support that the ancient Greeks played a version of football they called harpaston. This game apparently took place on a rectangular field with goal lines on both ends. Two teams of equal number, but varying player size, were divided by a center line. The game began by throwing the harpaston or handball into the air. The object of the game was to pass, kick, or run the ball past the opposing team's goal line.

The game next took to the streets. Participants from neighboring towns would meet at a designated point. Still without official rules or methods of keeping score, the bladder or ball would be kicked through the streets. This took place until protests from local shopkeepers forced players to confine their game to a vacant area.

It is here that the rules of the game first took shape. A field much like that used to play soccer was marked with boundaries. The team that kicked the ball over the opponent's goal line was awarded one point. It also was at this time that the game took on the name of futballe.

The game remained strictly a kicking game until American collegians blended soccer with rugby. In 1874, McGill University (Montreal, Canada) engaged Harvard University (Cambridge, Massachusetts) in two sports games. One game was played with Canadian rugby rules, which allowed players to run with the ball, as well as throw it. The other game followed U.S. soccer rules, which restricted players to only kicking the ball.

It seemed that Harvard preferred elements of both games and introduced them to Yale University in New Haven, Connecticut. Two years later, representatives from Harvard and Yale met in Massachusetts to create guidelines for this new game of football. Another new twist to the game was that it was played with an oval-shaped ball.

Spaulding Sports Worldwide, based in Chicopee, Massachusetts, takes credit for having produced the first American-made football in 1892.

Raw Materials

In the early stages of the game of football, a pig's bladder was inflated and used as the ball. By comparison, today's football is an inflated rubber bladder enclosed in a pebble-grained leather cover or cowhide. This material is used because it is both durable and easily tanned.

Design

The football's uneven shape makes it difficult to catch and hold and also causes unpredictable bounces. White laces sewn on the ball's surface help the players to grip it. There have been many attempts to alter the football's design; for example, dimples on footballs have been tried, but there was a tendency for dirt and mud to get caught in them.

The Manufacturing Process

1 After special tanning processes, the cowhide selected to be used for the foot-

A box containing 24 new balls is opened before each NFL game; 12 balls are put into play during each half. After the game, the balls are used for practices.

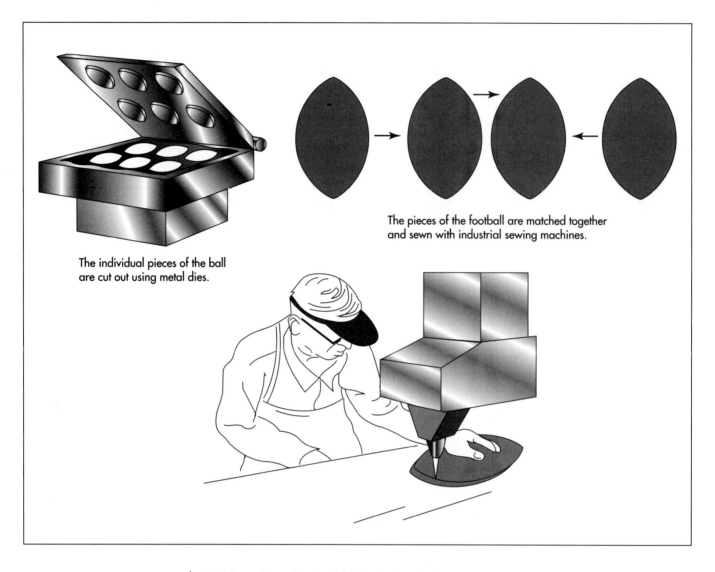

The individual pieces of the ball
are cut out using metal dies.

The pieces of the football are matched together
and sewn with industrial sewing machines.

ball is cut into a bend, which is the best and strongest part of the hide.

2 The bend is then die-cut into panels. Using a hydraulically-driven clicking machine, an operator cuts four panels into the precise shape required at the same time.

3 Next, each panel goes through a skiving machine in order to reduce it to a predetermined thickness and weight.

4 A synthetic lining is sewn to each panel. The lining, which is composed of three layers of cross-laid fabric firmly cemented together, prevents the panel from stretching or growing out of shape during use. The lining and panel are sewn together using an industrial size and strength version of a home sewing machine.

5 A facing is then applied to those areas that will carry the lacing holes as well as the hole for the inflating needle. The holes are then punched.

6 The four panels are sewn together by a hot-wax lock stitch machine to ensure that the seams are especially durable. Then, the ball is turned right side out.

7 Next, a two-ply butyl rubber bladder is inserted, the ball is laced, and then it is inflated with a pressure of not less than 12.5 lb (6 kg) but no more than 13.5 lb (6.1 kg). After inflation, the ball is checked to ensure it conforms to all size and weight regulations.

8 The ball is ready for branding with the manufacturer's name and number.

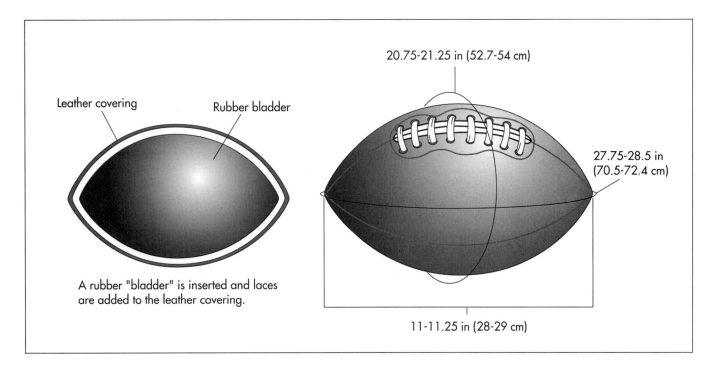

Leather covering

Rubber bladder

20.75-21.25 in (52.7-54 cm)

27.75-28.5 in (70.5-72.4 cm)

A rubber "bladder" is inserted and laces are added to the leather covering.

11-11.25 in (28-29 cm)

9 After final inspections, the balls are boxed and shipped to designated schools and ball clubs.

Quality Control

Since 1941, Wilson Sporting Goods Company, currently based in Chicago, Illinois, has been the official ballmaker for the National Football League (NFL). For all NFL games, the only sanctioned ball is a Wilson brand ball. The ball must measure 20.75-21.25 in (52.7-54 cm) around its middle (also called the girth, short axis, or belly); 27.75-28.5 in (70.5-72.4 cm) around its ends (the circumference, long axis); and 11-11.25 in (28-29 cm) from tip to tip (the length of the long axis). It also must weigh between 14-15 oz (397-425.25 g).

All balls designed for professional use are stamped with "NFL" on them for the National Football League and they also bear the signature of the League commissioner. A box containing 24 new balls is opened before each game; 12 balls are put into play during each half. After the game, the balls are used for practices.

Those balls that are used in the Super Bowl game also have the names of the participat-ing teams along with the date and location of the game.

The Future

Future changes to the football are more likely to occur in the area of materials rather than design. The goal is to "create a better feel right out of the box."

Spaulding Sports Worldwide currently is working on a proprietary material to create a composite-covered football. Two of the benefits of a composite cover compared with a leather cover are that it does not retain as much water; and that it is not as susceptible to becoming hard due to cold weather.

Where to Learn More

Books

Foehr, Donna Poole. *Football for Women and Men Who Want To Learn The Game.* National Press, Inc., 1988, pp. 94, 100, 101, 102, 127.

Ominsky, Dave and P.J. Harari. *Football Made Simple: A Spectator's Guide.* First Base Sports, Inc., 1994, pp. 1, 9.

—*Susan Bard Hall*

A two-ply butyl rubber bladder is inserted, the ball is laced, and then it is inflated with a pressure of not less than 12.5 lb (6 kg) but no more than 13.5 lb (6.1 kg). After inflation, the ball is checked to ensure it conforms to all size and weight regulations.

Football Helmet

Three out of every 1,000 helmets of every size and style are taken off the production line to the product testing lab, where they are placed on a quasi-humanoid head form and subjected to a battery of impact tests. Ten helmets are tested per day.

Background

Amateur and professional football players alike wear protective gear to reduce the likelihood of sustaining injury while playing the game of football. The football helmet with its chin strap, face mask, and optional mouth guard is one example of protective gear.

The football helmet serves an aesthetic purpose as well. Because the helmet bears the team's logo, it serves as a trademark. Credit goes to the Los Angeles Rams as being the first football team to design graphics for their helmets. The Rams horns still adorn their helmets, letting their opponents know they are not afraid to butt heads with them.

The first helmets, circa 1915, were basic, leather headgear without face masks. With their flat top design, they bore a strong resemblance to the soft leather headgear worn by today's wrestlers. The design of these helmets primarily protected the players' ears; yet, without ear holes, this type of helmet made on-field communication virtually impossible.

Helmets with harder leather to help protect the skull first started making an appearance during World War I. In the ensuing years, increasingly harder leathers were used to provide even greater protection. During the same time frame, the first fabric cushioning came on the scene to help absorb the shock brought upon by collisions. Helmet makers also began to phase out the flat top design, replacing it with a more oval shape. The advantage to this new shape was it allowed for blows to the head to be deflected to one side, rather than forcing the top of the head to absorb most of the impact.

Football helmet design took a giant step in 1939 when the John T. Riddell Company introduced plastic helmets. This also led the way for a redesign of helmet straps, which to this point, were designed to be affixed around the neck. The redesign called for the straps to attach to the chin.

Within 10 years, leather helmets became obsolete. Two other significant events took place in the 1940s. The National Football League (NFL) made football helmets required equipment, and the first face mask was developed.

Since the 1970s, football helmets have taken on another role — that of souvenir. Football fans have created demand for replica footballs of their favorite team, which can be found in virtually any store that specializes in sports memorabilia.

Raw Materials

Materials used for the production of football helmets have evolved from leather, to harder leather, to molded polycarbonate shells, which are used today because of their strength and weight.

Design

From the early flat top design without holes for ears to the more oval shape, probably the single innovation with the most impact on football helmet design took place in the early 1970s. Dr. Richard Schneider of the University of Michigan Hospital is reported to have believed that air was the most effective way to protect against blunt force. With this theory in mind, he invented an inflatable bladder for use inside a football helmet.

A prototype was developed and used by the University of Michigan team. It did not take long for the Bike Athletic Company to hire Dr. Schneider and begin mass producing the helmet, which today is known as Schutt Sport Group's Air™ Helmet.

The chin strap, which helps to secure the helmet to the player's head, began as straps designed to attach around the neck. The re-design of the straps to attach around the chin took place in 1939.

The face mask, which is usually made of plastic or metal bars, attaches to the front of the helmet. There are two types of face masks, the open cage and the closed cage. The open cage usually is preferred by quarterbacks, running backs, wide receivers and defensive backfield men because the open cage—with two or three horizontal bars and no vertical bar above the nose—enables better visibility. The closed cage usually is the choice of linesmen because the closed cage—vertical bar running the length of the mask over the nose with two, three, or four horizontal bars—helps to keep other players' fingers and hands out of their eyes. In the 1970s, vinyl coating was layered onto the bars to protect against chipping and abrasions. Soon, colors were added to the face masks as another way to distinguish players and teams.

The logo of a player's team usually adorns both sides of the helmet.

In the 1970s, a group known as NOCSAE (National Operating Committee on Standards for Athletic Equipment) established performance standards for football helmets, as well as prescribed verbage to go on the helmet itself. The NOCSAE warning label states that the helmet should not be used to strike an opponent. Such an action is against football rules and may cause severe brain or neck injury. Playing the game of football in itself can cause injury, and no helmet can prevent all such injuries. The warning also alerts players to use the helmet at their own risk. This NOCSAE warning was required to be placed inside every helmet. In 1983, the NOCSAE warning began to appear on the outside of every helmet.

Another design feature has been the use of radio receivers in the helmets so that coaches can relay plays to their signal callers. In order to bring the game closer to the fans, a "helmet-cam" also has been used so that fans get to see exactly what the players see on the field.

The Manufacturing Process

1 The helmet outer shell is constructed of a tough plastic called polycarbonate alloy. The polycarbonate alloy arrives at the manufacturing plant in pellet form — in boxes of thermoplastic pellets, the size of beebees. The pellets are loaded into an injection-molding machine, melted, and forced into a cavity the size of a football helmet. It takes approximately one minute to mold one shell. Shells come in small, medium, large, and extra large sizes.

2 The shell then drops out of the machine.

3 Next, a multi-drill fixture drills 14-15 holes into the mold, a process that takes approximately 12-15 seconds to complete.

4 Next, protective air liners are produced. Certain rotationally-molded, one-piece liners are inflatable and are used in the helmet for obtaining proper fit and to aid in dispersing the energy imparted by an impact. Other specifically-engineered liners contain special foams and energy-attenuating or elastic materials. Like air, these materials are designed to absorb kinetic energy of movement and slow or decrease the impact of a blow to the head. The foam-based liners are made in several pieces—one is for the back, neck, and sides of the helmet and another is for the crown.

To produce the special foams required for the liner, large sheets of foam are die-cut to size. Then, the vinyl encasement is die-cut to size. A piece of vinyl is loaded into a vacuum former. The pieces of the die-cut foam are put into the vinyl and thermo-formed to make an airtight seal. Another layer of vinyl is placed on top of the thermoform and the process is repeated.

5 The jaw pads, which are designed to fit below the earlobe, are affixed. Different sizes or thicknesses are available.

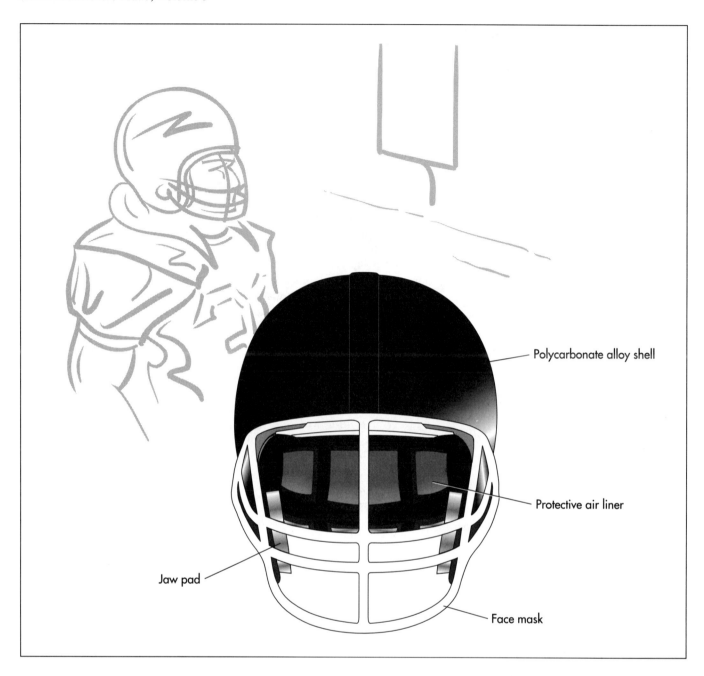

Polycarbonate alloy shell

Protective air liner

Jaw pad

Face mask

Materials used for the production of football helmets have evolved from leather, to harder leather, to molded polycarbonate shells, which are used today because of their strength and weight.

6 Face masks are then attached. There are several different styles of face masks. The face masks are made out of steel wire and coated with plastic. There also are three different versions of plastic face masks.

7 The chin straps are then attached.

8 The helmets can be painted in any one of the standard finished colors. There are over 50 standard colors to choose. More often however, the color finish is injection molded in at the time the shell is construct-ed. Decorative pieces such as decals are generally not applied by the manufacturer, but by the organization purchasing the helmets. The NFL does their own decaling as well.

9 At the end of the assembly line, each helmet is subjected to inspection to ensure that the workmanship standards have been met. Only then would each helmet be placed in a poly bag and into a compartmented carton for shipment to the warehouse. Each helmet has a serial number inside the shell, and the corresponding serial number is affixed to the outside of the carton.

Helmet Reconditioning Process

Regularly-scheduled helmet reconditioning helps to ensure that each athlete is protected to the full extent of their equipment. This reconditioning process also helps to prolong the effective life of the helmet and reduce replacement costs.

- High-pressure nozzles spray cleaning and sanitizing solutions onto the helmet to dislodge dirt and disinfect it. Separately, interior protective linings and accessories are cleaned and sanitized.

- Using glass beads through a carefully controlled pressure sand-blaster, loose and chipping paint is removed. Air buffers and cotton-buffing wheels are used to remove decals and the adhesive residue that still remains.

- Pressure and flow control nozzles are used to apply paint uniformly to maximize paint adhesion.

- Face masks are removed and inspected, then reinstalled on the reconditioned helmet using corrosion-resistant hardware.

- After thorough cleaning and sanitizing, jaw pads and chin straps are inspected, then reinstalled.

- Each helmet is hand-buffed and wiped, both inside and outside, to maximize helmet shine and cleanliness.

- Each helmet is placed in a poly bag to keep it dust-free.

- Helmets are then placed in compartmented cartons, which are designed to protect the helmets during transit.

Quality Control

The material used for the helmet shell must meet the approved standard guidelines created by the NOCSAE. All incoming raw materials that are to be used in the manufacture of football helmets are subject to inspection. Once the helmets have been produced, three out of every 1,000 of every size and style are taken off the production line to the product testing lab where they are placed on a quasi-humanoid head form and subjected to a battery of impact tests. Approximately 10 to 15 helmets are tested per day.

The Future

A new helmet design that is being tested is a one-piece helmet/shoulder pad combination which may help to protect players by distributing force through the entire torso, not just the head and neck. This product is still in the testing stages. Protective Sports Equipment has developed a polyurethane safety accessory that is designed to attach to the football helmet to reduce the impact that can cause concussions. Upon impact, the ProCap returns to its original form. The design and material used in the manufacture of the ProCap allows for the absorption of more of the shock from a collision. Initial tests of the polyurethane safety accessory have had inconclusive results. Significantly more testing and evaluation will be done before this product is accepted.

Riddell said its research and development department listens to suggestions and demands made by those with a vested interest in the game of football. They are continually investigating new raw materials that will help to spread out or extend the decceleration time of impact when a helmet contacts another object. The round/teardrop configuration currently used slides off another helmet and as such, helps to guard against rotational injuries as opposed to the helmet shape worn by hockey players that can lock together.

Where to Learn More

Books

Foehr, Donna Poole. *Football for Women and Men Who Want To Learn The Game.* National Press, Inc., 1988, pp. 97, 100, 101.

Ominsky, Dave and P.J. Harari. *Football Made Simple: A Spectator's Guide.* First Base Sports, Inc., 1994, pp. 10.

Periodicals

"A Symbol: Football's Most Prominent Tool has Evolved Along with the Sport." *American Football Quarterly,* October/November/December.

"Aging Helmets to be Sidelined." *The Physician and Sportsmedicine,* December 1990, p. 15.

"Football Caps Reduce Impact." *Machine Design,* January 8, 1993, pp. 16.

—*Susan Bard Hall*

Garbage Truck

The size of a front loader, rear loader, or side loader body is measured in cubic yards of garbage that it can contain. The size of a roll-off truck is measured in pounds of hoist capacity.

Background

You can call it garbage, trash, refuse, or solid waste. It's all the same thing, and getting rid of it has been a problem since the beginning of civilization. The earliest method of getting rid of garbage consisted of simply throwing it in a pile. As the population of an area grew, so did the need to move the garbage pile further and further away. In 500 B.C., the Greek city of Athens established the first municipal dump in the Western world when it required that garbage be disposed of at least one mile (1.6 km) from the city walls. Other cities were not so advanced. For example, in 1400, garbage had piled up so high outside the gates of Paris, France, that it interfered with the city's defenses.

The first vehicles for hauling garbage were probably two-wheeled carts drawn by animals or slaves. In the 1800s horse-drawn, four-wheeled wagons moved slowly down alleyways as garbagemen hoisted reeking barrels filled with wet garbage and dumped them into the open wagon bed. By the 1920s, motor power had replaced horse power, but little else had changed. The cry of "Here comes the garbage truck" was still the signal to go inside and close your windows.

The postwar consumer boom of the 1950s in the United States led to a significant increase in trash. After years of restrictions and shortages during World War II, people eagerly replaced old products with new ones. Many of the new products were meant to be used once and thrown away. Paper plates, plastic cups, paper towels and napkins, disposable diapers, and brown paper lunch bags all clogged the trash cans. The refuse vehicle industry responded in the late 1950s with the development of the first enclosed refuse trucks, utilizing hydraulic rams to compress the trash as it was collected. This allowed each truck to carry more trash per load.

Today, many municipalities in the United States have contracted with private firms to pick up their trash and dispose of it, rather than do it themselves. Out of this trend have emerged two or three giant refuse companies, each owning thousands of trucks. In order to remain competitive, these companies have designed trucks that are highly specialized and automated in an effort to deal with an ever-increasing amount of trash at the lowest cost.

Raw Materials

Most of the body components on a garbage truck are made of steel. The body floor, sides, top, and ends are made of steel sheet or plate and are reinforced with formed steel channels. Different thicknesses of sheet or plate are used for different areas of the body, depending on the stresses expected in that area. This helps minimize the weight of the body, and therefore, maximize the weight of trash the truck can carry.

The lift arms and forks on a front loader are cut from thick steel plates, and the torque tubes are made from thick-walled, seamless steel tubing. The packer blade, or head, is used to periodically compress the garbage inside the body. It is made from steel plate and slides on plastic, steel, or bronze shoes.

Purchased components include the vehicle cab and chassis, lights, warning labels, electrical wiring, and the hydraulic fluid, cylinders, hoses, and controls.

Design

There are five common kinds of garbage trucks: front loader, rear loader, side loader, recycling, and roll-off. Each is used for a different type of garbage collection. The size of a front loader, rear loader, or side loader body is measured in cubic yards of garbage that it can contain. The size of a roll-off truck is measured in pounds of hoist capacity.

A front loader has two long, hydraulically raised lift arms that are pivoted behind the truck cab and extend forward of the front bumper. Forks on the ends of these arms slip into slots on the sides of a large metal trash container. The hydraulic arms lift the entire container up and over the cab and tip the contents into an opening at the forward portion of the body top. An internal packer blade periodically compresses the trash and moves it to the rear of the body. Front loaders are generally used to pick up trash at commercial businesses, and the containers are commonly called dumpsters, although that is a proprietary name for one manufacturer's design. Front loaders have capacities of about 30-40 cu yds (23.0-30.6 cu m) and can be operated with a crew of one.

A rear loader has an opening at the lower rear portion of the body. Individual trash containers are manually dumped into this opening. A hydraulic paddle or blade is activated periodically to push the trash forward into the body. Rear loaders are usually used to pick up trash in residential areas. They have capacities of about 20-30 cu yd (15.3-23.0 cu m) and require a crew of two or three.

A side loader is operated in a similar manner as a rear loader, but the opening for the trash is on the side, just behind the cab, where the driver or loader can reach it quickly. With a manual side loader, the trash is manually dumped into the opening. With an automated side loader, a hydraulic arm with a gripping claw on the end grabs the trash container and quickly dumps the contents into the opening. Side loaders are used to pick up trash in residential areas. They have capacities of about 15-30 cu yd (11.5-23.0 cu m) and can be operated with a crew of one or two.

A recycling truck is designed to accept two or more recyclable commodities, such as newspapers, glass containers, metal cans, or other materials. It is equipped with a separate loading opening and bin for each material. Recycling trucks are used in residential areas and can be operated with a crew of one or two.

A roll-off truck carries an enclosed trash container on a tilting ramp attached to the truck frame. The container is rolled down the ramp and set on the ground at construction sites and other locations where a large amount of trash and debris needs to be removed. When the container is full, the truck returns and winches the loaded container up the ramp and onto the truck frame again. Roll-off trucks typically have a 60,000 lb (27,300 kg) hoist capacity and are operated with a crew of one.

The Manufacturing Process

Garbage truck bodies are built in a fixed location within a plant, rather than moving down an assembly line or moving from one work station to another. Component parts are fabricated in a machine shop and are then welded or assembled into subassemblies. The subassemblies are brought together and are welded or assembled into the finished body. The body is then lifted and mounted on a truck chassis.

Here is a typical sequence of operations for the assembly of a front-loader garbage truck:

Forming the body shell

1 The pieces for the body floor, sides, top, and front end are cut to size in a machine shop using band saws, metal shears, and cutting torches. Some flat pieces are bent in press brakes or curved in roller benders. Mounting holes are punched or drilled.

2 The pieces for each of the body components—floor, sides, top, and front—are moved to separate sub-assembly areas where they are welded together. Welding the long reinforcing channels on the sides is often done on a flat welding table with an automatic welder that is programmed to make welds in the correct areas. Other welding is done manually. Templates are sometimes used to position the pieces correctly, while clamps hold the pieces in position.

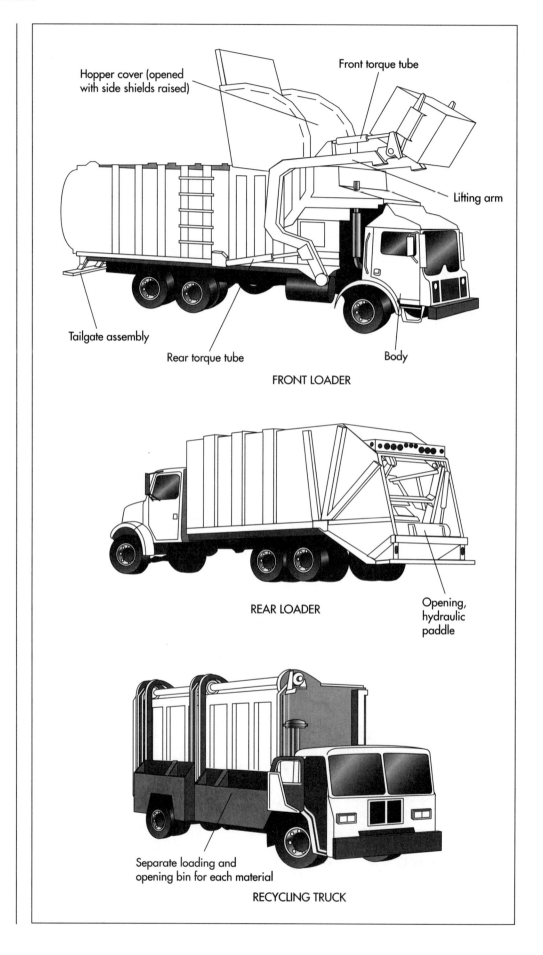

Hopper cover (opened with side shields raised)

Front torque tube

Lifting arm

Tailgate assembly

Rear torque tube

Body

FRONT LOADER

REAR LOADER

Opening, hydraulic paddle

Separate loading and opening bin for each material

RECYCLING TRUCK

3 Each of the body subassemblies is brought to the body assembly area, sometimes called a cell. First the floor is brought in and positioned on supports to make it level and stable. Then the sides are lifted in and are braced while they are welded to the floor. Then the top is lowered and welded into place, followed by the front. When the body shell is finished, it looks like an empty shoe box with one end missing.

Installing the operating subassemblies

4 While the body shell is being formed, the operating parts of the body are being fabricated and welded into subassemblies. These include the lift arms, fork assembly, hopper cover, tailgate, and packer blade. They are formed in the same sequence as described in steps 1 and 2. When the body shell is complete, the operating subassemblies are brought to the body assembly area in sequence.

5 The packer blade is installed first because it goes inside the body. It is lifted into position, and the packer hydraulic cylinder is attached between the packer blade and the body shell. Some body designs use two hydraulic cylinders for increased packing force.

6 The hopper cover is installed next. It is welded over the loading opening on the forward portion of the body top. The cover is hydraulically opened as the lift arms bring the trash container up and over the top, and it includes shields to prevent the trash from spilling over the sides. The hydraulic cylinder is attached to the mounting points.

7 The lift arms, fork assembly, and front and rear torque tubes are assembled and attached to the body shell. The various hydraulic cylinders are attached to their mounting points.

8 The tailgate is installed last. It is hung from its top-mounted hinges, and the hydraulic cylinders are attached to their mounting points.

Finishing the body

9 The body lights are installed and the electrical wiring is routed and connected. The hydraulic hoses are routed and connected to the various hydraulic cylinders.

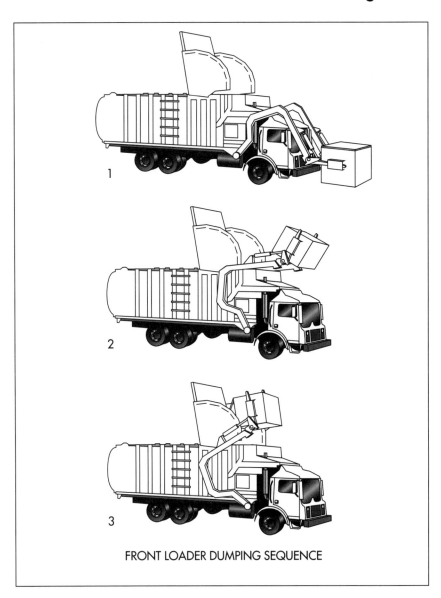

FRONT LOADER DUMPING SEQUENCE

10 At this point, some manufacturers paint the body before it is mounted on the truck, while others wait until the body is mounted on the truck before they paint it. If the body is to be painted at this point, the lights and the exposed portions of the hydraulic cylinder rods are masked off with paper and tape.

11 After the body is painted, any name plates, decorative striping, and warning labels are applied as required.

Modifying the truck cab and chassis

12 A truck cab and chassis is delivered to the garbage truck body builder. The most popular style is known as a low-cab-forward, in which the cab sits slightly

ahead of the engine. Many of the components required to work with a particular kind of garbage body are available directly from the truck manufacturer. Other components must be installed by the garbage body builder. This work is done in a separate area of the plant. It often includes mounting a hydraulic fluid tank and filters on the truck frame, installing a hydraulic pump and power-take-off on the side of the transmission, and mounting controls and instruments in the cab to operate the various hydraulic cylinders on the body.

Mounting the body

13 When the body is completed and the truck cab and chassis have been properly modified, the body is lifted and bolted to the truck frame rails. The rear mounting brackets are usually bolted tight, while the front mounting brackets are usually attached through springs which permit the frame rails to move slightly relative to the body. This allows the truck frame to flex slightly when traveling over uneven ground at landfill sites. If the rails were held rigidly against the body, the resulting stress forces could cause the frame rails to break.

14 The electrical wires and hydraulic hoses are connected between the truck and the body. The hydraulic tank and lines are filled with hydraulic fluid.

15 As a final test, all the lights are checked, and the hydraulic cylinders are actuated through their full cycle. The truck is then driven outside to await delivery to the customer.

Quality Control

Each component part is checked for dimensional accuracy before it is assembled. During welding, parts are located by templates or jigs and are clamped in place. After the body is mounted on the truck, all lights and hydraulic components are given an operational test to ensure they are functioning properly.

The Future

As disposal sites near urban areas fill up, there will be a two-pronged effort to deal with trash. One prong will be the transportation of trash to even more remote sites for disposal. These sites may be hundreds of miles away, and the trash will have to be transferred from the collection vehicles to larger vehicles for the final leg of its journey. Some western cities are even considering daily trash trains to haul refuse to abandoned open-pit mines far out in the desert.

The other prong will be continued efforts to reduce the amount of trash going to the landfill. Recycling efforts will be an important part of this effort, and will require additional separation of discarded materials and additional specialized trucks to pick up these materials. In some neighborhoods, each house currently has one container for yard wastes, such as lawn clippings, leaves, and small prunings; another container for newspapers; a third container for bottles, cans, and milk cartons; and a fourth container for magazines, junk mail, and other paper. A fifth container is provided for all non-recyclable trash destined for the landfill.

In order to encourage people to cut back on non-recyclable materials, some cities are considering the use of an electronic device to make each household "pay-by-the-pound" for this trash. A tiny electronic chip with the name and address of the household would be embedded in each trash can. When the trash is collected, a robot arm would grasp the can, and a sensor would read the information from the chip. The can would then be dumped into a hopper, which would weigh the trash. The weight and the name and address would then be recorded in a computer onboard the truck. At the end of each day, this information would be downloaded to a central computer, which would accumulate it and generate the monthly or quarterly bill for that household.

Where to Learn More

Books

Hadingham, Evan and Janet. *Garbage!: Where It Comes From, Where It Goes.* Simon & Schuster Inc., 1990.

Murphey, Pamela. *The Garbage Primer.* Lyons & Burford, 1993.

—*Chris Cavette*

Gas Mask

Background

A gas mask is a device designed to protect the wearer from noxious vapors, dust, and other pollutants. Masks may be designed to carry their own internal supply of fresh air, or they may be outfitted with a filter to screen out harmful contaminants. The latter type, known as an Air Purifying Respirator (APR), consists of a tight-fitting face piece that contains one or more filter cartridges, an exhalation valve, and transparent eye pieces. The first APR was patented in 1914 by Garret Morgan of Cleveland, Ohio, an African American inventor also credited with major improvements in the traffic signal. When the Cleveland Waterworks exploded in 1916, Morgan showed the value of his invention by entering the gas-filled tunnel under Lake Erie to rescue workers. Morgan's device later evolved into the gas mask, used in World War I to protect soldiers against chemicals used in warfare.

Since that early time, there have been significant advances in gas mask technology, particularly in the area of new filtration aids. In addition, masks have been made more comfortable and tighter fitting with modern plastics and silicone rubber compounds. Today APRs are used to filter many undesirable airborne substances, including toxic industrial fumes, vaporized paint, particulate pollution, and some gases used in chemical warfare. These masks are produced in several styles, some that cover only the mouth and nose and others that cover the entire face, including the eyes. They may be designed for military as well as industrial use but, even though the two types are similar in design, the military masks must meet different standards than those used in industry. This article will focus on manufacture of the full face type of mask used for industrial applications.

Raw Materials

A full-face gas mask consists of a filter cartridge, flexible face covering piece, transparent eye lenses, and a series of straps and bands to hold the device snugly in place. The filter cartridge is a plastic canister 3-4 inches (8-10 cm) across and 1 inch (2.5 cm) deep, which contains a filtration aid. Carbon based filtrants are commonly used because they can adsorb large quantities of organic gases, especially high molecular weight vapors like those used in chemical warfare. However, inorganic vapors are not usually strongly adsorbed on carbon. The adsorptive properties of carbon can be enhanced by impregnating the particles with specific reactants or decomposition catalysts. Such chemically treated carbon is known as "activated carbon." The type of activated carbon employed in a given filter cartridge depends on the specific type of industrial contaminant to be screened. For example, carbon treated with a combination of chromium and copper, known as "Whetlerite carbon," has been used since the 1940s to screen out hydrogen cyanide, cyanogen chloride, and formaldehyde. Today, due to concerns about chromium toxicity, a combination of molybdenum and triethylenediamine is used instead. Other types of activated carbon employ silver or oxides of iron and zinc to trap contaminants. Sodium-, potassium- and alkali-treated carbon are used to absorb sewage vapors (hydrogen sulfide), chlorine, and other harmful gases.

Today Air Purifying Respirators (APRs) are used to filter many undesirable airborne substances, including toxic industrial fumes, vaporized paint, particulate pollution, and some gases used in chemical warfare.

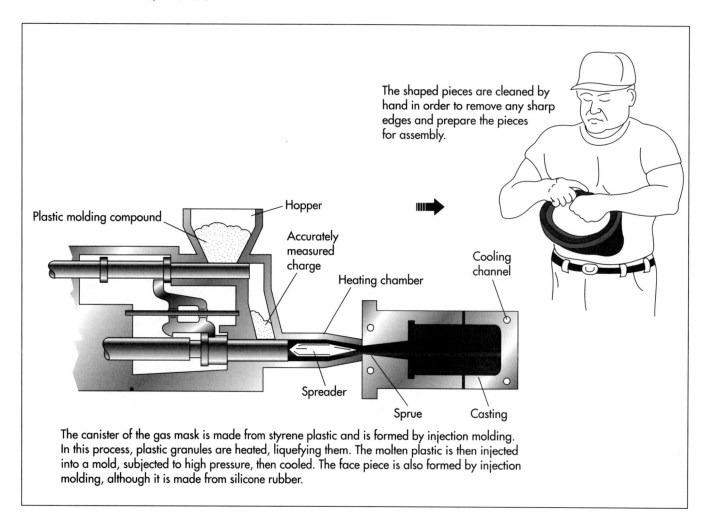

The shaped pieces are cleaned by hand in order to remove any sharp edges and prepare the pieces for assembly.

Plastic molding compound

Hopper

Accurately measured charge

Heating chamber

Cooling channel

Spreader

Sprue

Casting

The canister of the gas mask is made from styrene plastic and is formed by injection molding. In this process, plastic granules are heated, liquefying them. The molten plastic is then injected into a mold, subjected to high pressure, then cooled. The face piece is also formed by injection molding, although it is made from silicone rubber.

The "skirt," or face-covering piece, of the mask is used to hold the other components in place and to provide a secure seal around the face area. Depending upon mask design, an exhalation valve may be inserted in the face piece. This one-way valve allows exhaust gases to be expelled without allowing outside air into the mask.

The eyepieces used in gas masks are chemically resistant, clear plastic lenses. Their main function is to ensure the wearer's vision is not compromised. Depending on the industrial environment in which the mask is to be used, the eyepieces may have to be specially treated to be shatterproof, fog resistant, or to screen out certain types of light. Most gas mask manufacturers do not make their own eyepieces; instead they are molded from polycarbonate plastic by an outside supplier and shipped to the manufacturers for assembly.

The elastic straps that hold the mask on the face are typically made of silicone rubber.

Supplementary straps may be added to allow the mask to be comfortably hung around the neck during breaks in work.

Design

The design of the mask itself varies by the industrial application. Some masks are designed with speech diaphragms, some are built to accept extra filters, and others are made to be connected to an external air supply. Although the fundamental design does not vary for a given type of mask, the kind of filtrant used will vary depending on the product's intended use. Manufacturers stock a variety of mask styles and cartridge filtrants. When they receive orders for a specific type of mask, they can custom design a mask that has the appropriate features.

The Manufacturing Process

1 The canister is made from styrene plastic, which is resistant to water and other

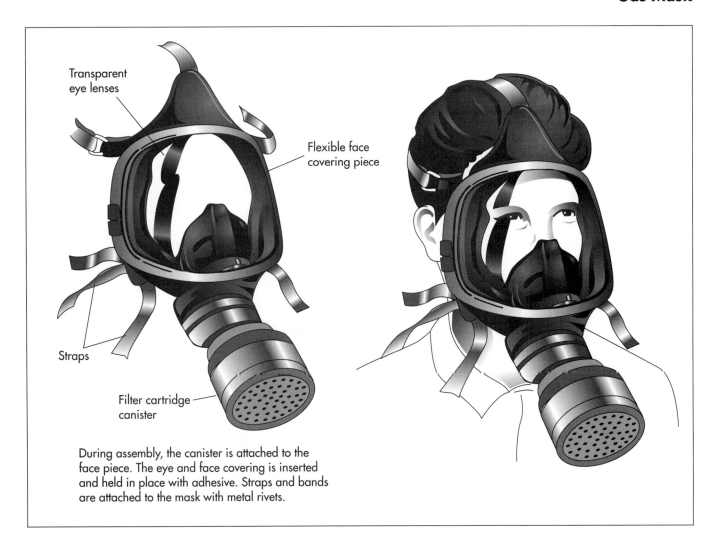

Transparent
eye lenses

Flexible face
covering piece

Straps

Filter cartridge
canister

During assembly, the canister is attached to the
face piece. The eye and face covering is inserted
and held in place with adhesive. Straps and bands
are attached to the mask with metal rivets.

chemicals, has good dimensional stability, and is specially designed for injection molding. Injection molding is a process by which molten plastic is injected into a mold under high pressure. The mold used for gas mask canisters consists of two disk-shaped pieces of metal that are clamped together. The plastic resin is liquefied by heating and then injected into the mold via an injection plunger. The mold is then subjected to high pressure. Most injection machines compress the mold with a pressure ranging from 50-2,500 tons (51-2,540 metric tons). After the molten plastic has been compressed, cooling water is forced through channels in the mold to cool and harden the plastic. The pressure is released, the two halves of the mold are separated, and the finished canister is ejected.

Styrene is a thermoplastic resin, which means it can be repeatedly remelted, so the scrap pieces can be reworked to make additional canisters. Therefore, there is very lit-

tle wasted plastic in this process. A similar molding process is conducted to create small circular screens that fit inside the canister. The screens are designed to hold the activated carbon in place inside the cartridge. As the canisters travel down the assembly line, one screen is inserted, the canister is filled with the appropriate filtrant, then the second screen is put into place.

2 The face piece is injection molded from silicone rubber. Silicone rubber has outstanding stability, is resistant to high temperatures, and can conform to curves in the face and head. It is also thermoplastic and can remolded as necessary. The molding process is very similar to the one described above. After molding the skirt must be removed from the mold, and any rough edges must be cleaned off by hand before the other components can be attached.

3 The pieces are assembled on a partially automated assembly line with two to

four line workers supervising the process. The completed filter canister is attached to the face piece and the eyepieces are inserted and held in place with adhesive. Finally the straps and bands are attached to the face piece with metal rivets. When assembly is complete, the mask is given a final quality check. When the masks pass inspection, they are identified with the appropriate markings in accordance with the American National Standard for Identification of Air Purifying Respirator Cartridges and Canisters. The finished masks are packaged for shipping. The containers used to package the masks must also designate the identity of the mask. Furthermore, they must be designed for easy access if the masks might be used in the event of an emergency.

Byproducts/Waste

Depending on the type of chemical treatment the activated carbon has been exposed to, it may be classified as chemical waste. This is the case with some filtrants, such as chromium-treated carbon. The injection molding process used for the canisters and the face pieces generates little waste since any lost resin can be remelted and used again. The lenses are manufactured by an outside vendor, so gas mask manufacturers do not have to address the issue of waste polycarbonate.

Quality Control

Gas masks, and air purifying respirators in general, are regulated by the Code of Federal Regulations (CFR). These regulations specify the type of masks to use for a specific application. Examples of the different mask types recognized by the CFR include self-contained breathing apparatus, non-powered air purifying particulate respirators, chemical cartridge respirators, and dust masks. The regulations stipulate the exact kind of testing that must be done to ensure the quality of the finished product. The type of testing depends on the masks' final application, that is, what kind of contaminants it will be expected to filter. The CFR specifies the types of contaminants that the gas must be tested with, and it also stipulates the conditions under which the testing must be conducted. For example, some masks must be exposed to the contaminant for long periods of time. Others

must be tested under specific temperature and humidity conditions. This is done by drawing an air stream contaminated with a known amount of poison through the mask. The amount of time required for the contaminant to saturate the filter and begin to pass through is then measured.

Testing is done at several points in the manufacturing process. There is an initial inspection of incoming goods to ensure they meet minimum quality specifications. This includes the filtrants, the resins used for molding, and the finished eyepieces as they are received. The canister must be tested after assembly to ensure it has proper seal and that the carbon filter works. The mask is tested once again after all componentry has been assembled. The final mask may be placed on a mannequin head to ensure that the seal is tight and that the mask maintains its seal in movement.

The Future

Over the last 80 years, the basic technology of gas masks has been tested repeatedly, and so is not likely to change in the future. The challenge for the APR industry will be to develop products for special purposes, such as infant respirators or masks for persons with head wounds and other disabling injuries. The future of these products also relies on advances in the material sciences, which allows production of smaller, more lightweight products. In fact, current research efforts in carbon chemistry are anticipated to result in the development of a filter canister that is only half the size of the current standard and is more effective. These and other improvements in materials will result in new generations of respirator devices for industrial use, as well as for medical and military applications.

Where to Learn More

Books

Ahmstead, B.H. *Manufacturing Processes.* John Wiley and Sons, 1977.

Other

Laboratory for National Testing of Gas Masks. http://www.niih.go.jp:80/guide/english/ profile/gasmask/gasmask.htm (July 9, 1997).

—*Randy Schueller*

Golf Ball

Background

Golf, a game of Scottish origin, is one of the most popular sports in the world. In the United States alone more than 24 million people play golf, including over 8,000 professional players. Golf tournaments around the world are popular with spectators, as well as with players, and since the 1960s, they have received wide television coverage. There is now even a cable channel devoted to golf, as well as numerous computer games.

The basic game involves using a variety of clubs to drive a small ball into a succession of either nine or 18 holes, over a course designed to present obstacles, in as few strokes as possible. A player is permitted to carry a selection of up to 14 clubs of varying shapes, sizes, and lengths. The standard golf ball used in the United States is a minimum of 1.68 in (4.26 cm) in diameter; the British ball is slightly smaller.

A golf course generally has 18 holes spread over a landscaped area that includes a number of hazards, including water, sand traps or bunkers, and trees. Difficulty is increased by varying distances among holes. Play on each hole is begun at the tee area, from which players drive the ball into the fairway. Each hole can vary in length from about 150-600 yards (135-540 m); successful players are those who are able to drive the ball more than 200 yards (180 m) from the tee, approaching most holes with fewer than three shots. At the end of the hole is the putting green, where the ball must be putted into the hole or cup to complete the hole.

Golf is usually played by groups of two to four people who move throughout the course together. The ball must be played from where it lies, except in specific circumstances. In stroke competition, the total number of strokes used to move the ball from the tee to the hole is recorded as the players' score for that individual hole. The player who uses the fewest strokes to complete the course is the winner. In match play, scores are compared after every hole, and a player wins, loses, or halves (ties) each hole.

Each hole must be reached in a specific number of shots (par), which usually depends on length. A birdie is a score on any one hole that is one stroke less than par, and an eagle is a score on a hole that is two less than par. A hole in one is scored when the player drives the ball into the hole with only one stroke.

Today, the golf ball market is worth around $550 million in annual sales, with over 850 million golf balls being manufactured and shipped every year. Currently, balls are made in two or three parts. A two-piece ball is made of rubber and plastic, and is mostly used by the casual golfer. These balls last a lot longer than the three-piece balls the pros use and hence make up 70% of all golf ball production. A three-piece ball consists of a plastic cover, windings of rubber thread, and a core that contains a gel or liquid (sugar and water) or is solid. A dimple pattern on the surface results in good flight performance.

The most common dimple patterns are the icosahedral, the dodecahedral, and the octahedral. The icosahedral pattern is based on a polyhedral with 20 identical triangular faces, much like a 20-sided die. Similarly, a dodecahedral is based on a polyhedral with 12 identical faces in the shape of pentagons. The octahedral is based on an eight-sided polyhedral with triangular faces. Some balls are based on the icosahedral with 500 dimples. As a general rule, the more dimples a ball has the better it flies, provided those dimples are about 0.15 in (0.38 cm) in diameter.

The size and depth of the dimples also affect performance. Shallow dimples generate more spin on a golf ball than deep dimples, which increases lift and causes the ball to rise and stay in the air longer and roll less. Deep dimples generate less spin on a golf ball than shallow dimples, which decrease lift and causes the ball to stay on a low trajectory, with less air time and greater roll. Small dimples generally give the ball a lower trajectory and good control in the wind, where as large dimples give the ball a higher trajectory and longer flight time.

Technological advances in materials and aerodynamics now allow the manufacturer to custom-fit a golf ball for a players' particular game, for weather conditions, and even for specific course conditions. Golf balls can be separated into four basic performance categories: distance and durability; control and maneuverability; distance and control; and slow clubhead speed. Within these categories there are more than 80 different balls of varying construction materials and design.

The United States Golf Association (USGA) has established rules for the ball in regard to maximum weight, minimum size, spherical symmetry, initial velocity, and overall distance. The weight of the ball must not be greater than 1.62 oz (45.93 g) and must be spherically symmetrical. The velocity shall not be greater than 250 feet (75 m) per second (255 feet [76.5 m] per second maximum) when measured on apparatus approved by the USGA. The overall distance standard states that the ball shall not cover an average distance in carry and roll exceeding 280 yards (84 m) (296.8 yards [89 m] maximum). These rules are updated every year.

Currently, there are around 850 models of balls that conform to these standards. Recently, balls that are about 2% larger than ordinary balls have been introduced that still conform to USGA rules. These balls have softer cores and thicker, harder covers, which leads to a straighter, longer shot.

History

The game of golf goes back as far as 80 B.C. when the Roman emperors played a game called paganica using a bent stick to drive a soft, feather-stuffed ball (or feathery). This ball was up to 7 in (17.5 cm) in diameter, much larger than the Scottish version. By the middle ages, the sport had evolved into a game called bandy ball, which still used wooden clubs and a smaller ball about 4 in (10 cm) in diameter.

Over the next five centuries the game developed on several continents and eventually evolved into the popular Scottish game known as golfe. Other European countries played similar games and a variation from the Netherlands was played in the American colonies as early as 1657. Although various types of wood, ivory, linen, and even metal balls were tried during the sport's early development in Europe, the feathery remained the ball of choice.

The Scottish game, however, is the direct ancestor of the modern game. The first formal golf club was established in Edinburgh in 1744. It established the first set of rules, which helped eliminate local variations in play. A decade later the Royal and Ancient Golf Club was established at Saint Andrews, Scotland, which became the official ruling organization of the sport. Its rules committee, along with the United States Golf Association (USGA), still rules the sport. A British player, Harry Vardon, helped popularize the sport in the United States during the late 1880s, although legend has it that a Scotsman named Alex McGrain was the first to play golf on the North American continent in eastern North Carolina over a hundred years earlier. The first American-made golf ball was produced by Spalding in 1895.

The first golf ball similar in size to today's came into existence around five or six hundred years ago, when the Dutchmen stuffed feathers into an 1.5 in (3.75 cm) leather pouch. This type of ball lasted for about 450 years. To make a feathery, the ballmak-

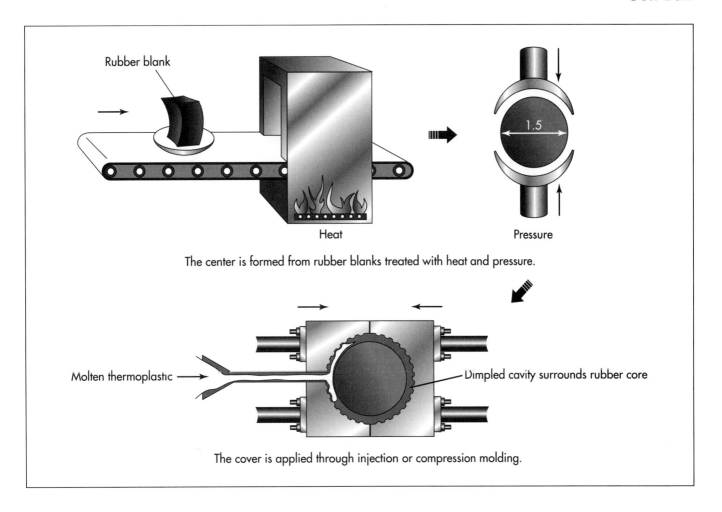

Rubber blank

1.5

Heat

Pressure

The center is formed from rubber blanks treated with heat and pressure.

Molten thermoplastic → Dimpled cavity surrounds rubber core

The cover is applied through injection or compression molding.

er stitched together a round pod made from strips of bull or horse hide that had been softened into leather. The pod was turned inside out, carefully leaving a small opening into which goose or chicken feathers were stuffed. In order to retain a spherical shape, the ballmaker used a leather cup as a crude mold. The opening was stitched up, the ball dried, hammered into a round shape, and rubbed with oil and chalk.

Finished featheries were made in different diameters and weights and were graded according to weight (measured in drams). Ballmakers determined the size and weight of each ball by adjusting the lengths and thickness of the leather used for the cover. Typically, feathery balls were made in the range of 20-29 drams. The featheries were first numbered according to their size and later according to diameter rather than weight. This numbering system has continued into the twentieth century.

The feathery was replaced when a much cheaper ball made out of gutta-percha, a

natural gum from Southeast Asia, was developed around 1850 in Italy. To make a gutta percha ball or gutty, a slice of resin rope that had been pre-mixed with a stabilizer was heated to make it pliable and then shaped into a sphere. Despite being rounder and smoother than the feathery, this ball had poorer flight performance. However, the new ball's affordability (dozens could be made per day instead of just a handful) made it practical for the working class to take up the sport in large numbers and this ball remained popular until about 1910.

The gutty ball went through several transformations during this time. Once ballmakers discovered that a rough surface was better aerodynamically, grooves were cut in the balls with a knife to simulate the stitching of the feathery. Next, the ballmakers pounded the ball with a chisel-faced hammer to produce nicks and bruises on the surface.

Further experimentation with the gutty through the mid-nineteenth century sought to improve the ball's flight performance.

The dimple pattern on the golf ball surface results in good flight performance. As a general rule, the more dimples a ball has the better it flies, provided those dimples are about 0.15 in (0.38 cm) in diameter.

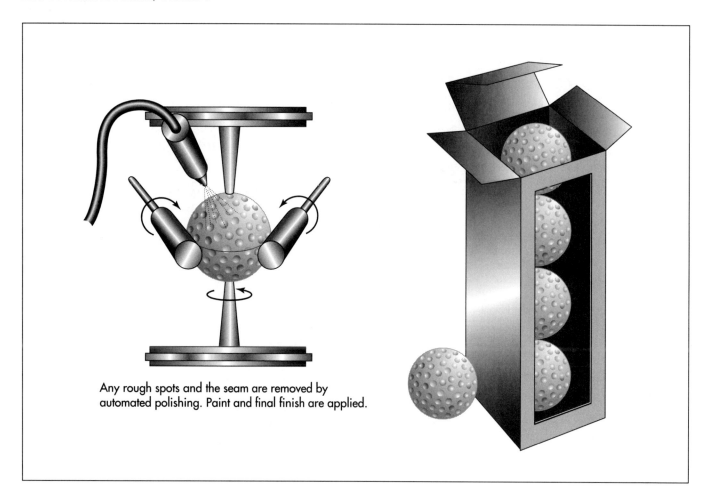

Any rough spots and the seam are removed by automated polishing. Paint and final finish are applied.

Ballmakers tried incorporating other substances such as cork dust, India rubber, bits of leather, and other materials into the pure gutta percha before shaping the ball. Though these balls were more durable, they lacked capacity for distance.

By the end of the 1870s, machined iron molds that had regular patterns inscribed on their inside were developed. One of the most popular of these was the brambleberry design with raised dimples. These molds created a regular pattern over the surface, eliminating hammering by hand. This refinement began a revolution in aerodynamic design for the golf ball. The rate of manufacture improved even further.

The game changed considerably in the early twentieth century when the B. F. Goodrich Company in Akron, Ohio, invented a lighter, tightly wound, rubber-threaded ball. The recessed dimpled ball was introduced by Spalding in 1908 and proved to be both aerodynamically and cosmetically a success. By 1930, it dominated the market,

with the spherical dimple becoming the standard. Other dimple shapes have since evolved, including truncated cone and elliptical dimples.

Raw Materials

A golf ball is made up of mostly plastic and rubber materials. A two-piece ball consists of a solid rubber core with a durable thermoplastic (ionomer resin) cover. The rubber starts out as a hard block, which must be heated and pressed to form a sphere.

The three-piece ball consists of a smaller solid rubber or liquid-filled center with rubber thread wound around it under tension, and an ionomer or balata rubber cover.

During the 1970s the interior of the ball improved further, thanks to a material called polybutadiene, a petroleum-based polymer. Though this material produced more bounce it was also too soft. Research at Spalding determined that zinc strengthened the material. This reinforced polybutadiene soon became widely used by the rest of the manufacturers.

The Manufacturing Process

Three-piece golf balls are more difficult to make and can require more than 80 different manufacturing steps and 32 inspections, taking up to 30 days to make one ball. Two-piece balls require about half of these steps and can be produced in as little as one day.

Forming the center

1 The center of the two-piece ball is a molded core. It is a blend of several different ingredients, all of which are chemically reactive to give a rubber type compound. After heat and pressure is applied, a core of about 1.5 inches (3.75 cm) is formed.

Forming the cover and dimples

2 Injection molding or compression molding is used to form the cover and dimples on a two-piece ball using a two-piece mold. In injection molding, the core is centered within a mold cavity by pins, and molten thermoplastic is injected into the dimpled cavity surrounding the core. Heat and pressure cause the cover material to flow to join with the center forming the dimpled shape and size of the finished ball. As the plastic cools and hardens, the pins are retracted and the finished balls are removed.

3 With compression molding, the cover is first injection molded into two hollow hemispheres. These are positioned around the core, heated and then pressed together, using a mold which fuses the cover to the core and also forms the dimples. Three-piece balls are all compression molded since the hot plastic flowing through would distort and probably cause breaks in the rubber threads.

Polishing, painting, and final coating

4 "Flash" or rough spots and the seam on the molded cover are removed. Two coats of paint are applied to the ball. Each ball sits on two posts, which spins so that the paint is applied uniformly. Spray guns that are automatically controlled are used to apply the paint. Next, the ball is stamped with the logo. The final step is the application of a clear coat for high sheen and scuff resistance.

Drying and packaging

5 After the paint is applied, the balls are loaded into containers and placed in large dryers. After drying, the balls are ready for packaging in boxes and other containers.

Quality Control

In addition to monitoring the manufacturing process using computers and monitors, three-piece balls are x-rayed to make sure the centers are perfectly round. Compression ratings are also used to measure compression-molded, wound golf balls. These ratings have no meaning when applied to two-piece balls, however. Instead, these balls are measured by a coefficiency rating, which is the ratio of initial speed to return speed after the ball has struck a metal plate. This procedure measures the coefficient of restitution.

Mechanical testing is also used to verify that the ball's performance meets the USGA's standards. Special equipment has been developed and some manufacturers even use wind tunnels to determine wind resistance and lift action. A machine called the True Temper Mechanical Golfer or Iron Byron, modeled after the swing of golf legend Byron Nelson, can be fitted for any club and can be set up at various swing speeds. For normal testing, the Iron Byron is configured using a driver, 5 iron, and 9 iron.

Another machine called the Ball Launcher provides the capability to propel balls through the air at any velocity, spin rate, and launch angle. This has the advantage of using launch conditions typical of a wide cross-section of golfers. Using both types of equipment, performance data associated with the flight of a golf ball can be measured and analyzed. These include the apogee angle, carry distance, total distance, roll distance, and statistical accuracy area.

The apogee angle indicates the height the trajectory of a ball reaches. It is measured using a camera with a telescopic lens pointing down range in conjunction with a gridded monitor. Carry distance is the distance a golf ball travels in the air and is measured using a grid system with markers in the landing zone. Total distance is the distance a golf ball travels in the air plus the roll distance. Roll distance is the total distance minus the carry distance.

The statistical accuracy area (SAA) or dispersion area is used as a measure of a golf ball's accuracy. For a given ball, the SAA value is based on the deviations of the ball's performance in the directions of carry and left/right of the centerline. These deviations are used to calculate an equivalent elliptical landing area.

The Future

As improvements in aerodynamic design continue, golf balls will be able to go even further. In fact, one golf ball manufacturer is already advertising that its balls can be driven 400 yards. However, some professional players are complaining that golf balls go too far and want the ball adjusted back about 10%. This means the USGA would have to tighten current requirements for carry and roll and for velocity in its ball-testing procedure. A 10% cutback would reduce drives by most tour pros by approximately 25 yards (22.5 m).

On the other hand, some experts believe that golf balls have reached their limit on distance and will not improve in this area over the next 20 years. Golf manufacturers will be challenged to achieve the ultimate consistency from one ball to the next, make balls that feel softer and stop faster on the greens, develop balls with greater durability, and invent the perfect dimple pattern. Space age materials may achieve some of these goals and metal matrix composites based on titanium are being considered. In addition, golf ball companies will have to manufacture more balls for specific categories of golfers. For example, four or five different types of trajectories might become available.

Where to Learn More

Books

New Trends in Golf Balls. Wilson Sporting Goods Co., Golf Division, 1997.

Periodicals

Achenbach, James. "Golf not ready to succumb to technology." *Golfweek,* February 22, 1997.

"Ancient spheres." *Golf Magazine,* April 1992, p. 178.

Braham, James. "All this for a golf ball?" *Machine Design,* December 12, 1991, p. 121.

Robinson, Bob. "Some PGA tour players renew call for shorter golf balls." *The Oregonian,* May 24, 1995.

Stogel, Chuck. "Big time wars in golf balls drive still-thriving industry." *Brandweek,* January 24, 1994, p. 30.

—*Laurel M. Sheppard*

Graham Cracker

Background

Graham crackers and related animal crackers are whole wheat crackers made with a special type of flour. They are slightly sweetened with sugar and honey and are sold in a variety of sizes and shapes. First developed in 1829, they remain a popular snack food, and millions of crackers are sold each year.

The development of the graham cracker is attributed to Sylvester Graham, an American clergyman. In 1829, he concocted the recipe for a cracker whose main ingredient was an unsifted, coarsely ground whole wheat flour. Touting his product as a health food, he produced and sold it locally. Over time, it became known the graham cracker. Due to its popularity and innovation, other bakeries copied his recipe and eventually developed methods for its mass production. Since then, graham crackers have been a popular snack food. They have also become an important ingredient in pie crust recipes.

From a recipe standpoint, animal crackers are very much like graham crackers. The primary difference between the two is the shape of the final product. Whereas graham crackers are typically square, animal crackers come in the shape of lions, tigers, camels, bears, and giraffes, to name a few. They were developed in England in the late 1800s and were initially imported to the United States. As their popularity grew, American bakeries began making them. A true innovation in the development of this product came from the National Biscuit Company, who packaged the crackers in a colorful box made to look like a circus wagon. This method of selling the product proved popular and spawned hundreds of variations on this theme. In the late 1950s, production technology improved, and the level of detail on animal crackers greatly increased.

Raw Materials

The recipe for graham crackers has remained essentially unchanged since its invention in 1829. The primary ingredients include whole-wheat flour, fat, and sugar. These, combined with other ingredients, provide the essential graham cracker characteristics.

Flour

The main component of most cracker recipes is wheat flour, which is obtained by grinding wheat seeds into a powder. Whole-wheat flour is composed of the three main parts of the wheat seed, the outer coat or bran, the germ, and the endosperm. The bran and germ are larger particles which add flavor, fiber, and color to the flour. The endosperm is responsible for the important baking characteristics. It is primarily composed of starch and protein, which when combined with water creates a mass, called gluten, that can be stretched and rolled without breaking. This property allows dough to be formed into various sizes and shapes.

The distinctive flavor and texture of graham cracker flour comes from the size of the flour particles used. For the correct taste, the flour must have the correct combination of small, medium, and large particles. If this combination is not right, the crackers will either turn out crumbly or have lumps.

Fats and oils

Fats and oils are another primary ingredient used in cracker manufacturing. They can be

The development of the graham cracker is attributed to Sylvester Graham, an American clergyman. In 1829, he concocted the recipe for a cracker whose main ingredient was an unsifted, coarsely ground whole wheat flour.

A 1959 graham cracker magazine advertisement. (From the collections of Henry Ford Museum & Greenfield Village.)

The graham crackers we enjoy today are a far cry from the whole wheat graham cracker that Sylvester Graham developed over 150 years ago. Graham was an early nineteenth century health reformer who lectured on the evils of meat, alcohol, fat, and processed grains such as refined flour. He urged housewives to bake using whole meal wheat flour (with the bran left in the flour) because whole meal or Graham flour rendered first-rate digestive foods. Thus, Graham's name was given to the biscuits, breads, and crackers made from this Graham flour. The true Graham crackers were baked without fat and refined sugar.

Sylvester Graham, born in Connecticut in 1794, was always sickly, small, and suffered from mental breakdowns. He believed a healthful diet would mend his body and mind. By the late 1820s, Graham had turned from foods he considered unhealthy, preferring unprocessed foods, vegetables, and water to refined grains, meat, and alcohol. Graham publicly denounced commercial bakeries, proclaiming their products tainted. Instead, he suggested mothers return to the kitchen and bake bread from whole meal wheat flour rather than purchase inferior products made from refined white flour. Mid-nineteenth century recipe books instructed housewives in testing for fresh Graham flour and featured baked goods of Graham flour— griddle cakes, Graham biscuits, and crackers. The graham cracker was only one such recipe, which later was mass produced and sold by large bakeries, the kind of bakery Graham would have denounced.

Graham crackers now contain whole wheat flour and other ingredients such as sugar and shortening.

Nancy EV Bryk

derived from a variety of plant and animal sources. Graham cracker recipes typically require hydrogenated vegetable shortening composed of soybean and cottonseed oil. Most of the naturally strong flavor of these oils is removed during the refining process. Butter can also be used. However, its flavor is retained during manufacturing.

There are many characteristics which make fats and oils important in graham cracker recipes. One characteristic is their insolubility in water. When water is added to flour, gluten is typically formed. But when fats and oils are present, they act as a barrier between the flour and water, and gluten formation is prevented. This "shortened" batter results in products that have a soft, crumbly texture. Using fats and oils improve the appearance of crackers and contribute to the taste.

Sweeteners

Graham crackers have a slightly sweet flavor. The primary sweetener is sugar, or sucrose, that is derived from sugar cane or sugar beet. It typically makes up about 5-15% of the recipe. Other sweetening ingredients used are dextrose, corn syrup, molasses, and honey. In addition to adding flavor, these ingredients have the extra benefits of improving the texture, affecting the color, contributing to the aroma, and preserving the product.

Other ingredients

Beyond the primary cracker ingredients, many other materials are added to give graham crackers their unique taste and texture. Cinnamon and salt contribute to the taste of the crackers. Whey is often added to ameliorate flavors without adding much flavor of its own. Leavening ingredients like sodium bicarbonate or sodium acid pyrophosphate give off carbon dioxide when mixed in the dough and are responsible for the air pockets throughout the cracker. Lecithin, which is derived from soybean oil, is used to make manufacturing easier by reducing the stickiness of the batter.

The Manufacturing Process

Graham crackers are made through a series of steps which convert the raw ingredients

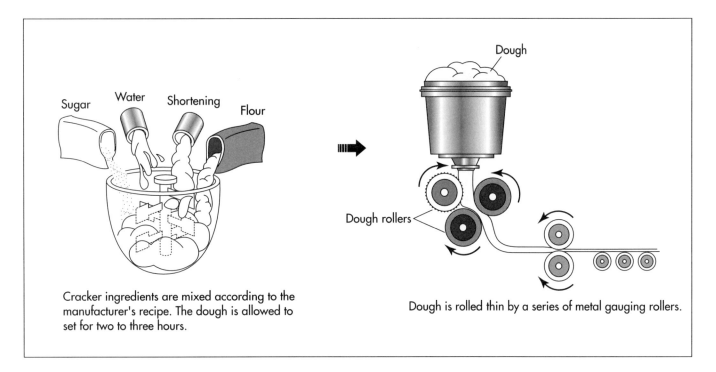

Cracker ingredients are mixed according to the manufacturer's recipe. The dough is allowed to set for two to three hours.

Dough is rolled thin by a series of metal gauging rollers.

into finished products. Important steps include ingredient handling, compounding, forming or machining, baking, post conditioning, and packaging.

Ingredient handling

1 Most of the major ingredients like flour, vegetable shortening, and sugar are delivered to cracker manufacturers in large quantities and stored in bulk tanks. Depending on the ingredient, these tanks may be fashioned with special equipment to control their internal environments. For example, a liquid such as vegetable shortening must be stored at a specific temperature until it is ready to be used. If the temperature varies too much, the shortening could be difficult to pump or could adversely affect the dough. Therefore, this tank has controls which can maintain the appropriate temperature. Other tanks may have refrigeration capabilities.

2 At the start of cracker production, ingredients are transferred to mixing tanks. Liquid bulk ingredients are transferred using metered pumps, which can move a specific quantity of material. Bulk solid materials are pumped via a method called pneumatic transfer, which involves fluidizing the powders with a stream of air and then pumping them with metered pumps. Minor ingredients that are supplied in boxes, bags, and drums are weighed and poured into the mixing tank by hand.

Compounding

3 The graham cracker ingredients are mixed together in specific quantities as specified in the recipe. Cracker doughs are mixed with either vertical spindle mixers or high-speed horizontal drum mixers. The order that ingredients are added to the mixture is important. Typically the process begins with sugar, water, and shortening. This forms a mixture with a cream-like consistency. Next the remaining ingredients are added, and a "short" dough is obtained. This dough is allowed to set for two to three hours for the leavening agents to work.

Machining

4 Graham crackers are usually sold in two forms, as squares or as animal crackers. The dough used for both is ostensibly the same. In the machining process, the dough is delivered from a hopper onto a conveyor belt and rolled thin by a series of metal gauging rolls. The thickness of the sheet is reduced by each of these rollers. Some manufacturers stack multiple sheets on top of each other in a process known as laminating. They are rolled out further, allowed to relax, and then sent along a conveyor belt to the cutting machines.

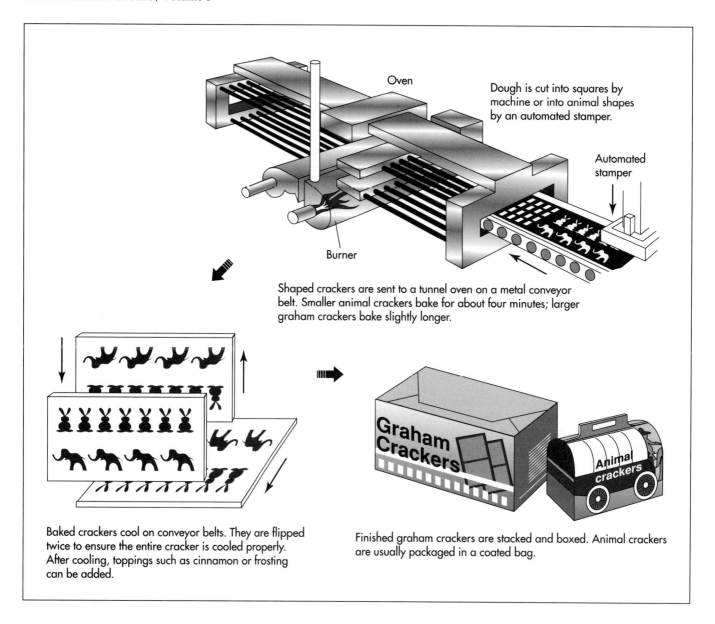

Oven

Dough is cut into squares by machine or into animal shapes by an automated stamper.

Automated stamper

Burner

Shaped crackers are sent to a tunnel oven on a metal conveyor belt. Smaller animal crackers bake for about four minutes; larger graham crackers bake slightly longer.

Graham Crackers

Animal crackers

Baked crackers cool on conveyor belts. They are flipped twice to ensure the entire cracker is cooled properly. After cooling, toppings such as cinnamon or frosting can be added.

Finished graham crackers are stacked and boxed. Animal crackers are usually packaged in a coated bag.

5 The edges of the dough sheets are cut smooth by rotary cutting machines, and excess dough is sent back to the hopper for reuse. When animal crackers are made, the sheet of dough is cut into the various animal shapes by cutting machines called stampers. These stampers, or rotary dies, have the animal shapes fashioned on them with intricate details. After this stage, sugar, cinnamon, or honey are applied to the top of the dough if the recipe requires it.

Baking

6 The crackers are baked in a tunnel oven. The dough is first transferred to a metal conveyor belt and then moved through the oven, which can be 100-300 feet (30-90 m) long. Baking takes place in three stages

called development, drying, and coloring. In the development stage the dough sets, taking on the size and shape of the final product. The greatest amount of water is lost in the drying stage. In the coloring stage, the dough is changed from pale white to a light golden brown. The amount of time a product spends baking is controlled by the speed of the moving conveyor belt. Animal crackers bake for as little as four minutes. Graham crackers bake slightly longer.

Post conditioning

7 After the crackers come out of the oven, they travel on a series of conveyors to cool. At some point in this process they are flipped over and then flipped back to ensure

that cooling is throughout. The total cooling time can be twice as long as the baking time.

8 Depending on the recipe, other coatings can be put on the crackers after they have cooled. These include such things as icings, chocolate coatings, or sugar. These can be applied using either stenciling, extruders, or depositors. Excess toppings are removed using forced air and vibrating shaker devices.

Packaging

9 The final step in the manufacturing process is packaging. Because of the fragile nature of some crackers, the packaging must be rigid and airtight. For square graham crackers, the crackers are cut and stacked individually and wrapped in flexible films. Animal crackers, which are less fragile, are typically packaged in a coated bag. Both types of crackers are then put inside boxes which are appropriately decorated to make the product appealing. These boxes are put into larger case boxes, which can be stacked on pallets and shipped to stores.

Quality Control

Quality control begins with the evaluation of incoming raw materials. Before they are allowed for use, these ingredients are tested in the Quality Control lab to ensure they conform to product specifications. Various sensory characteristics are checked, including appearance, color, odor, and flavor. Many other characteristics, such as the particle size of solids, viscosity of oils, and pH of liquids, are also studied. Each bakery relies on these tests to certify that the ingredients will produce a consistent, quality batch of graham crackers.

Various characteristics of each batch of final product is also carefully monitored to ensure that every graham cracker or animal cracker shipped to stores is of the same quality as the batches developed in the food laboratory. Quality control chemists and technicians check physical aspects of the crackers, including appearance, flavor, texture, and odor. The usual method of checking these characteristics is by comparing them to an established standard. For example, the color of a random sample is compared to a standard set during product development. Other qualities, such as taste, texture, and odor are evaluated by sensory panels. These are made up of a group of people who are specially trained to notice small differences in these characteristics. In addition to sensory tests, many specialized instrumental tests are also performed.

The Future

The trend in graham cracker products has been toward products which contain premium ingredients, are healthier for the consumer, or have unusual shapes. Graham cracker marketers have tended to tout organic ingredients in the recipes. Others have begun to use a low-fat recipe and make other "healthy" claims. Additionally, new flavors of graham crackers are constantly being introduced in the hopes that they will catch on and sustain sales over many years.

Where to Learn More

Books

Almond, N. *Biscuits, Cookies and Crackers: The Biscuit Making Process.* Elsevier Applied Science, 1989.

Booth, R. Gordon. *Snack Food.* Van Nostrand Reinhold, 1990.

Faridi, Hamed. *The Science of Cookie and Cracker Production.* Chapman & Hall, 1994.

Periodicals

Dornblaser, Lynn. "Everything they're cracked up to be. (new cracker products)" *Bakery Production and Marketing,* August 15, 1996, p. 26.

—*Perry Romanowski*

Gummy Candy

First developed in Germany in the early 1900s, gummy candy gained great popularity in the United States during the 1980s.

Background

Gummy candy is a unique candy composed of gelatin, sweeteners, flavorings, and colorings. Because of its nature it can be molded into literally thousands of shapes, making it one of the most versatile confection products ever. First developed in Germany in the early 1900s, it gained great popularity in the United States during the 1980s. Today, it continues to be popular, with sales totaling over $135 million in 1996 in the United States alone.

History

Gummy candy represents a more recent advance in candy technology. The technology, derived from early pectin and starch formulations, was first developed in Germany in the early 1900s by a man named Hans Riegel. He began the Haribo company, which made the first gummy bears in the 1920s. While gummy candy has been manufactured since this time, it had limited worldwide distribution until the early 1980s. It was then when Haribo began manufacturing gummy bears in the United States. The fad caught on, causing other companies to develop similar products. The gummy bears led to other types of gummy candy entries from companies such as Hershey, Brach's, and Farley's. Now, the candy is available in various different forms, from dinosaurs to fruit rolls. According to one gelatin manufacturer, nearly half of all gelatin made worldwide currently goes to making gummy candies.

Raw Materials

Gummy candy recipes are typically developed by experienced food technologists and chemists. By blending together different ingredients, they can control the various characteristics of gummy candy, such as texture, taste, and appearance. The primary ingredients include water, gelatin, sweeteners, flavors, and colors.

The main ingredient responsible for the candy's unique, gummy characteristics is gelatin. This is a protein derived from animal tissue that forms thick solutions or gels when placed in water. When used at an appropriate concentration, the gels take on the texture of the chewy, gummy candy. However, since these gels are thermoreversible, which means they get thinner as they are heated, gummy candies have a "melt in the mouth" characteristic. Both the texture and the amount of time it takes the candy to dissolve in the mouth can be controlled by the amount of gelatin used in a recipe.

Since gelatin is a tasteless and odorless compound that contains no fat, sweeteners and flavorings are added to give gummy candy its taste. Various sugars are added as sweeteners. Sucrose, derived from beets or sugar cane, provides a high degree of sweetness to the gummy candy. Fructose, which is significantly sweeter than common sucrose, is another sugar that is often used. Corn syrup is also used because it helps prevent the other sugars from crystallizing and ruining the gummy texture. Also, corn syrup helps add body to the candy, maintain moisture, and keep costs lower. Another sweetener is sorbitol, which has the added benefit of helping the candy maintain its moisture content. In addition to flavor, some of these sweeteners have the added benefit of preserving the gummy candy from microbial growth.

The sweetness of gummy candy is only one of its characteristics. Artificial and natural flavors are also used to create a unique taste. Natural flavors are obtained from fruits, berries, honey, molasses, and maple sugar. The impact of these flavors can be improved by the addition of artificial flavors that are mixtures of aromatic chemicals and include materials such as methyl anthranilate and ethyl caproate. Also, acids such as citric acid, lactic acid, and malic acid are added to provide flavor.

Gelatin gels have a natural faint yellow color, so dyes are added to create the wide array of colors found in gummy candy. Typical dyes include Red dye #40, Yellow dye #5, Yellow dye #6, and Blue dye #1. Using these federally regulated dyes, gummy manufacturers can make the candy almost any color they desire.

The textural characteristics of gelatin gels depends on many factors, such as temperature, method of manufacture, and pH. While the manufacturing method and temperature can be physically controlled, the pH is controlled chemically by the addition of acids. These include food grade acids such as citric acid, lactic acid, fumaric acid, and malic acid. Other ingredients are added during the manufacturing process as flavorants, lubricating agents, and shine enhancing agents. These include materials like beeswax, coconut oil, carnauba wax, mineral oil, partially hydrogenated soybean oil, pear concentrate, and confectioner's glaze, which are often added during the filling phase of manufacture.

The Manufacturing Process

Gummy manufacturing uses a starch molding process. First the candy is made, then it is filled into starched lined trays. The filled trays are then cooled overnight and the resulting formed candy is emptied from the trays. In the mass production of gummy candy, significant improvements have been made to increase the speed and efficiency of this process.

Compounding

1 The manufacture of gummy candy begins with compounding. Factory workers, known as compounders, follow instructions outlined in the recipes and physically pour the appropriate amount of gummy raw materials into the main mixing tanks. These tanks, which are equipped with mixing, heating, and cooling capabilities, are quite large. Depending on the size of the batch, gummy candy compounding can take from one to three hours. When the batch is complete, it is sent to the Quality Control (QC) laboratory to make sure that it meets the required specifications.

Forming candy

2 After the gummy candy is compounded and passes QC testing, it is either pumped or transferred to a starch molding machine known as a Mogul. This machine can automatically perform the multiple tasks involved in making gummy candy. It is called a starch molding machine because starch is a main component. In this machine, starch has three primary purposes. First, it prevents the candy from sticking to the candy molds, which allows for easy removal and handling. Second, it holds the gummy candy in place during the drying, cooling, and setting processes. Finally, it absorbs moisture from the candies, giving them the proper texture.

3 Making gummy candy in a Mogul is a continuous process. At the start of the machine, trays that contain previously filled, cooled, and formed gummy candy are stacked. The trays are then removed from the stack one-by-one and move along a conveyor belt into the next section of the machine, known as the starch buck.

4 As they enter the starch buck, the trays are inverted and the gummy candy falls out into a vibrating metal screen known as a sieve. The vibrating action of the sieve, in concert with oscillating brushes, removes all of the excess starch that adheres to the gummy candy. These pieces then move along a conveyor belt to trays, where they are manually transferred to other machines by which they can be decorated further and placed into appropriate packaging. A more recent advance, called the pneumatic starch buck, further automates this step. In this device, a tightly fitting cover is placed over the filled trays. When it is inverted, the candies adhere to the cover and remain in their

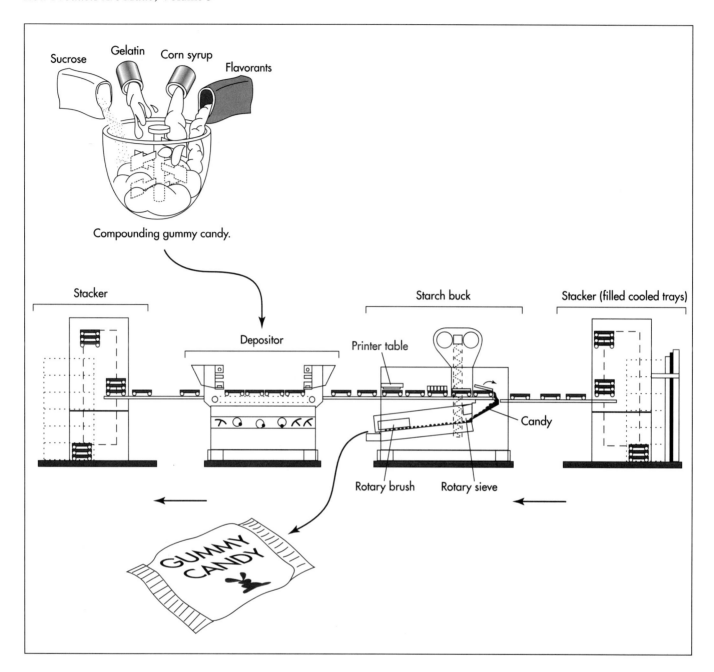

Sucrose Gelatin Corn syrup Flavorants

Compounding gummy candy.

Stacker   Starch buck   Stacker (filled cooled trays)

Depositor   Printer table

Candy

Rotary brush   Rotary sieve

GUMMY CANDY

Gummy candy is manufactured in a machine called a Mogul. Cooled trays of gummy candy are inverted in the starch buck. This candy is ready for packaging. The trays are then filled with starch to keep the candy from sticking and sent to the printer table, which imprints a pattern into the starch. The depositor fills the trays with the hot candy mixture, and the trays are sent back to the stacker to cool for 24 hours. Then the machine can start the process again.

ordered position. The excess starch is then removed by fast-rotating compressed-air jets. The candy can then be conveyed for further processing.

5 The starch that is removed from the gummy candy is reused in the process, but first it must be cleaned, dried, and otherwise reconditioned. Candy particles are first removed by passing the starch through a metal screen known as a sieve. It is then conveyed to a recirculating starch conditioning system. As it enters this machine, it is dried by being passed through hot, moving air. After drying, the starch is cooled

by cool air jets and conveyed back out to the Mogul to be reused in the starch molding process.

6 The starch returns from the drier via a conveyor belt to the Mogul, where it is filled into the empty trays and leveled. These were the same trays that were inverted and emptied in step two. These starch-filled trays then move to a printer table. Here, a board that has the inverse of the mold printed on it presses the starch down so the mold has an indent in it. From here, the trays are moved to the depositors.

7 The gummy candy, compounded in step 1, is transferred to the depositors. This is the part of the mogul that has a filling nozzle and can deliver the exact amount of candy needed into the trays as they pass under it. The depositor section of the mogul can contain 30 or more depositors, depending on how many imprints there are on the trays. In more modern depositors, the color, flavor, and acids can be added to the gummy base right in the depositor. This allows different colors and flavors to be made simultaneously, speeding up the process.

8 The filled trays are moved along to a stacking machine and then sent to a cooling room, where they stay until they are appropriately cooled and formed. This part of the process can take over 24 hours. After this happens, the trays are moved back to the Mogul, and the process starts all over again.

Quality Control

Quality control begins with the evaluation of the incoming raw materials. Before they are used, these ingredients are tested in the QC lab to ensure they conform to specifications. Various sensory characteristics are checked, including appearance, color, odor, and flavor. Many other characteristics, such as the particle size of the solids, viscosity of oils, and pH of liquids, are also studied. Each manufacturer depends on these tests to certify that the ingredients will produce a consistent, quality batch of gummy candy.

The characteristics of each batch of final product is also carefully monitored. Quality control chemists and technicians check physical aspects of the candy that include appearance, flavor, texture, and odor. The usual method of testing is to compare them to an established standard. For example, the color of a random sample is compared to a standard set during product development. Other qualities such as taste, texture, and odor are evaluated by sensory panels. These are made up of a group of people who are specially trained to notice small differences. In addition to sensory tests, many instrumental tests that have been developed by the industry over the years are also used to complement tests performed by humans.

The Future

Increasing the safety, speed, and efficiency of the manufacturing process are the major improvements being investigated for the future of the gummy candy industry. In any starch molding process, safety is a major concern because starch dryers represent an explosion hazard. Currently the U.S. government recommends minimizing these hazards by using spark-proof switches, blast walls, and other such mechanisms. Newer starch drying machines represent a reduced explosion hazard and improved microbiological killing. Additionally, moguls are being constructed that operate faster and more efficiently.

Since new products are the lifeline of any company in the candy business, new gummy flavors and colors are constantly being added to the base formula. Also, unique shapes are being molded, creating a plethora of new gummy candy. New forms of gummy candy are also being developed, most recently, a combination of gummy candy and marshmallow.

Where to Learn More

Books

Traxler, Hans. *The Life and Times of Gummy Bears*. Harper Collins, 1993.

Periodicals

Gelatin. Gelatin Manufacturers Institute of America, Inc., 1993.

Lepree, Joy. "Gelatin market softening offset by feedstock crimp." *Chemical Marketing Reporter,* July 18, 1994, p. 16.

Tiffany, Susan. "Infant Gummi Bear takes giant steps." *Candy Industry*, January 1995, p. 44.

—*Perry Romanowski*

Hair Dye

With intensive marketing of the first one-step hair-coloring treatment, the percentage of women in the United States who dyed their hair grew from approximately 8% in 1950 to almost 50% by 1973.

Background

Hair dye is one of the oldest known beauty preparations, and was used by ancient cultures in many parts of the world. Records of ancient Egyptians, Greeks, Hebrews, Persians, Chinese, and early Hindu peoples all mention the use of hair colorings. Early hair dyes were made from plants, metallic compounds, or a mixture of the two. Rock alum, quicklime, and wood ash were used for bleaching hair in Roman times, and herbal preparations included mullein, birch bark, saffron, myrrh, and turmeric. Henna was known in many parts of the world; it produces a reddish dye.

Many different plant extracts were used for hair dye in Europe and Asia before the advent of modern dyes. Indigo, known primarily as a fabric dye, could be combined with henna to make light brown to black shades of hair dye. An extract of the flowers of the chamomile plant was long used to lighten hair, and this is still used in many modern hair preparations. The bark, leaves, or nutshells of many trees were used for hair dyes. Wood from the brazilwood tree yielded brown hair dyes, and another hair dye known in antiquity as *fustic* was derived from a tree similar to the mulberry. Other dyes were produced from walnut leaves or nut husks, and from the galls, a species of oak trees. Some of these plant-derived dyes were mixed with metals such as copper and iron, to produce more lasting or richer shades.

The golden red hair captured by many Renaissance painters was artificially produced by some women. The Italian recipe was to comb a solution of rock alum, black sulfur, and honey through the hair and then let the hair dry in sunlight. Other hair dyes, dating from the sixteenth century, were preparations of lead, quicklime, and salt, or silver nitrate in rose water. Another early method of coloring hair was to apply powder. Pure white powder for hair or wigs was the mark of aristocratic dress in Europe during the seventeenth and eighteenth centuries. White powder was made of wheat starch or potato starch, sometimes mixed with plaster of paris, flour, chalk, or burnt alabaster. Similarly colored powders were sometimes used as well. These were made by adding natural pigments such as burnt sienna or umber to white powder to make brown, and India ink was sometimes used to make black powder. In Biblical times, people used powdered gold on their hair. The use of powdered gold and silver returned briefly as a fad in Europe among the wealthy in the mid-nineteenth century. Other hair colorants were blocks similar to crayons made with wax, soap, and pigments. These could be wetted and rubbed on the hair, or applied with a wet brush.

Preparations such as these were the only hair dyes available until the late nineteenth century. Hydrogen peroxide was discovered in 1818, but it was not until 1867 that it was exhibited at the Paris Exposition as an effective hair lightener. A London chemist and a Parisian hairdresser began marketing a 3% hydrogen peroxide formula at the Exposition as *eau de fontaine de jouvence golden* (golden fountain of youth water), and this was the first modern chemical hair colorant. Advances in chemistry led to the production of more hair dyes in the late nineteenth century. The first synthetic organic hair dye developed was pyrogallol, a

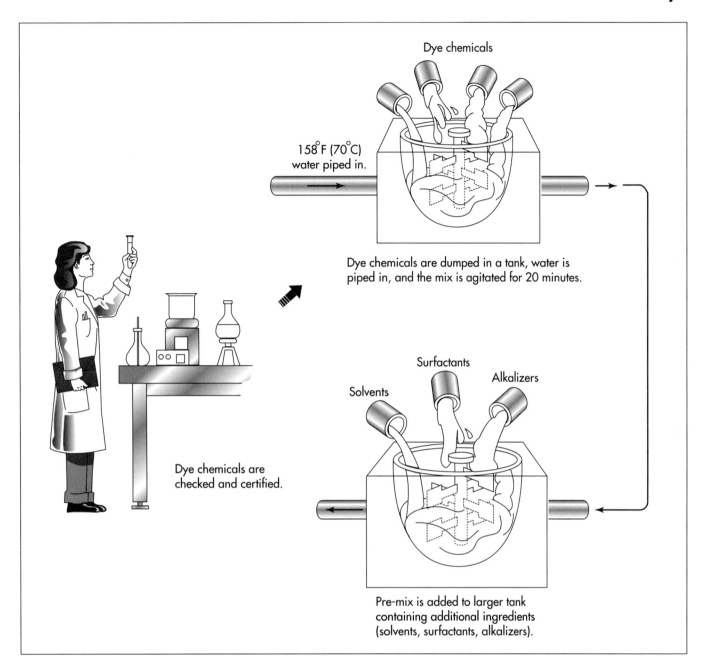

Dye chemicals

158°F (70°C) water piped in.

Dye chemicals are dumped in a tank, water is piped in, and the mix is agitated for 20 minutes.

Dye chemicals are checked and certified.

Solvents

Surfactants

Alkalizers

Pre-mix is added to larger tank containing additional ingredients (solvents, surfactants, alkalizers).

substance that occurs naturally in walnut shells. Beginning in 1845, pyrogallol was used to dye hair brown, and it was often used in combination with henna. So-called amino dyes were developed and marketed in Europe in the 1880s. The earliest was p-phenylenediamine, patented in Germany by E. Erdmann in 1888 as a dye for fur, hair, and feathers. To dye hair with p-phenylene-diamine and related dyes, a weak solution of the chemical, mixed with caustic soda, sodium carbonate, or ammonia, was applied to the hair. Then hydrogen peroxide was applied, which brought out the color. The amino dyes produced a more natural-look-ing black than previous dyes, and could make shades of red and brown as well.

A French hairdresser, Gaston Boudou, first marketed a standardized range of hair dyes in 1910. Whereas earlier hair colors had been mixed on the spot by hair dressers, and the colors produced were variable, Boudou's dyes produced a predictable color. Sold in a range of 18 colors, from black to light blond, these became very popular both in Europe and in the United States. The amino dyes, however, caused allergic reactions in a significant portion of users. Researchers in the United States are credited with creating a modified, less toxic amino-based hair dye, for standardizing the method of applying the dye, and for establishing strict specifications for the purity and strength of the raw materials. Further advances in hair dye chemistry were made by the makers of Clairol. Clairol produced the first one-step hair dye in 1950. This eliminated the time-consuming preliminary shampoo and pre-lightening that was the established hair-dying protocol. With intensive marketing of this easy-to-use product, the percentage of women in the United States who dyed their hair grew from approximately 8% to almost 50% by 1973.

Raw Materials

Most commercial hair dye formulas are complex, with dozens of ingredients, and the formulas differ considerably from manufacturer to manufacturer. In general, hair dyes include dyes, modifiers, antioxidents, alkalizers, soaps, ammonia, wetting agents, fragrance, and a variety of other chemicals used in small amounts that impart special qualities to hair (such as softening the texture) or give a desired action to the dye (such as making it more or less permanent). The dye chemicals are usually amino compounds, and show up on hair dye ingredient lists with such names as 4-amino-2-hydroxytoluene and m-Aminophenol. Metal oxides, such as titanium dioxide and iron oxide, are often used as pigments as well.

Other chemicals used in hair dyes act as modifiers, which stabilize the dye pigments or otherwise act to modify the shade. The modifiers may bring out color tones, such as green or purple, which complement the dye pigment. One commonly used modifier is resorcinol, though there are many others. Antioxidants protect the dye from oxidizing with air. Most commonly used is sodium sulfite. Alkalizers are added to change the pH of the dye formula, because the dyes work best in a highly alkaline composition. Ammonium hydroxide is a common alkalizer. Beyond these basic chemicals, many different chemicals are used to impart special qualities to a manufacturer's formula. They may be shampoos, fragrances, chemicals that make the formula creamy, foamy, or thick, or contribute to the overall action of the formula.

Hair dyes are usually packaged with a developer, which is in a separate bottle. The developer is most often based on hydrogen peroxide, with the addition of small amounts of other chemicals depending on the manufacturer.

The Manufacturing Process

Checking ingredients

1 Before a batch of hair dye is made, the ingredients must be certified. That is, the chemicals must be tested to make sure they are what they are labeled, and that they are the proper potency. Certification may be done by the manufacturer in-house. In many cases, the ingredients arrive from a reputable distributor who has provided a Certificate of Analysis, and this satisfies the manufacturer's requirements.

Weighing

2 Next a worker weighs out the ingredients for the batch. For some ingredients, only a small amount is necessary in the batch. But if a very large batch is being made, and several ingredients are needed in large amounts, these may be piped in from storage tanks.

Pre-mixing

3 In some hair dye formulas, the dye chemicals are pre-mixed in hot water. The dye chemicals are dumped in a tank, and water which has been already heated to 158°F(70°C) is pumped in. Other ingredients or solvents may also be added to the pre-mix. The pre-mix is agitated for approximately 20 minutes.

Mixing

4 The pre-mix is then added to a larger tank, containing the other ingredients of the hair dye. In a small batch, the tanks used may hold about 1,600 lbs (725 kg), and they are portable. A worker wheels the pre-mix tank to the second mix tank and pours the ingredients in. For a very large batch, the tanks may hold 10 times as much as the portable tanks, and in this case they are connected by pipes.

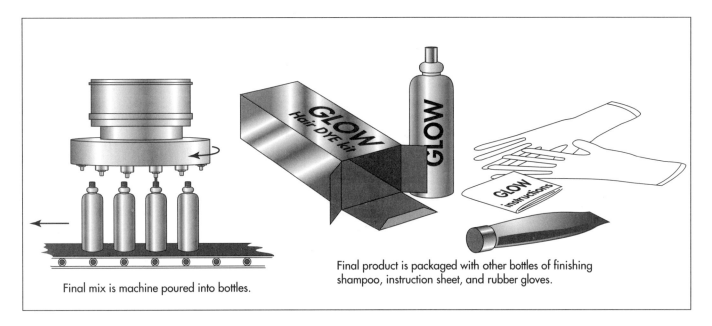

Final mix is machine poured into bottles.

Final product is packaged with other bottles of finishing shampoo, instruction sheet, and rubber gloves.

In a formula in which no pre-mixing is required, after checking and weighing, the ingredients go directly to the mixing step. The ingredients are simply mixed in the tank until the proper consistency is reached.

If a heated pre-mix is used, the second mix solution must be allowed to cool. The ingredients that follow the pre-mix may be additional solvents, surfactants, and alkalizers. If the formula includes alcohol, it is not added until the mix reaches 104°F(40°C), so that it does not evaporate. Fragrances too are often added at the end of the mix.

Filling

5 The finished batch of hair dye is then piped or delivered to a tank in the filling area. A nozzle from this tank lets a measured amount of hair dye into bottles, moving beneath it on a belt. The filled bottles continue on the belt to machines, which affix labels and cap them.

Packaging

6 From the filling area, the bottles are taken to the packaging line. At the packaging line, the hair dye bottle is put in a box, together with any other elements such as a bottle of developer or special finishing shampoo, instruction sheet, and gloves and cap, or any other tools provided for the consumer. After the package is complete, it is put in a shipping carton. The full cartons are then taken to the warehouse to await distribution.

Quality Control

Government regulations control what ingredients may be used in hair dyes, as many of them are toxic. Industry researchers will have already tested a formula numerous times in the laboratory before it reaches the manufacturing stage, to make sure a formula is non-irritating, works well, performs consistently, etc. As part of the manufacturing process, workers check their chemicals before they go into a batch, to make sure only the correct chemicals at the correct potency are used. After the batch is mixed, samples are taken, and these are subjected to a series of standard tests. Lab technicians make sure that the batch is the required viscosity and pH balance, and they will also test the dye's action on a swatch of hair. If a hair dye formula is being made for the first time, or if a formula has been altered, technicians will also test samples of the dye after the filling stage.

The Future

Hair dye manufacturers are increasing their use of computers to control and automate the manufacturing process. Computers can be used to weigh and measure ingredients, to control reactions, and to regulate equipment such as pumps. The future may see more fully automated manufacturers and increased efficiency.

Where to Learn More

Books

Balsam, M.S. and Edward Sagarin.*Cosmetics Science and Technology.* John Wiley & Sons, 1972.

Periodicals

Foltz-Gray, Dorothy. "Declare Your Right to Dye." *Health*, May-June 1996, pp. 54-57.

"Hair Dye Study." *FDA Consumer*, May 1994, p. 4.

—*Angela Woodward*

Harmonica

Background

The harmonica, or mouth organ, is a hand-held rectangular musical instrument. As the musician inhales and exhales into evenly spaced air channels, the metal reeds within produce musical tones. The length and thickness of the reed determines the note that is heard. Descended from the Jew's harp and Chinese sheng of ancient times, the harmonica has engendered various nicknames, including blues harp, pocket piano and Mississippi saxophone. Since its beginnings in the early 1800s, the harmonica has been used in variety of musical forms, from classical to folk to country to rock to blues to jazz.

History

Although it is impossible to pinpoint the exact day that the harmonica was invented, the first patent was issued to the teenaged Christian Friedrich Buschmann of Thuringer (now Germany) for his aura, a 4 in (10 cm) mouth organ that featured 21 blow notes arranged chromatically. It was quickly imitated throughout Europe and went by many names, such as mundharmonika, mundaeoline, psallmelodikon and symphonium. In 1826, Joseph Richter, a Bohemian instrument maker created a variation that was to become the standard. Richter's version featured 10 holes with 20 reeds on two separate plates that allowed both blow notes and draw notes. The plates were mounted on either side of a cedar comb. He tuned it to a diatonic, or seven-note, scale.

Several decades later, a young German clockmaker named Matthias Hohner learned to make a harmonica and consequently changed professions. Starting his new company in his kitchen in 1857, Hohner turned out 650 harmonicas in his first year with the help of family members and one paid worker. In 1862, Hohner, an astute marketer who had his name engraved on the plates of his harmonicas, introduced the instrument to North America, where its portability and affordability made it a favorite of the Western cowboy. African-American blues musicians also found the harmonica an affordable alternative to a piano or horn. Sonny Terry, James Cotton, Charles Musselwhite, and William Clarke are just a few of the blues legends who have lent their talents to the harmonica.

The harmonica soon entered the mainstream. In the period just before World War II, boys' harmonica bands were a popular vaudeville act. Larry Adler made a name for himself playing the harmonica with major symphony orchestras. In the late 1940s, the three-man Harmonicats sold 20 million copies of their rendition of "Peg o' My Heart." At the beginning of the 1960s, a group of 105 amateur harmonica players in Levittown, Pennsylvania, dubbed themselves the "Largest Uniformed Harmonica Band in the United States." Borrowing heavily from the African-American blues legacy, numerous white rock-and-roll musicians picked up the harmonica. Folk singer Bob Dylan popularized the practice of placing the harmonica on a neck frame to free the hands for playing the guitar, piano, or other instrument at the same time.

Today, five major types of harmonicas are produced: diatonic, diatonic tremolo-tuned, diatonic octave-tuned, chromatic, and or-

Folk singer Bob Dylan popularized the practice of placing the harmonica on a neck frame to free the hands for playing the guitar, piano, or other instrument at the same time.

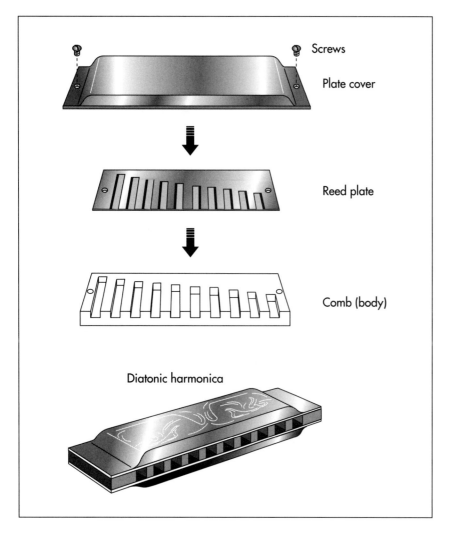

Screws

Plate cover

Reed plate

Comb (body)

Diatonic harmonica

chestral accompaniment. The single-reed diatonic harmonic is the most popular and can be heard in rock, country, blues, and folk music. It features 10 holes with 20 reeds, 10 for blow notes and 10 for draw notes. The tremolo has double holes, each of which contains a reed cut to the same key. Each hole allows both blow and draw notes. In the octave-tuned diatonic, the reeds in the double holes are an octave apart. Chromatic harmonicas play a 12-note octave, including all sharps and flats. The orchestral model can feature all blow notes or a combination of blow and draw notes. Some are designed to play chords.

Raw Materials

Originally, the body, or comb, of all harmonicas was constructed of wood. Now, most are made from injection-molded plastic. Some high-end models are made from metal alloys, lucite, or silver. Each of these

materials produces a distinct type of sound. Marine band and blues harp types continue to be made from moisture-resistant soft wood. The semi-hardness of the wood produces a rich sound while resisting swelling.

Reeds are cut from precision-tapered strips of brass alloy (a mixture of copper and zinc) material. Reed and cover plates are also machined from brass.

Screws and rivets are used to fasten the comb, reeds, reed plate, and cover plate.

The Manufacturing Process

While the individual parts are produced by machinery, the assembly is done by hand.

Creating the comb

1 Wooden combs are cut from a block of wood. Channels are carved out in descending lengths across the comb. Plastic combs are injection molded. The plastic compound is heated to a semi-fluid state and then mechanically injected into a mold. The compound hardens quickly, the mold is popped open, and the new comb is expelled.

Making the reed plate and reeds

2 The reed plate is stamped and machined, creating slits that correspond to the channels on the comb. Reeds are cut and tapered by machine. One end of each reed is riveted to the reed plate so that a reed lays over each slit. The opposite end of the reed is left free.

Tuning the reed plate

3 The reed plate is manually tuned. The tuner strikes the appropriate tuning fork and then files each reed to the correct tone. Filing the base end lowers the pitch; filing the free end raises the pitch.

Attaching the reed plate to the comb

4 The reed plate is attached to the comb with nails or screws. The assembly is done manually at a workbench similar to that used by a shoemaker. The nails are inserted into the holes with needle-nosed pliers and then tapped in gently with a small hammer.

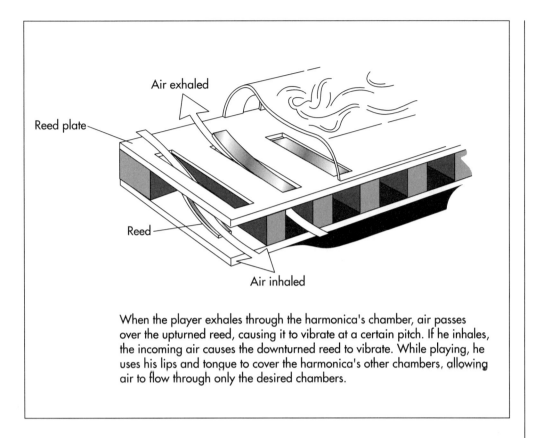

Reed plate

Air exhaled

Reed

Air inhaled

When the player exhales through the harmonica's chamber, air passes over the upturned reed, causing it to vibrate at a certain pitch. If he inhales, the incoming air causes the downturned reed to vibrate. While playing, he uses his lips and tongue to cover the harmonica's other chambers, allowing air to flow through only the desired chambers.

Attaching the plate cover

5 The plate cover, which has been machined, shaped, and stamped with the company name and harmonica type, is attached to the reed/comb assembly with screws or nails.

Packaging

6 The harmonicas are inserted into boxes and packed for shipment to retailers.

The Future

In the twenty-first century, less of the assembly will be done manually as the process becomes more automated and calibrated by computers. Manufacturers claim that the computerized process will increase the life span of the reeds and produce a harmonica that is more airtight.

Where to Learn More

Periodicals

Chelminski, Rudolph. "Harmonicas are... Hooty, Wheezy, Twangy and Tooty." *Smithsonian*, November 1995.

Other

Hohner Homepage. http://www.hohnerusa.com/html/history.html (January 29, 1997).

—*Mary F. McNulty*

Harp

The earliest harps probably developed from hunting bows and consisted of a few strings attached to the ends of a curved wooden body.

Background

A harp is a musical instrument consisting of a triangular frame open on both sides which contains a series of strings of varying lengths that are played by plucking. The length of the string determines how high or low a sound it makes. A modern concert harp stands about 70-75 in (1.8-1.9 m) high, is about 40 in (1 m) wide, weighs about 70-90 lb (32-41 kg), and has 47 strings, ranging in size from a few inches to several feet in length.

Smaller instruments similar to the harp include the lyre, which has strings of the same length but of varying thickness and tension; the psaltery, which has a frame open only on one side; and the dulcimer, which is similar to the psaltery but which is played by striking the strings with a hammer rather than plucking them.

History

The earliest harps probably developed from hunting bows and consisted of a few strings attached to the ends of a curved wooden body. A harp used in Egypt about five thousand years ago consisted of six strings attached to this kind of body with small wooden pegs. By 2500 B.C., the Greeks used large harps, consisting of strings attached to two straight pieces of wood which met at an angle.

By the ninth century, frame harps, which enclosed wire strings within a triangular wooden frame, appeared in Europe. They were fairly small [2-4 ft (0.6-1.2 meters) high] and were used by traveling musicians, particularly in Celtic societies. Many performers of traditional music (who are usually known as harpers rather than harpists) still use this type of instrument today.

The inability of these harps to play accidentals (notes a half-tone higher or lower than the notes of the scale to which the strings were tuned) led to a number of experiments. Harps were built with extra strings to play accidentals, either by increasing the number of strings in a single row or by adding a second row of strings parallel to the first to form double strung harps. In Wales, some harps had three rows of strings.

Instead of increasing the number of strings, some harpmakers devised mechanisms for changing the length of the strings, thereby adjusting the pitch. By the end of the seventeenth century, hooks were used in the Tyrol region of Austria to shorten strings as needed, providing two notes from each string. In 1720, Celestin Hochbrucker added seven pedals to control these hooks. In 1750, Georges Cousineau replaced the hooks with pairs of metal plates and doubled the number of pedals to produce three notes per string.

In 1792, Sébastien Érard replaced the metal plates with rotating brass disks bearing two studs, each of which gripped the string like a fork when the disk turned. He also reduced the number of pedals back to seven by devising pedals which could occupy three different positions each. Érard's design is still used in modern concert harps today. In the late nineteenth century and throughout the twentieth century, innovations were made in harpmaking by the American harp manufacturing company Lyon and Healy. These innovations included redesigning the stave back and the sound chamber of the harp.

Raw Materials

A harp is basically a large wooden triangle, usually made primarily of maple. The front, vertical side of the triangle is known as the column or the forepillar. The upper, curved side of the triangle is known as the neck. The third side of the triangle is known as the body. White maple is the best wood for these three sides because it is strong enough to withstand the stress of the strings. The soundboard, which is contained within the body and which amplifies the sound of the strings, is usually made of spruce. Spruce is used because it is light, strong, pliable, and evenly-grained, enabling it to respond uniformly to the vibrations of the strings to produce a rich, clear sound. The middle of the soundboard, known as the centerstrip, is attached to the base of the strings and is usually made of beech. Beech is used because it is tough enough to bear the tension of the strings.

The curved plate on the neck of the harp, to which the strings are attached, is made of brass. The disks which control the length of the strings are also brass, as are the pedals which control the disks. These external metal parts are often plated with gold for appearance and to resist tarnishing. The complex internal mechanism which connects the pedals to the disks, known as the action, is made of brass and stainless steel, with some parts such as washers made of a hard plastic such as nylon.

The strings of a harp are made of a variety of materials, including steel, gut (derived from the intestines of sheep), and nylon. Each material has different properties which make it suitable for a particular length of string.

The surface of a harp may be treated with clear lacquers or wood stains of various colors such as ebony or mahogany. It may also be inlaid with decorative woods such as walnut or avodire (a pale yellow West African wood). Some harps are gilded with 23 karat gold leaf. The soundboard may be decorated with paint or gold decals.

Design

Each harp is a unique work of art. The design of the harp depends on the needs of the performer. Traditional harpers require small, light instruments with strings controlled by levers. Classical harpists require much larger instruments with strings controlled by pedals. The exterior design of harps varies from simple curves with natural finishes to intricate carvings with a wide variety of decorations ranging from abstract geometric designs to romantic floral displays.

The Manufacturing Process

Making the wooden components

1 Boards of spruce, maple, beech, and other woods are received by the harpmaker and inspected. In order to perfectly match the grain of harp with a natural finish, boards of wood all from the same tree may be received together. The boards are then stored for about six months to become adjusted to the local climate in order to avoid any future problems with splitting or cracking.

2 Power woodcutting machines cut the boards into rough approximations of the pieces needed. More detailed shaping of these pieces is done with hand held woodcutting tools. Harpmakers learn their craft in a series of apprenticeships. New workers build the base of the harp, then go on to learn the skills needed to build the body and the soundboard. Only the most experienced harpmakers work on the column and the neck. Many thin layers of wood are glued together under pressure to form wooden parts which are stronger than solid wood. The various wooden parts are then stored to await assembly.

Making the metal components

3 Metalworkers use a wide variety of power and hand held tools to shape brass and steel into the nearly 1,500 pieces needed to make up the action of the harp. Some simple parts may be purchased from outside manufacturers. The metal components are then stored to await assembly.

Decorating the wooden components

4 Before assembly the wooden components are decorated as desired. The col-

A modern concert harp stands about 70-75 in (1.8-1.9 m) high, is about 40 in (1 m) wide, weighs about 70-90 lb (32-41 kg), and has 47 strings, ranging in size from a few inches to several feet in length.

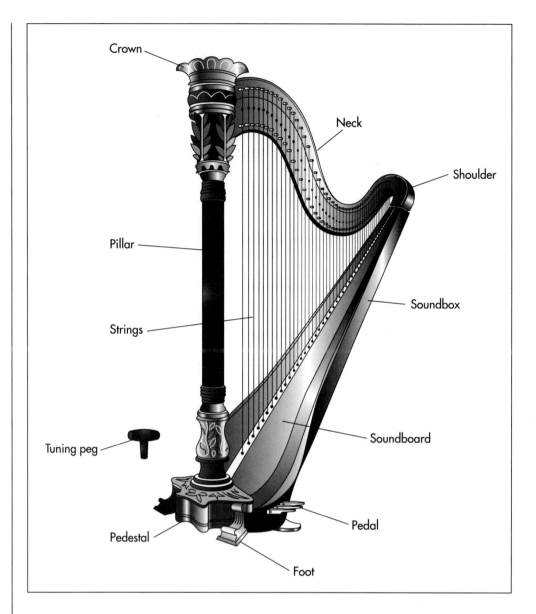

Crown

Neck

Shoulder

Pillar

Soundbox

Strings

Soundboard

Tuning peg

Pedestal

Pedal

Foot

umn may be hand carved with complex designs which take several weeks to complete. All wooden parts are sanded smooth in preparation for finishing. They are then sprayed with clear lacquer or colored wood stain. After one coat of lacquer or stain is applied, it is allowed to dry and then sanded smooth again. This process is repeated up to 10 times over as long as two weeks. The soundboard may then be painted with elaborate designs.

5 Some harps have gilded columns and bases. The gilder begins by sanding unfinished wooden parts to remove all imperfections. Layers of gesso (a special mixture of glues) are applied to the smooth wood. After the gesso sets, layers of clay are applied and sanded smooth. Glue is applied to

a small area of the smooth clay. Gold leaf 0.000004 inches thick (0.1 microns) is applied with a brush. (The gold is so thin that it cannot be handled directly by human hands.) The process is repeated on other small areas until an entire component is gilded. Excess gold is wiped away and another layer of gold leaf is applied. Some portions of the gold are burnished to a brilliant sheen by rubbing them with a tool made of polished agate. Clear lacquer is applied to protect the gold.

Assembling the harp

6 Master harpmakers begin the slow, painstaking process of bringing the wood and metal components together to form the harp. The parts of the neck, body,

soundboard, base, and column are brought together to form the frame. The complex mechanism of the action is fitted within the column and connected to the disks on the brass plate below the neck and the pedals on the base. Strings are attached to brass pegs on the neck, fed through the disks, and attached to the centerstrip of the soundboard. At first the strings are very loose. They are slowly tightened to the correct level of tension and tuned to the correct pitch.

7 After a final inspection, the harp is packed in close-fitting foam within a cardboard box to be shipped to the purchaser. The harp manufacturer also makes special protective wooden cases with wheels which allow the harp to be moved with relative ease.

Quality Control

Every step in the harpmaking process requires extreme attention to quality. Lumber is inspected for flaws. In particular, the spruce used for the soundboard is tested for its acoustic properties to ensure the quality of the sound it will produce. Each wooden component is individually inspected by a master harpmaker, then again after it has been sanded smooth for finishing. Metal components are also individually inspected. Those purchased from outside companies are inspected to ensure that they match the blueprints supplied by the harpmaker.

The strings are carefully tuned during the assembly process by an expert tuner. The action is tested to ensure that it is silent to avoid interfering with the music. The approximately 400 holes in the brass plate which holds the disks may be drilled by computer-controlled equipment to ensure accurate alignment. The harpmaker may choose to have a professional musician test each completed harp to ensure the quality of its sound.

The Future

Two seemingly contradictory trends hint at the future of the harp industry. Sparked by an increasing interest in Celtic music, more musicians are using harps similar to those used 1,000 years ago. On the other hand, many rock and jazz musicians are turning to electric harps, which produce amplified sounds in a manner similar to electric guitars. Despite these trends, it seems likely that harps similar to those designed by Sébastien Érard will continue to dominate the industry.

Where to Learn More

Books

Gammond, Peter. *Musical Instruments in Color*. Macmillan, 1976.

Rensch, Rosalyn. *Harps and Harpists*. Indiana University Press, 1989.

Other

Lyon and Healy. http://www.lyonhealy.com (July 9, 1997).

Strohmer, Shaun. "What Makes a Harp a Harp." http://harp.column.com/feature.html (September 25, 1996).

—Rose Secrest

Heat Pump

Heat pumps demonstrate remarkable versatility, providing both air conditioning and heating in the same system by simply reversing the direction of flow of the working fluid.

Background

As a result of society's increasing concern for ecological and environmental issues, the demand for more efficient ways to utilize heat and energy is rising. The heat pump industry uses technological advances such as year-round space heating to displace heat energy to a more useful location and purpose. This concept is accomplished by providing localized or redirected heat, while exchanging cool air with heated air.

The principles of heat pumps are actually the reverse of the technological and thermodynamic principles of an air conditioner unit. The majority of heat pumps give the added benefit of providing both heating in the winter and cooling in the summer. This can be accomplished simply by reversing the flow of the working fluid circulating through the coils. The heat pump is an entire thermodynamic system whereby a liquid and/or gas medium is pumped through an assembly where it changes phases as a result of altering pressure. Although relatively costly to setup, the heat pump system provides a more economical and efficient way to control temperatures and reuse existing heat energy.

Raw Materials

The manufacturing of heat pumps involves the use of large iron castings with stainless steel components and aluminum tubing. The castings, used in the pump and motor, will often have small amounts of nickel, molybdenum, and magnesium to improve the mechanical and corrosion-resisting characteristics of the casting. In smaller heat pumps, some components re-

quire the use of alloy steel to reduce weight. Depending on what type of working fluid is used (ammonia, water, or chlorofluorocarbons), the piping in the heat pump system may require corrosion resistant stainless steel or aluminum. In systems where consistency of thermodynamic properties are more critical, copper tubing may improve efficiency. Housing most of the components of the heat pump, the encasements are fabricated out of mild carbon sheet steel. The rest of the piping, fittings, valves, and couplings are stainless steel.

All heat pumps require a working fluid to transfer excess energy from one heat source to another. Traditionally, chlorofluorocarbons (CFCs) have been used as working fluids because of their superior thermodynamic properties. Because of the harmful effects CFCs are now known to have on the environment, they have been gradually phased out of production. Instead, water, hydrocarbons, and ammonia are frequently utilized in heat pump systems despite their lack of efficiency in some heat pump designs.

Design

Heat pumps all have the same basic components. These components consist of a pump, a condenser, an evaporator, and an expansion valve. Despite the relative similarities of these components, heat pump designs vary greatly depending on the specific application of the pump. The two major designs, vapor compression and absorption, utilize different thermodynamic principles, yet both include similar components and provide similar system efficiencies.

Heat pumps demonstrate remarkable versatility in providing both air conditioning and heating in the same system by simply reversing the direction of flow of the working fluid. In this regard, heat pumps eliminate the need for dual systems in order to maintain a desired temperature. However, this will be costly as it requires a system that is able to pump in both directions. In extremely adverse climates, heat pumps lose some of their effectiveness and may require an additional heat source. This supplemental heat can come from geothermally heated water or electric heaters.

The typical heat pump operation uses the working fluid to receive heat from a source positioned close to the evaporator. At the evaporator, the fluid vaporizes into a low pressure vapor. Upon entering the pump, the vapor is compressed to high pressure and enters a condenser which returns the vapor to a liquid and ultimately gives off its stored heat to the desired source. An expansion valve then allows the system to return to its low pressure liquid state, and the cycle begins again.

The Manufacturing Process

The pump is usually procured as a finished unit and installed into the system by integrating it with coupling and piping components. Designed for the specific size and fluid requirements of the system, the pump may be shipped, depending upon its size, directly to the installation site. This usually occurs with large commercial heat pumps supplying heat and/or refrigeration to office buildings. Smaller residential models may have the pump installed into an assembly that includes the condenser, evaporator, and various piping. These units, encased in a sheet metal box, will be comprised of various subassemblies for the condenser and evaporator in order to bolt every component to the box or to one another. Some of the brackets used will form the base of the unit where the pump will be bolted down to a metal pan and connected to an AC motor.

Encasements

1 Assembled from several different sheets of metal, encasement units are sheared to size in a shear press. After they are cut to the proper dimensions, small assembly holes are punched in the metal using a Computer Numerically Controlled (CNC) punch press. These punch presses have either a moveable table to move the sheet metal or a moveable die which is able to punch holes in different spots of the metal. Punch presses are often directed where to punch by a computer-aided design (CAD) program. Different shaped punching tools are stored within the machine, providing it with the ability to punch all of the necessary holes by simply changing the computer program.

2 After punching, the sheet will move to a Numerically Controlled (NC) press brake, where it will be bent in different shapes and configurations. The press brake bends the metal into many different shapes by using dies or tooling. Unlike the CNC punch press, the press brake will require a manual change in tooling to perform a different bend. The sheet is then ready to be welded, riveted, or bolted to the other sheets and brackets. Once assembled, these sheets provide most of the stability of stand-alone units.

Condenser and evaporator

3 The condenser and evaporator are made of many small, thin copper or aluminum tubes, which are bent around curved dies by tube bending machines. NC tube bending machines will be programmed to provide the same exact bend on each of the tubes, allowing them to be stacked one on top of the other. These tubes will then be attached to plates or fins through which the tubes will pass and be joined through tube expansion or joint welding. This creates a tightly sealed system. The tube and plate assembly will act as a heat exchanger by allowing the working fluid to pass through the system inside the tubes, while giving off the heat in the condenser to another fluid medium passing between the plates and acquiring the heat given off through the tubes.

4 In order to provide strength or connectivity to the components, small brackets are punched out of mild carbon steel. The brackets are usually punched out of steel coil that is continuously fed first through a decoiler. Once it is decoiled, it is sheared, bent, and formed in one continuous process.

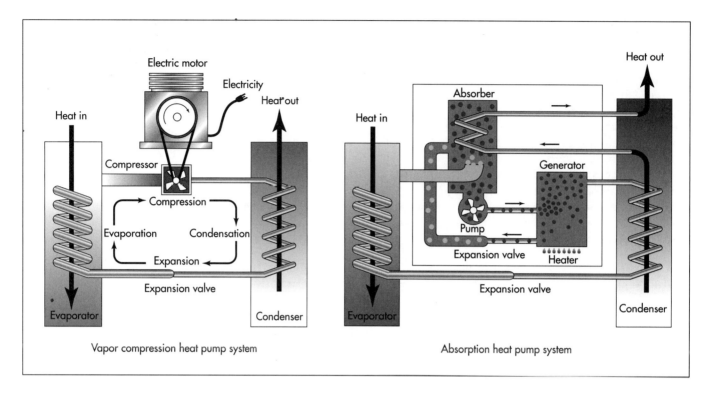

Vapor compression heat pump system

Absorption heat pump system

This is done with a progressive die configuration, where the bracket remains attached to the coil as it moves from station to station. Each station adds something to the bracket, either a hole or notch, and sends it to the next station, until finally it is sheared from the coil. This process may be outsourced to vendors who specialize in progressive die or transfer press operations and can provide better cost control.

Tubing

5 More tubing is fabricated and bent to provide the rest of the piping needed to connect the pump with the condenser and evaporator. Various fittings and connection components are utilized. The expansion valve, which is contained within some of the piping lines, is another component purchased as a whole unit. The expansion valve is a designed fitting that provides for the expansion of the working fluid and connection of smaller diameter tubing with larger diameter tubing. In small residential units, the valve is contained within the main box, while in larger commercial units, it may be installed on site in the piping system.

Painting/coating

6 Components, subassemblies, brackets, and/or plates are painted or powder coated for corrosion resistance. Before painting, however, some parts are treated with a special solvent to remove any grease or oil left from the manufacturing operations. This is usually done by submersing the parts in large tanks filled with solvent and then drying them in a special oven. Some parts, which are specially coated with zinc, nickel, or chrome, will be fed through an acid bath before being dipped into tanks of coating solution. Once cleaned, the parts are manually loaded onto trays or hung on specially designed racks and fed into a paint booth. The paint is applied with a pressurized paint dispenser that will spray paint into each crevice.

Packaging

7 After passing vigorous inspections, the heat pump is sent to packaging, where the system will be boxed and shipped to the installation site.

Installation

8 Generally, heat pumps will be installed at the construction site. The compressor and evaporator will be constructed of massive 3 in (7.5 cm) diameter tubing and have larger chambers, where the working fluid will change phases. The pump itself will be bolted to a concrete pad and connected

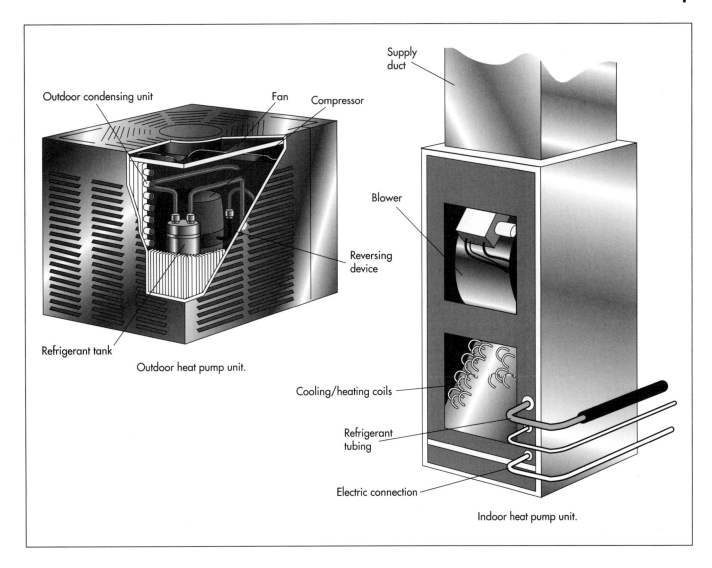

Outdoor condensing unit   Fan   Compressor

Supply duct

Blower

Reversing device

Refrigerant tank

Cooling/heating coils

Outdoor heat pump unit.

Refrigerant tubing

Electric connection

Indoor heat pump unit.

with a large DC motor or natural gas generator. The fittings and valves will be shipped and installed into the piping system, while supported by brackets and braces anchored to existing walls. These installations exhibit significant engineering challenges and often require cooperation between the contractor and heat pump manufacturer.

Quality Control

Each component that is procured from an outside supplier will usually be inspected for dimensional compliance before being assembled. Other components will be checked during their fabrication to ensure quality. The final assembly will then be tested by filling it with the appropriate working fluid and connecting the system to a power source to turn the pump. By measuring, with transducers or switches, the temperature and pressure levels of the fluid in different stages, the final system can be checked against predetermined criteria.

The Future

With the rising energy costs, the demand for the efficient heat pump will increase. The high initial cost will be returned in full as overall energy use decreases. The versatile heat pump will benefit organizations that aim to increase their exposure to new technological developments. As technology improves, the heat pump will ultimately produce more cost effective heating and cooling. Product development will generate competition among industries, causing the high manufacturing costs to decrease. Working fluid technology will continue to expand due to several experimental studies designed to meet future environmental concerns.

Where to Learn More

Other

"HydroHeat Geothermal Systems." October 4, 1996. http://www.njhpc.org/njh_uses.html (July 9, 1997).

"Heat Pump Working Fluids." October 1996. http://www.heatpumpcentre.org/hpcwrkf.htm (July 9, 1997).

"Heat Pump Technology." October 1995. http://www.heatpumpcentre.org/hpctek.htm (July 9, 1997).

"Heat Pumps in Industry." October 1996. http://www.heatpumpcentre.org/hpciapp.htm#industry systems (July 9, 1997).

—*Jason Rude*

Heavy-Duty Truck

Background

Trucks are divided into light-duty, medium-duty, and heavy-duty classifications depending on their weight. Heavy-duty trucks have a gross vehicle weight of 33,000 lb (15,000 kg) or more (i.e. the weight of the vehicle plus the weight of the payload is 33,000 pounds or more). When a heavy-duty truck is pulling a trailer, it may have a gross combination weight of 80,000 lb (36,360 kg) or more.

Technically, a vehicle that carries the load by itself, without a trailer, is known as a truck, or a straight truck. Examples include certain dump trucks, concrete mixers, and garbage trucks. A vehicle that pulls the load in a trailer is known as a tractor. The tractor is coupled to the trailer through a pivot point, known as the fifth wheel, which is mounted on top of the tractor frame. Most of the big rigs on highways are tractors pulling trailers.

History

The first gasoline-engine trucks were developed in the United States in the 1890s. During World War I, trucks played an important role moving supplies at home and overseas. With the development of a system of paved roads in the United States during the 1920s, the number of truck manufacturers grew. By 1925, there were more than 300 brands of trucks on the road. Some manufacturers came and went quickly. The Great Depression of the 1930s finished many more. By the 1990s, there were only nine heavy-duty truck manufacturers left in the United States. Together they build about 150,000-200,000 trucks a year.

Raw Materials

Trucks use steel for strength and durability, aluminum for light weight and corrosion resistance, polished stainless steel for bright finishes, and molded plastics for complex shapes.

Frame rails and crossmembers are usually formed from high-tensile steel. Suspension components, axles, and engine mounts are also made from steel. Some are cast and some are fabricated and welded.

The cab structure and outer skin may be made from steel or aluminum. If steel is used, the metal is coated with one or more layers of corrosion barriers such as zinc. On some cabs the roof may be made of fiberglass to form the complex curves required at the corners.

The hood and front fenders are usually molded in plastic or fiberglass because of the complex aerodynamic shapes. The front bumper may be stamped and drawn from steel or aluminum, or it may be molded in plastic and backed with a steel substructure.

Bright trim pieces—such as outside mirrors, sun visors, radiator grilles, and grab handles—are often made from polished stainless steel to give a long-lasting bright finish that will not crack or corrode.

The cab interior is finished with vinyl or cloth upholstery. The floors are covered with synthetic fiber carpeting or rubber mats. The dashboard and interior trim pieces are molded from plastic. The windows are made of laminated safety glass.

Fluids used in heavy-duty trucks include diesel fuel, petroleum-based or synthetic lu-

Heavy-duty trucks have a gross vehicle weight of 33,000 lb (15,000 kg) or more (i.e. the weight of the vehicle plus the weight of the payload is 33,000 pounds or more). When a heavy-duty truck is pulling a trailer, it may have a gross combination weight of 80,000 lb (36,360 kg) or more.

A 1911 Ford Model-T/Smith Form-A Truck Conversion tractor coupled to Fruehauf's 1914 flatbed semi-trailer. (From the collections of Henry Ford Museum & Greenfield Village.)

Today's most recognized form of heavy-duty truck, the tractor-trailer, or semi truck, was commercially developed in the 1910s. Some truck designers believed tractors, motor trucks designed only to pull separable trailers, could make truck operation profitable. If tractors easily connected to trailers, the more costly motorized tractors could remain busy hauling full trailers, while leaving less expensive trailers idle during loading or unloading.

In 1911, truck designer Charles Martin built a gasoline powered tractor to pull modified horse-drawn commercial wagons. His most significant innovation, however, was the fifth wheel coupler. A round plate with a central hole, it attached to the top of tractor frames to connect and support trailers. Buyers converted wagons into semi-trailers by raising them with jacks and removing their front axles. Lowering and locking a trailer's bottom mounted kingpin into a tractor's fifth wheel coupled the vehicles. Martin's Rocking Fifth Wheel handled the period's rough roads. It allowed tractor-trailers to bend when turning, but also accommodated the ups and downs of uneven surfaces. Nearly every truck manufacturer purchased Martin's popular device.

August Fruehauf, a Detroit blacksmith, launched an early trailer manufacturing company in 1914, by building a boat trailer for local lumberman Frederick Sibley. Sibley pulled it with a Model-T Ford car that he turned into a one ton truck with a Smith Form-A Truck conversion kit. Impressed that pivoting tractor-trailers maneuvered long, heavy loads through tight quarters, Sibley ordered more trailers for his business. By 1916, Fruehauf was a noted trailer manufacturer.

Erik R. Manthey

bricants, antifreeze, power steering fluid, and an environmentally safe, non-fluorocarbon gas known as R134A, which replaces freon in the air conditioning system.

Design

Truck manufacturers usually design a new model about every five to seven years. The new design incorporates advances in technology and materials, as well as changes desired by the customers. The design team will use a clay model to determine the overall styling, then build a prototype cab and hood for review and durability testing. As the design progresses, they will build an entire prototype vehicle for road testing. Just before the new truck goes into production, they will build one or more pilot models using actual production parts to spot any last-minute assembly problems.

In addition to the basic model, the engineers must also design all the options required by customers for different truck applications. Some manufacturers have as many as 12,000 options for their line of heavy-duty truck models.

The Manufacturing Process

Heavy-duty trucks are assembled from component parts. Each truck manufacturer usually builds its own cabs, and a few also build their own engines, transmissions, axles, and other major components. In most cases, however, the major components (and many of the other components) are built by other companies and are shipped to the truck assembly plant.

In most plants, the trucks move along an assembly line as components are added by different groups of workers at successive workstations. The truck starts with a frame assembly that acts as the "backbone" of the truck and finishes with the completed, fully operational vehicle being driven off the end of the assembly line under its own power.

Here is a typical sequence of operation for the assembly of a heavy-duty truck:

Assembling the frame

1 A pair of frame rails are selected from stock lengths of C-channel. They are laid side-by-side and fed through an automatic drilling machine or punch to make holes for connecting crossmember brackets, engine mounts, and other frame-mounted components. A computer tells the machine the size and location of the required holes along the length of the frame rails.

2 Small threaded studs are spot welded inside the C-section of the frame rails. The air lines for the brakes and the electrical wires for the lights and sensors are placed inside the frame rails and are secured with rubber-cushioned clamps fastened to the studs.

3 The brackets for the frame crossmembers are bolted in place using high-strength bolts or self-clinching fasteners. The left and right frame rails are then positioned opposite each other, and the crossmembers are added. The frame now resembles a long ladder with the rails as the sides and the crossmembers as the rungs.

4 Other frame-mounted components — such as engine mounts, suspension brackets, and air tanks — are bolted in place.

Installing the axles and suspensions

5 The front and rear axles are fitted with the proper hubs (the round ends to which the wheels are attached), brakes, and brake drums. The axles are clamped to the suspensions by means of long u-bolts. Some suspensions use long leaf springs while others use inflated rubber air bags.

6 The front and rear axles and suspensions are lifted into place and attached to the suspension brackets on the frame. The shock absorbers are attached between the axles and the frame.

Finishing the frame

7 Up until this point the frame assembly is usually moved from station to station either manually or with overhead hoists. The frame is now placed on a moveable support and begins moving down the assembly line. The air tanks and brake chambers are connected to the air lines, and the lights and sensors are connected to the proper wires.

8 If the vehicle is to be a tractor, the fifth wheel is lifted onto the frame and bolted into place. From this point on the frame assembly with the axles, suspensions, and frame-mounted components is referred to as the chassis.

Painting the chassis

9 All components that are not to be painted are covered with masking tape or paper. The chassis then moves into a paint booth where it is painted with compressed air spray guns. Most truck manufacturers require that all component parts be received with a primer coat of paint, so priming is not necessary.

10 After the chassis has been thoroughly painted and visually checked, it moves into a drying oven where a flow of hot air dries the paint. As it emerges from the oven, the masking tape and paper are removed.

Installing the engine and transmission

11 The engine and transmission are brought into the plant alongside the assembly line. Almost all trucks now use diesel engines. The clutch is installed and the transmission is bolted onto the rear of the engine. The fan, alternator, and other engine components are installed and connected with hoses and electrical wiring.

12 The finished engine/transmission package is then hoisted using lifting eyes that are part of the engine and is lowered onto the engine mounts in the chassis, where it is bolted in place. The radiator assembly is bolted onto its brackets ahead of the engine. The fuel lines, air hoses, starter cables, and coolant hoses are connected to the engine.

Finishing the chassis

13 The fuel tanks are secured to their frame brackets and connected to the fuel lines. Batteries are secured in the battery box, but are not connected to prevent accidental sparking.

14 The tires are mounted on the wheels at a workstation adjacent to the as-

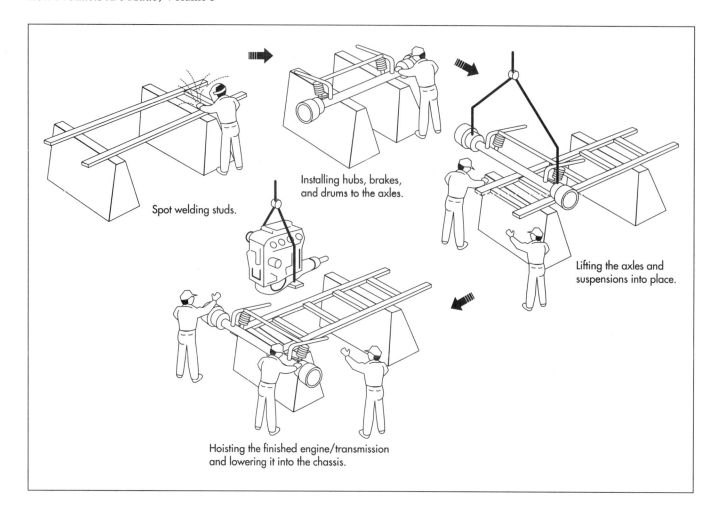

Spot welding studs.

Installing hubs, brakes, and drums to the axles.

Lifting the axles and suspensions into place.

Hoisting the finished engine/transmission and lowering it into the chassis.

sembly line. Aluminum wheels are left natural or may be polished. Steel wheels are painted before the tires are mounted. The tires and wheels are mounted on the axle hubs, and the lug nuts are tightened. At this point, the truck is taken off its moveable supports and sits on its own tires.

Assembling the cab, hood, and sleeper

[Steps 15-23 are performed in a separate area off the assembly line]

15 The cab and sleeper substructures are welded or fastened together in jigs to hold the pieces in place. The substructures give the cab and sleeper their strength and provides fastening points for the outer skin and the inner upholstery and trim.

16 The outer skin pieces are welded or fastened in place. This includes the sides, back, floor, and roof pieces. The joints between pieces are overlapped and

sealed to prevent leaks. The cab and sleeper doors are secured to the hinges.

17 The hood is usually a molded plastic piece and is shipped to the plant without any hardware attached. The hood is checked for rough surfaces and is sanded as required.

Painting the cab, hood, and sleeper

18 The cab, hood, and sleeper for each truck are painted at the same time. The surfaces are cleaned and the areas that are not to be painted are masked off with paper or tape. If a paint design such as a different color stripe is specified, the stripe area is painted first, then the stripe is masked off and the main body color is applied on a second pass through the paint booth. After each pass, the cab, hood, and sleeper go through a drying oven. After the final pass, the masking is removed and the paint is visually inspected.

Heavy-Duty Truck

In most plants, the trucks move along an assembly line as components are added by different groups of workers at successive workstations. The truck starts with a frame assembly that acts as the "backbone" of the truck and finishes with the completed, fully operational vehicle being driven off the end of the assembly line under its own power.

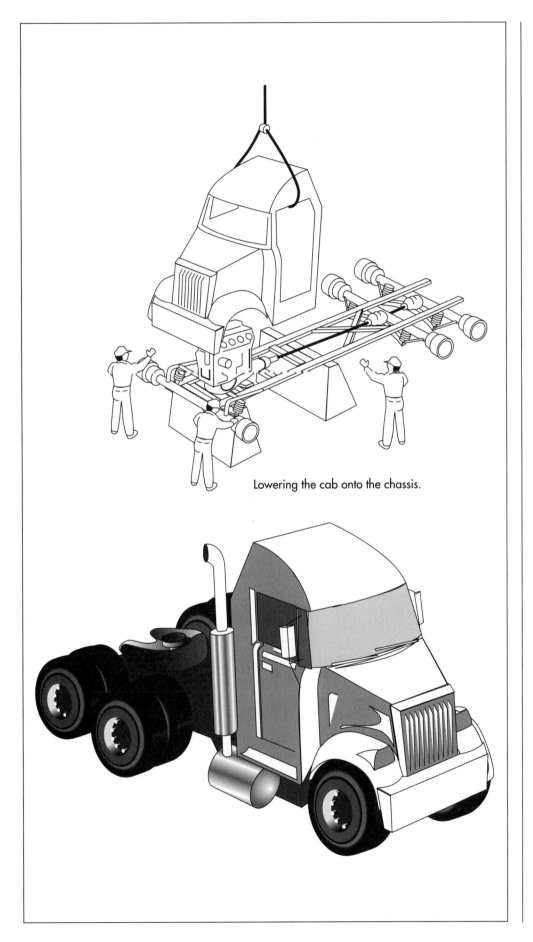

Lowering the cab onto the chassis.

Finishing the cab, hood, and sleeper

19 The grille, headlight brackets, hood hinges and latches, and the manufacturer's emblem or name are installed on the hood. The finished hood is then stored alongside the assembly line.

20 The exterior components of the cab and sleeper — the grab handles, mirrors, visors, etc. — are mounted before any work on the interior begins.

21 The instrument panel is attached to the dashboard. The gauges, warning lights, and switches are installed and hooked up to the appropriate wires and hoses. The entire dashboard assembly is then installed in the cab along with the cab heater system and steering column.

22 Pads of foam insulation are placed in the cab and sleeper walls, and the interior upholstery pieces are secured in place on the walls and ceiling. Plastic trim pieces are screwed in place to cover exposed edges and seams. The floor is covered with a rubber mat or fabric carpet laminated to a sound-absorbing pad, and the edges are secured. The seats are installed on top of the floor covering and secured with bolts into the main cab structure.

23 The windshield and rear windows are carefully pressed into place. A rubber gasket seals the edges between the glass and the cab structure.

Installing the cab, hood, and sleeper

24 The completed cab is lowered onto the chassis and bolted to its mounts. The sleeper is bolted in place behind the cab. The steering column is connected to the steering box. The transmission shift lever is installed through the floorboard, and the clutch pedal is attached to the clutch linkage.

25 After all the cab connections are made, the hood is lowered onto the chassis and secured to its pivot point. The bumper is attached to brackets on the frame. Wire connections are made for the headlights and front turn signals.

Adding fluids

26 The engine, radiator, and other reservoirs are filled, and the air conditioning system is charged. A small amount of diesel fuel is added to the tanks to allow a short road test. The steering wheel, which had been left out to give working room in the cab, is now installed, and the batteries are connected. The completed truck is then driven off the end of the assembly line.

Aligning the front and rear axles

27 To make sure that the front and rear axles are parallel to each other and perpendicular to the centerline of the frame, the truck is placed on a laser alignment machine and the axle positions are adjusted as required. The angle of the wheels is also adjusted. This ensures that the truck will handle properly and have satisfactory tire life.

Testing the completed truck

28 The truck is driven onto a dynamometer and secured with chains. The rear wheels of the truck sit on rollers set into the ground and connected to the dynamometer. As the truck engine spins the rear wheels on the rollers, the dynamometer measures the engine power to ensure it is operating correctly.

29 The truck is driven slowly through a water spray booth as the driver checks for cab leaks. The driver then takes the truck out for a short drive to check out the overall operation. If the truck passes all the tests, it is parked on "ready row" to be delivered to the dealer.

Quality Control

In addition to testing the completed truck, each component part and assembly operation is inspected. Parts are checked for correct dimensions before they reach the assembly line. Assembly operations are checked by the production workers themselves and are double-checked by quality control inspectors. The instrument panel is tested to make sure all the gauges and switches are working before it is installed in the truck. Even the thickness of the paint is checked with an electronic meter to ensure it meets the standard.

The Future

Heavy-duty trucks have evolved slowly over the last 100 years and will probably continue a slow evolution in the future. An increased concern about fuel efficiency has led to more aerodynamic designs. Likewise an increased concern about exhaust emissions has led to cleaner combustion engines. Heavy-duty trucks are still one of the most economical ways to ship the wide variety of raw materials and finished goods needed in our complex society, and they will probably remain one of our principal forms of transportation for many decades to come.

Where to Learn More

Books

Karolevitz, Robert F. *This Was Trucking*. Superior Publishing Company, 1966.

Rasmussen, Henry. *Mack: Bulldog of American highways*. Motorbooks International, 1987.

Rasmussen, Henry. *Peterbilt: The class of the industry*. Motorbooks International, 1989.

Other

Freightliner home page. 1996. http://www.freightlinertrucks.com (July 9, 1997).

Kenworth home page. 1996. http://www.paccar. com (July 9, 1997).

Peterbilt home page. 1996. http://www.peterbiltmotors.com (July 9, 1997).

Volvo GM Heavy Trucks home page. March 18, 1997. http://www.volvotrucks.volvo.com (July 9, 1997).

—*Chris Cavette*

Hologram

Rather than repeatedly handling the fragile skull of the 2,300 year old Lindow Man, researchers studied its holographic image.

Background

A hologram is a flat surface that, under proper illumination, appears to contain a three-dimensional image. A hologram may also project a three-dimensional image into the air—a lifelike image that can be photographed although it cannot be touched. Because they cannot be copied by ordinary means, holograms are widely used to prevent counterfeiting of documents such as credit cards, driver's licenses, and admission tickets. The word hologram comes from the Greek roots *holos* meaning whole and *gramma* meaning message. The process of making a hologram is called holography. When a hologram is made, light from a laser records an image of the desired object on film or a photographic plate.

There are basically two types of holograms. A reflection hologram is viewed when lit from the front, while a transmission hologram is viewed by shining a light through it from the back side. An embossed hologram is made by backing a transmission hologram with a mirror-like substance, which allows it to be viewed when lit from the front. Holograms can also be made that show moving objects; these sequences, called stereograms, are typically three to 20 seconds long.

Although a hologram is a visual image of a physical object, it is quite different from a photograph. For instance, when an object is photographed, each portion of the photo contains an image of the corresponding portion of the original object. Each section of a hologram, however, contains a complete image of the original object, viewed from a vantage point that corresponds to the section's position on the hologram. Thus, if the transparent plate containing a transmission hologram is broken, each piece will still be able to project the entire image, albeit from a different point of view. Using a piece from near the top of the holographic plate will produce an image as seen from above, while using a piece from near the bottom of the plate will create the impression of looking upward toward the object.

Another interesting property of holograms is that they preserve the optical properties of objects such as lenses. For instance, consider making a hologram of a magnifying glass placed in front of a butterfly. When viewing the holographic image of those objects, an observer will find that the portions of the butterfly seen through the image of the magnifying glass will be enlarged.

Holographic packaging has been shown to increase the sales of certain products. Projection holograms are especially eye-catching and are used at trade shows and retail stores. They can be used to display extremely delicate or valuable objects. A classic example was an image of a diamond-adorned hand that was projected over the sidewalk outside the Cartier jewelry store in New York City in 1970. Not only did it catch the attention of people walking by it, it attracted television news crews. In fact, it was even attacked by an umbrella-wielding pedestrian who thought it was the "work of the devil." In another instance, rather than repeatedly handling the fragile skull of the 2,300 year old Lindow Man, researchers studied its holographic image. Scotland Yard's Forensic Science Department used this holographic image to construct a physical model of the remains of the prehistoric man. As yet another application of holography, former Chicago Bears football coach Mike Ditka displayed a holographic portrait of himself in his restaurant to create a somewhat personal image when he could not be there in person.

Holograms can be made at home by hobbyists for a modest investment in equipment. The process requires a laser and an isolation table to prevent movement of the equipment while the film is being exposed. Holograms are also produced commercially and can be reproduced in large quantities. Using stock artwork, a master hologram for mass production can be created for as little as $2,500, whereas using custom artwork can cost $5,000 to $10,000. Reproducing the image costs from 1 to 4 cents per inch (2.5 cm), depending on the volume; this represents a 40% decrease since embossed holograms were first marketed in the late 1970s. Finished holograms can be attached to other objects as pressure-sensitive labels (0.5 to 1.5 cents each) or by hot stamping (2 to 5 cents each). Once the artwork is finalized, it takes about three months to create and reproduce a batch of commercial holograms. It is estimated that more than $200 million worth of embossed holograms were manufactured in 1995.

History

The first hologram was made in 1947 by Dennis Gabor, a Hungarian-born scientist who was working at the Imperial College of London. Gabor was attempting to refine the design of an electron microscope. He devised a new technique, which he decided to test with a filtered light beam before trying it with an electron beam. Gabor made a transmission hologram by carefully filtering his light source, but the process did not become practical until technology provided a way to produce coherent light—light that consists of a single frequency and a single wavelength. Hologram production took off with the invention of the laser in 1960, as a laser generates light that is of a single color (frequency) and produces waves that travel in phase with one another.

In 1962, using a laser to replicate Gabor's holography experiment, Emmett Leith and Juris Upatnieks of the University of Michi-gan produced a transmission hologram of a toy train and a bird. The image was clear and three-dimensional, but it could only be viewed by illuminating it with a laser. That same year Uri N. Denisyuk of the Soviet Union produced a reflection hologram that could be viewed with light from an ordinary bulb. A further advance came in 1968 when Stephen A. Benton created the first transmission hologram that could be viewed in ordinary light. This led to the development of embossed holograms, making it possible to mass produce holograms for common use.

Nearly a quarter century after he had made the first hologram, Gabor was awarded the Nobel Prize for Physics for this achievement in 1971. The following year, Lloyd Cross made the first recording of a moving hologram by imprinting sequential frames from ordinary moving picture film onto holographic film.

Raw Materials

Holograms made by individuals are usually exposed on very high resolution photographic film coated with a silver halide emulsion. Holograms made for mass production are exposed on a glass plate pretreated with iron oxide and then coated with photoresist. The photoresist material will chemically react to the specific wavelength of light that will be used to create the hologram. Because of their availability at a relatively low cost, helium-neon lasers are most commonly used by individuals who make their own holograms. Commercial hologram manufacturers use different laser types such as ruby, helium-cadmium, or krypton-argon ion.

After exposure, the film or photoresist plate is processed in chemical developers like those used in photography. Both nickel and silver are used to make the production masters that will be used to stamp multiple copies of the holograms onto polyester or polypropylene film. Aluminum is used to create the reflective coating on the back of embossed holograms.

Design

A three-dimensional, physical object can be used to create a hologram. The holographic image is normally the same size as the orig-

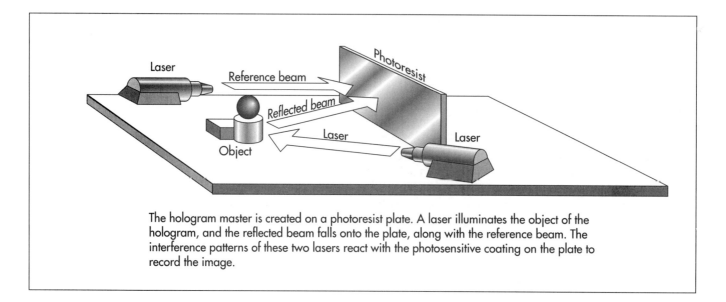

The hologram master is created on a photoresist plate. A laser illuminates the object of the hologram, and the reflected beam falls onto the plate, along with the reference beam. The interference patterns of these two lasers react with the photosensitive coating on the plate to record the image.

inal object. This may require construction of a detailed scale model of the actual subject in a size suitable for the holographic image. Alternatively, the artwork that is to be reproduced as a hologram can be computer generated, in which case software controls the laser exposure of the image file, one pixel at a time. (Pixels are the individual dots that comprise a graphic image on a computer screen or printout.)

The Manufacturing Process

Various manuals are available that explain to amateur holographers how to make holograms at home. The following steps describe the commercial mass production of a holographic image of an actual, three-dimensional object.

Mastering

1 A laser is used to illuminate the physical object, with the reflected light falling on the photoresist plate. Simultaneously, a reference beam from the laser also falls directly on the photoresist plate. The interference patterns of these two light beams react with the photo-sensitive coating to record a holographic image of the object. Common exposure times are between one to 60 seconds. In photography, slight motion of the object or the film results in a blurred image. In holography, however, the exposed plate will be blank (contain no image at all) if during the exposure there is movement as small as one

fourth the wavelength of the laser light (wavelengths of visible light range from 400 to 700 billionths of a meter).

A typical photoresist plate has a 6 in (15.24 cm) square working area; an extra half-inch (1.25 cm) of space on two edges allows the plate to be clamped into position. Because many holograms are smaller than this, several different images can be "ganged" (clustered) onto one plate, just as numerous individual photographs are exposed on one roll of film.

2 The plate on which the original hologram is recorded is called the master. After being exposed, the master is processed in a chemical bath using standard photographic developers. Before proceeding with production, the master is inspected to confirm that the image has been properly recorded. Because of the chemical reactions caused by the laser and the developer on the photoresist, the developed plate's surface resembles the surface of a phonograph record; there are about 15,000 grooves per inch (600 per cm), reaching a depth of about 0.3 microns (1 micron is a thousandth of a millimeter).

Electroforming

3 The master is mounted into a jig (frame) and sprayed with silver paint to achieve good electrical conductivity. The jig is lowered into a tank along with a supply of nickel. An electric current is introduced, and the master is electroplated with nickel. The jig is removed from the tank and washed with

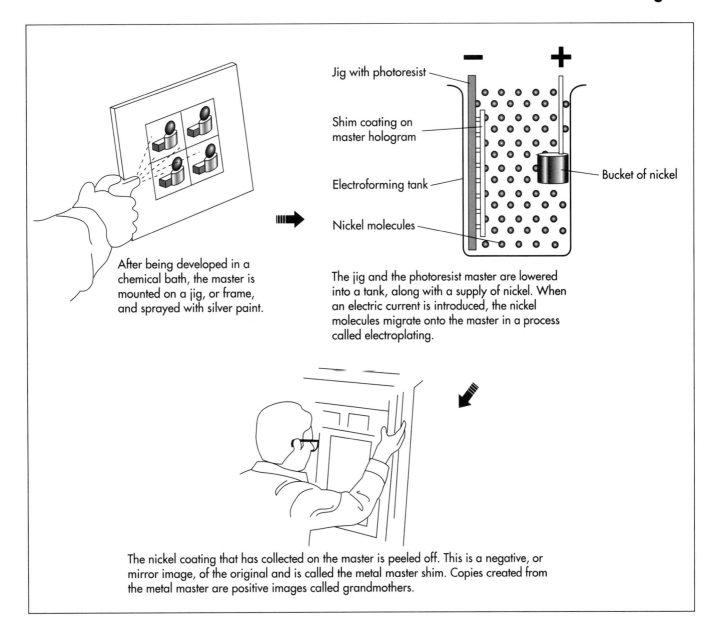

Jig with photoresist

Shim coating on master hologram

Electroforming tank

Nickel molecules

Bucket of nickel

After being developed in a chemical bath, the master is mounted on a jig, or frame, and sprayed with silver paint.

The jig and the photoresist master are lowered into a tank, along with a supply of nickel. When an electric current is introduced, the nickel molecules migrate onto the master in a process called electroplating.

The nickel coating that has collected on the master is peeled off. This is a negative, or mirror image, of the original and is called the metal master shim. Copies created from the metal master are positive images called grandmothers.

deionized water. The thin, nickel coating, which is called the metal master shim, is peeled off the master plate. It contains a negative image of the master hologram (the negative is actually a mirror image of the original hologram).

Using similar processes, several generations of shims are created. Those made from the metal master shim are known as "grandmothers," and they contain positive images of the original hologram. At this stage, numerous copies of the original image are "combined" (duplicated in rows) on one shim that can be used to print multiple copies with a single impression. Successive generations of shims are known as "moth-

ers," "daughters," and "stamper shims." Because these generations alternate between negative and positive images of the original, the stamper shims are negative images that will be used during actual production runs to print the final product holograms.

Embossing

4 Stamper shims are mounted in embossing machines. A roll of polyester film (or a similar material) that has been smoothed with an acrylic coating is run through the machine. Under intense heat and pressure, the shim presses the holographic image onto the film, to a depth of 25 millionths of a millimeter. The embossed film is rewound onto a roll.

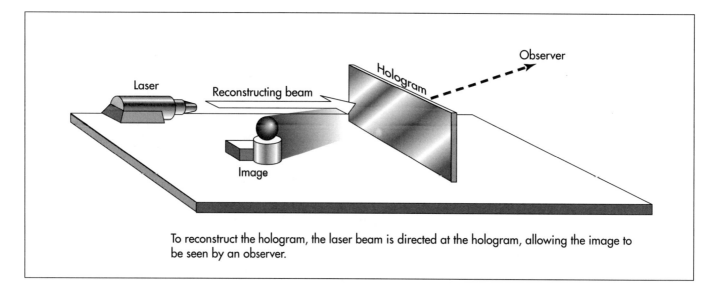

To reconstruct the hologram, the laser beam is directed at the hologram, allowing the image to be seen by an observer.

Metallizing

5 The roll of embossed film is loaded into a chamber from which the air is removed to create a vacuum. The chamber also contains aluminum wire, which is vaporized by heating it to 2,000°F (1,093°C). The sheet is exposed to the vaporized aluminum as it is rewound onto another roll, and in the process it becomes coated with aluminum. After being removed from the vacuum chamber, the film is treated to restore moisture lost under the hot vacuum condition. A top coating of lacquer is applied to the film to create a surface that can be imprinted with ink. The roll of film, which may be as wide as 92 in (2.3 m), is sliced into narrower rolls.

Converting

6 Depending on what type of film was used and what kind of product is being made, one or more finishing steps may be done. For instance, the film may be laminated to paper board to give it strength. The film is also cut into shapes desired for the final product and may be printed with messages. Heat-sensitive or pressure-sensitive adhesive is applied to the back of holograms that will be affixed to other objects or used as stickers.

Finishing

7 The holograms are either attached to other products or are counted and packaged for shipment.

The Future

Today, the most common use of holograms is in consumer products and advertising materials. There are some unusual applications too. For example, in some military aircraft, pilots can read their instruments while looking through the windshield by using a holographic display projected in front of their eyes. Automobile manufacturers are considering similar displays for their cars.

Holograms can be created without visible light. Ultraviolet, x-ray, and sound waves can all be used to create them. Microwave holography is being used in astronomy to record radio waves from deep space. Acoustical holography can look through solid objects to record images, much as ultrasound is used to generate images of a fetus within a woman's womb. Holograms made with short waves such as x rays can create images of particles as small as molecules and atoms.

Holographic television sets may project performers into viewers' homes within the next decade. Fiber optic communications systems will be able to transmit holographic images of people to distant homes of friends for realistic visits. Just as CD-ROM technology used optical methods to store large amounts of computer information on a relatively small disk, three-dimensional holographic data storage systems will further revolutionize storage capacities. It is estimated that this technology will store an amount of information equivalent to the contents of the Library of Congress in a space the size of a sugar cube.

Where to Learn More

Books

Kasper, Joseph E. and Steven A. Feller. *The Complete Book of Holograms: How They Work and How To Make Them.* John Wiley & Sons, 1987.

Saxby, Graham. *Practical Holography.* Prentice Hall, 1994.

Unterseher, Fred, Jeannene Hansen, and Bob Schlesinger. *Holography Handbook: Making Holograms the Easy Way.* Ross Books, 1996.

Other

DeFreitas, Frank. "On-Line Intro. To Holography Lessons." *Holostudio.* http://www.enter.net/~holostudio/intro.html (7 May 1997).

"History and Development of Holography." *Holophile, Inc.* http://www.connix.com/~barefoot/history.htm (7 May 1997).

Holographic Dimensions, Inc. http://www.hmt.com:80/holography/hdi/index.html (7 May 1997).

Outwater, Christopher and Van Hamersveld. *Practical Holography.* http://hmt.com/holography/hdi/holobook.htm (7 May 1997).

Pennsylvania Pulp and Paper Co. *Prismatic Illusions.* http://www.holoprism.com (13 May 1997).

—*Loretta Hall*

Hot Air Balloon

Balloons rise because of the displacement of air, applying the principle that the total upward buoyant force is equal to the weight of the air displaced.

Background

A hot air balloon is a nonporous envelope of thin material filled with a lifting gas that is capable of lifting a suspended payload into the atmosphere. Balloons rise because of the displacement of air, applying the principle that the total upward buoyant force is equal to the weight of the air displaced.

Over the course of history, balloon envelopes have been made of paper, rubber, fabric, and various plastics. The shapes of balloons have also varied over time, but today the most common ones are spheres, oblate spheroids, and aerodynamic configurations. Lifting gases have also varied. Today, the most commonly used gases are helium, hydrogen, and heated air.

History

In the late eighteenth century, Joseph and Jacques Montgolfier pioneered hot air ballooning in France. In 1782, they discovered that heated air in a lightweight bag caused it to rise. In 1783, they demonstrated their discovery publicly in Annonay, France. A few months later they repeated the experiment at Versailles, this time sending up a sheep, a rooster, and a duck as passengers.

The first manned balloon flight took place in Paris, France, on November 21, 1783. Coordinated by the Mongolfier brothers, Pilatre de Rozier and the Marquis Francois Laurent d'Arlandes were launched into the air in a balloon made of paper and linen. Smoke and heated air were used for the lifting gas.

Early balloons were used in war and sport. In the nineteenth century, the balloon was honed for use in war to spy behind enemy lines. Peacetime uses included taking the earliest aerial photographs. These balloons were useful but they were not steerable. Serious scientific balloon experiments only began late in the twentieth century.

Between 1934 and 1961, crewed balloon flights conducted research in the stratosphere [the atmosphere 6-15 mi (9.7-24.14 km) above Earth's surface]. Pressurized capsules allowed crews to go as high as 100,000 ft (30 km). In 1961, the advent of space flight made many of these experiments obsolete.

Since the early 1960s, the hot air balloon has been used as a free balloon (i.e. released into the atmosphere) to carry people aloft as a sport. In the 1970s, consumer hot air balloon sales soared, and in 1973 the first World championships were held in the United States. Today, there are various ballooning events around the world, but the main objective of most serious ballooners is to make and break records.

Other types of balloons currently used include the meteorological balloon, the zero-pressure balloon, the superpressure balloon (a constant level balloon), the military tethered balloon, and the powered balloon.

Modern hot air ballooning

Today's hot air balloons have two main parts, the envelope (or gas bag) and the basket. The gas bag is usually spherical and constructed of a nonextensible material. The heated air that lifts the balloon comes from a hydrocarbon gas burner attached above the basket.

The basket (also called the gondola) carries the passengers. Lift is controlled by adjusting the burning rate of the gas. A valve at the top of the balloon has a rope attached so that passengers can control descent. A rip cord and rip panel allow the rapid release of gas on landing to prevent the dragging of the load on impact.

Balloons can only go so far up into the atmosphere. The current limit for practical ballooning is 34 miles (55 km).

One of the most interesting aspects of hot air ballooning is the inflation of the hot air balloon. Under normal conditions, a four-man balloon can be inflated and launched with a crew of four to five people. To inflate such a large object, a large space is needed. The basket is laid out on its side. The envelope is connected to the basket and spread out over the ground. A few crew members hold open the mouth of the balloon and while a fan partially inflates a balloon with cold air.

The balloon's pilot enters the envelope at this point to make preliminary pre-flight checks on operational lines, rigging, pulleys, Velcro™ tab, flying wires, parachute, and the fabric of the envelope.

One fuel tank is used when the burner is turned on. A rush of trapped air surges into the envelope. The mouth tends to close behind it because of the rush. It takes about 60 seconds to fill a 20,000 cu ft (6,096 cu m) balloon. As the envelope fills, it rises above the basket.

The crew dealing with the crown of the balloon hold it steady downwind and prevent it from rolling side to side. As the lift increases, the crew walks the crown line to the basket.

After the inflation is completed, the pilot and passengers come aboard to make final checks. The balloon continues heating until the balloon becomes "light" (ready for take off).

Raw Materials

Envelope

Envelopes balance their load with load tapes or cords. Americans prefer a heavy fabric to share the load. Their European counterparts prefer lighter fabric and balance it with more load tape structures.

The fabric is woven from two kinds of yarn, nylon and Dacron (polyester). There are advantages and disadvantages to both. Nylon is lighter and stronger, but Dacron can withstand higher temperatures. The woven fabric is actually a mesh structure that allows air to pass through it. Most fabrics have a tensile strength of 40-100 lb (18.16-45.4 kg) per inch-wide strip.

To contain the air, the woven fabric is coated with a sealant. The most common is polyurethane, plus additives like neoprene (synthetic rubber) or silicone, and an ultra-violet inhibitor to protect the coating from breaking down because of the sun. The number of coats is determined by air tightness balanced with material fragility.

Two other important parts of the envelope are the parachute and the rip panel, both of which help control the balloon, especially when descending. The parachute is fail-safe, so it has become the dominant control. It is made of fabric similar to the envelope. The rip panel is sealed by Velcro™, and has a secondary opening called a vent that is also made with fabric similar to the envelope.

A net supports the basket and distributes the weight of the basket evenly over the balloon.

Basket

The basket's body is usually made of rattan and willow woven together. The floor can be made of plywood. The edges of the basket are commonly bound in leather, suede, or rawhide. Stainless steel wires and/or upright rigid supports attach the basket to the burner frame. Some manufacturers suspend the basket from a load ring that is itself suspended from the envelope. This load ring can also work as the frame for the burner.

Burner

The burner is a single-unit propane burner powered by two or more fuel tanks. The fuel tanks are joined to the burner with a permanent hose coupling. All the burners are constant-burning pilot flames.

There are a couple of components to the burner. The liquid valve regulates the

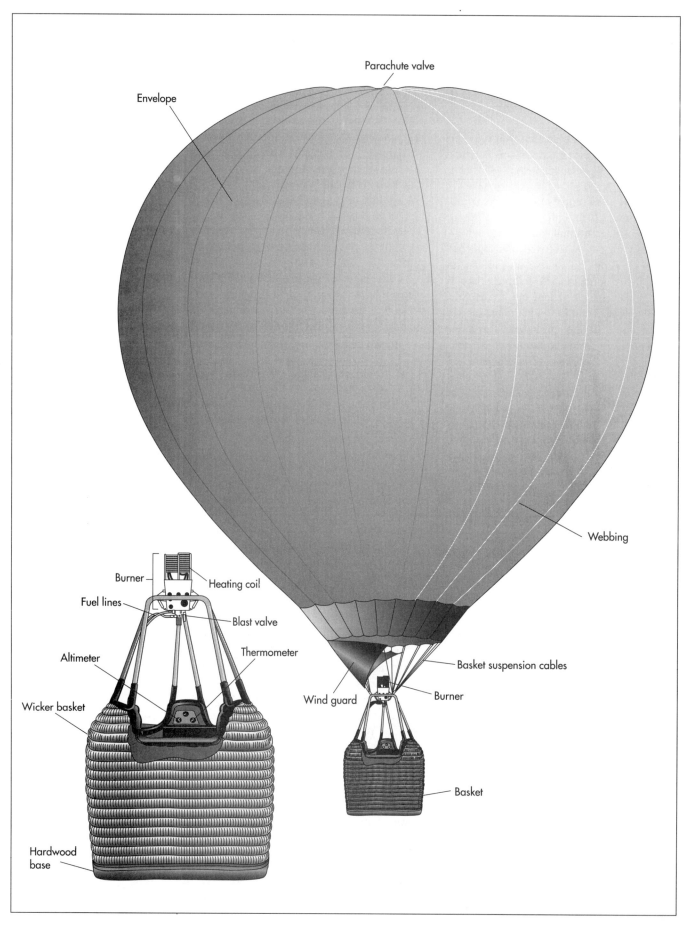

amount of fuel used by the burner. The pressure gauge shows the amount of gas pressure as it enters the burner coil indicating the amount of power available. The pilot light burns the vapor off the top of the tank. The vaporized propane emerges through jets that produce a flame with a fat base and a long tip. The stainless steel vaporizing coil passes the propane to produce heat.

Design

Envelope

Envelopes are designed to reduce fabric stress by producing a lightly curved gore (the sectional panels that are sewn together to make the envelope). Curved gores are longer in the center than they are in their ends. There are three main types of gores: vertical, horizontal and diagonal. Diagonal panels are the most economic because they waste the least amount of fabric. Many gores are computer designed.

There are two prevalent balloon shapes: teardrop and round. The teardrop can climb faster using less energy, but the round shaped balloon uses less fabric, thereby having less surface area to heat. An envelope can expect to last 400 flying hours.

Basket

Baskets have remained essentially the same since the 1700s. They are made of wicker and are square. They flex on impact. In the early 1970s, one company manufactured a gondola that was made of aluminum and fiberglass, but it tended to shatter on hard landings. The other big change is a triangle-shaped basket which allowed for some innovations with instrument panel placement. Baskets last about 800 flying hours.

The Manufacturing Process

Envelope

1 Envelope construction basically involves the sewing of the gores together. Whether done by hand or industrial sewing machine, there are three stitches. The double lap seam features two rows of parallel stitching along the folded over fabric seam. Preferred by manufacturers for its strength and lightness, the seam features about eight stitches in every inch (3 per cm). A few manufactures use a flat seam (straight parallel stitching holds two pieces of fabric together) and the zigzag (zigzag parallel stitching with a double lap of fabric). The load tapes and cords are also sewn in.

2 After the envelope is stitched, it is coated. The coating is applied mechanically and under pressure.

3 Finally, if the envelope is to be used for advertising purposes, an applique is attached with a slogan or name. It can be applied with acrylic spray paint, or ready-made adhesive letters or banners can be attached. If the artwork is large, it can be sewn into the envelope proper by being cut directly into the gores. This can be a complex, demanding process.

Basket

4 Baskets are manufactured base-first. On top of the plywood base with runners, a frame of cane is built up to 1 in (2.5 cm) in diameter. At the corner, the frame surrounds stainless steel wires and load frames. Around the frame, the rattan or willow is woven. Holes are left in the body of the basket for cylinder straps. The finished basket is coated with varnish to help maintain its shape and set the cane together. Finally, the edges are protected with sewn in rawhide, leather, or suede. The instrument/dashboard is built in as are the propane tanks for the burner unit.

Burner

5 Many balloon manufacturers outsource these components and assemble them between the basket and the envelope after the other parts are put together.

The Future

Innovations that will allow hot air balloons to go higher, for a longer period of time, and under more control will continue to happen. Many of the innovations center on improving the burner and the deflation system.

Where to Learn More

Books

Gibbs-Smith, C.H. *Ballooning*. Penguin Books, 1948.

Hildebrant, A. *Airships Past and Present.* Archibald Constable & Co., Ltd., 1908.

Other

"A Brief Look at the History of Ballooning." http://www.geocities.com/TheTropics/4744/hist.htm (April 30, 1997).

Jervis, Mark. "A Short History of Ballooning." http://www.bris.ac.uk/Depts/Union/BUHABS/first.html (April 30, 1997).

—*Annette Petrusso*

Ice Cream

History

Our love affair with ice cream is centuries old. The ancient Greeks, Romans, and Jews were known to chill wines and juices. This practice evolved into fruit ices and, eventually, frozen milk and cream mixtures. In the first century, Emperor Nero reportedly sent messengers to the mountains to collect snow so that his kitchen staff could make concoctions flavored with fruit and honey. Twelve centuries later, Marco Polo introduced Europe to a frozen milk dessert similar to the modern sherbet that he had enjoyed in the Far East. The Italians were especially fond of the frozen confection that by the sixteenth century was being called ice cream. In 1533, the young Italian princess Catherine de Médici went to France as the bride of the future King Henry II. Included in her trousseau were recipes for frozen desserts. The first public sale of ice cream occurred in Paris at the Café Procope in 1670.

Frozen desserts were also popular in England. Guests at the coronation banquet of Henry V of England in the fourteenth century enjoyed a dessert called *crème frez*. By the seventeenth century, Charles I was served crème ice on a regular basis. Eighteen-century English cookbooks contained recipes for ice cream flavored with apricots, violets, rose petals, chocolate, and caramel. Other early flavorings included macaroon and rum. In early America, George Washington and Thomas Jefferson were especially fond of ice cream. Dolley Madison was known to serve it at White House state dinners.

Because ice was expensive and refrigeration had not yet been invented, ice cream was still considered a treat for the wealthy or for those in colder climates. (In a note written in 1794, Beethoven described the Austrians' fear that an unseasonably warm winter would prevent them from enjoying ice cream.) Furthermore, the process of making ice cream was cumbersome and time-consuming. A mixture of dairy products, eggs, and flavorings was poured into a pot and beaten while, simultaneously, the pot was shaken up and down in a pan of salt and ice.

The development of ice harvesting and the invention of the insulated icehouse in the nineteenth century made ice more accessible to the general public. In 1846, Nancy Johnson designed a hand-cranked ice cream freezer that improved production slightly. The first documented full-time manufacturing of ice cream took place in Baltimore, Maryland, in 1851 when a milk dealer named Jacob Fussell found himself with a surplus of fresh cream. Working quickly before the cream soured, Fussell made an abundance of ice cream and sold it at a discount. The popular demand soon convinced him that selling ice cream was more profitable than selling milk.

However, production was still cumbersome, and the industry grew slowly until the industrialization movement of the early twentieth century brought electric power, steam power, and mechanical refrigeration. By the 1920s, agricultural schools were offering courses on ice cream production. Trade associations for members of the industry were created to promote the consumption of ice cream and to fight proposed federal regulations that would call for selling ice cream by weight rather than volume, and the disclosure of ingredients.

After World War II, with raw materials readily available again, the ice cream industry produced over 20 qt (19 l) of ice cream for each American per year.

The Prohibition era proved to be very profitable for the ice cream industry. Denied alcoholic beverages, many people ate ice cream instead. Breweries were often converted to ice cream factories, although it is likely that some of the plants were merely fronts for illegal liquor sales. Although the repeal of Prohibition in 1933 and the ensuing depression slowed ice cream sales, the industry continued to grow. The movie industry was especially instrumental in the promotion of ice cream and scenes depicting stars enjoying the frozen concoctions were plentiful. Ice cream parlors sprang up in every town and the parlor employee, the so-called soda jerk, developed into a cultural icon.

After World War II, with raw materials readily available again, the ice cream industry produced over 20 qt (19 l) of ice cream for each American per year. During the 1950s, competition sprang up between the ice cream parlor and the drug store that sold packaged ice cream. It was during this time that usage of lesser quality ingredients increased. Many producers were adding very low percentages of butterfat and pumping large quantities of air into the ice cream to fill out the carton.

The 1970s saw the development of gourmet ice cream manufacturers with an emphasis on natural ingredients. People also became interested in making ice cream at home. Upscale restaurants offer homemade ice cream on their dessert lists.

Raw Materials

Today, ice cream is made from a blend of dairy products (cream, condensed milk, butterfat), sugar, flavorings, and federally approved additives. Eggs are added for some flavorings, particularly French vanilla. The broad guidelines allow producers to use ingredients ranging from sweet cream to nonfat dry milk, cane sugar to corn-syrup solids, fresh eggs to powdered eggs. Federal regulations do stipulate that each package of ice cream must contain at least 10% butterfat.

The additives, which act as emulsifiers and stabilizers, are used to prevent heat shock and the formation of ice crystals during the production process. The most common additives are guar gum, extracted from the guar bush, and carrageenan, derived from sea kelp or Irish moss.

Ice cream flavors have come a long way from the standard vanilla, strawberry, and chocolate. By the 1970s, the International Association of Ice Cream Manufacturers had recorded over 400 different flavors of ice cream. In an ever-expanding array of combinations, fruit purees and extracts, cocoa powder, nuts, cookie pieces, and cookie dough are blended into the ice cream mixture.

Air is added to ice cream to improve its ability to absorb flavorings and to facilitate serving. Without air, ice cream becomes heavy and soggy. On the other hand, too much air results in ice cream that is snowy and dry. The federal government allows ice cream to contain as much as 100% of its volume in air, known in the industry as overrun.

Makers of high-quality ice cream (sometimes known as gourmet ice cream) use fresh whole dairy products, a low percentage of air (approximately 20%), between 16-20% butterfat, and as few additives as possible.

The Manufacturing Process

Although ice cream is available in a variety of forms, including novelty items such as chocolate-dipped bars and sandwiches, the following description applies to ice cream that is packaged in pint and half-gallon containers.

Blending the mixture

1 The milk arrives at the ice cream plant in refrigerated tanker trucks from local dairy farms. The milk is then pumped into 5,000 gal (18,925 l) storage silos that are kept at 36°F (2°C). Pipes bring the milk in pre-measured amounts to 1,000 gal (3,785 l) stainless steel blenders. Premeasured amounts of eggs, sugar, and additives are blended with the milk for six to eight minutes.

Pasteurizing to kill bacteria

2 The blended mixture is piped to the pasteurization machine, which is composed

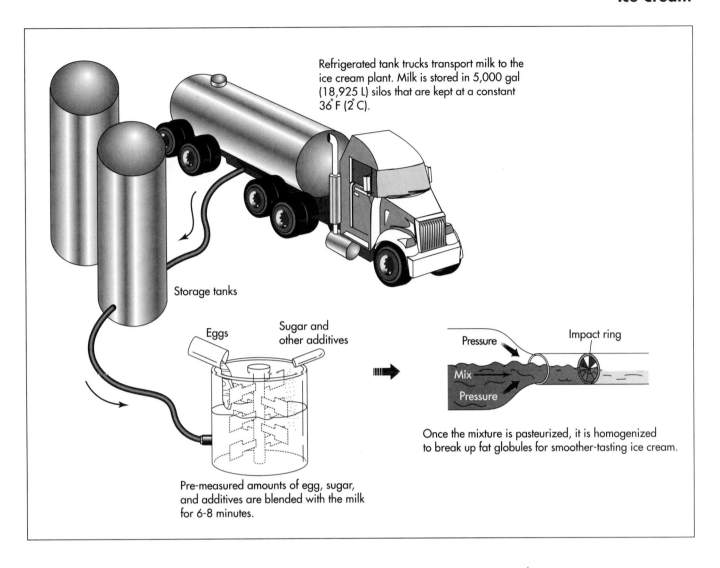

Refrigerated tank trucks transport milk to the ice cream plant. Milk is stored in 5,000 gal (18,925 L) silos that are kept at a constant 36°F (2°C).

Storage tanks

Eggs

Sugar and other additives

Pre-measured amounts of egg, sugar, and additives are blended with the milk for 6-8 minutes.

Pressure

Impact ring

Mix

Pressure

Once the mixture is pasteurized, it is homogenized to break up fat globules for smoother-tasting ice cream.

of a series of thin stainless steel plates. Hot water, approximately 182°F (83°C), flows on one side of the plates. The cold milk mixture is piped through on the other side. The water warms the mixture to a temperature of 180°F (82°C), effectively killing any existing bacteria.

Homogenizing to produce a uniform texture

3 By the application of intensive air pressure, sometimes as much as 2,000 pounds per square inch (141 kg per sq cm), the hot mixture is forced through a small opening into the homogenizer. This breaks down the fat particles and prevents them from separating from the rest of the mixture. In the homogenizer, which is essentially a high-pressure piston pump, the mixture is further blended as it is drawn into the pump cylinder on the down stroke and then forced back out on the upstroke.

Cooling and resting to blend flavors

4 The mixture is piped back to the pasteurizer where cold water, approximately 34°F (1°C), flows on one side of the plates as the mixture passes on the opposite side. In this manner, the mixture is cooled to 36°F (2°C). Then the mixture is pumped to 5,000 gal (18,925 l) tanks in a room set at 36°F (2°C), where it sits for four to eight hours to allow the ingredients to blend.

Flavoring the ice cream

5 The ice cream is pumped to stainless steel vats, each holding up to 300 gal (1,136 l) of mixture. Flavorings are piped into the vats and blended thoroughly.

Freezing to soft-serve consistency

6 Now the mixture must be frozen. It is pumped into continuous freezers that

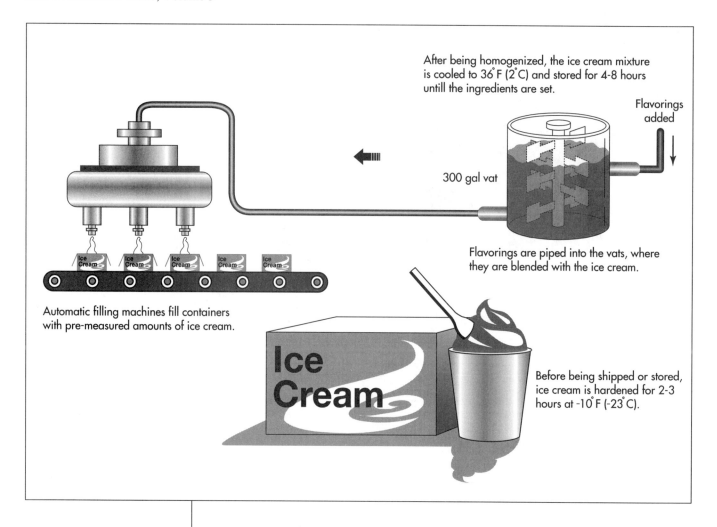

After being homogenized, the ice cream mixture is cooled to 36°F (2°C) and stored for 4-8 hours untill the ingredients are set.

Flavorings added

300 gal vat

Flavorings are piped into the vats, where they are blended with the ice cream.

Automatic filling machines fill containers with pre-measured amounts of ice cream.

Ice Cream

Before being shipped or stored, ice cream is hardened for 2-3 hours at -10°F (-23°C).

can freeze up to 700 gal (2,650 l) per hour. The temperature inside the freezers is kept at -40°F(-40°C), using liquid ammonia as a freezing agent. While the ice cream is in the freezer, air is injected into it. When the mixture leaves the freezer, it has the consistency of soft-serve ice cream.

Adding fruit and sweetened chunks

7 If chunks of food such as strawberry or cookie pieces are to be added to the ice cream, the frozen mixture is pumped to a fruit feeder. The chunks are loaded into a hopper at the top of the feeder. Another, smaller hopper, fitted with a starwheel, is located on the front of the feeder. An auger on the bottom of the machine turns the hoppers so that the chunks drop onto the starwheel in pre-measured amounts. As the mixture passes through the feeder, the starwheel pushes the food chunks into the ice cream. The mixture then moves to a blender where the chunks are evenly distributed.

Packaging and bundling the finished product

8 Automatic filling machines drop preprinted pint or half-gallon-sized cardboard cartons into holders. The cartons are then filled with premeasured amounts of ice cream at the rate of 70-90 cartons per hour. The machine then places a lid on each cartons and pushes it onto a conveyer belt. The cartons move along the conveyer belt where they pass under a ink jet that spray-paints an expiration date and production code onto each carton. After the imprinting, the cartons move through the bundler, a heat tunnel that covers each cup with plastic shrink wrapping.

Hardening

9 Before storage and shipping, the ice cream must be hardened to a temperature of -10°F (-23°C). The conveyer system moves the ice cream cartons to a tunnel set at -30°F (-34°C). Constantly turning ceiling

fans create a wind chill of -60°F (-51°C). The cartons move slowly back and forth through the tunnel for two to three hours until the contents are rock solid. The cartons are then stored in refrigerated warehouses until they are shipped to retail outlets.

Quality Control

Every mixture is randomly tested during the production process. Butterfat and solid levels are tested. The bacteria levels are measured. Each mixture is also taste-tested.

Ice cream producers also carefully monitor the ingredients that they purchase from outside suppliers.

The Future

Ice cream manufacturers continue to develop new flavorings. Ironically, given the industry's experiences during Prohibition, one of the more recent innovations has been the introduction of liqueur-flavored ice creams.

Where to Learn More

Books

Dickson, Paul. *The Great American Ice Cream Book*. Atheneum, 1972.

Lager, Fred. *Ben and Jerry's: The Inside Scoop*. Crown Publishers, 1994.

Periodicals

"Centrifugal pumps handle chocolate: overcoming the challenges of pumping heavy products." *Dairy Foods*, September 1994.

Gorski, Donna. "A cordial challenge." *Dairy Foods*, January 1995.

O'Donnell, Claudia D. "The story behind the story: two dairy processors tell a tale of fruits, flavors and nuts." *Dairy Foods*, May 1993.

—*Mary F. McNulty*

Imitation Crab Meat

First invented in the mid 1970s, imitation crab meat has become a popular food in the United States, with annual sales of over $250 million.

Background

Imitation crab meat is a seafood product made by blending processed fish, known as surimi, with various texturizing ingredients, flavorants, and colorants. First invented in the mid 1970s, imitation crab meat has become a popular food in the United States, with annual sales of over $250 million.

Surimi is the primary ingredient used to create imitation crab meat. It is mostly composed of fish myofibrillar proteins. These proteins are responsible for the quintessential characteristic of surimi that makes imitation crab meat manufacture possible, namely the ability to form a sturdy gel. The gel can be shaped and cut into thin strips which, when rolled together, mimic the texture of real crab meat.

Although imitation crab meat was introduced in the United States in the early 1980s, the Japanese have been using surimi-based products for over 800 years. Traditionally called *kamaboko*, the first recorded surimi manufacturing procedure was found in a Japanese cookbook written in 1528. Commercial production of *kamaboko* products began on a small scale in the nineteenth century. However, modern manufacturing did not start until the twentieth century, when efficient methods of bulk fishing were developed.

The basic manufacturing technology that is used today was primarily developed in Japan between 1945 and 1960. During this time, scientists developed techniques that made large-scale surimi production possible. For example, better methods of preservation were developed, and consequently the shelf life of surimi products was extended. Additionally, the science behind the gel forming properties of the myofibrillar proteins was worked out, and many factors that contributed to its texture were discovered.

One of the major problems with surimi was that when it was frozen, it lost its gel forming properties. As scientists investigated this problem, they discovered that the incorporation of cryoprotectant materials such as sucrose and sorbitol protected the surimi from degradation during freezing. This was important because it allowed imitation crab meat manufacturers to use surimi which was produced days earlier. This development of the mid 1960s resulted in a tremendous growth of the surimi-based seafood industry.

The process of making imitation crab meat from surimi was invented independently by Y. Sugino and K. Osaki by 1975. Early production of this product in the United States began in 1983 by the Japanese company Yamasa Enterprises. As popularity of this product increased, other companies also began production, and by 1986 the market for imitation crab meat was $250 million. Ultimately, U.S.-based corporations took market share away from imported products and now export imitation crab meat to Japan.

Raw Materials

Various ingredients are mixed together to make a product which has the color, taste, and texture of crab meat. The surimi used in the manufacture of imitation crab meat is most commonly processed from the Alaska pollock or walleye pollock. To a lesser extent, the New Zealand hoki is also used.

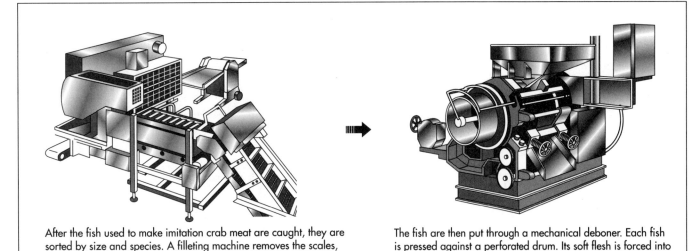

After the fish used to make imitation crab meat are caught, they are sorted by size and species. A filleting machine removes the scales, head, tail, and viscera of the fish.

The fish are then put through a mechanical deboner. Each fish is pressed against a perforated drum. Its soft flesh is forced into the drum, while the scales and bones are left on the outside. Waste matter is scraped off and collected in a bin.

These fish are particularly useful because they are abundant, have little flavor, and are inexpensive to process. Other fish that have been used include the blue whiting, croaker, lizardfish, and Pike-conger. However, these fish currently present some problems during surimi manufacture, which limits their use.

During the manufacture of surimi, various processing ingredients are added. Cryoprotectant materials such as sugar and sorbitol are added prior to freezing to prevent the degradation of the gel-forming properties of surimi. These ingredients also have an impact on the taste of the final product and help extend its shelf life.

While surimi gels provide structure, other ingredients are needed to help stabilize and modify its texture. One important ingredient is starch. It improves the texture and stabilizes the gel matrix. This is particularly important for the stability of the product when it is frozen. The amount of starch is usually about 6% of the recipe. Egg white is also added to the surimi to improve the gel structure. It has the ability to increase the gel strength and improve its appearance by making the surimi more glossy and whiter. Vegetable oil is also used to improve the appearance of surimi and modify its texture.

Flavoring is added to surimi to make it taste like crab meat. These flavorants can be nat-ural or artificial, but typically a mixture of both is used. Natural flavoring compounds include amino acids, proteins, and organic acids, which are obtained through aqueous extraction of edible crabs. Artificial flavors can be made to closely match crab meat flavor and are typically superior to naturally derived flavorants. Artificial flavoring compounds include esters, ketones, amino acids, and other organic compounds. Additionally, seasonings and secondary flavorants are added to the meat to improve the overall flavor. Common ingredients include nucleotides, monosodium glutamate, vegetable proteins, and mirin.

The coloring for imitation crab meat is typically made using water insoluble compounds like carmine, caramel, paprika, and annato extract. By combining these and other ingredients, various shades of red, orange, and pink can be obtained. Before using the colorants, they are mixed in a surimi paste. This allows them to be easily applied to the imitation crab meat bundles.

The Manufacturing Process

Sorting, cleaning, and filleting

1 The manufacture of imitation crab meat begins by preparing the fish that will be converted to surimi. When the fish are caught in large nets, they must be manually sorted by species and cleaned. They are fur-

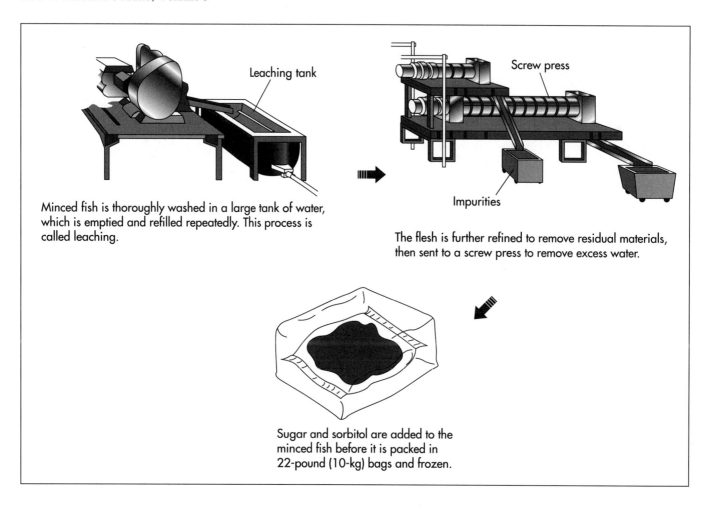

Minced fish is thoroughly washed in a large tank of water, which is emptied and refilled repeatedly. This process is called leaching.

Leaching tank

Screw press

Impurities

The flesh is further refined to remove residual materials, then sent to a screw press to remove excess water.

Sugar and sorbitol are added to the minced fish before it is packed in 22-pound (10-kg) bags and frozen.

ther mechanically sorted by size to optimize the yield of fillets. Scales are mostly removed from the fish after sorting. The fish are then conveyored to a filleting machine, which removes the head, tail, and viscera. Water washing is done next to remove excess fluids. This whole process can be done either on the fishing boat or in land-based manufacturing plants.

Preparing surimi

2 The prepared fish fillets can then be minced, or mechanically deboned, and made into surimi. This is done using a mechanical deboner, which removes the skin, scales, fins, and bones. This machine is made up of a thick rubber belt and a perforated drum. As the fish passes through this machine, the belt presses it against the drum, forcing the soft flesh particles to the interior of the drum while leaving the harder scales and bones on the outside. The drum is constantly rotated and the excess outer material is scraped off and collected in a waste bin.

3 The minced fish is next thoroughly washed with water in a process called leaching. This is done in a large tank which is emptied and refilled with water repeatedly. Leaching removes many undesirable water soluble materials such as fats, inorganic salts, and some proteins. After the final leaching cycle, the mince is partially dewatered before moving to the refining phase of manufacture.

4 Refining machines are made up of a cylindrical screen and a rotor. The mince is selectively separated with the soft, white meat in the front of the machine and the harder, browner meat in the back. This refining step removes any residual materials such as skin, bones, and scales. The refined mince is sent to a screw press that removes all excess water.

5 After dewatering, cryoprotective compounds such as sugar and sorbitol are added to the mince to help protect the fish proteins from breaking down during the final, freezing stage of manufacture. The

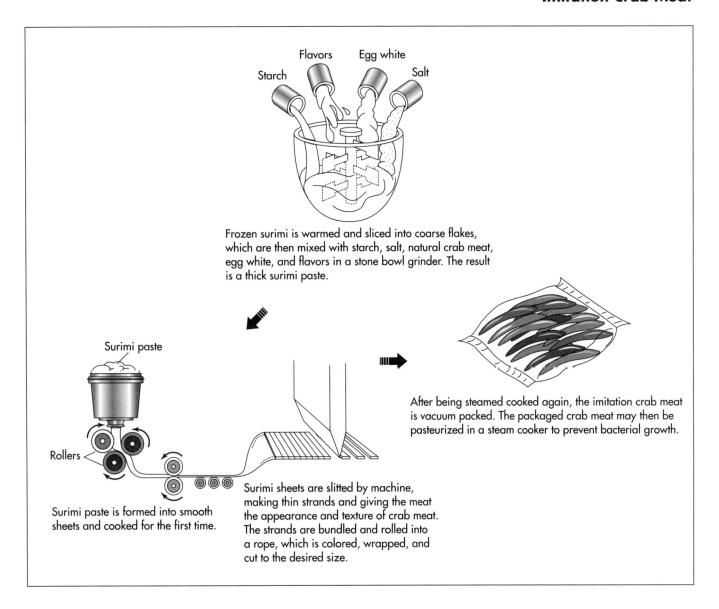

Starch Flavors Egg white Salt

Frozen surimi is warmed and sliced into coarse flakes, which are then mixed with starch, salt, natural crab meat, egg white, and flavors in a stone bowl grinder. The result is a thick surimi paste.

Surimi paste

Rollers

Surimi paste is formed into smooth sheets and cooked for the first time.

Surimi sheets are slitted by machine, making thin strands and giving the meat the appearance and texture of crab meat. The strands are bundled and rolled into a rope, which is colored, wrapped, and cut to the desired size.

After being steamed cooked again, the imitation crab meat is vacuum packed. The packaged crab meat may then be pasteurized in a steam cooker to prevent bacterial growth.

final step in surimi production involves packing it in polyethylene bags in 22-pound (10-kg) blocks and quickly freezing it to below -4°F (-20°C). The surimi is stored at this temperature until it is ready to be used.

Forming the crab meat

6 The frozen surimi is converted to imitation crab meat through various steps. First, it is warmed to about 25°F (-4°C), then sliced into coarse flakes. In a process known as comminution, the surimi flakes are then mixed together in a stone bowl grinder with other ingredients in the crab meat recipe. These ingredients include starch, salt, natural crab meat, egg white, and flavors. This mixture results in a thick surimi paste, which is then transferred to a holding tank.

7 The paste is pumped from the holding tank to the sheet-forming equipment. Here, continuous sheets of surimi, about 10 inches (25 cm) wide and 0.05 inch (1.2 mm) thick are produced. Due to the chemical nature of the surimi protein, these sheets are very smooth. After the sheets are formed, they are sent to machines for the initial cooking. This cooking helps set the sheets and prepares them for the slitting operation, which gives the meat the appearance and texture of crab meat.

8 The slitting is done by a machine which is composed of two steel rollers that cut the surimi sheet into thin 0.1 inch (1.5 mm) wide strands. These thin strands are then bundled and rolled into a rope. This rope is given the appropriate color, wrapped, and

cut to the desired size. It is then steamed cooked, forming a product that looks and tastes very much like the crab meat it is designed to imitate.

Packaging

9 Imitation crab meat is mechanically vacuum packed in thermoformed trays. This protects the meat from contamination and provides an appealing look. Some common plastics used for packing include polyethylene, nylon, and polyester. After packing, the imitation crab meat is typically pasteurized in a steam cooker. This step helps prevent bacterial growth and increases shelf life.

Quality Control

In the manufacture of imitation crab meat, quality control tests are performed at various points. For example, the characteristics of the incoming raw materials are analyzed. Specific properties such as pH, percentage of moisture, odor, taste, and appearance are all evaluated. The quality of the incoming fish is also checked. Most important is the test for rancidity.

The quality of the surimi is also examined by testing various characteristics. The chemical composition is tested using laboratory methods. Such things as protein content, moisture, and lipid content are all checked. Also, visual assessment of the color and texture of the surimi gives an indication of quality, as does a pH test. Since the gel-forming ability of the surimi is paramount to its use in imitation crab meat, various tests are run to ensure that it meets minimum standards before it is used. Finally, imitation crab meat is susceptible to microbial attack. Therefore, manufacturers routinely test whether their products are contaminated.

Byproducts/Waste

The water left over from the manufacture of surimi is characterized as waste water. It is composed of many water-soluble substances, fats, and suspended particles. Environmental regulations require that manufacturers treat this water before returning it to the environment. This is done using such things as filters, centrifuges, and chemical treatments.

The Future

Future developments in the imitation crab meat industry are likely to be found in a few key areas. One important area of research has focused on the development of surimi from different kinds of fish. These would include fish that currently have low economic value and are quite abundant. Many of these new fish have more fat and different body chemistries than the fish currently used, so the challenge will be to improve the surimi that can be made using them. In the manufacturing area, a more continuous process is being developed. These processes result in better yields of surimi. Also, environmental concerns will lead to new technologies that will minimize the amount of waste involved in manufacture. Finally, new crab meat recipes aimed at improving the nutritional value of the product will be developed.

Where to Learn More

Books

Lanier, Tyre, and Chong Lee, eds. *Surimi Technology.* Marcel Dekker, 1993.

Sikorski, Zdzislaw. *Seafood Proteins.* Chapman and Hall, 1994.

Periodicals

Okada, Minoru. "A fish story. What is the 'plastic food' really made of?" *Chemtech,* October 1991, pp. 588-591.

—*Perry Romanowski*

Instant Coffee

Background

Instant (or soluble) coffee has been widely used for decades because of its convenience. During the height of its popularity in the 1970s, nearly a third of the roasted coffee imported into the United States was converted into an instant product, resulting in annual sales of more than 200 million pounds. Today, about 15% of the coffee consumed in the United States is prepared by mixing instant granules with hot water, either at home, in offices, or in vending machines. Furthermore, development of good quality instant products has helped popularize coffee in cultures that historically drank tea.

Since its invention, researchers have sought to improve instant coffee in a variety of ways. For example, some of the early powdered versions did not dissolve easily in water, leaving clumps of damp powder floating in the cup. Coffee aroma dissipates easily, and manufacturers have tried to develop treatments that will make a jar of instant coffee smell like freshly ground coffee when it is opened. More modern manufacturing processes make instant coffee granules that look more like ground coffee. Finally, a major goal has been to produce an instant coffee that tastes as much as possible like the freshly brewed beverage.

The primary advantage of instant coffee is that it allows the customer to make coffee without any equipment other than a cup and stirrer, as quickly as he or she can heat water. Market researchers have also found that consumers like making coffee without having to discard any damp grounds. Some

coffee drinkers have become so used to drinking instant coffee that at least one manufacturer found in taste tests that their target audience did not even know what fresh-brewed coffee tastes like.

History

The desire to make coffee instantly by simply mixing a liquid or dry concentrate with hot water goes back hundreds of years. The earliest documented version of instant coffee was developed in Britain in 1771. The first American product was developed in 1853, and an experimental version (in cake form) was field tested during the Civil War. In 1901, the first successful technique for manufacturing a stable powdered product was invented in Japan by Sartori Kato, who used a process he had developed for making instant tea. Five years later, George Constant Washington, a British chemist living in Guatemala, developed the first commercially successful process for making instant coffee.

Washington's invention, marketed as "Red E Coffee," dominated the instant coffee market in the United States for 30 years, beginning around 1910. During the 1930s, the Brazilian coffee industry encouraged research on instant coffee as a way of preserving their excess coffee production. The Nestlé company worked on this effort and began manufacturing Nescafé in 1938, using a process of co-drying coffee extract along with an equal amount of soluble carbohydrate. Instant coffee was enormously popular with American soldiers during World War II; one year, the entire production from the U.S. Nescafé plant (in excess of one million cases) went solely to the mil-

Instant coffee was enormously popular with American soldiers during World War II; one year, the entire production from the U.S. Nescafé plant (in excess of one million cases) went solely to the military.

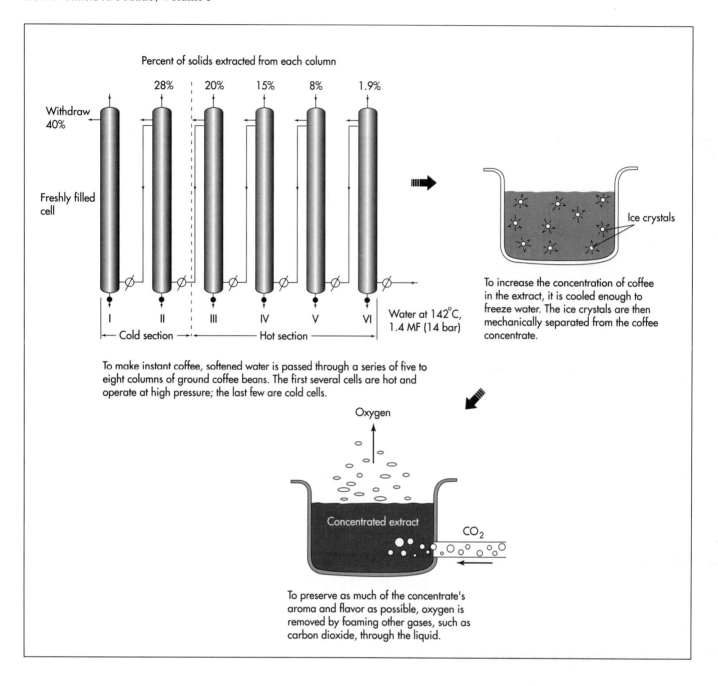

Percent of solids extracted from each column

28%   20%   15%   8%   1.9%

Withdraw 40%

Freshly filled cell

I   II   III   IV   V   VI

Water at 142°C, 1.4 MF (14 bar)

├── Cold section ──┤├────── Hot section ──────┤

To make instant coffee, softened water is passed through a series of five to eight columns of ground coffee beans. The first several cells are hot and operate at high pressure; the last few are cold cells.

Ice crystals

To increase the concentration of coffee in the extract, it is cooled enough to freeze water. The ice crystals are then mechanically separated from the coffee concentrate.

Oxygen

Concentrated extract

CO_2

To preserve as much of the concentrate's aroma and flavor as possible, oxygen is removed by foaming other gases, such as carbon dioxide, through the liquid.

itary. By 1950, Borden researchers had devised methods for making pure coffee extract without the additional carbohydrate component. This improvement boosted instant coffee use from one out of every 16 cups of coffee consumed domestically in 1946 to one out of every four cups in 1954. In 1963, Maxwell House began marketing freeze-dried granules, which reconstituted into a beverage that tasted more like freshly brewed coffee. During the next five years, all of the major manufacturers introduced freeze-dried versions, and by the mid-1980s, 40% of the instant coffee used in the United States was freeze dried.

Raw Materials

Two of the 50 known species of coffee beans dominate the beverage coffee industry. *Coffee arabica* varieties, grown primarily in Latin America, India, and Indonesia, are relatively mild in flavor and, consequently, bring a higher price. They are also relatively expensive to harvest, since individual coffee cherries must be hand picked at their peak of ripeness. *Coffee robusta* varieties, grown mainly in Africa, India, and Indonesia, have a harsher flavor, but they are cheaper to grow since they can be harvested over a range of ripeness and are more resistant to diseases and insects. Because of their more attractive price, the *robustas* are widely used in the manufacture of instant coffees.

Roasting at temperatures above 300°F (180°C) drives the moisture out of coffee beans. Beans destined for use in instant products are roasted in the same way as beans destined for home brewing, although the moisture content may be left slightly higher (about 7-10%). The beans are then ground coarsely to minimize fine particles that could impede the flow of water through the industrial brewing system.

The Manufacturing Process

Extraction

1 The manufacture of instant coffee begins with brewing coffee in highly efficient extraction equipment. Softened water is passed through a series of five to eight columns of ground coffee beans. The water first passes through several "hot" cells (284-356°F, or 140-180°C), at least some of which operate at higher-than-atmospheric pressure, for extraction of difficult components like carbohydrates. It then passes through two or more "cold" cells (about 212°F, or 100°C) for extraction of the more flavorful elements. The extract is passed through a heat exchanger to cool it to about 40°F (5°C). By the end of this cycle, the coffee extract contains 20-30% solids.

Filtration and concentration

2 After a filtering step, the brewed coffee is treated in one of several ways to increase its concentration. The goal is to create an extract that is about 40% solids. In some cases, the liquid is processed in a centrifuge to separate out the lighter water from the heavier coffee extract. Another technique is to remove water by evaporation before cooling the hot, brewed extract. A third alternative is to cool the extract enough to freeze water, and then mechanically separate the ice crystals from the coffee concentrate.

Recovery of aromatic volatiles

3 Part of the enjoyment of making and drinking coffee is smelling the aroma.

During the several steps of the manufacturing process, volatile aromatic elements are lost; they must be returned in a later step to produce an attractive instant coffee product. Aromatics can be recovered during several stages of the manufacturing process. For instance, gases released during the roasting and/or grinding processes can be collected. Ground, roasted coffee can be heated to release additional aromatic gases. Passing steam or appropriate solvents through a bed of ground, roasted coffee can strip and capture aromatic components. Aromatic oils can be expressed from spent coffee grounds by exerting pressure of at least 2,000 lb per sq in (14,000 kPa). Gases can also be distilled from coffee extract after the brewing process is complete.

4 To preserve as much of the aroma and flavor as possible, oxygen is removed from the coffee extract. This is accomplished by foaming other gases, such as carbon dioxide or nitrogen, through the liquid before it enters the dehydration phase of the manufacturing process.

Dehydration

Two basic methods are available for converting the liquid coffee extract to a dry form. Spray drying is done at a higher temperature, which affects the taste of the final product, but it is less costly than freeze drying.

Spray drying

5 Cooled, clarified liquid concentrate is sprayed through a nozzle at the top of a drying tower. The tower is at least 75 ft (23 m) tall. Air that has been heated to about 480°F (250°C) is blown downward through the mist to evaporate the water. The air is diverted out of the tower near the bottom, and it is filtered to remove fine particles, which can be recirculated back through the tower or reintroduced during the agglomeration step. The dry coffee powder collects in the bottom of the tower before being discharged for further processing. The resulting powder contains 2-4% moisture and consists of free-flowing, non-dusty particles.

6 Spray drying may be followed by a step to form the powder into coarser particles that will dissolve more completely in the consumer's cup. The agglomeration process

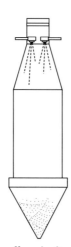

To evaporate the coffee, the liquid concentrate is sprayed through a nozzle at the top of a drying tower. Warm air blows downward, evaporating the water. Dry coffee powder collects in the bottom of the tower.

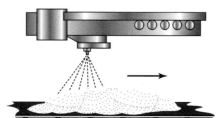

Volatile aromas that have been recovered from earlier steps in the manufacturing process are sprayed on the dry coffee particles.

Coffee is packaged in a low-oxygen and low-humidity atmosphere to optimize product flavor and consistency.

basically involves rewetting the surfaces of the coffee powder particles and bringing the particles into contact, so that they will adhere to each other and form larger, more granular particles. This is accomplished by exposing the powder to steam or a fine mist, while tumbling it in the air.

Freeze drying

7 Freeze drying may be used instead of spray drying. The process involves four steps, beginning with "primary freezing." Coffee extract is chilled to a slushy consistency at about 20°F (-6°C).

8 The prechilled slush is placed on a steel belt, trays, or drums and further cooled in a series of steps, until it reaches a temperature of -40-(-50)°F (-40-[-45]°C). Quick cooling processes (taking 30-120 seconds) result in smaller, lighter colored products, while slower processes (taking 10-180 minutes) generate larger, darker granules.

9 The slabs of ice are broken into pieces and ground into particles of the proper size for the drying step. The particles are sieved to ensure proper sizing, and those that are too small are melted and returned to the primary freezing stage.

10 The frozen particles are sent into a drying chamber where, under proper conditions of heat and vacuum, the ice vaporizes and is removed.

Aromatization

11 Volatile aromas that have been recovered from earlier steps in the manufacturing process are sprayed on the dry coffee particles. This may be done during the packaging operation.

Packaging

12 Instant coffee particles are hygroscopic—that is, they absorb moisture from the air. Consequently, they must be packaged under low humidity conditions in a moisture-proof container to keep the product dry until purchased and opened by the consumer. Also, to prevent loss of aroma and flavor, the product is packaged in a low-oxygen atmosphere (usually carbon dioxide or nitrogen).

Byproducts/Waste

Spent coffee grounds from the brewing process are the primary waste product. At least one manufacturer burns these grounds to heat water and generate steam that is used in the manufacturing process. The process is designed to be environmentally friendly, minimizing waste products by maximizing the use of the raw materials.

The Future

Since the introduction of General Foods International Coffees in the 1970s, instant coffees have been available in flavored varieties. Recent innovations include instant mixes for latte and mocha beverages. Maxwell House is test marketing an instant iced coffee product in vanilla, mocha, and original coffee flavors.

Where to Learn More

Books

Pintauro, Nicholas. *Soluble Coffee Manufacturing Processes.* Noyes Development Corp., 1969.

Pintauro, Nicholas. *Coffee Solubilization: Commercial Processes and Techniques.* Noyes Data Corp., 1975.

Ullmann, Fritz. "Coffee." *Ullmann's Encyclopedia of Industrial Chemistry.* VCH, 1985, pp. 315-339.

Periodicals

McCormack, Tim. "To Brew or Not To Brew." *Fancy Food Magazine,* January 1996.

—*Loretta Hall*

Iron-On Decal

The term decal is from the French decalquer, meaning to trace or to copy.

Background

An iron-on decal is an image printed on special paper that allows it to be transferred to fabric by applying heat and pressure. Iron-ons are one of the four primary types of decals; the other three types are slide off, varnish, and pressure sensitive. The term decal is from the French *decalquer*, meaning to trace or to copy. Decals are traced, or more accurately, stenciled, onto a fabric screen. Ink is forced through the stencil and onto a printing substrate in a process is known as screen printing. Iron-on decals are printed on a temporary substrate that is designed to release the image when exposed to pressure and heat. The act of ironing the decal causes the image to move from its paper backing to the fabric, hence the name "iron on."

The concept of using stencils to print images has been employed throughout recorded history. There is even evidence to suggest that the first stencils were made from large leaves and were used by prehistoric people to create cave paintings. Over time, stencil printing was used for decorating such items as furniture, walls, and fabrics. A drawback of this technique was that holes in the stencils are connected by "bridges," which show up as interrupting gaps or "islands" in the final printed image. This problem was solved in the early 1900s by Samuel Simons, who devised a way to print through a silk fabric screen. The silk allowed the ink to slip around the fine fibers and give a smoother looking image. This modified method of stencil printing is known today as "silk screening." Simon's invention was first officially discussed in his 1907 British patent, but unfortunately it

was slow to catch on commercially. Not until 1938 did a group of practicing artists under the federal Works Projects Administration began to seriously explore the technique. By the end of World War II, the silk screening process was recognized as a valuable printing method, and today it is one of the most widely used of all printing technologies. In the last 40 years, printers have developed techniques to screen print different types of ink on a wide variety of substrates. Heat-transferred graphic decals are made using these advanced techniques.

Raw Materials

Production of iron-on decals requires stencil-making materials, inks, a porous printing screen, and a printing substrate to receive the final image.

Stencil materials

Stencils are made from nonporous paper or plastic coated with lacquer, gelatin, or a combination of glue and tusche, a heavy ink-like substance. These materials are either oil soluble or water soluble, depending on the type of printing. The stencil blocks portions of the screen during the printing process, so that ink only touches the paper in designated spots.

Inks

The inks commonly used for iron-on decals fall into two categories, plastisol type and sublimation type. Both types use pigments created from a variety of metals, clays, plants, and synthetic chemicals to provide color. These pigments are suspended or dissolved in a liquid solvent such as mineral spirits, alcohol, or water. Plastisol ink is a lacquer-based ink that is specially designed for use on fabrics. It is dried by heating it to 300°F (149°C) for several minutes. This type of ink is also quite thick and requires some special handling to produce a good image. Sublimation ink is not really an ink at all but rather a dye-like pigment. When exposed to heat, the pigment vaporizes and moves from its temporary support to the fabric, where it becomes permanently bonded to the fibers. Synthetic fibers such as polyester and nylon are particularly effective at bonding with sublimation inks. Both of these ink types are available in a wide array of colors, so that almost any image can be reproduced.

Printing screen

The screens used in this process are typically finely woven fabrics, like silk, nylon, and dacron, or stainless steel meshes that are stretched tightly over a rigid frame. Small scale printing screens may be wood or plastic. Large commercial screens are typically made of metal.

Printing substrates

The printing substrate used for heat transfers depends on the type of ink employed. Plastisol inks can only be printed on a specially coated paper stock that is designed to absorb the inks. When the paper is placed face down on fabric, and heat and pressure are applied, the coating melts and allows the image to transfer to the fabric. The coating produces a rubbery feel on the finished garment. On the other hand, sublimation dyes are printed on uncoated paper stock and are transferred as a result of the chemical change from solid to gas.

The Manufacturing Process

The process of creating iron-on decals involves three key steps: preparation of the stencil to be used to print the image; the screen printing process itself; and transfer of the image to the fabric substrate.

Stencil preparation

1 The first step is to create a stencil of the image. As described above, stencils are

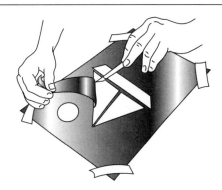

Iron-on decals are made using stencils, which consist of a paper or plastic backing sheet coated with a thin film of lacquer or gelatin. First, an outline of the image to be printed is cut into the film layer of the stencil. This portion is then peeled away.

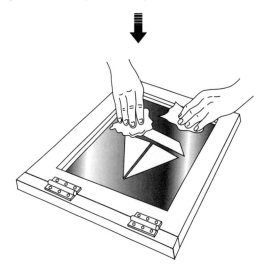

A solvent is then used to adhere the stencil to the underside of a fabric screen.

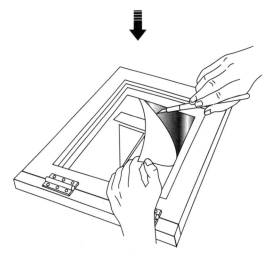

Once the stencil has dried, the backing layer is removed, leaving the film layer attached to the screen.

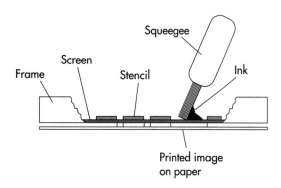

To transfer the image onto release paper, ink is placed on top of the screen and forced through the openings in the stencil using a rubber squeegee. This process can be done by hand or it can be automated. Different colors are laid down one at a time.

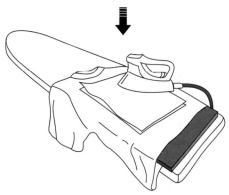

To transfer the image onto cloth, heat and pressure are applied to the decal using a household iron. Cardboard is inserted underneath the garment to provide a stable backing, and the decal is placed on the garment with the image facing down. Scrap cloth may be placed on the decal to protect it from the direct heat of the iron.

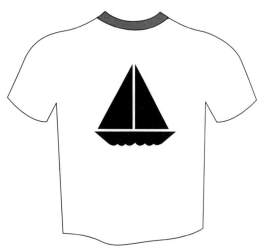

After the decal is cooled, the release paper can be peeled off, leaving the finished image on the garment.

made from paper or plastic sheets, which are coated with a thin layer of lacquer or gelatin. An outline of the image to be printed is cut into this thin layer and the lacquer or gelatin is peeled away. A solvent is then used to adhere the stencil to the underside of the fabric screen. Once the stencil has dried on the screen, the paper or plastic backing sheet is removed, leaving only the film layer. The portions of the stencil that were cut away expose a section of the screen through which ink can be forced. Such stencils may be produced from photographic images by a similar process, but instead of cutting away the portions of the image, a light-sensitive material is used to mark an image on the screen. Alternatively, the stencil may be painted directly on the screen using tusche.

Screen printing

2 Image transfer is accomplished by forcing various inks, which are placed on top of the screen fabric, through the stencil and onto the printing substrate. A rubber squeegee is used to force ink through openings in the stencil. Ink application may be done by hand or by automated printing processes. The different colors are laid down sequentially, one at a time. The plastisol type must be heated to cure after each application. This curing is done at temperatures of 225-250° F (107-121°C). The ink must be heated to this temperature for about one minute and then cooled before the next color can be added. (The sublimation type requires no heat curing.) It should also be noted that the colors are laid down in reverse order, from last to first. Therefore, the finished decal resembles a multi-layer sandwich. The bottom layer is the release paper, followed by what are called the detail colors. Then the background colors are laid down one at a time.

Transfer process

3 Upon completion of the printing process, the image is ready to be transferred to a T-shirt or other garment. Before transferring the image, the fabric must be laid out on a smooth hard surface. For a T-shirt, a piece of heavy cardboard is inserted between its front and back to provide firm support and prevent the transfer from sticking to the back of the garment. Heat and

pressure are then applied using an ordinary household iron or a special piece of equipment called a dry mounting press. The latter consists of two flat, electrically heated metal plates that provide even pressure and heat to the fabric. Using the dry press allows for better image transfer with less chance of scorching. For iron-on decals made with plastisol inks, the dry press is set to about 300°F (149°C). This temperature is adequate to melt the lacquer layer holding the inks. Continued heat and pressure for one to three minutes cause the molten ink/lacquer film to migrate to the fabric. When the heat source is removed, the ink/lacquer combination cools and binds to the fabric. After cooling is complete, the paper backing is peeled away. A similar process is used for sublimation inks, except no carrier layer is needed because the pigments vaporize and transfer directly to the fabric as result of heat and pressure. A temperature of 350-375°F (177-191°C) is usually sufficient to affect this change. Sublimation inks transfer more cleanly than the plastisol type since there is no lacquer coating or film to be transferred.

Quality Control

The quality control measures employed for decals are meant to ensure that the image transfer is clean and crisp. The following factors have been determined to be critical to image quality:

- The screen fabric must be properly adhering to the stencil.

- The right type of adhesive must be used to adhere the stencil to the screen.

- Contact time for the adhesive must be limited to avoid softening the stencil.

- The stencil must not be cut with poor or dull tooling.

- The screen must be properly cleaned after each ink application.

With plastisol inks, care must be taken during image transfer to ensure that residual lacquer layer does not stick to the paper and cause smearing of the image.

Byproducts/Waste

The decal manufacturing process generates waste in the form of excess materials used in stencil production (lacquer, gelatin, and paper). The very nature of the stencil dictates that some material will be wasted, because unused portions of the image are cut away. Excess inks that are wiped off the screen also contribute waste products, as do solvents used for cleaning equipment. Depending on the chemistry of the specific materials employed, the waste may be flammable and considered hazardous. To a large extent this depends on whether the waste material is water or solvent based.

The Future

Methods of producing iron-on decals stand to be improved as advances are made in ink and paper coating chemistry. The development of quicker drying inks with a wider range of colors and better adherence to paper and fabric substrates would substantially increase the efficiency with which these products are made. The current trend toward increased regulations to protect the environment will likely impact the types of inks, solvents, and lacquers that are used in decal production. Development of environmentally safe, or "green," products would be an asset to the industry.

For example, a recent improvement in computer technology has led to a better screen printing process, known as FM screening. This process allows printers to use smaller, more uniform screens, resulting in smoother images. Another interesting advance in decal printing now allows those with a personal computer to make their own iron-ons. Several manufacturers offer specially treated iron-on paper that can be printed with a standard color printer. An image on this paper can then be simply and easily ironed on to shirts or other garments. This process can be used by anyone to create unique and memorable garments, but it cannot replace the commercial printing processes currently used for screen printing iron-on decals.

Where to Learn More

Books

Grattaroti, Rosalie, ed. *Great T-shirt Graphics*. Rockport Publishers, 1993.

Swerdlow, Robert. *The Step by Step Guide to Screen-Process Printing.* Prentice-Hall, 1985.

Periodicals

McDougall, Paul. "FM Screening: Big Gains from Tiny Dots?" *Folio: the Magazine for Magazine Management*, January 15, 1994, p. 25.

Strashun, Joann. "Screening Options Gain Momentum." *Graphic Arts Monthly*, February 1994, P. 55.

Toth, Debora. "Printers Ponder Screening Options." *Graphic Arts Monthly*, August 1995, p. 45.

—*Randy Schueller*

Latex

Background

A latex is a colloidal suspension of very small polymer particles in water and is used to make rubber.

Natural

Dipped goods (medical and surgical items, household and industrial gloves, boots, and balloons) utilize more than half of all natural latex consumed in the United States. The adhesives industry is the second largest user of natural latex in products such as shoes, envelopes, labels, and pressure sensitive tape.

Natural latex with a high solids content is also used for making molds for casting plaster, cement, wax, low temperature metals, and limited run polyester articles. Natural latex has the ability to shrink around the object to be reproduced, so that the smallest detail will be reproduced in the cast. Latex is even being used to help stabilize desert soils to make them suitable for agricultural uses.

Natural latex is produced from the *Hevea brasilienesis* rubber tree and is the protective fluid contained beneath the bark. It is a cloudy white liquid, similar in appearance to cow milk. It is collected by cutting a thin strip of bark from the tree and allowing the latex to exude into a collecting vessel over a period of hours.

Hevea trees mature at five to seven years of age and can be tapped for up to 30 years. Rubber yields range around a ton per acre (2.5 tons per ha) on the larger plantations, but yields four times as much are theoretically possible. Trees often are rested for a period after heavy tapping.

Natural latex was once commercially produced in the Amazon in great quantities. In recent times, production of natural latex has moved to Malaysia, Indonesia, and other Far Eastern regions. More than 90% of the total world production of natural rubber now comes from Asia, with well over half of that total originating in these countries. Other leading Asian producers include Thailand, India, and Sri Lanka. China and the Philippines both have substantially increased their rubber production as well.

Synthetic

Most synthetic rubber is created from two materials, styrene and butadiene. Both are currently obtained from petroleum. Over a billion pounds (454,000,000 kg)of this type of rubber was manufactured in the United States in 1992. Other synthetic rubbers are made from specialty materials for chemical and temperature resistant applications.

Tires account for 60-70% of all natural and synthetic rubber used. Other products containing rubber include footwear, industrial conveyor belts, car fan belts, hoses, flooring, and cables. Products such as gloves or contraceptives are made directly from rubber latex. Latex paints are essentially a solution of colored pigment and rubber latex. Latex foam is made by beating air into the latex before coagulating it.

History

The Indians of Central and South America used rubber as early as the eleventh century to coat fabric or to make into balls, but it was not until the French scientist Charles de la Condamine visited South America during

More than 90% of the total world production of natural rubber now comes from Asia, with well over half of that total originating in Malaysia and Indonesia.

Buds from high-grade rubber trees are grown and then grafted to year-old seedlings in order to create new, high-grade trees.

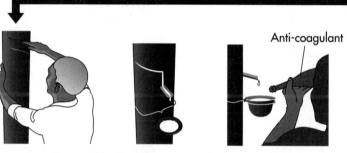

Slanted strips of bark 1/3 in (0.8 cm) thick are cut from the tree to start the tap. The thickness of the cut determines both the flow of latex and the amount of damage done to the tree. Precise skill is necessary in order to successfully tap a tree.

the 1700s that the first samples were sent back to Europe. Rubber was given its present English name by the British chemist Joseph Priestley in about 1770. The first modern use for rubber was discovered in 1818 by a British medical student named James Syme. He used it to waterproof cloth in order to make the first raincoats, a process patented in 1823 by Charles Macintosh. Thomas Hancock devised methods for mechanically working rubber so it could be shaped, and he built England's first rubber factory in 1820. Also during this period, Michael Faraday discovered that natural rubber is composed of units of a chemical compound called isoprene.

During the mid-nineteenth century, Charles Goodyear discovered vulcanization, a process that retains the rubber's elasticity under temperature changes. This process heats rubber with sulfur, which causes cross linking, decreasing rubber's tackiness and sensitivity to heat and cold.

In 1882, John Boyd Dunlop of Ireland was granted a patent for his pneumatic tire. As the demand for tires began to deplete natural rubber supplies, the British cultivated huge rubber plantations in Singapore, Malaysia, and Ceylon (Sri Lanka). Seeds were taken from Brazil and first germinated in England and then shipped to these countries. Today, all natural rubber produced in Asia comes from trees that are descendants of the Brazilian seeds.

By the early 1900s, various countries sought ways to improve rubber compounds and to develop synthetic materials. In 1910, sodium was found to catalyze polymerization. When the Germans were cut off from natural rubber supplies during World War I, they used this discovery to make about 2,500 tons (2,540 metric tons) of rubber made from dimethylbutadiene.

During World War II, the Japanese gained control of the major sources of natural rubber in Asia. In response, the United States' synthetic rubber industry increased its production by an astonishing 10,000%, from 7,967 tons (8,130 metric tons) in 1941 to more than 984,000 tons (1 million metric tons) in 1944. Following the war, other countries developed their own synthetic rubber factories to avoid having to rely on overseas rubber supplies.

Improvements in synthetic rubber have continued, and in addition, higher yielding hybrid trees have been developed that yield twice as much natural latex as the conventional ones. In 1971, a tree stimulant was developed that resulted in an average increase of 30% in latex production with no apparent harm to the trees.

Raw Materials

The composition of latex sap consists of 30-40% rubber particles, 55-65% water, and small amounts of protein, sterol glycosides, resins, ash, and sugars. Rubber has high elasticity and a polymer molecular structure. This structure consists of a long chain made up of tens of thousands of smaller units, called monomers, strung together. Each monomer unit has a molecular size comparable with that of a simple substance such as sugar. Other special chemicals are used as preservatives or stimulants during the harvesting process.

Both synthetic and natural rubber production require the use of vulcanizing chemicals, primarily sulfur. Fillers such as carbon black are also added to provide extra strength and stiffness. Oil is often used to help processing and reduce cost.

The Manufacturing Process

Growing and processing natural rubber is one of the most complex agricultural industries and requires several years. It combines botany, chemistry, and sophisticated machinery with dexterous skills of the people who harvest the trees. Contrast this with synthetic rubber production, which involves chemical reactions and sophisticated chemical processing machinery that is automatically controlled by computers. The production of natural latex is described below.

Planting

1 Seeds from high-grade trees are planted and allowed to grow for about 12 to 18 months in the nursery before a new bud is grafted to the seedling. After bud grafting, the year-old seedling tree is cut back and is ready for transplanting. The bud sprouts shortly after transplanting, resulting in a new tree with better properties. Approximately 150 trees are planted per acre (375 per ha), which are cultivated and cared for until they are ready for tapping in about six to seven years.

Tapping

2 To harvest latex, a worker shaves off a slanted strip of bark halfway around the tree and about one third in (0.84 cm) deep. Precise skill is required for if the tree is cut too deeply, the tree will be irreparably damaged. If the cut is too shallow, the maximum amount of latex will not flow. The latex then bleeds out of the severed vessels, flows down along the cut until it reaches a spout, and finally drops into a collection cup that will later be drained.

3 Tapping is repeated every other day by making thin shavings just below the previous cut. When the last scar created by the cuts is about 1 ft (0.3 m) above the ground, the other side of the tree is tapped in similar fashion, while the first side renews itself. Each tapping takes about three hours and produces less than a cup of latex.

4 A tapper first collects the cut lump, which is coagulated latex in the cup, and tree lace, which is latex coagulated along the old cut. Next, the tapper makes a new cut. The latex first flows rapidly, then declines to a steady rate for a few hours, after which it slows again. By the next day, the flow has nearly stopped as the severed vessel becomes plugged by coagulated latex.

5 To prevent most of the liquid latex from coagulating before it can be conveniently pooled and transported, the tapper adds a preservative such as ammonia or formaldehyde to the collection cup. Both the liquid and coagulated latex is sent to factories for processing.

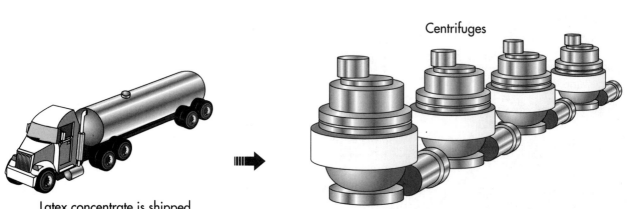

Latex concentrate is shipped by tank truck to the centrifuge operation.

Centrifuges

A liquid concentrate is formed from 10% of the collected latex by removing water from the sap and, thereby, increasing the rubber content to 60%.

Quality control is maintained throughout the manufacturing process by skilled technicians.

Acid is added to the latex and then water is removed to form bales of dry stock rubber.

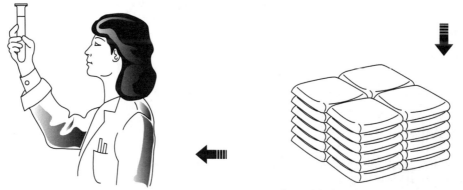

Raw rubber is extruded, molded, rolled, whipped, bonded, and mixed with other materials to make thousands of products.

6 To increase tree yields and reduce tapping times, chemical stimulants are used. Puncture tapping, in which the bark is quickly pierced with sharp needles, is another method that can improve productivity, since it enables the same worker to tap more trees per day.

Producing liquid concentrate

7 About 10% of the latex is processed into a liquid concentrate by removing some of the water and increasing the rubber content to 60%. This is achieved either by spinning the water out of the latex through centrifugal force, by evaporation, or by a method known as creaming. In this method, a chemical agent is added to the latex that causes the rubber particles to swell and rise to the liquid's surface. The concentrate is shipped in liquid form to factories, where it is used for coatings, adhesives, and other applications.

Producing dry stock

8 Other rubber and field latex is coagulated with acid. A giant extrusion dryer that can produce up to 4,000 lbs (1,816 kg) per hour removes the water, creating a crumb-like material. The dried rubber is then compacted into bales and crated for shipment.

Forming sheets

9 Ribbed smoked sheets are made by first diluting the latex and adding acid. The acid makes rubber particles bunch together above the watery serum in which they are suspended. After several hours, roughly one pound (0.45 kg) of soft, gelatinous rubber coagulates for every three pounds (1.35 kg) of latex.

10 The rubber is allowed to stand for one to 18 hours, then the slabs are pressed into thin sheets through a system of rollers that wrings out excess liquid. The final set of rollers leaves a ribbed pattern on the sheets that increases the surface area and hastens drying. The sheets are dried for up to a week in smoke houses before being packed and shipped.

Producing other products

11 To make rubber products, the mix is shaped by placing it in a heated mold, which helps shape and vulcanize the material. For more complex products, such as tires, a number of components are made, some with fiber or steel-cord reinforcement, which are then joined together. Surgical gloves are made by dipping a ceramic form into latex, withdrawing the form, and then drying the latex shape.

Quality Control

A number of quality checks are made after the latex is harvested. After tapping, the latex is checked for purity and other properties. After each step of the production process, technicians check physical properties and chemical composition, using a variety of analytical equipment.

The Future

The production of natural rubber has failed to meet the growing demand for rubber, and hence, today two-thirds of the world's rubber is synthetic. However, developments, such as the invention of epoxidized natural rubber which is produced by chemically treating natural rubber, may reverse this trend. The synthetic rubber industry is also continuing to make processes more efficient, less costly, and less polluting, as well as developing new additives, compounds, and applications.

Though there are as many as 2,500 other plants that produce rubber, it is not made fast enough to be profitable. United States Department of Agriculture researchers are looking at ways to speed up the process by genetically engineering a plant to make larger initiator molecules. These molecules start the rubber-making process, and if such molecules were larger, rubber could be produced up to six times faster.

Where to Learn More

Periodicals

Amato, Ivan. "History of Rubber Production and Use." *News Service of the American Chemical Society.*

"New Ways to Make Rubber Tried." *The Associated Press News Service*, March 24, 1995.

Other

Firestone Synthetic Rubber & Latex Company, PO Box 26611, Akron, OH 44139-0008. http://www.firesyn.com.

CEMENTEX Latex Corp., 121 Varick St., New York NY 10013. (212) 741-1770. http://www.cementex.com.

—*Laurel M. Sheppard*

Lumber

Background

Lumber is a generic term that applies to various lengths of wood used as construction materials. Pieces of lumber are cut lengthwise from the trunks of trees and are characterized by having generally rectangular or square cross sections, as opposed to poles or pilings, which have round cross sections.

The use of wood as a construction material predates written history. The earliest evidence of wood construction comes from a site near Nice, France, where a series of post holes seems to indicate that a hut 20 ft (6m) wide by 50 ft (15 m) long was built there 400,000 years ago using wood posts for support. The oldest wood construction found intact is located in northwest Germany, and was built about 7,300 years ago. By 500 B.C. iron axes, saws, and chisels were commonly used to cut and shape wood. The first reference to cutting wood in a sawmill, rather than using hand tools, comes from northern Europe and dates from about 375. The sawmill was powered by the flow of water.

In North America, European colonists found vast forests of trees, and wood became the principal building material. The circular saw, which had been developed in England, was introduced in the United States in 1814 and was widely used in sawmills. A large-scale bandsaw was developed and patented by Jacob R. Hoffman in 1869 and replaced the circular saw for many sawmill operations.

Lumber produced in early sawmills had varying dimensions depending on the customer's specific order or the mill's standard practice. Today, lumber pieces used in construction have standard dimensions and are divided into three categories, depending on the thickness of the piece. Lumber with nominal thicknesses of less than 2 in (5 cm) are classified as boards. Those with nominal thicknesses of 2 in (5 cm) but less than 5 in (13 cm) are classified as dimension. Those with nominal thicknesses of 5 in (12.5 cm) and greater are classified as timbers. The nominal widths of these pieces vary from 2-16 in (5-40 cm) in 1 in (2.5 cm) increments. Most rough-cut lumber pieces are dried and then finished, or surfaced, by running them through a planer to smooth all four sides. As a result, the actual dimensions are smaller than the nominal dimensions. For example, a standard two-by-four piece of dried, surfaced dimension lumber actually measures 1.5 in (3.8 cm) by 3.5 in (8.9 cm).

Pieces of lumber that are not only surfaced, but also machined to produce a specific cross sectional shape are classified as worked lumber or pattern lumber. Decorative molding, tongue-and-groove flooring, and shiplap siding are examples of pattern lumber.

Today, processing wood products is a billion-dollar, worldwide industry. It not only produces construction lumber, but also plywood, **fiberboard**, paper, cardboard, turpentine, rosin, textiles, and a wide variety of industrial chemicals.

Raw Materials

The trees from which lumber is produced are classified as hardwoods or softwoods. Although the woods of many hardwoods are hard, and the woods of many softwoods are soft, that is not the defining characteristic. Most hardwood trees have leaves,

Many of the trees being logged today are second-generation or third-generation trees that are younger and smaller in diameter than the original old-growth trees. These younger trees also contain a higher percentage of juvenile wood, which is less dimensionally stable than older wood.

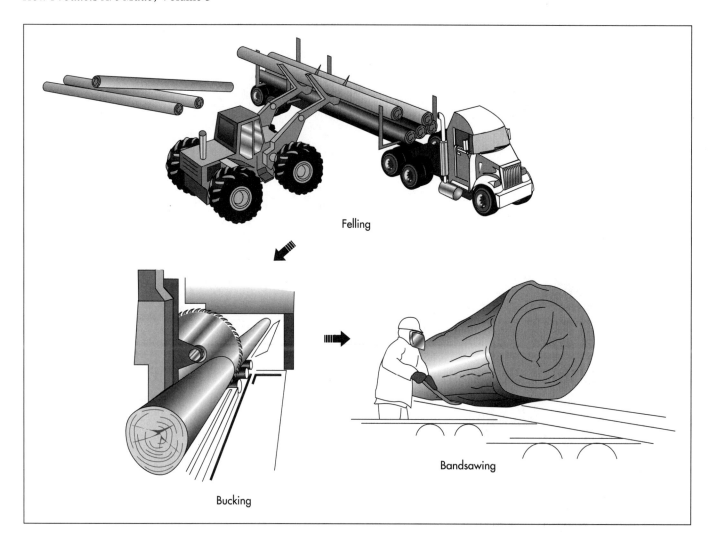

Felling

Bucking

Bandsawing

During felling, the trees are cut down with chain saws and the limbs are removed. At the mill, the logs are debarked and bucked, or cut to a predetermined length. Then they proceed to the bandsaw for further processing.

which they shed in the winter. Hardwood trees include oaks, maples, walnuts, cherries, and birches, but they also include balsa, which has one of the softest and lightest of all the woods. Softwood trees, on the other hand, have needles instead of leaves. They do not shed their needles in the winter, but remain green throughout the year and are sometimes called evergreens. Softwood trees include pines, firs, hemlocks, spruces, and redwoods.

Hardwoods are generally more expensive than softwoods and are used for flooring, cabinetry, paneling, doors, and trimwork. They are also extensively used to manufacture furniture. Hardwoods are available in lengths from 4-16 ft (1.2-4.8 m). Softwoods are used for wall studs, joists, planks, rafters, beams, stringers, posts, decking, sheathing, subflooring, and concrete forms. They are available in lengths from 4-24 ft (1.2-7.3 m).

Both hardwood and softwood lumber pieces are graded according to the number and size of defects in the wood. Defects include knots, holes, pitch pockets, splits, and missing pieces on the edges or corners, called wanes. These defects primarily affect the appearance, but may also affect the strength of the piece. The higher grades are called select grades. Hardwoods may also be graded as firsts or seconds, which are even higher than select. These grades have very few defects and are used for trim, molding, and finish woodwork where appearance is important. The higher the grade, the fewer the number of defects. The lower grades are called common grades and are used for general construction where the wood will be covered or where defects will not be objectionable. Common grades are designated in descending order of quality by a number such as #1 common, #2 common, and so on. Pieces of softwood common grade lumber may also be designated by an equivalent name, such as select merchantable, construction, and so on. Lumber intended for uses other than construction, such as boxes or ladders, are given other grading designations.

The Manufacturing Process

In the United States, most trees destined to be cut into lumber are grown in managed forests either owned by the lumber company or leased from the government. After the trees have reached an appropriate size, they are cut down and transported to a lumber mill where they are cut into various sizes of lumber.

Here is a typical sequence of operations for processing trees into lumber.

Felling

1 Selected trees in an area are visually inspected and marked as being ready to be cut down, or felled. If a road does not already exist in the area, one is cut and graded using bulldozers. If operations are expected to extend into the rainy season, the road may be graveled, and culverts may be installed across streams to prevent washouts.

2 Most tree felling is done with gasoline-powered chain saws. Two cuts are made near the base, one on each side, to control the direction the tree will fall. Once the tree is down, the limbs are trimmed off with chain saws, and the tree is cut into convenient lengths for transportation.

3 If the terrain is relatively level, diesel-powered tractors, called skidders, are used to drag the fallen tree sections to a cleared area for loading. If the terrain is steep, a self-propelled yarder is used. The yarder has a telescoping hydraulic tower that can be raised to a height of 110 ft (33.5 m). Guy wires support the tower, and cables are run from the top of the tower down the steep slopes to retrieve the felled trees. The tree sections, or logs, are then loaded on trucks using wheeled log loaders.

4 The trucks make their way down the graded road and onto public highways on their way to the lumber mill. Once at the mill, giant mobile unloaders grab the entire truck load in one bite and stack it in long piles, known as log decks. The decks are periodically sprayed with water to prevent the wood from drying out and shrinking.

Debarking and bucking

5 Logs are picked up from the log deck with rubber-tired loaders and are placed on a chain conveyor that brings them into the mill. In some cases, the outer bark of the log is removed, either with sharp-toothed grinding wheels or with a jet of high-pressure water, while the log is slowly rotated about its long axis. The removed bark is pulverized and may be used as a fuel for the mill's furnaces or may be sold as a decorative garden mulch.

6 The logs are carried into the mill on the chain conveyor, where they stop momentarily as a huge circular saw cuts them into predetermined lengths. This process is called bucking, and the saw is called a bucking saw.

Headrig sawing large logs

7 If the log has a diameter larger than 2-3 ft (0.6-0.9 m), it is tipped off the conveyor and clamped onto a moveable carriage that slides lengthwise on a set of rails. The carriage can position the log transversely relative to the rails and can also rotate the log 90 or 180 degrees about its length. Optical sensors scan the log and determine its diameter at each end, its length, and any visible defects. Based on this information, a computer then calculates a suggested cutting pattern to maximize the number of pieces of lumber obtainable from the log.

8 The headrig sawyer sits in an enclosed booth next to a large vertical bandsaw called the headrig saw. He reviews the suggested cutting pattern displayed on a television monitor, but relies more on his experience to make the series of cuts. The log is fed lengthwise through the vertical bandsaw. The first cut is made along the side closest to the operator and removes a piece of wood called a slab. The outer surface of the slab has the curvature of the original tree trunk, and this piece is usually discarded and ground to chips for use in paper pulp.

9 The carriage is returned to its original position, and the log is shifted sideways

Depending on the size of log, it may be cut in different ways to optimize the size and number of resulting boards. After boards are cut, they are dried and planed.

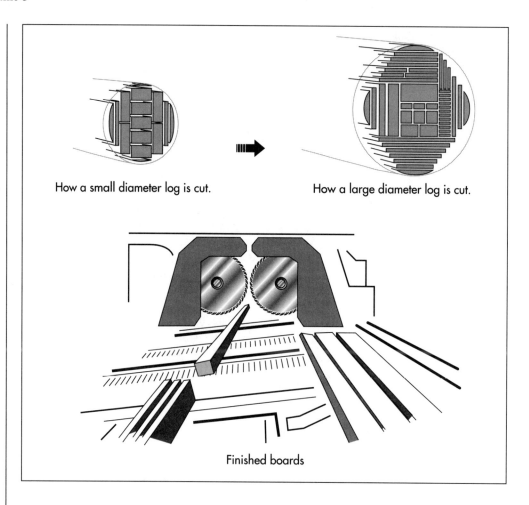

How a small diameter log is cut.   How a large diameter log is cut.

Finished boards

or rotated to make subsequent cuts. The headrig sawyer must constantly review the log for internal defects and modify the cutting pattern accordingly as each successive cut opens the log further. In general, thinner pieces destined to be made into boards are cut from the outer portion of the log where there are fewer knots. Thicker pieces for dimension lumber are cut next, while the center of the log yields stock for heavy timber pieces.

Bandsawing small logs

10 Smaller diameter logs are fed through a series of bandsaws that cut them into nominal 1 in (2.5 cm), 2 in (5 cm), or 4 in (10 cm) thick pieces in one pass.

Resawing

11 The large cut pieces from the headrig saw, called cants, are laid flat and moved by chain conveyor to multiple-blade bandsaws, where they are cut into the required widths and the outside edges are

trimmed square. The pieces that were cut from smaller logs may also pass through multiple-blade bandsaws to cut them to width. If the pieces are small enough that they do not need further cutting, they may pass through a chipper, which grinds the uneven edges square.

Drying or seasoning

12 The cut and trimmed pieces of lumber are then moved to an area to be dried, or "seasoned." This is necessary to prevent decay and to permit the wood to shrink as it dries out. Timbers, because of their large dimensions, are difficult to thoroughly dry and are generally sold wet, or "green." Other lumber may be air dried or kiln dried, depending on the required moisture content of the finished piece. Air-dried lumber is stacked in a covered area with spacers between each piece to allow air to circulate. Air-dried woods generally contain about 20% moisture. Kiln-dried lumber is stacked in an enclosed area, while 110-180°F (44-82°C) heated air is circulated through the

stack. Kiln-dried woods generally contain less than 15% moisture and are often specified for interior floors, molding, and doors where minimal shrinkage is required.

Planing

13 The dried pieces of lumber are passed through planers, where rotating cutting heads trim the pieces to their final dimensions, smooth all four surfaces, and round the edges.

Grade stamping and banding

14 Each piece of lumber is visually or mechanically inspected and graded according to the amount of defects present. The grade is stamped on each piece, along with information about the moisture content, and a mill identification number. The lumber is then bundled according to the type of wood, grade, and moisture content, and the bundle is secured with steel bands. The bundle is loaded on a truck or train and shipped to a lumber yard for resale to customers.

Quality Control

There are very few pieces of perfect lumber. Even though great care is taken to avoid or minimize defects when sawing the wood to the required sizes, there are almost always some defects present. The number and location of these defects determines the grade of the lumber, and the purchaser must choose the grade that is appropriate for each specific application.

The Future

As the number of older trees available for logging diminishes, so does the lumber industry's ability to selectively cut pieces of lumber to the sizes needed for construction. Many of the trees being logged today are second-generation or third-generation trees that are younger and smaller in diameter than the original old-growth trees. These younger trees also contain a higher percentage of juvenile wood, which is less dimensionally stable than older wood.

To counter this trend, the lumber industry is literally taking trees apart and putting them back together again to manufacture the sizes, strengths, and stability required for construction. Actually, they have been doing this for decades in the form of plywood and glue-laminated beams, and some of the new products use similar technology.

One of the new manufactured lumber products is called parallel strand lumber. It begins much like plywood with a thin veneer of wood being peeled off a log. The veneer passes under a fiber-optic scanner that spots defects and cuts them out, sort of like an automated cookie cutter. The veneer is then dried and cut into 0.5 in (1.3 cm) wide strips. The strips are fed into one end of a machine, which coats them with a phenolic resin glue and stacks them side-to-side and end-to-end to form a solid 12 in by 17 in (30 cm by 43 cm) beam of wood. The beam is zapped with 400,000 watts of microwave energy, which hardens the glue almost instantly. As the beam emerges from the other end of the machine, it is cut into 60 ft (18.3 m) lengths. It is then further cut into various sizes of lumber, and sanded smooth. The resulting pieces are significantly stronger and more dimensionally stable than natural wood, while being attractive enough to be used for exposed beams and other visible applications.

Where to Learn More

Books

Bramwell, Martyn, ed. *The International Book of Wood.* Simon and Schuster, 1976.

Forest Products Laboratory. *Wood Handbook: Wood as an Engineering Material.* United States Department of Agriculture, 1987.

Hoadley, R. Bruce. *Understanding Wood: A Craftsman's Guide to Wood Technology.* The Taunton Press, 1980.

Hornbostel, Caleb. *Construction Materials, 2nd Edition.* John Wiley and Sons, Inc., 1991.

Vila, Bob. *This Old House Guide to Building and Remodeling Materials.* Warner Books, Inc., 1986.

Periodicals

Crosby, Bill. "The New Lumber." *Sunset* (Central West edition), November 1995, pp. 72-76.

McCafferty, Phil. "Reinventing Wood." *Popular Science,* May 1990, pp. 96-99, 117.

McCafferty, Phil. "New Strengths For Lumber." *Popular Science,* January 1992, pp. 68-69, 95.

Wardell, C. "Engineered Lumber From the Top Down." *Popular Science,* October 1995, p. 53.

—*Chris Cavette*

M & M® Candy

Background

M&Ms® chocolate candies have two possible origins. Some sources say that M&Ms® were invented in the 1930s, based on a suggestion by soldiers fighting in the Spanish Civil War. The soldiers protected chocolate candy from melting in their pockets and packs by covering the candy with a sugary coating.

The M&M®/Mars Company adapted this coating to create its most popular product.

M&M®/Mars, the company that currently manufactures M&Ms®, first manufactured the candies in 1940. Forrest Mars, Sr., who formed M&M Limited in Newark, New Jersey, wanted to sell chocolates that could be sold year round, especially during the summer months when sales traditionally decreased. Since air conditioning was not available, retailers did not have the means to keep chocolate from melting and consumers did not buy it. By putting his chocolate inside the candy shell, the chocolate did not melt and M&Ms® could be sold any time. They could be eaten neatly in almost any climate. In fact, they were made part of service rations during World War II.

History

At the heart of every M & M® candy is its chocolate, and chocolate has a long history. The cocoa tree is a native plant of South America's river valleys. It was brought north to Mexico by the seventh century A.D. The Mayans and Aztecs made a drink from the beans of the cocoa tree. The upper classes drank this concoction, called *cacahuatl*, which was a blend of the cocoa beans, red pepper, vanilla, and water.

In 1528, the drink was brought back to Europe by Spanish explorers returning from the New World. They found the drink, as drunk by the natives, bitter and unpalatable, and they blended the cocoa beans with other ingredients including sugar, cinnamon cloves, anise, almonds, hazelnuts, vanilla, orange-flower water, and musk. They ground their new concoction and heated it to create a paste. The paste was smoothed onto the leaves of a plantain tree and allowed to hardened. The slab of chocolate was then removed.

To make *chocalatal*, the ancestor of our hot chocolate, the Spanish dissolved a bit of the chocolate slab in hot water and thin corn broth. It was stirred to distribute the fats from the chocolate paste more evenly. By the mid-seventeenth century, choclatal had replaced cachuatl in its original areas for all except the lower classes of Mexico.

The Spanish were also the first to consume chocolate in its solid form, though it was nothing like the center of a M&M®. After conquering the unfounded and widespread fear that the consumption of chocolate would lead to bowel obstructions, Europeans were able to enjoy the treat, and cookbooks including chocolate candy recipes sprang up. As made in the mid 1700s, the chocolate candy was a combination of the chocolate paste described above and sugar held together by plant gums. The candy had a coarse, crumbly texture that did not hold the sugar well, and it was not popular.

It was not until 1828 that a method was developed to produce the solid chocolate that we know today. A Dutch chocolate maker, Conrad van Houten, invented a screw press

One hundred million individual M&Ms® can be manufactured per day.

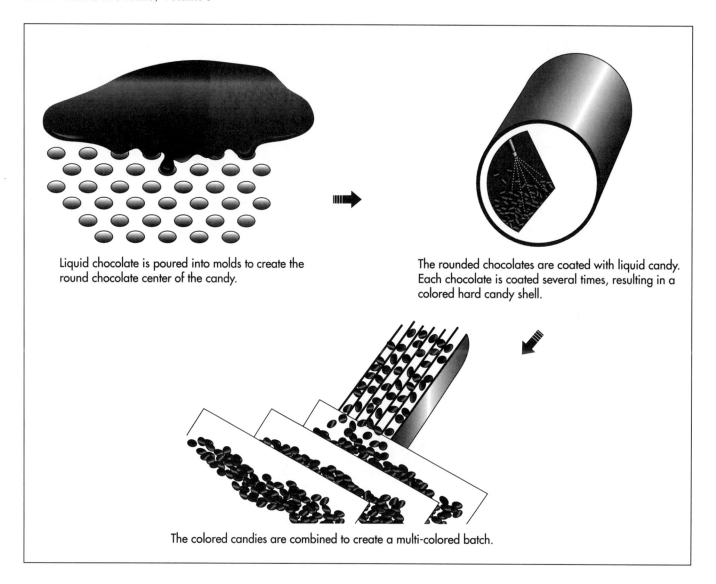

Liquid chocolate is poured into molds to create the round chocolate center of the candy.

The rounded chocolates are coated with liquid candy. Each chocolate is coated several times, resulting in a colored hard candy shell.

The colored candies are combined to create a multi-colored batch.

that squeezed most of the butter out of cocoa beans. This process separated cocoa powder from cocoa butter. The powder was more useful on its own, providing a better base for hot chocolate. When cocoa butter was blended with regular ground cocoa beans, the resulting paste was smoother and easier to blend with sugar. Within 20 years, an English company introduced the first commercially successful prepared hard chocolate.

In 1876, Swiss candymaker Daniel Peter used dried milk to make solid milk chocolate. In 1913, another Swiss candymaker, Jules Sechaud, developed a technique for making chocolate shells filled with other confections. Chocolate was indeed publicly popular by this time, though still expensive.

Hershey Foods marketed one of the first chocolate bars that was both widely afford-

able and available. Milton Hershey, the company's founder, became fascinated by a chocolate statue displayed at the 1893 World's Exposition in Chicago, Illinois. When he decided to manufacture chocolate, he used fresh milk and mass production techniques. The latter insured he could sell large quantities of individually wrapped chocolate at inexpensive prices. Hershey began to manufacture chocolate bars in 1904.

In 1940, Forrest Mars, Sr., and an associate whose name has been lost to history, started to manufacture M&Ms®. The M in M&Ms® comes from the first initial of the last names of Mars and his associate. Peanut-centered M&Ms® were first introduced in 1954; almond-centered M&Ms®, 1988; mint chocolate M&Ms®, 1989; peanut butter chocolate (peanut butter creme center surrounded by chocolate surrounded by a candy shell)

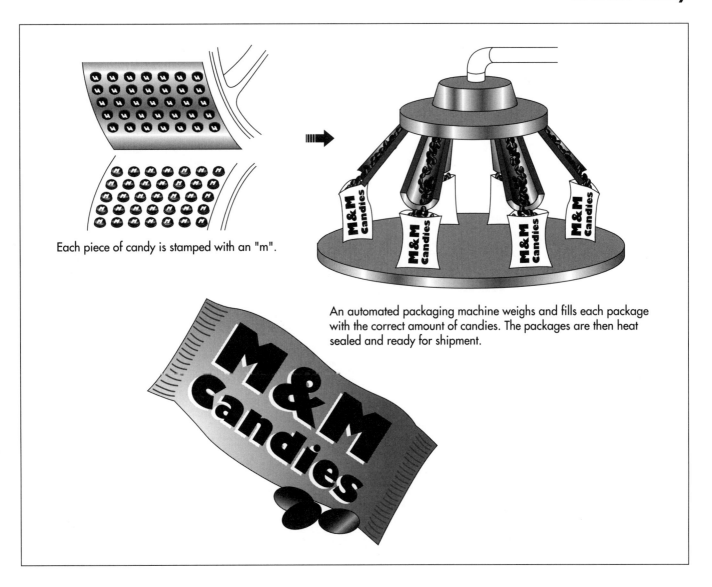

Each piece of candy is stamped with an "m".

An automated packaging machine weighs and fills each package with the correct amount of candies. The packages are then heat sealed and ready for shipment.

M&Ms®, 1990. The M&M®/Mars Company claims that combined sales of all the M&M® varieties makes it the best-selling snack brand in the United States.

In 1976, red-colored M&Ms® were discontinued due to a health concern associated with a certain red food coloring. This controversial coloring was not used in the M&Ms®, but the company did not want to confuse the consumer. Red reappeared in M&Ms® in 1987. (In response to a consumer survey, blue M&Ms® replaced tan M&Ms® in 1995.)

Raw Materials

M&Ms® have two main components, hardened liquid chocolate and the hard candy shell. Liquid chocolate comes from a blend of whole milk, cocoa butter, sugar, and chocolate liquor, among other ingredients. The chocolate liquor is a thick syrup that is made from the grinding of cocoa beans. Roasted cocoa nibs (nibs are the meat of the bean) undergo the process of broyage, in which they are crushed by a grinder made of revolving granite blocks. Chocolate liquor is actually composed of small particles of roasted nibs suspended in oil. The candy shell is made from a blend of sugar and corn syrup.

The Manufacturing Process

Molding

1 The liquid chocolate is poured into tiny molds to create the chocolate centers of the candy. (If they are peanut or almond

M&Ms®, the chocolate surrounds a whole peanut or almond. For peanut butter chocolate M&Ms®, the peanut butter center is made first, and then surrounded by the chocolate. The production process that follows is essentially the same for all varieties.)

2 After the candies are formed, they are "tumbled" to make the chocolate center smooth and rounded. Then they are allowed to harden.

Coating

3 When the chocolates are hard, they are transported via conveyor belt to the coating area, where the process called panning is performed.

4 During panning, the chocolates are rotated in large containers as liquid candy made of sugar and corn syrup is sprayed onto them. The coats are sprayed on rotating chocolates at timed intervals. These intervals allow each coat to dry. Each coat leaves an even layer, a shell, of dry candy substance. The chocolate centers receive several coatings to ensure a uniform, complete coat on every piece.

5 The color is added to a finishing syrup and applied as the final coat. Each batch is a different color. Finally, the liquid candy dries into the hardened shell.

Printing

6 The single-colored batches are combined into the mixtures of red, yellow, blue, green, brown, and orange. They are then transported to the machine that stamps the "m" on the shells. A special conveyor belt carries the pieces to this machine. Each piece rests in its own indentation. The piece runs under rubber etch rollers that gently touch each candy to print the "m." This machine is specially designed to imprint the "m" without cracking the thin candy shell. The process used is similar to the off-set printing process. Approximately 2.6 million M&Ms® are transported to the etching machine per hour. One hundred million individual M&Ms® can be manufactured per day.

Packaging

7 A special packaging machine weighs the candies, pours the proper amount into individual bags, and heat-seals the package. Plain M&Ms® are proportioned (approximately) as follows: 30% brown; 20% yellow; 20% red; 10% green; 10% orange; 10% blue. Peanut M&Ms® are 20% brown; 20% yellow; 20% red; 20% blue; 10% green; and 10% orange. Peanut Butter Chocolate M&Ms® and Almond M&Ms® have even proportions (20% each) of yellow, red, green, blue, and brown.

8 The finished packages are moved along a conveyor belt to a machine that assembles the shipping cartons and fills them with the appropriate number of candy packages. The machine also seals the cartons shut.

Quality Control

After the single color M&M® batches are combined, the pieces are sifted to eliminated any misshapen pieces. If an M&M® is missing an imprinted "m," it is not considered a reject. Because of the minor variations in shape from piece to piece, it is impossible to guarantee an "m" on each piece of candy.

Where to Learn More

Books

Hirsch, Sylvia Balser and Morton Gill Clark. *A Salute to Chocolate.* Hawthorn Books, 1968.

The Story of Chocolate. Chocolate Manufacturers' Association of the U.S.A.

Periodicals

Cavendish, Richard. "The Sweet Smell of Success." *History Today,* July, 1990, pp. 2-3.

Galvin, Ruth Mehrtens. "Sybaritic to Some, Sinful to Others, but How Sweet it is!" *Smithsonian,* February 1986, pp. 54-64.

Marshall, Lydia and Ethel Weinberg. "A Fine Romance." *Cosmopolitan,* February 1989, pp. 52-54.

Other

"M&Ms® Factory Tour." 1997. http:\\www.m-ms.com/tour/index.html (July 9, 1997).

—Annette Petrusso

Magnetic Resonance Imaging (MRI)

Magnetic resonance imaging (MRI) is a medical device that uses a magnetic field and the natural resonance of atoms in the body to obtain images of human tissues. The basic device was first developed in 1945, and the technology has steadily improved since. With the introduction of high-powered computers, MRI has become an important diagnostic device. It is noninvasive and is capable of taking pictures of both soft and hard tissues, unlike other medical imaging tools. MRI is primarily used to examine the internal organs for abnormalities such as tumors or chemical imbalances.

History

The development of magnetic resonance imaging (MRI) began with discoveries in nuclear magnetic resonance (NMR) in the early 1900s. At this time, scientists had just started to figure out the structure of the atom and the nature of visible light and ultraviolet radiation emitted by certain substances. The magnetic properties of an atom's nucleus, which is the basis for NMR, were demonstrated by Wolfgang Pauli in 1924.

The first basic NMR device was developed by I. I. Rabi in 1938. This device was able to provide data related to the magnetic properties of certain substances. However, it suffered from two major limitations. Firstly, the device could analyze only gaseous materials, and secondly, it could only provide indirect measurements of these materials. These limitations were overcome in 1945, when two groups of scientists led by Felix Bloch and Edward Purcell independently developed improved NMR devices. These new devices proved useful to many researchers, allowing them to collect data on many different types of systems. After further technological improvements, scientists were able to use this technology to investigate biological tissues in the mid 1960s.

The use of NMR in medicine soon followed. The earliest experiments showed that NMR could distinguish between normal and cancerous tissue. Later experiments showed that many different body tissues could be distinguished by NMR scans. In 1973, an imaging method using NMR data and computer calculations of tomography was developed. It provided the first magnetic resonance image (MRI). This method was consequently used to examine a mouse and, while the testing time required was more than an hour, an image of the internal organs of the mouse resulted. Human imaging followed a few years later. Various technological improvements have been made since to reduce the scanning time required and improve the resolution of the images. Most notable improvements have been made in the three-dimensional application of MRI.

Background

The basic stages of an MRI reading are simple. First the patient is placed in a strong constant magnetic field and is surrounded by several coils. Radiofrequency (RF) radiation is then applied to the system, causing certain atoms within the patient to resonate. When the RF radiation is turned off, the atoms continue to resonate. Eventually, the resonating atoms return to their natural state and, in doing so, emit a radiofrequency radiation that is an NMR signal. The signal is

In 1973, an imaging method using NMR data and computer calculations of tomography was developed. It provided the first magnetic resonance image (MRI).

then processed through a computer and converted into a visual image of patient.

The NMR signals that are emitted from the body's cells are primarily produced by the cells' protons. Early MR images were constructed based solely on the concentration of protons within a given tissue. These images, however, did not provide good resolution. MRI became much more useful for constructing an internal image of the body when a phenomena known as relaxation time, the time it takes for the protons to emit their signal, was taken into consideration. In all body tissues, there are two types of relaxation times, T1 and T2, that can be detected. Different types of tissues will exhibit different T1 and T2 values. For example, the gray matter in the brain has a different T1 and T2 value than blood. Using these three variables (proton density, T1, and T2 value), a highly resolved image can be constructed.

MRI is most used for creating images of the human brain. It is particularly useful for this area because it can distinguish between soft tissue and lesions. In addition to structural information, MRI allows brain functional imaging. Functional imaging is possible because when an area of the brain is active, blood flow to that region increases. When the scans are taken with sufficient speed, in fact, blood can be seen moving through organs. Another application for MRI is muscular skeletal imaging. Injuries to ligaments and cartilage in the joints of the knees, wrists, and shoulder can be readily seen with MRI. This eliminates the need for traditional invasive surgeries. A developing use for MRI is tracking chemicals through the body. In these scans NMR signals from molecules such as carbon 13 and phosphorus 31 are received and interpreted.

Raw Materials

The primary functioning parts of an MRI system include an external magnet, gradient coils, RF equipment, and a computer. Other components include an RF shield, a power supply, NMR probe, display unit, and a refrigeration unit.

The magnet used to create the constant external magnetic field is the largest piece of any MRI system. To be useful, the magnet must be able to produce a stable magnetic field that penetrates throughout a certain volume, or slice, of the body. There are three different kinds of magnets available. A resistive magnet is made up of thin aluminum bands wrapped in a loop. When electricity is conducted around the loop a magnetic field is created perpendicular to the loop. In an MRI system, four resistive magnets are placed perpendicular to each other to produce a consistent magnetic field. As electricity is conducted around the loop, the resistance of the loop generates heat, which must be dissipated by a cooling system.

Superconducting magnets do not have the same problems and limitations of the resistive type of magnet. Superconducting magnets are ring magnets, made out of a niobium-titanium alloy in a copper matrix, which are supercooled with liquid helium and liquid nitrogen. At these low temperatures, there is almost no resistance, so very low levels of electricity are needed. This magnet is less expensive to run than the resistive type, and larger field strengths can be generated. The other type of magnet used is a permanent magnet. It is constructed out of a ferromagnetic material, is quite large, and does not require electricity to run. It also provides more flexibility in the design of the MRI system. However, the stability of the magnetic field the permanent magnet generates is questionable, and its size and weight may be prohibitive. While each of these different kind of magnets can produce magnetic fields with varying strength, an optimum field strength has not been discovered.

To provide a method for decoding the NMR signal that is received from a sample, magnetic field gradients are used. Typically, three sets of gradient coils are used to provide data in each of the three dimensions. Like the primary magnets, these coils are made of a conducting loop that creates a magnetic field. In the MRI system, they are wrapped around the cylinder that surrounds the patient.

The RF system has various roles in an MRI machine. First, it is responsible for transmitting the RF radiation that induces the atoms to emit a signal. Next, it receives the emitted signal and amplifies it so it can be manipulated by the computer. RF coils are the primary pieces of hardware in the RF

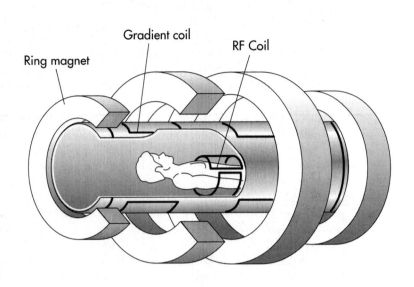

Ring magnet

Gradient coil

RF Coil

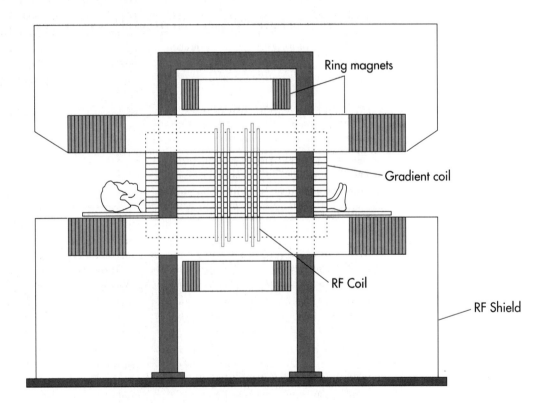

Ring magnets

Gradient coil

RF Coil

RF Shield

Magnetic resonance imaging (MRI) is a medical diagnostic tool used to create cross-section images of the hard and soft tissues of the body. The patient is placed on a table that slides inside a machine containing large electromagnets. An MRI system with magnets mounted vertically creates a horizontal magnetic field. Horizontally mounted magnets create a vertical magnetic field. By using magnets arranged perpendicularly to each other, a consistent magnetic field is created. RF coils then emit radio waves, which are absorbed by the atoms in the body. The excited atoms emit signals that are gathered by gradient coils. These signals are interpreted by a computer to create images of the body from any angle.

system. They are constructed to create an oscillating magnetic field. This field induces atoms in a defined area to absorb RF radiation and then emit a signal. In addition to sending the RF signal, the coils can also receive the signal from the patient. Depending on the type of MRI system, either a saddle RF coil or a solenoid RF coil is used. The coil is usually positioned alongside the subject and is designed to fit the patient. To reduce RF interferences, an aluminum sheet is used.

The final link in the MRI system is a computer, which controls the signals sent and processes and stores the signals received. Before the received signal can be analyzed by the computer, it is translated through an analog-digital convertor. When the computer receives signals, it performs various reconstruction algorithms, creating a matrix of numbers that are suitable for storage and building a visual display using a Fourier transformer.

The Manufacturing Process

The individual components of an MRI system are typically manufactured separately and then assembled into a large unit. These units are extremely heavy, sometimes weighing over 100 tons (102 metric tons).

Magnet

1 The most frequently used magnets in an MRI system are superconducting electromagnets. These can be made using various materials, but the basic design involves a coil of conductive wire, a cooling system, and a power supply. The coils are made by wrapping wire, constructed from filaments of a niobium titanium alloy embedded in copper, in a large loop. To create the necessary magnetic field, a number of coils are used. In one type of system eight coils are used, six to create the primary magnetic field and two to compensate for the excess field.

2 The coils are immersed in a vessel containing liquid helium. This reduces the temperature to a level that makes them superconductive. To help keep the temperature stable, the vessel is surrounded by two more vessels containing other coolants like liquid nitrogen. This construction is then

suspended with thin rods in a vacuum-sealed container. A power supply is hooked up to the magnetic coils and is used only when the magnet needs to be energized. The magnet is attached to the patient support, which is a sliding table that brings the patient into the magnetic field.

Gradient coils

3 The gradient coils are resistant type electromagnets. In an MRI system, there are typically three sets of gradient coils. Each coil is made by winding thin strips of copper or aluminum in a specific pattern. The coils are given strength by introducing an epoxy into their structure. The size of these coils determines the width of the opening into which the patient is placed. Since a smaller coil requires less energy, this width must be large enough to prevent claustrophobia in the patient but small enough to require a reasonable amount of electricity. These gradient coils are typically shielded to prevent interfering eddy currents.

RF system

4 The electronic components of the RF system may be provided by outside suppliers and assembled by the MRI manufacturer. These components are attached to the RF coils, which are made with varying designs. The transmitter and receiver coils are composed of the same type of materials as the gradient coils. They are also constructed much like the main magnet. However, they are made up of a loop of conducting material, such as copper, that can create an oscillating magnetic field. One type of RF coil is a surface coil, which is shaped in a circle and is applied directly on the patient. Another type is the saddle coil. These can either be fitted right into the magnet bore or shaped into a birdcage coil and placed just inside the gradient coils. Each type of coil is attached to a power source.

Computer

5 The computer is supplied by computer manufacturers and modified and programmed for use in an MRI system. Attached to it is the user interface, the Fourier transformer, the signal converter, and a preamplifier. A display device and a laser printer are also included.

Final assembly

6 Each of the components of the MRI are assembled together and placed into an appropriate frame. Assembly can take place at the plant or on-site, where the system will be used. In either case, the nature of the magnet typically requires special handling precautions, such as transporting it in an air-suspended vehicle.

Quality Control

The quality of each MRI system being manufactured is ensured by making visual and electrical inspections throughout the entire production process. The performance of the MRI is tested to be sure it is functioning properly. These tests are done under different environmental conditions, such as excessive heat and humidity. Most manufacturers set their own quality specifications for the MRI systems that they produce. Standards and performance recommendations have also been proposed by various medical organizations and governmental agencies.

The Future

The focus of current MRI research is in areas that include improving the scan reso-lution, reducing scan time, and improving MRI design. The methods for improving resolution and decreasing scan time involve reducing the signal to noise ratio. In an MRI system, noise is caused by randomly generated signals that interfere with the signal of interest. One method for reducing it is by using a high magnetic field strength. Improved designs for MRI systems will also help reduce this interference and decrease the noise associated with electromagnets. In the future, real time MRI scans should be available.

Where to Learn More

Books

Boer, Jacques and Marinus Vlaardinger-broek. *Magnetic Resonance Imaging Theory and Practice*. Springer, 1996.

Brown, J. and J. Heiken. *Manual of Clinical Magnetic Resonance Imaging*. Raven Press, 1991.

Rinck, P. *Magnetic Resonance in Medicine*. Blackwell Scientific Publications, 1993.

—*Perry Romanowski*

Maple Syrup

The northeastern section of North America is the only region in the world where the environmental conditions are capable of producing tappable sap that will yield maple syrup.

Background

The Algonquin Indians called it *sinzibuk-wud*, meaning drawn from wood. It was the Algonquins and the other Native American tribes of the northeastern United States and southeast Canada who first showed French and British settlers how to draw the sap of *Acer saccharum*, the sugar maple, and reduce it into a sweet, thick liquid known today as maple syrup.

In early March, when the days started to become warm but nights were still freezing, Native Americans would cut a vee in the trunk of a maple and insert a reed or curved piece of bark into the opening. Under the opening, they would set a larger piece of bark or a clay pot to catch the dripping sap. The sap was concentrated either by leaving it out overnight and then tossing out the water, which had frozen on top, or by placing hot stones into the sap to evaporate some of the water. The resulting product was used in cooking and sometimes as a sweet drink. European settlers introduced iron and copper pots into the process, which allowed the sap to be heated longer, removing more of the water and producing what we know as maple syrup today. Throughout the 1700s, both maple syrup and maple sugar served as an integral unit of trade for the early colonies, but they would soon be supplanted by another sweet crop from warmer climates, namely sugar cane.

Since the sixteenth century, Spain, England, and France had grown profitable sugar cane crops in their Caribbean island colonies. In 1803, the Louisiana Purchase gave U.S. investors direct access to the territory's burgeoning cane sugar industry, which had been developed by French growers fleeing slave revolts in the Caribbean. In 1849, large scale cane cultivation by U.S. growers began in Hawaii. Throughout the 1800s, improvements to production methods, combined with constant pressure from the now-powerful sugar industry for increased tariff protection from the federal government, as well as cane's naturally high yield (one acre of cane will produce 12.5 tons of raw sugar [31 metric tons per ha]) succeeded in making cane sugar the sweetener of choice.

The maple and its syrup remain an integral part of spring in northeastern North America. Many of the towns and villages in the area arrange a yearly festival centered around the maple harvest. The sugar maple is the state tree of New York, Vermont, Wisconsin, and West Virginia. The maple leaf flies at the center of Canada's national flag, which is entirely appropriate since the Canadian province of Quebec is by far the largest producer of syrup in the world. Today, the United States and Canada produce a combined average of a little over five million gal (18,920,000 l) of maple syrup annually.

Raw Materials

Of the over 200 different species of tree in the maple family, only a few produce sap of sufficient quality and sugar content to be used for maple syrup. The northeastern section of North America is the only region in the world where the environmental conditions are capable of producing tappable sap that will yield maple syrup, and where one is able to find the sugar maple tree. The sugar, or rock, maple can grow to 130 ft (40 m) with a diameter of 3 ft (1 m). A tree must be at least 12 in (30.5 cm) in diameter before it can be tapped; it will take 40 years to reach that size. The sugar maple is also highly prized for its hard, beautifully grained wood used in furniture making and as a veneer. Some sugar maples form intricate patterns such as bird's eye maple, which has small circles scattered throughout the wood resembling birds' eyes. Other North American native maples, the black, the red, and the silver maple, will also produce syrup. The black maple has a yield close to the sugar and is used alongside it in syrup production. The red and silver are used less often as they generally produce less sugar, have a shorter growing season, and may have sap that is slightly cloudy.

The sap itself is made up of 90% water. The remaining 10% of the sap is a mixture of sugars (mostly levulose, which is a variant of fructose and hexose, which is glucose plus fructose), calcium, potassium, phosphorous, iron, and trace amounts of B vitamins. In order to make the sap into syrup, most of the water will be removed, which is why it takes approximately 40 gal (151 l) of sap to make 1 gal (4 l) of syrup.

The Manufacturing Process

In the making of maple syrup, one could say that the farmer comes last and not be wrong. The sap of the maple will never become maple syrup until the farmer intercedes. The following five components make up that transformation.

Season

1 In the winter, the maple does not grow, and it stores its sap in its roots. By late spring when the trees begin to produce leaves and flowers, a chemical change will have occurred in the sap giving it an unpleasant taste, which farmers describe as "buddy." It is only in early spring, usually February through April, when warm days cause the sap to flow up from the roots to the branches to feed new growth and cold nights cause the sap to return to the roots, that the trees can be tapped.

Tapping

2 To tap a maple tree, farmers drill a 0.5 in (1.3 cm) hole about 2.5-3 in (6-8 cm)

A. W. Mayo and Son promoted their family maple syrup and confection business on a sleigh. (From the collections of Henry Ford Museum & Greenfield Village.)

Americans have always had a sweet tooth. Native Americans satisfied their taste for sugar through the production of maple syrup and European settlers of New England learned the practice. After the Revolutionary War, many farm families in the northern United States and territories produced maple syrup and maple sugar for home consumption. Surveyors of new territories noted the presence of "sugar" trees, and settlers named streams near maple groves "Sugar Creek."

Cane sugar production in America did not develop until the late eighteenth century. The growing season in the southern part of the United States did not allow sugar cane to mature, and no methods of processing immature cane existed. In 1795, in what was then French territory, Etienne de Bore perfected a means of crystallizing sugar from immature sugar cane in New Orleans. Other planters copied de Bore and the American cane sugar industry began. By 1802, 75 sugar plantations produced as much as eight million lb (four million kg) of brown sugar. When the United States acquired the Louisiana Purchase, it gained these sources of sugar production. Midwestern farmers, and later, Western producers who used irrigation, grew sugar beets to sell to sugar processors.

Sugars from cane and beets are generally used for baking, but maple syrup and maple sugar remain popular in spite of other sweeteners. Families with an affinity for this tasty confection continue to tap trees in the spring, and thousands of Americans purchase maple syrup to pour over their morning stack of pancakes.

Leo Landis

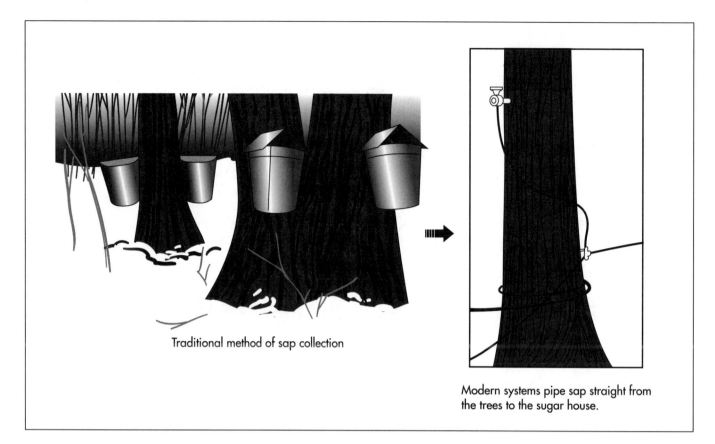

Traditional method of sap collection

Modern systems pipe sap straight from the trees to the sugar house.

To collect maple sap, holes are drilled into the trees and hollow spikes are inserted. Traditionally, pails collected the syrup that dripped out, but modern systems send the sap directly to the sugar house.

into the trunk. The hole is drilled at a slightly upward angle to prevent sap from collecting in the hole, freezing, and cracking the trunk. Care is taken only to drill into the light-colored sapwood, which will heal, and not into the darker heartwood, which will not. Trees from 12-15 in (30.5-38 cm) in diameter supports one tap, and the number of taps increases as the trees grow larger, with a maximum of four taps per tree. Each tap yields an average of 10 gal (38 l) of sap per season, which makes about a quart of syrup. New holes must be drilled at least 4 in (10 cm) above and 6 in (15 cm) to either side of previous holes. This prevents large areas of scar tissue from forming in the wood and leaving the tree open to disease. Properly cared for, the holes will heal completely in one or two years and will cause no damage to the tree.

Collecting

3 Once the tree is tapped, farmers gently drive a hollow spike called a spile into the hole. The spile is a round, hollow piece of wood about the diameter of a broom handle with a metal tip to help drive the end into the wood. Traditional methods use a canvas bag

or a bucket hung from the spile to catch the dripping sap. The bucket or bag is covered to keep out debris. Once a day, the farmer empties the buckets into a large gathering tank pulled through the fields by a horse or tractor. The tank is pulled to the building where the sap is processed, called the sugar house, and emptied into a holding tank.

While the sight of pails hung from maples and the sound of dripping sap is much more romantic than modern systems, this traditional method of collection has a few inherent flaws, especially for large-scale producers. The first is that collecting syrup bucket by bucket is both time and labor intensive, and is therefore costly. The second is that sap should be processed immediately after being collected in order to produce the best quality syrup. Under the traditional system, if collection exceeds production, the sap must be stored. And if collection falls behind, production must slow.

Modern systems eliminate the collection process altogether and send sap straight from the trees to the sugar house. To accomplish this, a plastic tube is attached to each spile; the tubes run together to a larger pipe, and the pipes, in turn, run from the

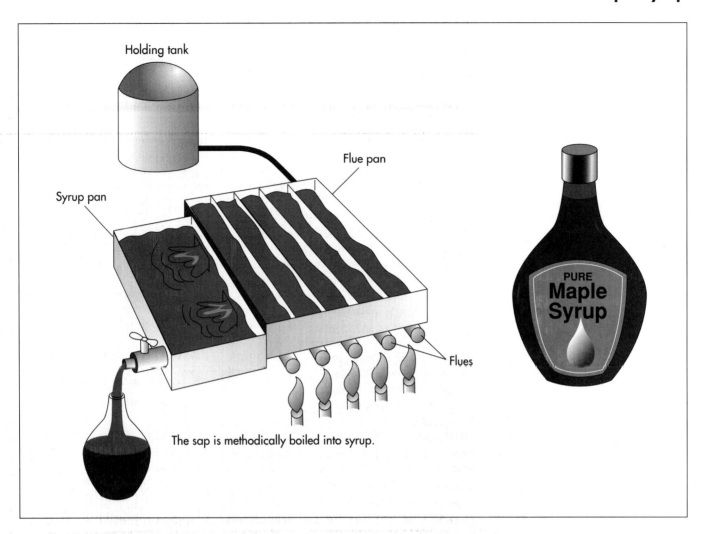

Holding tank

Syrup pan

Flue pan

PURE
Maple Syrup

Flues

The sap is methodically boiled into syrup.

various parts of the orchard directly to the sugar house. A pump maintains a constant, slight vacuum and keeps the sap constantly moving through the pipeline.

Sugaring

4 Possibly the most important piece of the maple syrup-making process takes place in the sugar house. The initial sugar content of the sap, its clarity, and to some extent, its taste are all determined by weather, soil, and ultimately, the tree, all of which are outside the control of the grower. It is only through the process of heating and condensing the sap that it takes on the distinctive sweet flavor of maple syrup.

Most commercial syrup producers use a continuous feed system to make syrup. Sap continuously, but very slowly, enters from a holding tank on one end, and finished syrup is continuously drained and bottled at the other end. The sap is heated over a wood- or oil-fueled fire in a series of long, shallow stainless steel pans. This causes the water in the sap to steam off, condensing the sap, and causes a chemical change in the sap, which brings out a flavor not present before it is heated. The sap first enters the flue pan, which has a corrugated bottom lined with a series of pipes called flues. The flues serve two purposes. First, the fire heats the air inside the flues, which provides a more even, gentle heat than the fire gives directly and which allows the sap to be heated longer and at a higher temperature, evaporating more water without the danger of scorching the sap. Second, the wavy pan bottom increases the surface area of the pan and allows more syrup to be in contact with the heat at the same time. From the flue pan, the sap flows into the flat-bottomed syrup pan where it is finished. The two pans are divided by partitions that allow a slow, constant flow of sap.

Maple sap becomes maple syrup at the moment when its concentration of sugar reach-

Collected sap enters the holding tank and then travels to the flue pan. Here it is heated, concentrating the sap and causing a chemical change. Maple sap becomes maple syrup at the moment when its concentration of sugar reaches 66%. The syrup is filtered and poured into containers.

es 66%. Producers test this concentration using two methods. The first uses the formula that syrup is finished when it reaches the temperature at which water boils plus 7.1°F (4.3°C). Because water boils at different temperatures depending on altitude, producers boil a sample of water and measure its temperature. When the syrup reaches the sample's temperature plus 7.1°F (4.3°C), it is done. The second, perhaps more scientific, method involves an instrument called a hydrometer. A hydrometer is a sealed glass tube with a small amount of weight in one end. Running horizontally around the outside of the tube are a series of lines which indicate concentration in percentage; the lines are specially calibrated for maple syrup. Once the sap has been heated, a sample is placed in a cylinder and the hydrometer is placed in the sap. The hydrometer sinks to the line that indicates the current sugar concentration in the sap. When the hydrometer sinks to the line that indicates 66%, the sap has become syrup. As the sugar approaches 66%, it is checked constantly. Finishing the syrup at the proper sugar concentration is critical: too high and the syrup will crystallize; too low and it will spoil.

Bottling

5 Once the syrup is properly finished, it is poured through a cloth filter to remove any sugar sand, called nitre. It is then poured into glass or metal containers while still hot. The container is filled to the very top and then tightly sealed. Because maple syrup contains no preservatives, this final step ensures that the container is sterile and airtight and prevents any spoiling.

Quality Control

In order to be considered pure maple syrup, a product must be 100% pure. Products containing other ingredients are labelled table or pancake syrup. The grades listed on the bottle have nothing to do with purity or sweetness (all maple syrup has the same sugar concentration) or even necessarily with quality or taste. The grades are based solely on the syrup's ability to transmit light. The more light that can shine through the syrup, the higher the grade. This system can be used because the opacity of syrup re-

lates directly to the way it tastes. Connoisseurs will say that syrups from different regions have distinctive tastes, but generally, the more opaque the syrup, the heavier the maple taste will be. The United States Department of Agriculture assigns grades A through C to syrup, with A being the lightest. Within each grade are three further divisions for color, light, medium, and dark amber. Grade A dark amber is the most commonly used for table syrup. It has a pleasant, full-bodied maple taste. Grade A light amber is considered the finest syrup. Its taste is very delicate and subtle. Occasionally, growers will produce syrup graded extra fancy, which is even subtler than grade A light amber. Grade B syrup has a much richer, more "mapley" taste; it is primarily used in cooking. Grade C has a strong, thick taste and is used almost exclusively as a commercial sweetener.

Where to Learn More

Books

Gemming, Elizabeth. *Maple Harvest: The Story of Maple Sugaring.* Coward, McCann, & Geoghehan, 1976.

Muir, Reginald. *The Vermont Maple Syrup Cook Book.* Phoenix Publishing, 1974.

Nearing, Helen. *The Maple Sugar Book.* Schocken Books, 1970.

Periodicals

Berk, Gesina. "Sugaring Time - How Maple Sugar is Made." *Humpty Dumpty's Magazine,* March 1996, pp. 16-20.

Clark, Edie. "Sap Bucket Blues." *Yankee,* March 1996, p. 18.

Martin, Rux. "Caught up in the Romance of Maple." *Yankee,* March 1994, pp. 132-139.

Robinshaw, Sue. "Spring in the Sugarbush." *Countryside & Small Stock Journal,* March-April 1993, pp. 63-66.

Other

"Camp 100 % Pure Maple Syrup, the Maple Syrup Experts!" http://www.ivic.qc.ca/abriweb/erable/faq_32.html#anchor709074 (March 29, 1997).

Heiligmann, Randall B. "Hobby Maple Syrup Production." http://www.ag.ohio-state.edu/~ohioline/forestry/f-36.html (January 16, 1997).

"Maple Syrup-A Sweet Natural Resource." http://monsterbit.com/touch/maple1.html (January 16, 1997).

"The Story of Maple Syrup." http://www.state.vt.us/agric/minfo.htm (January 16, 1997).

"What is Maple Syrup?" March 5, 1996. http://www.mkl.com/~jmitchell/maple/syrup.html#whatis (March 29, 1997).

—*Michael Cavette*

Marker

Background

Markers, or felt-tip pens, serve a variety of functions. Children use them to make bright, colorful drawings. The stereotypical teacher uses a glaring, unmistakable red felt-tip to grade papers. Retail employees, roadside vendors, performers, and protestors rely on the indelible, eye-catching shades and thick inking surface of these writing and drawing utensils to announce sales, prices, and productions, or to create strongly worded posterboard signs to convey dissatisfaction. Markers are also useful for permanently marking surfaces, which is often necessary for identification purposes—putting names on clothing tags, boxes, and tape which can be adhered to almost any item.

History

The felt-tip pen was invented by Sidney Rosenthal in 1953. This inventor from Richmond Hill, New York, placed a felt tip on the end of a small, stout bottle of permanent ink and discovered that the resulting marks saturated a heavy, absorbent surface, yielding rich color and permanence.

The felt-tip pen had many predecessors. Inks and dyes have been used throughout human evolution for marking objects, from cave writings fashioned from natural dyes of the earth applied with sticks to graffiti applied with paint from an aerosol can. Ink is a combination of a coloring agent, or pigment, and a liquid containing oils, resins, and chemical solvents. Initially, ink was fashioned from different colored juices and plant and animal extracts. Today, synthetic materials are used in addition to these natural substances.

The use of ink for writing and printing dates back to 3,200 B.C., when the Egyptians used a mixture of fine soot and vegetable gum to create a substance that could be used for writing and painting. Both Egyptians and Greeks used iron oxide (or, more commonly, rust) to make red ink. Around 2,000 B.C., the Chinese began making red ink from mercury sulfate and black ink from iron sulfur mixed with sumac tree sap. Like the Egyptians, they formed their ink into a solid block or stick that would be mixed with water when used. Europeans did not begin commonly using ink until the seventeenth century, using tannic acid from tree bark and iron salt to create the recipe that formed the blue and black inks still used today.

Like inks and dyes, pens have been used since antiquity. The earliest pens were made from hollow reeds and, later, hollow wing feathers of geese and swans, called quills, infused with ink. Steel pen nibs came into use in the early 1800s, and then fountain pens, which did not require a constant resupply of ink like previous incarnations, gained popularity.

House paint also came into popular use in the late 1800s when Edwin Binney and Harold Smith used red oxide pigments (a mixture of naturally occurring dyes and chemicals) to create a viscous coloring substance—the paint that was used to color America's first classic red barns. Binney and Smith, through their company Binney & Smith, then created a line of carbon black pigments that were used by the Goodrich company to color its white auto tires black.

Next, Binney & Smith acquired a water-powered stone mill in Easton, Pennsylva-

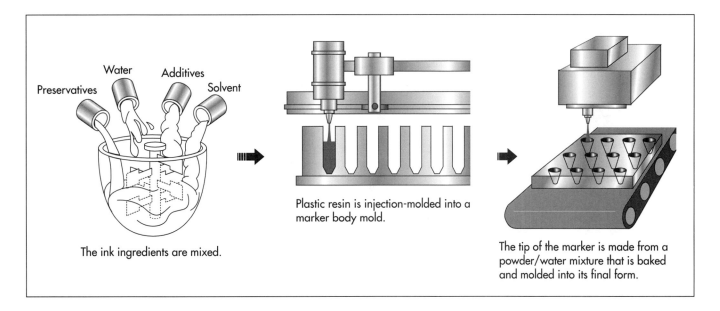

Preservatives Water Additives Solvent

The ink ingredients are mixed.

Plastic resin is injection-molded into a marker body mold.

The tip of the marker is made from a powder/water mixture that is baked and molded into its final form.

nia, and began fashioning slate culled from the area into pencils. From there, the company created dustless chalk in 1902. In 1903, the company fashioned a variant of its industrial wax marking crayons, which were smaller and came in a variety of colors created by colored pigments added to paraffin wax. Thus, Crayola crayons were born.

By the late 1950s while Binney & Smith was expanding its crayon business, Rosenthal was building up his company, Speedry Chemical Products, which manufactured and marketed his felt-tip pen invention. Rosenthal initially geared his product at the art supplies market, but soon thought to capitalize on its mass market appeal due to its suitability to poster-making, sign-lettering, and other marking purposes. Competitors threw their hats in the ring as early as 1958, when Carter's, Inc., came out with a more slender marker with an aluminum ink tube. Speedry sued Carter's for patent infringement, but lost. Other companies began marketing pens containing water-soluble inks that could be used on normal-weight paper (Rosenthal's invention required heavier paper to keep the ink from soaking straight through to the underlying surface) and with capillary flow technology, which enhanced the movement of the ink into the tip of the writing utensil. When Rosenthal changed the name of his company to Magic Marker Corporation in 1966, he was already suffering the effects of this increased competition.

Despite its name becoming synonymous with its product, regardless of the manufacturer, Magic Marker Corporation continued to lose money and filed for bankruptcy in 1980. In 1989, Binney & Smith, now a subsidiary of the Hallmark Corporation, purchased the rights to the Magic Marker name, stating that it was motivated to purchase the moniker due to remaining high consumer recognition of the brand name. The defunct Magic Marker Industries would be able to use the royalty income garnered from selling the use of its name to pay off creditors. Binney & Smith now manufacturers a wide variety of Crayola markers as well as Magic Markers.

Raw Materials

The marker body, cap, and plugs are formed from plastic resin. The marker reservoir, which holds the ink, is formed from polyester. Powder and water are used to form the felt writing tip. In addition, markers require ink, and the pigments and synthetic substances used to make it. Toluol and xylol used to be common synthetics used as solvents in dye, but due to their toxic nature these substances have largely been replaced with safer chemicals such as cyclic alkylene carbonates, although these chemicals are still used to make the indelible ink contained in permanent markers. The solvent is the substance into which the dye is diluted. Water also acts as a solvent in ink. Additives may also be used in an ink mixture to act as wetting agents.

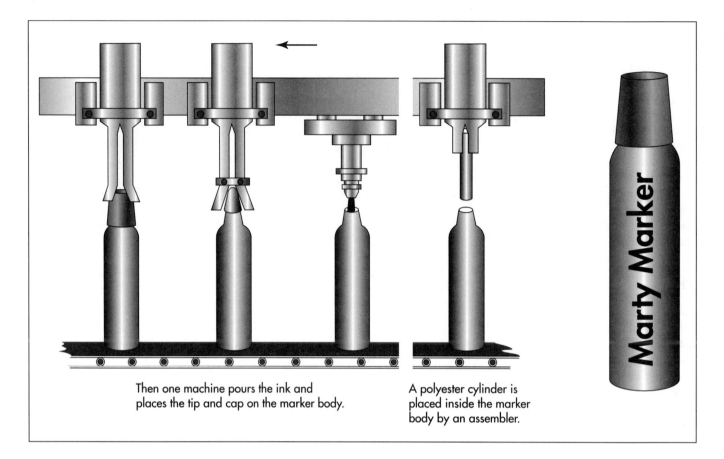

Then one machine pours the ink and places the tip and cap on the marker body.

A polyester cylinder is placed inside the marker body by an assembler.

Marty Marker

The Manufacturing Process

Making the ink

1 The ink mixture is concocted. Preferable ratios are 1-10 % water by weight (the water content must be sufficient for complete dissolution or dispersion of the dye), with the remainder of the weight being made up of a solvent such as alkyl or alkylene carbonate.

2 Conventional additives, such as nonylphenylpolyglycol ether, alkylpolyglycol ether, fatty acid polyglycol ester, or fatty alcohol ethoxalates, and preservatives, such as ortho-phenolphenyl and its sodium salt, ortho-hydroxydiphenyl, or 6-acetoxy-2,4-dimethhyl-m-dioxane, may also be added to the mixture.

Making the marker

3 To make the body of the marker, plastic resin is injection-molded into a marker body. Injection molding involves heating a substance, in this case plastic resin, into a molten state and forcing (injecting) it into a

mold of the desired shape, then allowing it to cool and harden. Marker caps and plugs are formed in the same manner as the barrel.

4 The nib, or tip, of the marker is made from powder which is mixed with water, molded, and baked into its pointed or flat form.

5 Using one machine for all the following functions, an assembler then places a polyester cylinder inside the marker barrel to form a reservoir for the ink, fills the reservoir with ink, and inserts the nib at the bottom and the cap at the top.

6 The markers are then placed into color assortment and packaged for retail marketing.

Byproducts/Waste

While individual markers can be disposed of like any other household waste, special concern must be given to the disposal of excess ink by the manufacturer. Typically, ink disposal would not be subject to any stringent environmental regulations because it does not contain toxic substances, but there is a special disposal requirement. It is against federal regulations to discharge anything to a sewer system that will alter the color of the contents of that system. Thus, liquid ink cannot be dumped down a drain, but must be combined with an absorbent material and disposed of like a solid. Furthermore, if any of the contents of the ink exhibit certain characteristics, such as toxicity, that qualify those contents as a hazardous waste, the substance must be disposed of according to federal and state hazardous waste regulations.

The Future

The traditional marker is continually being enhanced. Markers are available in both indelible and washable formulas. Markers are also produced that can yield changing colors, color over colors, emit scents, and be used to create three-dimensional drawings. One of the most popular spin-offs of the marker is the highlighting marker, which comes in a variety of translucent, fluorescent colors and can be used for marking text. This is a popular product with students and professionals. Ink companies are also at work developing products that are more environmentally sound, such as inks using vegetable oils or water instead of traditional petroleum or other solvent ingredients.

Where to Learn More

Other

Flint Ink Corporation. 1997. http://www.flintink.com/index2.html (July 14, 1997).

"How are Crayola Markers Made?" Binney & Smith, Inc. 1996. http://www.crayola.com:80/manufacture/markers/home.html (July 14, 1997).

U.S. Patent # 4,931,093 (Marker or Felt Tip Pen) (from the Great Lakes Patent and Trade Center at the Detroit Public Library).

—*Kristin Palm*

Marshmallow

Background

Marshmallows are one of the earliest confections known to humankind. Today's marshmallows come in many forms, from solid (soft pillows dropped in cocoa or roasted on a stick) to semi-liquid (covered in chocolate or formed into chicks for Easter) to the creme-like (used as a base in other candies or as an ice cream topping). In essence, all marshmallows are aerated candies.

History

Originally, however, marshmallows were made from the root sap of the marsh mallow (*Althaea officinalis*) plant. It is a genus of herb that is native to parts of Europe, north Africa, and Asia. Marsh mallows grow in marshes and other damp areas. The plant has a fleshy stem, leaves, and pale, five-petaled flowers. The first marshmallows were made by boiling pieces of the marsh mallow root pulp with sugar until it thickened. After it had thickened, the mixture was strained and cooled. As far back as 2000 B.C., Egyptians combined the marsh mallow root with honey. The candy was reserved for gods and royalty.

The marsh mallow root also has medicinal qualities. Marsh mallow roots and leaves can work as a laxative. It also was used by early Arab doctors as a poultice to retard inflammations. Marsh mallow roots were also used in treating chest pains, to soothe coughs and sore throats, and as an ointment. Whether used as a candy or for medicinal purposes, the manufacturing process of marsh mallows was limited to a small, almost individual, scale. Access to marsh mallow confections was limited to the wealthy until the mid-nineteenth century. Common people only tasted marsh mallows when they took pills; doctors sometimes hid the medicine inside the candy to cover the pill's undesirable taste.

Modern marshmallow confections were first made in France around 1850. This first method of manufacture was expensive and slow because it involved the casting and molding of each marshmallow. French candy makers used the mallow root sap as a binding agent for the egg whites, corn syrup, and water. The fluffy mixture was heated and poured onto the corn starch in small molds, forming the marshmallows. At this time, marshmallows were still not mass manufactured. Instead, they were made by confectioners in small stores or candy companies.

By 1900, marshmallows were available for mass consumption, and they were sold in tins as penny candy. Mass production of marshmallows became possible with the invention of the starch mogul system of manufacture in the late nineteenth century. In the starch mogul system, a machine automatically fills trays with starch about 2 in (5.08 cm) thick, which is then evened off and slightly compressed. Then a printing board, made of plaster, wood, or metal trays shaped to mold the marshmallow of the final product is pressed into the starch and withdrawn. Then the space created is filled with hot creme. The first moguls were wood, but all were steel by 1911. Gelatin and other whipping agents replaced the mallow root in the ingredient list.

In 1955, there were nearly 35 manufacturers of marshmallows in the United States. About this time, Alex Doumak, of Doumak, Inc., patented a new manufacturing method called the extrusion process. This invention changed the history of marshmallow production and is still used today. It now only takes 60 minutes to produce a marshmallow. Today, there are only three manufacturers of marshmallows in the United States, Favorite Brands International (Kraft marshmallows), Doumak, Inc., and Kidd & Company.

Raw Materials

Marshmallows are made from only a few ingredients, which fall into two main categories: sweeteners and emulsifying agents. Sweeteners include corn syrup, sugar, and dextrose. Proportionally, there is more corn syrup than sugar because it increases solubility (the ability to dissolve) and retards crystallization. Corn starch, modified food starch, water, gum, gelatin, and/or whipped egg whites are used in various combinations. The resulting combination gives the marshmallows their texture. They act as emulsifying agents by maintaining fat distribution and providing the aeration that makes marshmallows puffy. Gum, obtained from plants, also can act as an emulsifier in marshmallows, but it is also important as a gelling agent.

Most marshmallows also contain natural and/or artificial flavoring. If they are colored marshmallows, the color usually comes from an artificial coloring.

The Manufacturing Process

Cooking

1 A solution is formed by dissolving sugar and corn syrup in water and boiling it. Egg whites and/or gelatin is mixed with the sugar solution. Then the ingredients are heated in a cook kettle to about 240°F (115°C). The resulting mixture is passed through a strainer to remove extraneous matter.

2 In the pump, the mixture is then beaten into a foam to two or three times its original volume. At this stage, flavoring can be added.

Forming

3 The heated mixture is transferred to a heat exchanger. Air is pumped into the mixture. The mixture cools in a tempering kettle, passes through another filter, and

Paradise pudding from the recipe booklet "The Jell-O Girl Entertains," circa 1930. (From the collections of Henry Ford Museum & Greenfield Village.)

In the early twentieth century, marshmallows were considered a child's confection, dispensed as penny candy at general stores along with licorice whips and peppermint drops. But through a fortuitous connection with other popular foods and some clever marketing, marshmallows would soon become a staple ingredient at pot-luck dinners, family get-togethers, and even elegant parties.

A perusal through twentieth-century cookbooks and recipe booklets reveals that marshmallows usually served as an ingredient in cakes, candies, and desserts. They also became well-known as a topping for steaming cups of hot cocoa and as a roasted treat at cookouts and picnics. Increasingly, they served as a sweet addition to salads and side dishes, including their classic contribution to the Thanksgiving dinner table—atop a dish of baked sweet potatoes or yams.

The 1935 recipe booklet, "Campfire Marshmallow Cookery," expanded upon the usual marshmallow classics with 50 "perfect" recipes. These ranged from everyday dishes like marshmallow ice box loaf and campfire rice pudding to special occasion desserts, including a selection of dainty marlows and mallobets (or ice creams and sherbets).

Perhaps the greatest distinction for marshmallows occurred as a result of their advantageous connection with gelatin salads and desserts, which rose in popularity during the 1920s and 1930s. Recipe booklets for Jell-O and Knox Gelatine from that time include recipes that called for marshmallows on almost every page—recipes like banana fluff, lime mallow sponge, cocoa tutti frutti, and paradise pudding.

Donna R. Braden

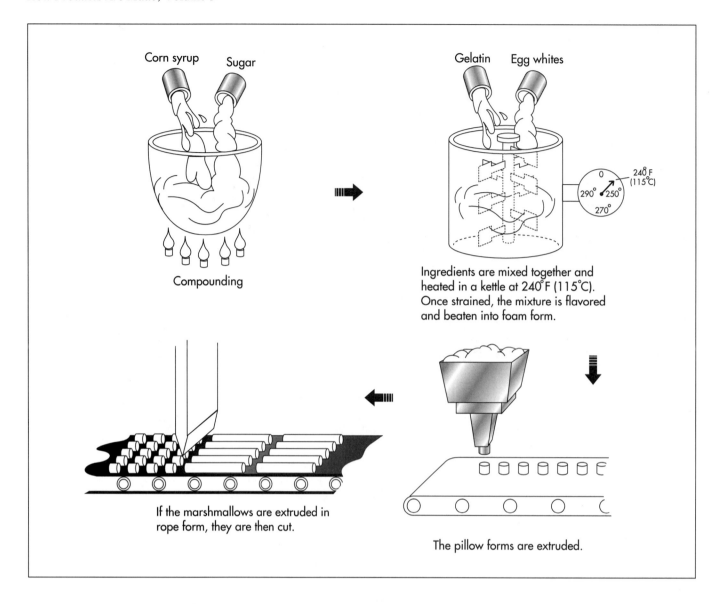

Corn syrup   Sugar

Compounding

Gelatin   Egg whites

240°F (115°C)
0
290°   250°
270°

Ingredients are mixed together and heated in a kettle at 240°F (115°C). Once strained, the mixture is flavored and beaten into foam form.

If the marshmallows are extruded in rope form, they are then cut.

The pillow forms are extruded.

continues on to the "hill." Marshmallows are extruded through a machine or deposited onto bands.

The extrusion process involves the foam being squeezed through a die to produce marshmallow's familiar pillow shape. Usually, they get a coating of corn starch to counter stickiness and help maintain their form after they have been extruded. Sometimes the pillows are formed into a rope of pillows. If so, they are cut and dried on a rubber conveyor belt.

Cooling

4 After the pillows are formed, they are sent through a cooling drum, where excess starch is removed. They also are cooled enough to be packaged.

Packaging

5 After the pillows have cooled, they are weighed and packaged. Before being put in cases, some manufacturers pass their product through a metal detector. The case is code dated and shipped to retail stores.

Quality Control

Throughout the manufacturing process, marshmallows are checked for extraneous matter. The mixture passes through strainers, screens, and metal detectors. Visual inspections are also used to ensure quality. Many marshmallow manufacturers have standards for many aspects of a marshmallow, including its size and texture. An ideal marshmallow should be light. This quality

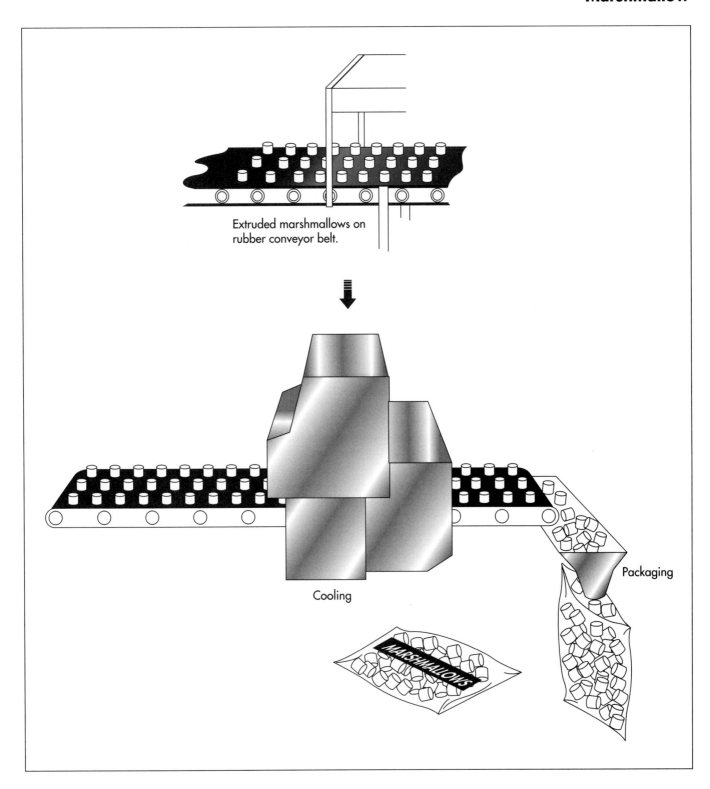

Extruded marshmallows on rubber conveyor belt.

Cooling

Packaging

is affected by how much air is beat into the marshmallow. The more air that is worked into the mixture, the lighter the resulting marshmallow.

An ideal marshmallow is created in a manufacturing environment where the size of the bubbles in the foam that forms the marsh-

mallow is controlled. These small bubbles should be evenly distributed throughout a stable foam. If several factors are controlled, including every factor of the beating process, marshmallows will be formed correctly. A substandard marshmallow has large bubbles. These bubbles are conducive to syrupy layers and voids.

Where to Learn More

Periodicals

Groves, Reg. "Process Control of Marshmallows." *Candy Industry*, July 1991, p.20.

Groves, Reg. "Technology and techniques in marshmallow production." *Candy Industry*, September 1995, pp. 46-53.

Other

Marshmallow fluff site. 1996. http://www.marshamallowfluff.com (July 9, 1997).

—*Annette Petrusso*

Mascara

Background

Mascara is a cosmetic applied to the eyelashes to make the lashes thicker, longer, and darker. It is one of the most ancient cosmetics known, having been used in Egypt possibly as early as 4000 B.C. Egyptians used a substance called kohl to darken their lashes, eyebrows, and eyelids. Egyptian kohl was probably made of galena or lead sulfite, malachite, and charcoal or soot. The Babylonians and ancient Greeks also used black eye cosmetics, as did the later Romans. Cosmetics of all sorts fell out of use in Europe after the fall of Rome, though eye cosmetics continued to be important in the Arab world. The use of cosmetics was revived in Europe during the Renaissance.

Early mascara from the modern era usually took the form of a pressed cake. It was applied to the lashes with a wetted brush. The ingredients typically were 50% soap and 50% black pigment. The pigment was sifted and combined with soap chips, run through a mill several times, and then pressed into cakes. A variation on this was cream mascara, a lotion-like substance that was packaged in a tube. To apply it, the user would squeeze a small amount of mascara out of the tube onto a small brush. This was a messy process that was much improved with the invention in the 1960s of the mascara applicator. This patented device was a grooved application rod that picked up a consistent amount of mascara when pulled from the bottle. The grooved rod was soon replaced with a brush. This new ease of application may have contributed to the increased popularity of mascara in the late 1960s.

Raw Materials

There are many different formulas for mascara. All contain pigments. In the United States, federal regulations prohibit the use of any pigments derived from coal or tar in eye cosmetics, so mascaras use natural colors and inorganic pigments. Carbon black is the black pigment in most mascara recipes, and iron oxides provide brown colors. Other colors such as ultramarine blue are used in some formulas. One common type of mascara consists of an emulsion of oils, waxes, and water. In formulas for this type of mascara, beeswax is often used, as is carnauba wax and paraffin. Oils may be mineral oil, lanolin, linseed oil, castor oil, oil of turpentine, eucalyptus oil, and even sesame oil. Some formulas contain alcohol. Stearic acid is a common ingredient of lotion-based formulas, as are stiffeners such as ceresin and gums such as gum tragacanth and methyl cellulose. Some mascaras include fine rayon fibers, which make the product more viscous.

The Manufacturing Process

There are two main types of mascara currently manufactured. One type is called anhydrous, meaning it contains no water. The second type is made with a lotion base, and it is manufactured by the emulsion method.

Anhydrous method

1 In this method, ingredients are mixed in tanks or kettles, which make a small batch of 10-30 gal (38-114 l). The ingredients are first carefully measured and weighed. Then a worker empties them into

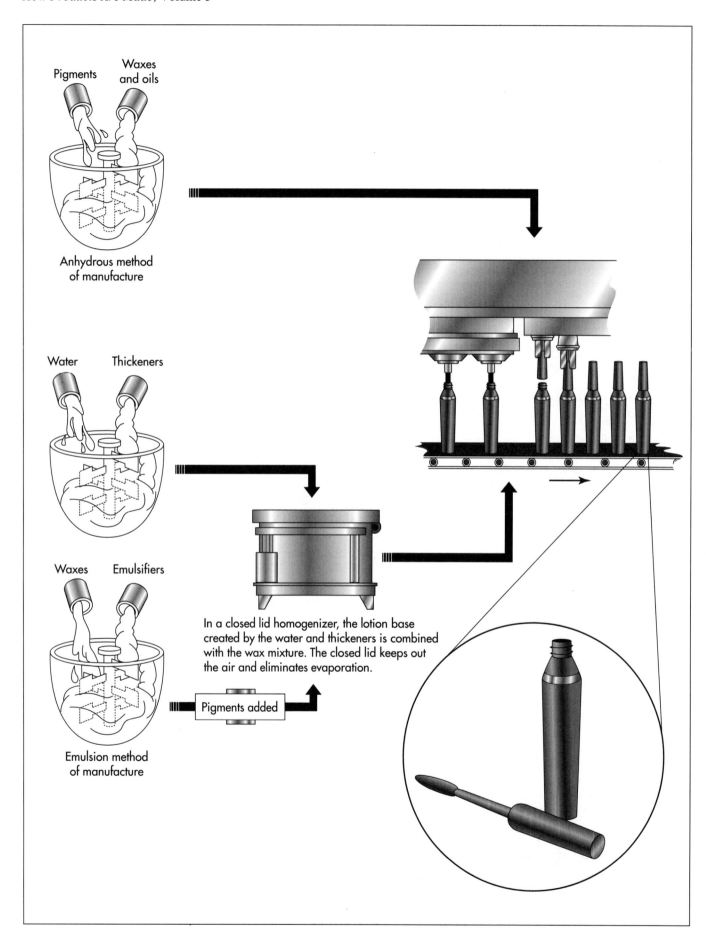

Pigments

Waxes
and oils

Anhydrous method
of manufacture

Water

Thickeners

Waxes

Emulsifiers

Emulsion method
of manufacture

Pigments added

In a closed lid homogenizer, the lotion base
created by the water and thickeners is combined
with the wax mixture. The closed lid keeps out
the air and eliminates evaporation.

the mixing tank. Heat is applied to melt the waxes, and the mixture is agitated, usually by means of a propeller blade. The agitation continues until the mixture reaches a semi-solid state.

Emulsion method

2 In this method, water and thickeners are combined to make a lotion or cream base. Waxes and emulsifiers are heated and melted separately, and pigments are added. Then the waxes and lotion base are combined in a very high speed mixer or homogenizer. Unlike the tank or kettle above, the homogenizer is enclosed and mixes the ingredients at very high speed without incorporating any air or causing evaporation. The oils and waxes are broken down into very small beads by the rapid action of the homogenizer and held in suspension in the water. The homogenizer may hold as little as 5 gal (19 l), or as much as 100 gal (380 l). The high-speed mixing action continues until the mixture reaches room temperature.

The following steps are common to both types of mascara.

Filling

3 After the mascara solution has cooled or reached the proper state, workers transfer it to a tote bin. Next, they roll the tote bin to the filling area and empty the solution into a hopper on a filling machine. The filling machine pumps a measured amount (typically about 0.175 oz [5 g]) of the solution into glass or plastic mascara bottles. The bottles are usually capped by hand. Samples are removed for inspection, and the rest are readied for distribution.

Quality Control

Checks for quality and purity are taken at various stages in the manufacture of mascara. The chemicals are checked in the tank before the mixing begins to make sure the correct ingredients and proper amounts are in place. After the batch is mixed, it is re-checked. After the batch is bottled, representative samples from the beginning, middle, and end of the batch are taken out. These are examined for chemical composition. At this point they are also tested for microbiological impurities.

The Future

Some mascaras on the market today boast all-natural ingredients, and their recipes vary little from products that might have been made at home 100 years ago. One development that may affect mascara manufacturing in the future, however, is the development of new pigments. Researchers in the plastics industry have developed bold, vivid pigments that have recently been introduced to lipsticks. Plastic-derived pigments may be of interest to mascara manufacturers as well.

Where to Learn More

Books

Angeloglou, Maggie. *A History of Make-up*. The Macmillan Company, 1970.

Aucoin, Kevyn. *The Art of Make Up*. Harper Collins, 1994.

Schemann, Andrew. *Cosmetics Buying Guide*. Consumer Reports Books, 1993.

Wetterhahn, Julius. "Eye Makeup," in *Cosmetics: Science and Technology*. M. S. Balsam and Edward Sagarin, ed. John Wiley & Sons, 1972.

Periodicals

Iverson, Annemarie. "Pigment of the Imagination." *Harper's Bazaar*, May 1995, pp. 160-164.

—*Angela Woodward*

Opposite page:
Mascara can be made in two different ways. In the anhydrous method, all the ingredients are mixed, heated, and agitated. With the emulsion method, water and thickeners are combined, while the waxes and emulsifiers are mixed and heated separately. Pigments are added before both mixtures are combined in a high-speed agitator called a homogenizer. The result of either method is a semi-solid substance that is ready to be packaged.

Match

Modern match manufacturing is a highly automated process using continuous-operation machines that can produce as many as 10 million matches in an eight-hour shift with only a few people to monitor the operation.

Background

A match is a small stick of wood or strip of cardboard with a solidified mixture of flammable chemicals deposited on one end. When that end is struck on a rough surface, the friction generates enough heat to ignite the chemicals and produce a small flame. Some matches, called strike-anywhere matches, may be ignited by striking them on any rough surface. Other matches, called safety matches, will ignite only when they are struck on a special rough surface containing certain chemicals.

History

The first known use of matches was in 577 during the siege of a town in northern China. Women in the town used sticks coated with a mixture of chemicals to start fires for cooking and heating, thus allowing them to conserve their limited fuel by putting the fires out between uses. The details of this technique were subsequently lost to history. It was not until 1826 that John Walker of England invented the first friction matches. Walker's matches were ignited by drawing the heads through a folded piece of paper coated with ground glass. He began selling them in 1827, but they were difficult to light and were not a success.

In 1831, Charles Sauria of France developed a match that used white phosphorus. These matches were strike-anywhere matches and were much easier to ignite. Unfortunately, they were too easy to ignite and caused many unintentional fires. White phosphorus also proved to be highly toxic. Workers in match plants who inhaled white phosphorus fumes often suffered from a horrible degeneration of the jawbones known as "phossy jaw." Despite this health hazard, white phosphorus continued to be used in strike-anywhere matches until the early 1900s, when government action in the United States and Europe forced manufacturers to switch to a nontoxic chemical.

In 1844 Gustaf Pasch of Sweden proposed placing some of the match's combustion ingredients on a separate striking surface, rather than incorporating them all into the match head, as an extra precaution against accidental ignition. This idea—coupled with the discovery of less-reactive, nontoxic red phosphorus—led J. E. Lundstrom of Sweden to introduce safety matches in 1855. Although safety matches posed less of a hazard, many people still preferred the convenience of strike-anywhere matches, and both types continue to be used today.

The first matchbook matches were patented in the United States by Joshua Pussey in 1892. The Diamond Match Company purchased the rights to this patent in 1894. At first, these new matches were not well accepted, but when a brewing company bought 10 million matchbooks to advertise their product, sales soared.

Early match manufacturing was mainly a manual operation. Mechanization slowly took over portions of the operation until the first automatic match machine was patented by Ebenezer Beecher in 1888. Modern match manufacturing is a highly automated process using continuous-operation machines that can produce as many as 10 million matches in an eight-hour shift with only a few people to monitor the operation.

Raw Materials

Woods used to make matchsticks must be porous enough to absorb various chemicals, and rigid enough to withstand the bending forces encountered when the match is struck. They should also be straight-grained and easy to work, so that they may be readily cut into sticks. White pine and aspen are two common woods used for this purpose.

Once the matchsticks are formed, they are soaked in ammonium phosphate, which is a fire retardant. This prevents the stick from smoldering after the match has gone out. During manufacture, the striking ends of the matchsticks are dipped in hot paraffin wax. This provides a small amount of fuel to transfer the flame from the burning chemicals on the tip to the matchstick itself. Once the paraffin burns off, the ammonium phosphate in the matchstick prevents any further combustion.

The heads of strike-anywhere matches are composed of two parts, the tip and the base. The tip contains a mixture of phosphorus sesquisulfide and potassium chlorate. Phosphorus sesquisulfide is a highly reactive, non-toxic chemical used in place of white phosphorus. It is easily ignited by the heat of friction against a rough surface. The potassium chlorate supplies the oxygen needed for combustion. The tip also contains powdered glass and other inert filler material to increase the friction and control the burning rate. Animal glue is used to bind the chemicals together, and a small amount of zinc oxide may be added to the tip to give it a whitish color. The base contains many of the same materials as the tip, but has a smaller amount of phosphorus sesquisulfide. It also contains sulfur, rosin, and a small amount of paraffin wax to sustain combustion. A water-soluble dye may be added to give the base a color such as red or blue.

The heads of safety matches are composed of a single part. They contain antimony trisulfide, potassium chlorate, sulfur, powdered glass, inert fillers, and animal glue. They may also include a water-soluble dye. Antimony trisulfide cannot be ignited by the heat of friction, even in the presence of an oxidizing agent like potassium chlorate, and it requires another source of ignition to start

the combustion. That source of ignition comes from the striking surface, which is deposited on the side of the matchbox or on the back cover of the matchbook. The striking surface contains red phosphorus, powdered glass, and an adhesive such as gum arabic or urea formaldehyde. When a safety match is rubbed against the striking surface, the friction generates enough heat to convert a trace of the red phosphorus into white phosphorus. This immediately reacts with the potassium chlorate in the match head to produce enough heat to ignite the antimony trisulfide and start the combustion.

Match boxes and match books are made from cardboard. The finned strips of cardboard used to make the matches in match books are called a comb.

The Manufacturing Process

Matches are manufactured in several stages. In the case of wooden-stick matches, the matchsticks are first cut, prepared, and moved to a storage area. When the matchsticks are needed, they are inserted into holes in a long perforated belt. The belt carries them through the rest of the process, where they are dipped into several chemical tanks, dried, and packaged in boxes. Cardboard-stick matches used in match books are processed in a similar manner.

Here is a typical sequence of operations for manufacturing wooden-stick matches:

Cutting the matchsticks

1 Logs of white pine or aspen are clamped in a debarking machine and slowly rotated while spinning blades cut away the outer bark of the tree.

2 The stripped logs are then cut into short lengths about 1.6 ft (0.5 m) long. Each length is placed in a peeler and rotated while a sharp, flat blade peels a long, thin sheet of wood from the outer surface of the log. This sheet is about 0.1 in (2.5 mm) thick and is called a veneer. The peeling blade moves inward toward the core of the rotating log until only a small, round post is left. This post is discarded and may be used for fuel or reduced to wood chips for use in making paper or chipboard.

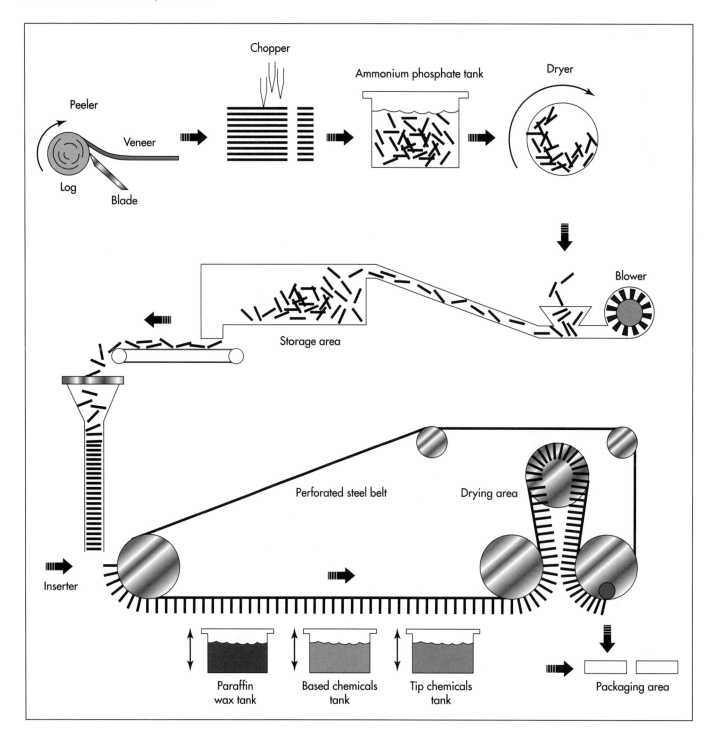

3 The sheets of veneer are stacked and fed into a chopper. The chopper has many sharp blades that cut down through the stack to produce as many as 1,000 matchsticks in a single stroke.

Treating the matchsticks

4 The cut matchsticks are dumped into a large vat filled with a dilute solution of ammonium phosphate.

5 After they have soaked for several minutes, the matchsticks are removed from the vat and placed in a large, rotating drum, like a clothes dryer. The tumbling action inside the drum dries the sticks and acts to polish and clean them of any splinters or crystallized chemical.

6 The dried sticks are then dumped into a hopper and blown through a metal duct to the storage area. In some operations the sticks are blown directly into the match-making facility rather than going to storage.

Forming the match heads

7 The sticks are blown from the storage area to a conveyor belt that transfers them to be inserted into holes on a long, continuous, perforated steel belt. The sticks are dumped into several v-shaped feed hoppers that line them up with the holes in the perforated belt. Plungers push the matchsticks into the holes across the width of the slowly moving belt. A typical belt may have 50-100 holes spaced across its width. Any sticks that do not seat firmly into the holes fall to a catch area beneath the belt and are transferred back to the feed hoppers.

8 The perforated belt holds the matchsticks upside down and immerses the lower portion of the sticks in a bath of hot paraffin wax. After they emerge from the wax, the sticks are allowed to dry.

9 Further down the line, the matchsticks are positioned over a tray filled with a liquid solution of the match head chemicals. The tray is then momentarily raised to immerse the ends of the sticks in the solution. Several thousand sticks are coated at the same time. This cycle repeats itself when the next batch of sticks is in position. If the matches are the strike-anywhere kind, the sticks move on to another tray filled with a solution of the tip chemicals, and the match ends are immersed in that tray, only this time not quite as deeply. This gives strike-anywhere matches their characteristic two-toned appearance.

10 After the match heads are coated, the matches must be dried very slowly or they will not light properly. The belt loops up and down several times as the matches dry for 50-60 minutes.

Packaging the matches

11 The cardboard inner and outer portions of the match boxes are cut, printed, folded, and glued together in a separate area. If the box is to contain safety matches, the chemicals for the striking strip are mixed with an adhesive and are automatically applied to the outer portion of the box.

12 When the matches are dry, the belt moves them to the packaging area, where a multi-toothed wheel pushes the finished matches out of the holes in the belt. The matches fall into hoppers, which measure the proper amount of matches for each box. The matches are dumped from the hoppers into the inner portions of the cardboard match boxes, which are moving along a conveyor belt located below the hoppers. Ten or more boxes may be filled at the same time.

13 The outer portions of the match boxes move along another conveyor belt running parallel to the first belt. Both conveyors stop momentarily, and the filled inner portions are pushed into the outer portions. This cycle of filling the inner portions and pushing them into the outer portions is repeated at a rate of about once per second.

14 The filled match boxes are moved by conveyor belt to a machine, which groups them and places them in a corrugated cardboard box for shipping.

Quality Control

The chemicals for each portion of the match head are weighed and measured exactly to avoid any variation in the match composition that might affect performance. Operators constantly monitor the operation and visually inspect the product at all stages of manufacture. In addition to visual inspection and other normal quality control procedures, match production requires strict attention to safety. Considering that there may be more than one million matches attached to the perforated belt at any time means that the working environment must be kept free of all sources of accidental ignition.

The Future

The use of matches in the United States has steadily declined in the last few decades. This decline is the result of several factors: the availability of inexpensive, disposable lighters; the decrease in the use of tobacco products by the general public; and the development of automatic lighting devices for gas-fired stoves. Of the matches that are sold, book matches far outsell wooden stick matches because of their advertising value.

Opposite page:
Stripped logs are placed in a peeler, which cuts a sheet about 0.1 in (2.5 mm) thick, called veneer, from the log. The veneer proceeds to the chopper, which cuts it into small sticks. The sticks are soaked in a dilute solution of ammonium phosphate and dried, removing splinters and crystallized solution. The matches are dumped into a feed hopper, which lines them up. A perforated conveyor belt holds them upside down while they are dipped in a series of three tanks. The matches are dried for 50-60 minutes before they are packaged.

Worldwide, matches will continue to be in demand for the foreseeable future, although their production will probably follow the demand and migrate to other countries.

Where to Learn More

Books

Bennett, H., ed. *The Chemical Formulary, Vol. XV.* Chemical Publishing Company, Inc., 1970.

Periodicals

Bean, M.C. "History of the Match," *Antiques and Collecting Hobbies.* September, 1992, pp. 42-44.

— Chris Cavette

Microscope

Background

A microscope is an instrument used to produce enlarged images of small objects. The most common kind of microscope is an optical microscope, which uses lenses to form images from visible light. Electron microscopes form images from beams of electrons. Acoustic microscopes form images from high-frequency sound waves. Tunneling microscopes form images from the ability of electrons to "tunnel" through the surface of solids at extremely small distances.

An optical microscope with a single lens is known as a simple microscope. Simple microscopes include magnifying glasses and jeweler's loupes. An optical microscope with two lenses is known as a compound microscope. The basic parts of a compound microscope are the objective, which holds the lens near the specimen, and the eyepiece, which holds the lens near the observer. A modern compound microscope also includes a source of light (either a mirror to catch external light or a light bulb to provide internal light), a focusing mechanism, and a stage (a surface on which the object being examined can be held in place). Compound microscopes may also include a built-in camera for microphotography.

Ancient peoples noted that objects seen through water appeared larger. The first century Roman philosopher Seneca recorded the fact that letters seen through a glass globe full of water were magnified. The earliest simple microscopes consisted of a drop of water captured in a small hole in a piece of wood or metal. During the Renaissance, small glass lenses replaced the water. By the late seventeenth century, the Dutch scientist

Antonie van Leeuwenhoek built outstanding simple microscopes using very small, high-quality lenses mounted between thin brass plates. Because of the excellence of his microscopes, and the fact that he was the first to make observations of microscopic organisms, Leeuwenhoek is often incorrectly thought of as the inventor of the microscope.

The compound microscope made its first appearance between the years 1590 and 1608. Credit for this invention is often given to Hans Janssen, his son Zacharias Janssen, or Hans Lippershey, all of whom were Dutch spectacle makers. Early compound microscopes consisted of pairs of lenses held in a small metal tube and looked much like modern kaleidoscopes. Because of the problem of chromatic aberration (the tendency of a lens to focus each color of light at a slightly different point, leading to a blurred image) these microscopes were inferior to well-made simple microscopes of the time.

The earliest written records of microscopic observations were made by the Italian scientist Francesco Stelluti in 1625, when he published drawings of a bee as seen through a microscope. The first drawings of bacteria were made by Leeuwenhoek in 1683. During the seventeenth and eighteenth centuries, numerous mechanical improvements were made in microscopes in Italy, including focusing devices and devices for holding specimens in place. In England in 1733, the amateur optician Chester Moor Hall discovered that combining two properly shaped lenses made of two different kinds of glass minimized chromatic aberration. In 1774, Benjamin Martin used this technique in a microscope. Many advances were made in the building of microscopes in the nine-

The earliest written records of microscopic observations were made by the Italian scientist Francesco Stelluti in 1625, when he published drawings of a bee as seen through a microscope.

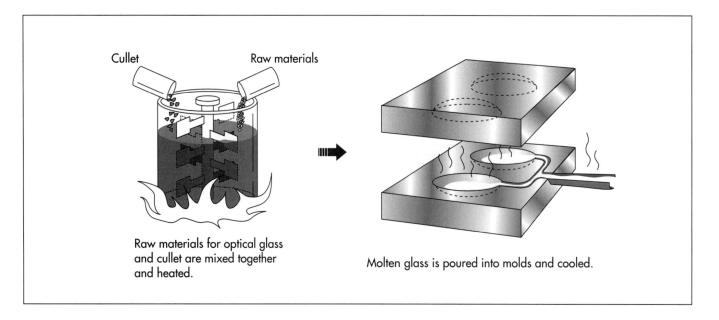

Cullet   Raw materials

Raw materials for optical glass and cullet are mixed together and heated.

Molten glass is poured into molds and cooled.

teenth and twentieth centuries. Electron microscopes were developed in the 1930s, acoustic microscopes in the 1970s, and tunneling microscopes in the 1980s.

Raw Materials

An optical microscope consists of an optical system (the eyepiece, the objective, and the lenses inside them) and hardware components which hold the optical system in place and allow it to be adjusted and focused. An inexpensive microscope may have a mirror as a light source, but most professional microscopes have a built-in light bulb.

Lenses are made of optical glass, a special kind of glass which is much purer and more uniform than ordinary glass. The most important raw material in optical glass is silicon dioxide, which must be more than 99.9% pure. The exact optical properties of the glass are determined by its other ingredients. These may include boron oxide, sodium oxide, potassium oxide, barium oxide, zinc oxide, and lead oxide. Lenses are given an antireflective coating, usually of magnesium fluoride.

The eyepiece, the objective, and most of the hardware components are made of steel or steel and zinc alloys. A child's microscope may have an external body shell made of plastic, but most microscopes have an body shell made of steel.

If there is a mirror included, it is usually made of a strong glass such as Pyrex (a

trade name for a glass made from silicon dioxide, boron dioxide, and aluminum oxide). The mirror has a reflective coating made of aluminum and a protective coating made of silicon dioxide.

If a light bulb is included, it is made from glass and contains a tungsten filament and wires made of nickel and iron within a mixture of argon and nitrogen gases. The base of the light bulb is made of aluminum.

If a **camera** is included, it contains lenses made of optical glass. The body of the camera is made of steel or other metals or of plastic.

The Manufacturing Process

Making the hardware components

1 Metal hardware components are manufactured from steel or steel and zinc alloys using precision metalworking equipment such as lathes and drill presses.

2 If the external body shell of an inexpensive microscope is plastic, it is usually a light, rigid plastic such as acrylonitrile-butadiene-styrene (ABS) plastic. ABS plastic components are made by injection molding. In this process the plastic is melted and forced under pressure into a mold in the shape of the final product. The plastic is then allowed to cool back into a solid. The mold is opened and the product is removed.

Making optical glass

3 The proper raw materials for the type of optical glass desired are mixed in the proper proportions, along with waste glass of the same type. This waste glass, known as cullet, acts as a flux. A flux is a substance which causes raw materials to react at a lower temperature than they would without it.

4 The mixture is heated in a glass furnace until it has melted into a liquid. The temperature varies with the type of glass being made, but is typically about 2550°F (1400°C).

5 The temperature is raised to about 2800°F (1550°C) to force air bubbles to rise to the surface. It is then slowly cooled and stirred constantly until it has reached a temperature of about 1800°F (1000°C). The glass is now an extremely thick liquid, which is poured into molds shaped like the lenses to be made.

6 When the glass has cooled to about 600°F (300°C), it is reheated to about 1000°F (500°C). This process, known as annealing, removes internal stresses which form during the initial cooling period and which weaken the glass. The glass is then allowed to cool slowly to room temperature. The pieces of glass are removed from the molds. They are now known as blanks.

Making the lenses

7 The blank is now placed in a vise and held beneath a rapidly rotating cylindrical cutter with a diamond blade. This cutter, known as a curve generator, trims the surface of the blank until a close approximation of the desired curve is obtained. The cut lens is inspected and cut again if necessary. The difficulty of this process varies widely depending on the type of glass being cut and the exact curvature required. Several cuttings may be required, and the time involved may be a few minutes or more than half an hour.

8 Several cut blanks are placed on the surface of a curved block in such a way that their curved surfaces line up as if they were all part of one spherical surface. This allows many lenses to be ground at the same time. A cast iron grinding surface known as a tool is placed on top on the lenses. The block of lenses rotates while the tool moves at random on top of it. A steady flow of liquid moves between the tool and the lenses. This liquid, known as a slurry, contains water, an abrasive (usually silicon carbide) to do the grinding, a coolant to prevent overheating, and a surfactant to keep the abrasive from settling out of the slurry. The lenses are inspected after grinding and reground if necessary. The grinding process may take one to eight hours.

9 The lenses are moved to a polishing machine. This is similar to the grinding machine, but the tool is made of pitch (a thick, soft resin derived from tar). A pitch tool is made by placing tape around a curved dish, pouring in hot, liquid pitch, and letting it cool back into a solid. A pitch tool can be used about 50 times before it must be reshaped. It works in the same manner as a grinding tool, but instead of an abrasive the slurry contains a polishing substance (usually cerium dioxide). The lenses are inspected after polishing and the procedure is repeated as necessary. Polishing may take from half an hour to five hours. The lenses are cleaned and ready to be coated.

10 The lenses are coated with magnesium fluoride. They are then inspected again, labeled with a date of manufacture and a serial number, and stored until needed.

Making the mirror

11 If a mirror is included, it is made in a way similar to the way in which a lens is made. Unlike a lens, it is cut, ground, and polished to be flat rather than curved. A reflective coating is then applied. Aluminum is heated in a vacuum to produce a vapor. A negative electrostatic charge is applied to the surface of the mirror so that it attracts the positively charged aluminum ions. This allows a thin, even coating of metal to be applied. A protective coating of silicon dioxide is then applied. Like a lens, the mirror is inspected, labeled, and stored.

Assembling the microscope

12 All of the final assembly of the microscope is done by hand. The workers wear gloves, masks, and gowns so that dirt does not damage the lenses or the internal mechanisms of the microscope. First the

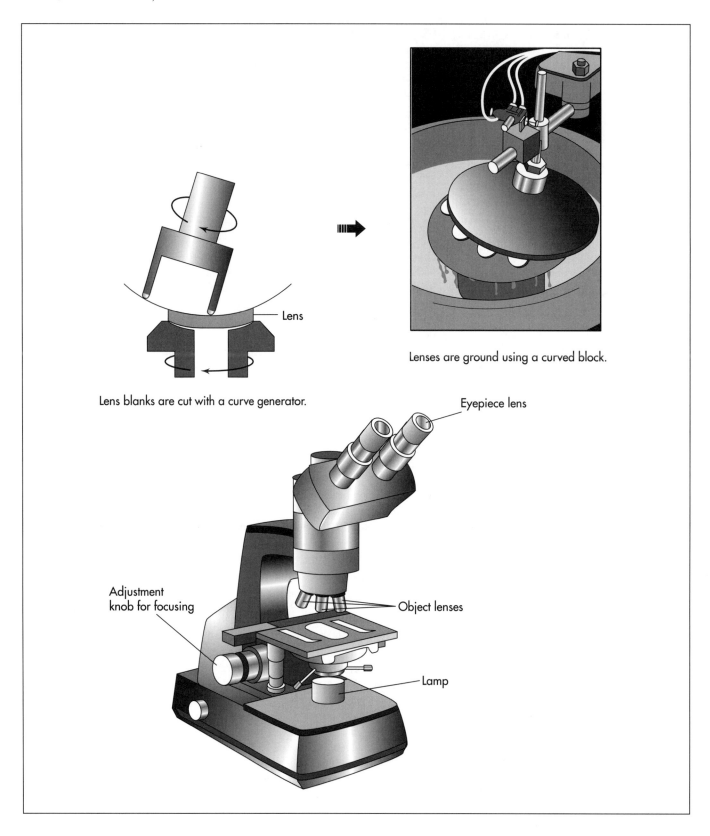

Lens blanks are cut with a curve generator.

Lens

Lenses are ground using a curved block.

Eyepiece lens

Adjustment knob for focusing

Object lenses

Lamp

lenses are placed in the steel tubes, which make up the bodies of the eyepiece and the objective. These tubes are manufactured in standard sizes, which allow them to be assembled into a standard size microscope.

13 The focusing mechanism of most microscopes is a rack and pinion system. This consists of a flat piece of metal with teeth on one side (the rack) and a metal wheel with teeth (the pinion), which controls the movement of the rack. The rack and pinion direct the objective so that its movement toward or away from the object being observed can be controlled. In many microscopes, the rack and pinion are attached to the stage (the flat metal plate on which the object being observed rests) and the objective remains stationary. After the rack and pinion system is installed, the knobs that control it are attached.

14 The external body shell of the microscope is assembled around the internal focusing mechanism. The eyepiece (or two eyepieces, for a binocular microscope) and objective (or a rotating disk containing several different objectives) are screwed into place. Eyepieces and objectives are manufactured in standard sizes that allow many different eyepieces and objectives to be used in any standard microscope.

15 If the microscope contains a mirror, this is attached to the body of the microscope below the opening in the stage. If it contains a light bulb instead, this may be attached in the same place (to shine light through the observed object) or it may be placed to the side of the stage (to shine light on top of the object). Some professional microscopes contain both kinds of light bulbs to allow both kinds of observation. If the microscope contains a camera, it is attached to the top of the body.

16 The microscope is tested. If it functions correctly, the eyepiece and objective are usually unscrewed before packing. The parts of the microscope are packed securely in close-fitting compartments lined with cloth or foam. These compartments are often part of a wood or steel box. The microscope is then placed in a strong cardboard container and shipped to consumers.

Quality Control

The most critical part of quality control for a microscope is the accuracy of the lenses. During cutting and polishing, the size of the lens is measured with a vernier caliper. This device holds the lens between two jaws. One remains stationary while the other is gently moved into place until it touches the lens. The dimensions of the lens are read off a scale, which moves along with the movable jaw.

The curvature of the lens is measured with a spherometer. This device looks like a pocket watch with three small pins protruding from the base. The two outer pins remain in place, while the inner pin is allowed to move in or out. The movement of this pin is connected to a scale on the face of the spherometer. The scale reveals the degree of curvature of the lens. A typical lens should vary no more than about one-thousandth of an inch (25 micrometers).

During polishing, these tests are not accurate enough to ensure that the lens will focus light properly. Optical tests must be used. One typical test, known as an autocollimation test, involves shining a pinpoint light source through a lens in a dark room. A diffraction grating (a surface containing thousands of microscopic parallel grooves per inch) is placed at the point where the lens should focus the light. The grating causes a pattern of light and dark lines to form around the true focal point. It is compared with the theoretical focal point and the lens is repolished if necessary.

The mechanical parts of the microscope are also tested to ensure that they function correctly. The eyepiece and the objective must screw firmly into their proper places and must be perfectly centered to form a sharp image. The rack and pinion focusing mechanism is tested to ensure that it moves smoothly and that the distance between the objective and the stage is controlled precisely. Rotating disks containing multiple objectives are tested to be sure that they rotate smoothly and that each objective remains firmly in place during use.

The Future

Amateur observers may soon be able to purchase microscopes with small, built-in video cameras, which allow the movements of microscopic organisms to be recorded. Computers may be built into the internal control mechanisms of the microscope to provide automatic focusing.

Where to Learn More

Books

Bradbury, Savile. *An Introduction to the Microscope*. Oxford University Press, 1984.

Jacker, Corrine. *Window on the Unknown: A History of the Microscope.* Charles Scribner's Sons, 1966.

Rochow, Theodore George, and Eugene George Rochow. *An Introduction to Microscopy by Means of Light, Electrons, X-rays, or Ultrasound.* Plenum Press, 1978.

Periodicals

Bardell, David. "The First Record of Microscopic Observations." *Bioscience*, January 1983, pp. 36-38.

Other

Ford, Brian J. "History of the Microscope." October 11, 1996. http://www.sciences.demon.co.uk/whistmic.htm

—*Rose Secrest*

Mosquito Repellent

Mosquito repellents are substances that are designed to make surfaces unpleasant or unattractive to mosquitos. They typically contain an active ingredient that repels mosquitos as well as secondary ingredients, which aid in delivery and cosmetic appeal. They are available in many forms, from creams to lotions to oils, but are most often sold as aerosol products.

History

Traditionally, various types of substances have been used to repel mosquitos. These include such things as smoke, plant extracts, oils, tars, and muds. As insect repellent technology became more sophisticated, individual compounds were discovered and isolated. This allowed the formulation of new, more efficient forms of mosquito repellents.

The first truly effective active ingredient used in mosquito repellents was citronella oil. This material is an herbal extract derived from the citronella plant, an Asian grass. While citronella had been used for centuries for medicinal purposes, its repellence was only accidentally discovered in 1901, when it was used as a hairdressing fragrance. Since citronella oil is a fragrant material, it is thought that the chemical terpenes of which it is composed are responsible for its repellent activity. Citronella oil does repel mosquitos, but it has certain characteristics which limit its effectiveness. For example, it is very volatile and evaporates too quickly from surfaces to which it is applied. Also, large amounts are needed to be effective.

The disadvantages of using citronella oil prompted researchers to study alternative synthetic compounds. Many of the early attempts at creating synthetic insect repellents were initiated by the United States military. Out of this research came the discovery of the repellent dimethyl phthalate in 1929. This material showed a good level of effectiveness against certain insect species, but it was ineffective against others. Two other materials were developed as insect repellents. Indalone was found to repel insects in 1937, and Rutgers 612 (2-ethyl-1,3-hexane diol) was synthesized soon after. Like dimethyl phthalate, these materials had certain limitations which prevented their widespread use.

Since none of the available materials were ideal repellents, research into new synthetic materials continued. In 1955, scientists synthesized DEET (n-n-diethylmetatoluamide), currently the most widely used active ingredient for mosquito repellents. After its discovery, repellent manufacturers developed many different forms in which to deliver DEET, such as creams, lotions, and aerosols.

Mode of Action

Most repellent chemicals work by interfering with the mosquito's homing system. This homing system, located on the antennae, is made up of a number of chemical receptors. Research has shown that these chemical receptors are activated by lactic acid, which naturally evaporates from the skin of warm-blooded animals. The mosquitos have the innate ability to follow the lactic acid emissions to their source. However, when a repellent ingredient such as DEET is applied to the skin, it also evaporates. It is thought that the chemical inhibits the binding of the lactic acid to the mosqui-

While citronella had been used for centuries for medicinal purposes, its repellence was only accidentally discovered in 1901, when it was used as a hairdressing fragrance.

to's chemical receptors. This essentially "hides" the protected person from the mosquito. Since the active ingredient must evaporate from the surface to work, the repellent activity lasts for a limited time.

Raw Materials

The active ingredient in a mosquito repellent is primarily responsible for its usefulness. For a material to be valuable as a mosquito repellent, it must meet certain criteria. First, it must effectively discourage insect attack on the treated area for many hours and on many different types of surfaces. Second, it must work under a variety of different environmental conditions. Next, it must not be toxic or cause irritation when applied to human or animal skin. Additionally, it must be cosmetically acceptable, having a pleasant odor, taste, and feel. It should also be harmless to clothing. Finally, it should have a relatively low cost and be effective against other common types of insects, such as flies.

While thousands of compounds have been studied for their use as insect repellents, DEET (n,n-diethyl-m-toluamide) has been used more than any other. DEET is the compound which results from a reaction of *m*-toluic acid with thionyl chloride followed by a reaction with diethyl amine. This material is isolated and purified before it is supplied to mosquito repellent manufacturers. Other repellent ingredients used include citronella oil, dimethyl phthalate, lavender, lemongrass oil, and peppermint oil. It has been found that mixtures of various repellent compounds often provide greater effectiveness than any one compound alone. The active ingredients contained in the mosquito repellents generally make up 5-30% of the final products.

The inert ingredients that are in a mosquito repellent depend on the form that the product will take. Currently, mosquito repellents are sold as aerosols, pumps, lotions, and oils. Mosquito repellents that are sold as lotions or creams are essentially skin creams which have DEET added at a certain level. They are primarily composed of water, surfactants, fatty alcohol, fragrance, and other emollients. When applied to the skin, these products have the dual benefit of repelling

mosquitos and moisturizing skin. These products are generally less effective than aerosol forms, however, because they do not allow the active ingredients to evaporate as easily.

Aerosols are the most common form for mosquito repellents. They are made up of a few different types of ingredients, including a solvent, a propellant, and miscellaneous ingredients. The solvent is usually an organic alcohol such as ethanol or propanol, whose primary responsibility is to dilute the active ingredient to an appropriate concentration. It also aids in keeping all of the raw materials mixed, ensuring that the product will remain effective even after long-term storage. The propellant is a volatile compound which creates the pressure that causes the rest of the product to be forced out of the container. Common propellants include liquified hydrocarbon gases like propane, butane, or isobutane, hydrofluorocarbons, and dimethyl ether. Other ingredients such as fragrances and emollients are added to aerosol mosquito repellents to make them more cosmetically appealing. Still other compounds are added to prevent corrosion and other stability problems.

In addition to the ingredients, the packaging components are also an important part of an aerosol mosquito repellent. The can is typically a metal container made up of tin-plate steel. The coating of tin keeps the steel from reacting with the ingredients used in the repellent formulation. The valve is another key packaging component. It has the dual task of sealing the pressurized contents in the can and controlling the dispensing of these contents. Valves have three sections: a diptube, which feeds the product from the can to the valve body; the valve body, which mixes the product and propellant; and the actuator button, which when pressed, allows the product to be released.

The Manufacturing Process

The production of mosquito repellents can be broken down into two steps. First a large batch of the repellent formulation is made, and then the batch is filled into the packaging. Since aerosols are the most common form of mosquito repellent, the following

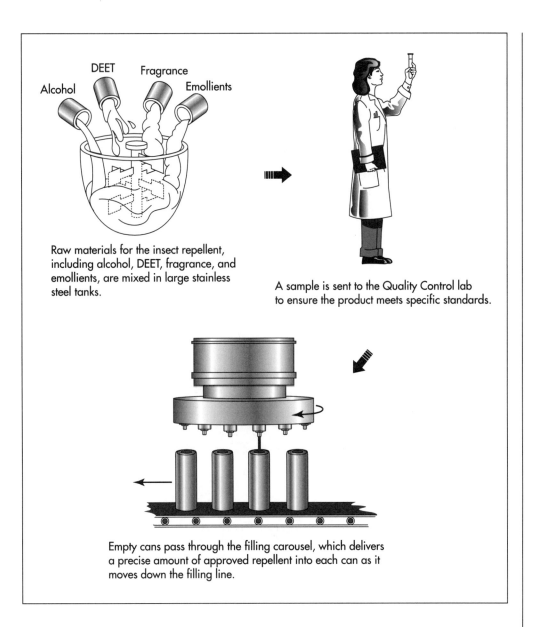

Raw materials for the insect repellent, including alcohol, DEET, fragrance, and emollients, are mixed in large stainless steel tanks.

A sample is sent to the Quality Control lab to ensure the product meets specific standards.

Empty cans pass through the filling carousel, which delivers a precise amount of approved repellent into each can as it moves down the filling line.

description details their production. Other forms of repellents like creams and lotions are produced in a similar way, except that the filling process is less involved.

Compounding

1 The first step in the manufacturing process is compounding. In the compounding area, raw materials are mixed together in large stainless steel tanks. For an aerosol, the alcohol is pumped into the tank, and the other materials, including DEET, fragrance, and emollients, are manually poured in and allowed to mix. All of the ingredients except the propellant are added at this phase of production. Since some of the materials in this process are flammable, special precautions are taken to prevent ex-

plosion, such as using spark-proof electrical outlets and blast-proof walls.

2 When the batch is finished, a sample is sent to the quality control lab and tested to make sure it meets the set standards for the product. After passing these tests it is pumped to the filling lines to make the finished product.

Filling

3 The filling line is a series of machines connected by a conveyor belt system that combine all of the components to make the finished mosquito repellent product. The first machine in the system feeds the empty cans onto the conveyor line. This machine has a large hopper that is filled with empty

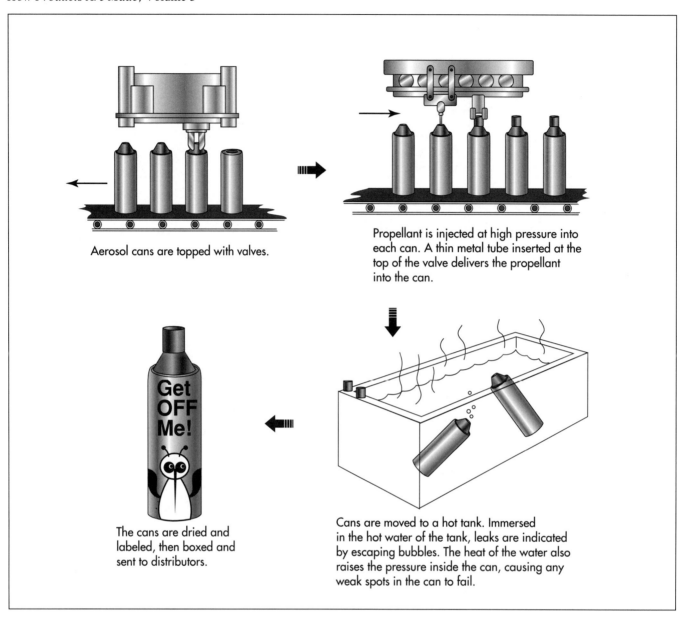

Aerosol cans are topped with valves.

Propellant is injected at high pressure into each can. A thin metal tube inserted at the top of the valve delivers the propellant into the can.

The cans are dried and labeled, then boxed and sent to distributors.

Cans are moved to a hot tank. Immersed in the hot water of the tank, leaks are indicated by escaping bubbles. The heat of the water also raises the pressure inside the can, causing any weak spots in the can to fail.

cans which are physically manipulated until they are standing upright and correctly oriented.

4 The metal cans are then automatically moved along the conveyor belt and cleaned with a jet of compressed air to remove any dust and debris. They next travel to the filling carousel. The filling carousel is made up of a series of piston filling heads that are calibrated to deliver exactly the correct amount of product into the cans. As the cans move through this section of the filling line, they are filled with product.

5 The next step in the filling process involves topping the cans with a valve, adding the propellant, and pressurizing the

cans. The valve is put on by the valve inserter machine. Much like the bin that holds the empty cans, the valves are also put in a hopper and then correctly sorted and aligned. As the cans pass by, the valves are put on. These valves are then tightly affixed to the can by the valve crimping machine. Depending on the type of filling technique, the propellant is either injected through the valve at high pressure or injected into the can before the valve is crimped.

6 After the cans are capped and filled, they are moved to a hot tank, a long trough filled with hot water. Here the cans are checked for escaping bubbles that would indicate a propellant leak. The high temperature of the waterbath also raises the

internal pressure of the can, which is intended to cause any weak spots in the can to fail. This is a crucial quality control step that prevents damaged cans from being sold to the public.

7 When the cans exit the waterbath, they are dried by high-pressure air jets. Other components are then added, such as the actuating button and the overcap. Any needed labels or printing are also added at this point.

8 The finished cans are then moved to the boxing area, where they are put into boxes, typically a dozen cans at a time. These boxes are then stacked onto pallets and hauled away in large trucks to distributors. High speed aerosol production lines like the one described can move at speeds of about 200 cans per minute or more.

Quality Control

Quality control is an essential step in the production of mosquito repellents. Tests are performed at various points in the manufacturing process to ensure that the finished products are consistent from run to run, remain effective over a long period of time, and are safe to use.

Before production begins, the incoming raw materials are checked to ensure they meet the previously set specifications. Tests such as pH, specific gravity, and moisture content are typically performed. Additionally, the cans are inspected for dents, corrosion, and other weaknesses. During manufacture, samples of the repellent are taken during different points along the filling line, and the characteristics or the product are tested. Some of the monitored parameters include the level of active ingredient, pressure, spray rate, and spray pattern. Other testing

is conducted to ensure that the cans evacuate properly. Also, long-term stability studies may be done to establish that the cans do not show undue signs of corrosion.

The Future

The use of many available mosquito repellents is not without its drawbacks. Products that use DEET or citronella oil as the primary active ingredients have been reported to causes rashes in some people. There have even been cases in which children who used DEET products have become very ill. For this reason, research has focused on finding new types of repellents and methods for improving the safety of the ones that are currently available. One recent advance in repellent technology is the use of chemicals to "encapsulate" DEET. It is thought that this product form will protect the user from the harmful effects of DEET while still maintaining its repellent activity. More investigation will have to be completed before this is verified.

Where to Learn More

Books

Knowlton, J. and S. Pearce. *Handbook of Cosmetic Science and Technology*. Elseveir Science Publications, 1993.

Periodicals

Holmes, Hannah. "The Battle of the Bug." *Backpacker*, April 1996, pp. 68-72.

Romanowski, P. and R. Schueller. "Aerosols for Apprentices." *Cosmetics & Toiletries,* May 1996, pp. 35-40.

—*Perry Romanowski*

Nicotine Patch

Nicotine is well suited for transepidermal delivery because it is a liquid which is known to penetrate skin easily. In fact, there are documented cases of tobacco workers suffering from nicotine overdose as a result of handling raw tobacco leaves, a condition known as Green Tobacco Sickness.

Background

A nicotine patch is a device designed to deliver nicotine through the skin and into the blood stream. It is used to help prevent the craving for nicotine that smokers experience when attempting to quit. This type of drug delivery device, generically known as a transepidermal patch, consists of a drug reservoir sandwiched between an occlusive back layer and a permeable adhesive layer that attaches to the skin. The drug slowly leaches out of the reservoir, travels through the skin, and then into the blood stream. Since skin is intended to keep most chemicals out of the body, only certain drugs have the proper chemical characteristics that allow them to be delivered in this manner. The drug molecules must be small enough to penetrate the skin's many layers. It must also be nonirritating to the skin and have a low melting point, so it can be incorporated in liquid form.

Originally transepidermal delivery of drugs required they be formulated into a topically applied cream or lotion. These vehicles tend to be messy, are difficult to administer in consistent amounts, and have unpredictable absorption rates. To overcome these problems, researchers developed a method to deliver an exact dosage by mixing the drug into a bandage adhesive. This early version of the patch successfully brought a known quantity of drug to a specified area of skin for a given length of time. However, it could not control the rate at which the drug was released. In the 1950s, technology was developed to create membranes which could be used to control the diffusion rate of drugs. In the late 1960s and 1970s, advances pioneered by the Alza Corporation

allowed drugs to be contained by these membranes. Thus modern controlled release patches was born. Transepidermal patches were developed to deliver precise quantities of a variety of drugs to the skin for a prolonged period of time. In early 1996 the FDA approved a patch containing nicotine for sale without a prescription. The first brand to be marketed under this new Over The Counter (OTC) regulation was Johnson & Johnson's Nicotrol®.

Raw Materials

Nicotine

The chemical nicotine is an addictive component of cigarettes. The body develops a physical, as well as psychological, craving for nicotine. The patch helps satisfy this craving while the smoker is attempting to quit. Nicotine is well suited for transepidermal delivery because it is a liquid which is known to penetrate skin easily. In fact, there are documented cases of tobacco workers suffering from nicotine overdose as a result of handling raw tobacco leaves, a condition known as Green Tobacco Sickness. Depending on the type of patch, the amount of nicotine compound employed varies between 5% and 50%. The drug may be used in its pure form, or it may be linked with other chemicals entities such as hydrochloride, dihydrochloride, sulfate, tartrate, bitartarate, zinc chloride, and salicylate to form derivatives.

When preparing patches with these chemicals, there are two key areas of concern. The first is dosage, since too high a dose can cause irregular heartbeat, palpitations, nausea, vomiting, dizziness, or weakness. In fact, 60 mg of nicotine (the equivalent of smoking 60 cigarettes at once) is considered to be a lethal dose. Therefore, it is critical that the patch be calibrated to deliver the prescribed amount. The second consideration is related to the solvent properties of nicotine. The drug will attack or dissolve many of the materials used to make patch components. Many adhesives, for example, become stringy and loose their tackiness when exposed to nicotine. Or, they may become so heavily loaded with the drug that they deliver an unacceptably large burst of nicotine when attached to the skin. The compatibility of all patch materials that contact nicotine must be carefully evaluated.

Delivery vehicle

The patch itself is a small disk approximately 1 in (2.5 cm) or less in diameter, which may be assembled in several different configurations. One type of patch consists of a plastic chamber that contains the drug and is covered by a selectively permeable membrane to control the rate at which the drug is delivered. This carrier layer can be made from a variety of plastics, including polyvinyl chloride, polystyrene, polyurethane, ethylene vinyl acetate, polyester, polyolefin, and polycarbonate. Alternately, the carrier may be of the matrix type, also known as the monolith type. In this configuration, the drug is dispersed or suspended in the solid plastic matrix of the carrier. In yet another patch design, the drug is mixed directly with the adhesive and applied to a plastic support layer. Regardless of which patch design is employed, the disk must deliver the drug at a controlled rate. It is also important that the device be made from a plastic material which is flexible enough to be applied and removed from skin without breaking or tearing.

Backing layer

All patch configurations feature an occlusive backing layer that is impermeable to the drug. This is typically a plastic sheet laminated with metallic foil to increase its barrier properties and prevent the drug from leaking.

Adhesive

The adhesive used to mount the patch on the skin is extremely important. There are a number of medical grade, pressure-sensitive adhesives, such as acrylate ester/vinyl pyrrolidone copolymers, dimethyl silicone polymers, and acrylate polymers. The latter dominate medical adhesive market, mainly because of their low level of allergenicity. In addition to being nonirritating to the skin, a patch adhesive must have good water resistance so it continues to adhere when the skin perspires. It must have sufficiently high cohesive strength to allow clean removal of adhesive from skin, and it must have properties that allow it to accommodate skin movement without losing the bond and without excessive skin irritation. As described above, care must be taken to ensure the adhesive will not degrade after prolonged contact with nicotine.

Other ingredients

Other ingredients, such as pigments, dyes, inert fillers, and processing aids, may be mixed in with the drug. Certain types of patches also include permeation enhancers to improve drug penetration. For example, one manufacturer of transepidermal patches includes low levels of alcohol to enhance skin penetration. Some nicotine patches contain antipruritic (anti-itch) agents to treat the pruritus associated with transdermal delivery of nicotine. These antipruritic drugs are selected from a group consisting of bisabolol, oil of chamomile, chamazulene, allantoin, D-panthenol, glycyrrhetenic acid, corticosteroids, and antihistamines.

The Manufacturing Process

Carrier preparation

1 The exact manufacturing process depends on the type of patch being constructed. In general, patch membranes are made by one of several techniques, all of which are designed to create a series of uniform diffusion pores. For example, in the polymer precipitation method, a polymer film is cast on a steel belt containing a solvent-water mixture. As the polymer film hardens, the solvent evaporates and creates a multitude of tiny holes. Porous membranes can also be created by stretching a thin melted film of polymer. As the film is uniformly stretched, small pores are formed. This method is typically used for

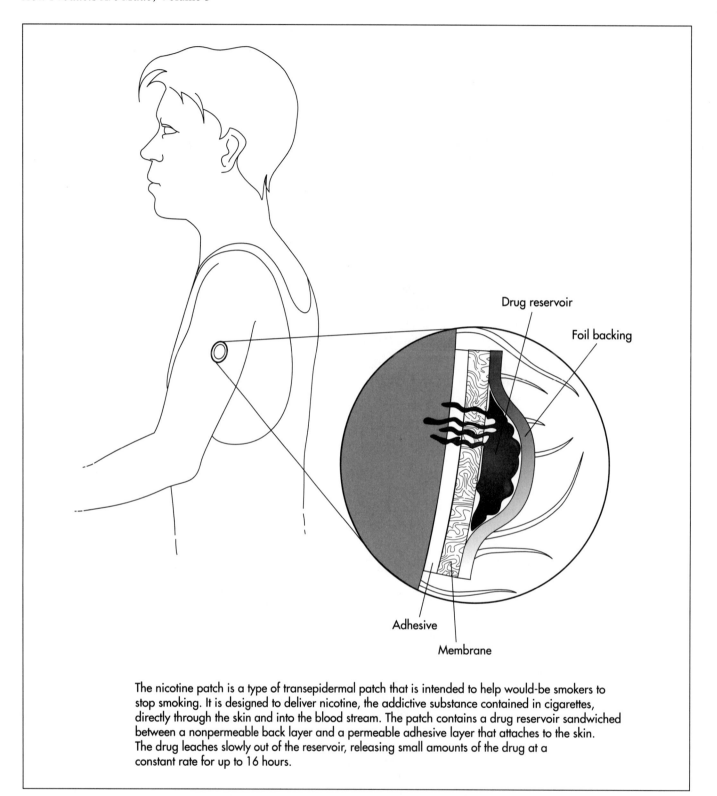

The nicotine patch is a type of transepidermal patch that is intended to help would-be smokers to stop smoking. It is designed to deliver nicotine, the addictive substance contained in cigarettes, directly through the skin and into the blood stream. The patch contains a drug reservoir sandwiched between a nonpermeable back layer and a permeable adhesive layer that attaches to the skin. The drug leaches slowly out of the reservoir, releasing small amounts of the drug at a constant rate for up to 16 hours.

polypropylene films. Polycarbonate films can be perforated by exposure to certain forms of nuclear particles. Regardless of the process, the goal is to create a thin plastic membrane with a multitude of microscopic channels through which the drug can diffuse. These membranes can be attached to the plastic housing containing the drug in subsequent operations.

2 In systems where the drug is intermingled with the patch material, like the matrix type and the mixed adhesive type, the process is somewhat different. The mix-

ture is prepared in a special type of mixer known as a Hockenmeyer mixer. The adhesive is added first and premixed at high speeds for a predetermined period of time. Next, other additives such as viscosity control agents may be added, and further mixing occurs. Then the drug component is slowly added and the mixing speed is increased. Final adjustments in pH or viscosity may be made at this point.

Processing and packaging

3 Final processing depends on the type of carrier. Reservoir type patches must be filled with the drug mixture. The drug-adhesive matrix described above is coated onto silicone-treated polyester film. The silicone ensures the patch can be easily removed to expose the adhesive layer. The completed patch is oven dried to remove solvents and then laminated to a carrier or backing strip. This backing strip can be further processed by die cutting and then packaged as a finished product.

Quality Control

All drugs must undergo stringent testing to ensure they are correctly synthesized and chemically pure. For drugs delivered via transepidermal patch, additional testing is necessary to determine the dosage rate of the product. This rate can be quantified by a method wherein a measured dose is applied to a sample of excised abdominal skin stretched across a small container known as a Franz-type diffusion cell. The amount of drug that diffuses through the skin sample and into the cell can be measured with a variety of analytical techniques, such as high-performance liquid chromatography. This value can be related to determine how much drug will be delivered during actual product usage.

Other key tests are done to ensure the patch adheres to skin properly. Skin is a highly unstable surface which is constantly expanding and contracting. Normally, the strength of an adhesive is evaluated by applying the product to a steel plate. This method is not effective for medical adhesives, however, because these adhesives bond to skin much differently than they bond to metal. To overcome this problem, researchers use a film of collagen (a skin protein material) for peel adhesion studies. The adhesive itself may be evaluated to ensure it does not wet too easily. This is accomplished by measuring the contact angle of a drop of water on the adhesive. The water drop should not wet the adhesive, and the angle should remain for 24 hours. There are a battery of other tests for adhesives, including static shear and the Polyken tack test. It should also be noted that these adhesives fall under government regulations for medical devices, which require certain safety testing such as eye irritation testing and allergic reaction screening.

The Future

The science of delivering nicotine and other drugs transepidermally is still evolving. Current technology can only deliver doses up to 16 hours in length, which stands to be improved. There are also opportunities for improvements to make membrane materials and adhesives that are more imperious to the solvent effects of nicotine. Furthermore, improved delivery systems that offer more advantages to the consumer should be developed. In fact, the Cygnus Corporation is developing a patch with adjustable doses, so the amount of nicotine can be varied depending upon the user's requirements. It is anticipated that recent approval to market these drugs without a prescription will result in increased market activity and improvements in many of the these areas.

Where to Learn More

Books

Satas, Donald, ed. *Handbook of Pressure Sensitive Adhesives*. Van Nostrand Reinhold, 1989.

Periodicals

"Habitrol manufacturer agrees to balance information." *FDA Consumer*, June 1993, p. 4.

"How the nicotine patch works." *Muscle & Fitness*, November 1992, p 34.

Starr, Cynthia. "Quitters or not, patients still like nicotine patches." *Drug Topics*, November 22, 1993, p. 24.

—*Randy Schueller*

Olive Oil

Cans or dark-tinted bottles will keep the deep-green color of the olive oil intact. Oil placed in clear-glass bottles will fade to a yellowish-green. However, the flavor is not affected.

Background

The olive and the tree on which it grows have been revered since ancient times. Archaeological digs have unearthed evidence that olive trees existed on the island of Crete in 3500 B.C. The Semitic peoples were cultivating the tree's fruit by 3000 B.C. They particularly liked to use the oil of the olive to anoint the body during religious ceremonies, and to light their lamps. An ancient Hebrew law prohibiting the destruction of any olive tree is still obeyed.

By the time of the Roman Empire, olives were a mainstay of the agricultural economy. The Romans also used the oil to grease the axles of wagons and chariots. The Greeks traded it for wheat; the elaborately decorated clay pots that they used to transport the oil became part of the civilization's burgeoning art industry.

The olive tree is mentioned frequently in the Koran and in the Bible. Noah receives the message that land is near when a dove arrives at the ark with an olive branch in its mouth. Greek mythology associates the goddess Athena with the olive tree and credits Acropos, the founder of Athens, with teaching the Greeks to extract oil from the tree's fruit.

A member of the evergreen family, the olive tree features a gnarled trunk and leaves with a silvery underside. Its strong root system is perfect for penetrating sand, limestone, or heavy, poorly aerated soil. The trees thrive best in regions with rainy winters and hot, dry summers. Although it may take up to eight years before a tree produces its first harvest, a single tree can live for centuries.

Early oil producers pressed the olives by crushing them between huge cone-shaped stones as they turned slowly on a base of granite. Today, most factories employ hydraulic presses, exerting hundreds of tons of pressure, to separate the oil from the olive paste. Spain and Italy are the primary commercial producers of olives and olive oil. Greece is close behind them. However, California, Australia, and South Africa are emerging as leaders in the industry. Some wineries are planting olives to offset poor wine harvests. Ironically, olive trees were planted in California by missionaries in the 1800s, which by the turn of the century were producing an excellent grade of olive oil. However, the market demand was weak so the trees were uprooted and grape vines were planted in their place.

In the late twentieth century, emphasis on good nutrition and a fascination with the so-called Mediterranean diet has resulted in a resurgence in the olive oil trade. Olive oil is touted as a monounsaturate that is healthier for human consumption than corn and vegetable oils. The oil is also promoted as a dandruff reliever and, when mixed with beeswax, a homemade lip balm. In the late 1990s, the United States and Canada consumed olive oil at a yearly rate of 147,600 tons (150,000 metric tons). The demand often exceeds the supply, and during the 1990s prices rose significantly.

Raw Materials

The primary ingredient of olive oil is the oil that is expressed from ripe olives. In the late spring, small flowers appear on the olive trees. Wind pollination results in the blossoming of the olives, which reach their

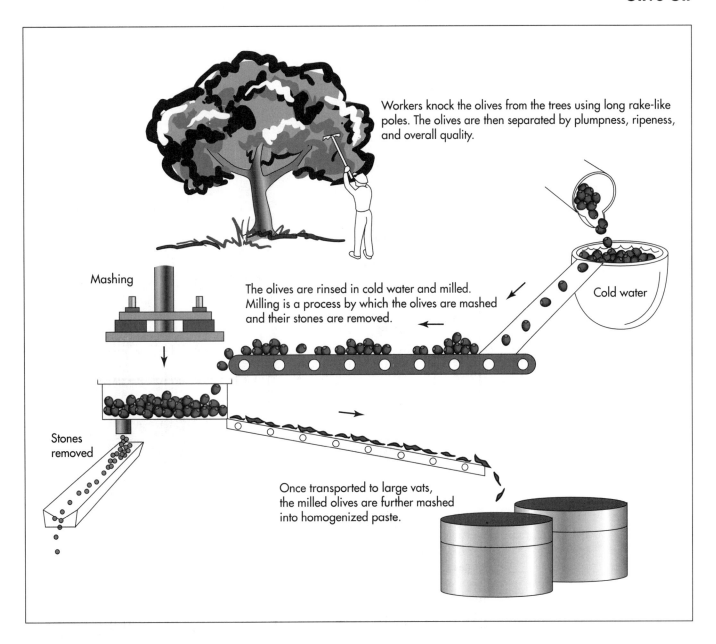

Workers knock the olives from the trees using long rake-like poles. The olives are then separated by plumpness, ripeness, and overall quality.

Cold water

Mashing

The olives are rinsed in cold water and milled. Milling is a process by which the olives are mashed and their stones are removed.

Stones removed

Once transported to large vats, the milled olives are further mashed into homogenized paste.

peak oil content approximately six months later. Thus, the olives are harvested from November to March, after they have progressed in color from green to reddish violet to black. It is often necessary to harvest olives from the same trees several times in order to gather olives at the same stage of maturation.

Since ancient times, workers have knocked the fruit from the trees with long-handled poles. The process has not changed significantly over the centuries. Modern poles resemble rakes. Originally, nets were spread under the tree to catch the falling olives. Many producers are now using plastic cov-

ers to cushion the fall and to allow for cleaner, faster gathering.

One quart (0.95 L) of extra virgin olive oil, the highest level of quality, requires 2,000 olives. The only added ingredient in extra virgin olive oil is the warm water used to flush away the bitterness of the olives, caused by the presence of oleuropein. Extra virgin olive oil contains not more than 1% oleic acid. Pure olive oil, that which results from the second pressing, is often mixed with extra virgin olive oil. The commercial, or non-edible, grades are put through a refining process that may leave traces of soda solutions and bleaching carbons.

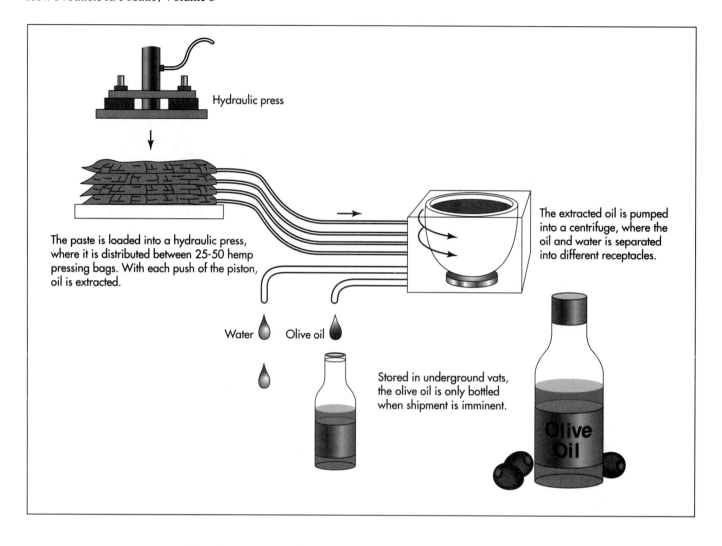

Hydraulic press

The paste is loaded into a hydraulic press, where it is distributed between 25-50 hemp pressing bags. With each push of the piston, oil is extracted.

Water

Olive oil

The extracted oil is pumped into a centrifuge, where the oil and water is separated into different receptacles.

Stored in underground vats, the olive oil is only bottled when shipment is imminent.

Olive Oil

The Manufacturing Process

Collecting and grading the olives

1 After the ripe olives have been combed from the trees, they are picked over by hand to weed out unsound olives. The olives are divided into categories according to their plumpness, state of ripeness, and quality. Then the olives are taken to the press and stored for a short period of time, from a few hours to several days. The period is short enough to prevent fermentation but long enough to allow the olives to get warm so that they release their oil easily.

Washing and milling the olives

2 The olives are rinsed in cold water and then passed along a conveyer belt between rollers or continuous hammers. This machinery, often called the olive crusher, breaks down the cells and de-stones the

olives. Depending on the resiliency of the olives' skin and the stage of maturation, it may be necessary to pass the fruit through the mill a second time.

Creating an olive paste through malaxation

3 In ancient times, the olives were mashed into a paste with a simple mortar and pestle. This principle was expanded upon until the stone mortars were large enough to require slaves or pack animals to operate them. In the modern process, the milled olives travel from the mill into vats in which slowly turning blades mash the olives into a homogenized paste.

Cold-pressing the olive paste to extract the oil

4 The oil is extracted by loading the paste into a hydraulic press. The olive paste is evenly spread over hemp pressing bags or disks covered with synthetic fibers. Each bag or disk is covered with approximately 9-13 lb (4-6 kg) of paste. Between 25 and 50 bags or disks are stacked onto a press plate. Plate guides are inserted at intervals of five to six bags. The plates serve to maintain the balance of the stack and to distribute the pressure evenly. A piston pushes up against the stack, and the oil seeps slowly through the pressing bags to attached tubes. The solid material remains inside the pressing bags.

5 The term cold-pressing refers to the fact that the oil is extracted without heating the paste, furthering insuring the purity of the oil. The oil that is expressed is a reddish mixture of the oil and the inherent vegetable water. This is the oil that receives the appellation of "extra-virgin" olive oil. The paste is removed from the bags and run through several more presses to obtain the lesser grades of oil that remain.

Separating the oil from the vegetable water

6 Originally, the oil and water mixture was stored in vats until the oil rose to the top and was skimmed off. Some fermentation was inevitable, affecting the taste and smell of the olive oil. Today, the separation is accomplished swiftly by pumping the mixture into a centrifuge. The centrifuge is comprised of a rotating drum and an auger that are spun on the same axis at great speed. Because the oil and the vegetable water are of differing densities, the centrifuge forces them apart and into separate receptacles.

Storing and packaging the oil

7 The oil is stored in underground vats until it is ready to be shipped. Then the oil is canned or bottled on an assembly line. Cans or dark-tinted bottles will keep the deep-green color of the olive oil intact. Oil placed in clear-glass bottles will fade to a yellowish-green. However, the flavor is not affected.

8 In many cases, olive oil distributors purchase the olive from the producers and rebottle it. Packaging has become more ornate as the popularity of olive oil has grown. It is not unusual to purchase olive oil in unusually shaped bottles topped with netting or rope. Some packagers also hire professional artists to design their labels.

Quality Control

The olive oil industry is regulated by government food agencies, such as the Food and Drug Administration (FDA) in the United States. By regulation, olive is classified into five grades. Virgin olive oil is that which is obtained from the first pressing. Pure is a mixture of refined and virgin oil. Refined, or commercial, consists of the lower grade lampante oil from which the acid, color, and odor have been removed through processing. Lampante is a highly acidic grade; its name is derived from its use as lamp oil. Sulfide olive oil is chemically extracted from the olives through the use of solvents and is refined many times.

The popularity of olive oil in the late twentieth century has spawned many bottlers who are combining various grades of olive oil and labeling them illegally as virgin or pure. A 1995 FDA report charged that only 4% of the 73 domestically produced or distributed olive oils it tested were pure. The North American Olive Oil Association disputed the findings, stating that of the 300 oils the association tests each year, only a handful are found to be impure. In any event, the situation has become one of "buyer beware."

The Future

Finding workers who are willing to perform the laborious task of picking olives is becoming more difficult. Therefore, the olive oil industry is pursuing methods for mechanizing the collecting process. Among the larger olive oil companies, centrifugation methods are becoming more popular for the pressing process as well as for separating the oil from the vegetable water. Although centrifugation requires more energy and water, the method takes up less space in the factory and requires a shorter set-up time. Centrifugation also eliminates the need for pressing bags, which must be washed after each pressing.

Where to Learn More

Periodicals

Benavides, Lisa. "For Olive Importers, It's All Greek to Them." *Boston Business Journal*, October, 25, 1996, p. 3.

Burros, Marian. "Eating Well." *The New York Times*, October 23, 1996, p. C3.

"From the Olive Tree to Olive Oil." Pompeian, Inc.

"Green, With Envy." *Prevention*, August 1996, p. 106.

Muto, Sheila. "Impurity of Olive Oil Is Raising Concerns." *The New York Times*, January 3, 1996, p. C2.

—*Mary F. McNulty*

Pacemaker

The pacemaker is an electronic biomedical device that can regulate the human heartbeat when its natural regulating mechanisms break down. It is a small box surgically implanted in the chest cavity and has electrodes that are in direct contact with the heart. First developed in the 1950s, the pacemaker has undergone various design changes and has found new applications since its invention. Today, pacemakers are widely used, implanted in tens of thousands of patients annually.

Background

The heart is composed of four chambers, which make up two pumps. The right pump receives the blood returning from the body and pumps it to the lungs. The left pump gets blood from the lungs and pumps it out to the rest of the body. Each pump is made up of two chambers, an atrium and a ventricle. The atrium collects the incoming blood. When it contracts, it transfers the blood to the ventricle. When the ventricle contracts, the blood is pumped away from the heart.

In a normal functioning heart, the pumping action is synchronized by the pacemaker region of the heart, or sinoatrial node, which is located in the right atrium. This is a natural pacemaker that has the ability to create electrical energy. The electrical impulse is created by the diffusion of calcium ions, sodium ions, and potassium ions across the membrane of cells in the pacemaker region. The impulse created by the motion of these ions is first transferred to the atria, causing them to contract and push blood into the ventricles. After about 150 milliseconds, the impulse moves to the ventricles, causing them to contract and pump blood away from the heart. As the impulse moves away from each chamber of the heart, that section relaxes.

Unfortunately, the natural pacemaker can malfunction, leading to abnormal heartbeats. These arrhythmias can be very serious, causing blackouts, heart attacks, and even death. Electronic pacemakers are designed to supplement the heart's own natural controls and to regulate the beating heart when these break down. It is able to do this because it is equipped with sensors that constantly monitor the patient's heart, and a battery that sends electricity, when needed, through lead wires to the heart itself to stimulate the heart to beat.

In addition to outer units, artificial pacemakers can be permanently implanted in a patient's chest. This is done by first guiding the lead through a vein and into a chamber of the heart, where the lead is secured. Fluoroscopic imaging helps facilitate this process. The pacemaker itself is next placed in a pocket, which is formed by surgery just above the upper abdominal quadrant. The lead wire is then connected to the pacemaker, and the pocket is sewn shut. This is a vast improvement over early methods, which required opening the chest cavity and attaching the leads directly to the outer surface of the heart.

History

The idea of using an electronic device to provide consistent regulation of the beating heart was not initially obvious to the early developers of the pacemaker. The first pacemaker, developed by Paul Zoll in 1952, was a portable version of a cardiac resuscitator. It had two lead wires that could be at-

First developed in the 1950s, the pacemaker has undergone various design changes and has found new applications since its invention. Today, pacemakers are widely used, implanted in tens of thousands of patients annually.

tached to a belt worn by the patient. It was plugged into the nearest wall socket and delivered an electric shock that stimulated the heart of a patient having an attack. This stimulation would usually be enough to cause the heart to resume its normal function. While moderately effective, this early pacemaker was primarily used in emergency situations.

Through 1957 and 1960 significant improvements were made to Zoll's original invention. In an attempt to reduce the amount of voltage needed to restart the heart and increase the length of time electronic pacing could be accomplished, C. Walton Lillehei made a pacemaker that had leads attached directly to the outer wall of the heart. Later, in 1958, a battery was added as the power source, making the pacemaker truly portable, which allowed patients to be mobile. This also enabled patients to use the pacemaker continuously instead of only for emergencies. Lillehei's pacemaker was external. William Chardack and Wilson Greatbatch invented the first implantable pacemaker. It was implanted in a living patient in 1960.

The modern technique for putting a pacemaker into a patient's heart was developed by Seymour Furman. Instead of cutting open the chest cavity, he used a method of inserting the leads into a vein and threading them up into the ventricles. With the leads inside the heart, even lower voltages were needed to regulate the heartbeat. This increased the length of time a pacemaker could be inside a person. Although his method was not widely used initially, by the late 1960s most cardiac specialists had switched to Furman's endocardial pacemakers. Since then improvements have been made in their design, including smaller pacemaker devices, longer lasting batteries, and computer controls.

Raw Materials

The materials used to construct pacemakers must be pharmacologically inert, nontoxic, sterilizable, and able to function in the environmental conditions of the body. The various parts of the pacemaker, including the casing, microelectronics, and the leads, are all made with biocompatible materials. Typically, the casing is made of titanium or a titanium alloy. The lead is also made of a metal alloy, but it is insulated by a polymer such as polyurethane. Only the metal tip of the lead is exposed. The circuitry is usually made of modified silicon semiconductors.

Design

Many types of pacemakers are available. The North American Society of Pacing and Electrophysiology (NASPE) has classified them by which heart chamber is paced, which chamber is sensed, how the pacemaker responds to a sensed beat, and whether it is programmable. Despite this vast array of models, all pacemakers are essentially composed of a battery, lead wires, and circuitry.

The primary function of a pacemaker battery is to store enough energy to stimulate the heart with a jolt of electricity. Additionally, it also provides power to the sensors and timing devices. Since these batteries are implanted into the body, they are designed to meet specific characteristics. First, they must be able to generate about five volts of power, a level that is slightly higher than the amount required to stimulate the heart. Second, they must retain their power over many years. A minimum time frame is four years. Third, they must have a predictable life cycle, allowing the doctor to know when a replacement is required. Finally, they must be able to function when hermetically (airtight) sealed. Batteries have two metals that form the anode and cathode. These are the battery components through which charge is transferred. Some examples include lithium/iodide, cadmium/nickel oxide, and nuclear batteries.

Pacemaker leads are thin, insulated wires that are designed to carry electricity between the battery and the heart. Depending on the type of pacemaker, it will contain either a single lead, for single chamber pacemakers, or two leads, for dual chamber pacemakers. With the constant beating of the heart, these wires are chronically flexed and must be resistant to fracture. There are many styles of leads available, with primary design differences found at the exposed end. Many of the leads have a screw-in tip, which helps anchor them to the inner wall of the heart.

The circuitry is the control center of the pacemaker. Located here are heart monitoring sensors, voltage regulators, timing cir-

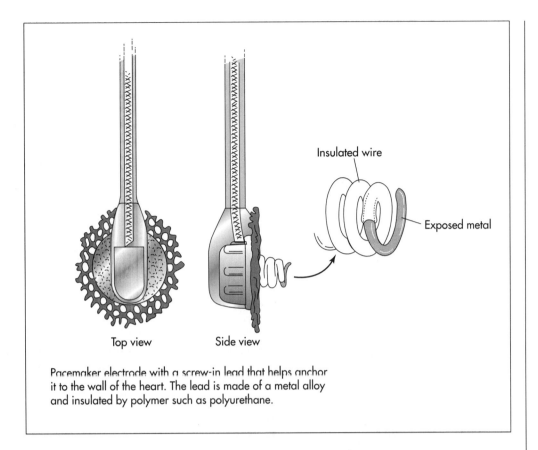

Insulated wire

Exposed metal

Top view Side view

Pacemaker electrode with a screw-in lead that helps anchor
it to the wall of the heart. The lead is made of a metal alloy
and insulated by polymer such as polyurethane.

cuits, and externally programmable controls. The circuitry is composed primarily of resistors, capacitors, diodes, and semiconductors. Modern pacemaker circuitry is a vast improvement over earlier models. With the application of semiconductors, circuit boards have become much smaller. They also require less energy, produce less heat, and are highly reliable.

The Manufacturing Process

Pacemakers are sophisticated electronic devices. Therefore, some manufacturers rely on outside suppliers to provide many of the component parts. The construction of a pacemaker is not a linear process but an integrated one. Component parts such as the battery, leads, and the circuitry are constructed individually, then pieced together to form the final product.

Making the battery

1 The primary type of battery used in pacemakers is a lithium/iodine cell. One method used by manufacturers to make these batteries involves first mixing togeth-er the iodine and a polymer such as poly2-vinylpyridine (PVP). They are heated together, forming a molten charge-transfer complex. This liquid is then poured into a half moon-shaped, preformed cell which contains the other components of the battery, including the lithium anode (positive charge) and a cathode collector screen. The iodine/polymer blend solidifies as it cools to form the cathode. After the cathode is formed, the battery is hermetically sealed to prevent moisture from entering.

Making the leads

2 The leads are typically composed of a metal alloy. The wire is made by an extrusion process in which the metal is heated until it is molten, then pushed through an appropriately sized opening. It is cut, then bundled with many other wires and treated with a polymeric insulator such as polyurethane. One end of the lead wires is fashioned with a shaped tip, and the other is fitted with a pacemaker connector.

Making the motherboard

3 The motherboard contains all the electrical circuitry of the pacemaker, including

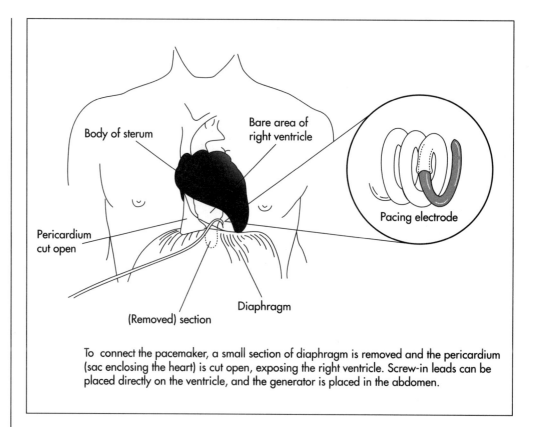

Body of sterum

Bare area of
right ventricle

Pacing electrode

Pericardium
cut open

Diaphragm

(Removed) section

To connect the pacemaker, a small section of diaphragm is removed and the pericardium (sac enclosing the heart) is cut open, exposing the right ventricle. Screw-in leads can be placed directly on the ventricle, and the generator is placed in the abdomen.

the semiconductor chips, resistors, capacitors, and other devices. Using a complex method known as hybridization, these components are combined to form a single complex circuit. Construction begins with a small board (less than 0.32 sq in [2 sq cm]) which has the electronic configuration mapped out. The appropriate components are put in place on the board. They are then affixed using a minimum number of soldering welds.

Final assembly and packaging

4 When all of the component pieces are available, final assembly takes place. The circuitry is connected to the battery, and both are inserted into the metal casing. The casing used for a pacemaker is typically formed using titanium or a titanium alloy. It is constructed in multiple pieces that are sealed together after the other pacemaker components are introduced. A fitting is also affixed to the casing, providing a connecting point for the leads.

5 The finished devices are then put into final packaging along with accessories. After being exhaustively tested, they are then sent out to distributors and finally to doctors.

Quality Control

The quality of each pacemaker is ensured by making visual and electrical inspections throughout the entire production process. These tests will detect most flaws. Since the batteries must be absolutely reliable, they are specially manufactured and exhaustively tested, thereby increasing the associated costs tremendously. The functionality of each finished pacemaker is also tested before it is sent out for sale. Many of these tests are done under varying environmental conditions, such as excessive humidity and stress.

Manufacturers set their own quality standards for the pacemakers that they produce. However, standards and performance recommendations are required by various medical organizations and governmental agencies. In the United States, pacemakers are classified as Class III biomedical devices, which means they require pre-market approval from the United States Food and Drug Administration (FDA).

The Future

With the increasing numbers of senior citizens in the United States, it is anticipated that a greater percentage of the population will require pacemakers. As research efforts continue, future devices promise to be longer lasting, more reliable, and more versatile. Advances in battery technology, such as using radioactive isotopes for power, will undoubtedly improve the longevity of implanted pacemakers. Developments in microelectronics should provide even smaller devices which are less prone to environmental interferences. A late-breaking development in the field is the application of cardiac pacemaking technology to the brain. In this system, scientists connect the lead wires to a specific site on the brain and stimulate it as needed to regulate heartbeat. This device has been shown to be particularly effective in calming the tremors associated with Parkinson's disease.

Where to Learn More

Books

Banbury, Catherine. *Surviving Technological Innovation in the Pacemaker Industry, 1959-1990*. Garland Pub., 1997.

Ellenbogen, Kenneth, ed. *Clinical Cardiac Pacing*. Saunders, 1995.

Fox, Stuart. *Human Physiology*. WCB Publishers, 1990.

Moses, H., J. Schneider, B. Miller, and G. Taylor. *A Practical Guide to Cardiac Pacing*. Little, Brown and Co., 1991.

Periodicals

Jeffrey, Kirk. "Many Paths to the Pacemaker." *Invention & Technology,* Spring 1997, pp. 28-39.

—*Perry Romanowski*

Paper Currency

Background

The existence of money as a means of buying or selling goods and services dates back to at least 3000 B.C., when the Sumerians began using metal coins in place of bartering with barley. The use of paper money began in China during the seventh century, but its uncertain value, as opposed to the more universally accepted value of gold or silver coins, led to widespread inflation and state bankruptcy. It was not until 1658, when Swedish financier Johann Palmstruck introduced a paper bank note for the Swedish State Bank, that paper money again entered circulation.

The first paper money in what is now the United States was issued by the Massachusetts Bay Colony in 1690. It was valued in British pounds. The first dollar bills were issued in Maryland in the 1760s. During the American Revolution, the fledgling Continental Congress issued Continental Currency to finance the war, but widespread counterfeiting by the British and general uncertainty as to the outcome of the revolution led to massive devaluation of the new paper money.

Stung by this failure, the United States government did not issue paper money again until the mid 1800s. In the interim, numerous banks, utilities, merchants, and even individuals issued their own bank notes and paper currency. By the outbreak of the Civil War there were as many as 1,600 different kinds of paper money in circulation in the United States—as much as a third of it counterfeit or otherwise worthless.

Realizing the need for a universal and stable currency, the United States Congress authorized the issue of paper money in 1861. In 1865, President Lincoln established the Secret Service, whose principal task was to track down and arrest counterfeiters. This early paper currency came in several different types, designs, and denominations, but had the common characteristic of being somewhat larger in size than today's money. It was not until 1929 that the current-sized bills went into circulation. In 1945, the government stopped printing bills in denominations greater than $100, and in 1963, they stopped printing silver certificate bills with the assurance that the dollar amount was available "...in silver payable to the bearer on demand."

Raw Materials

With paper money, the materials are as important as the manufacturing process in producing the final product. The paper, also known as the substrate, is a special blend of 75% cotton and 25% linen to give it the proper feel. It contains small segments of red and blue fibers scattered throughout for visual identification. Starting in 1990, the paper for $10 bills and higher denominations was made of two plies with a polymer security thread laminated between them. The thread was added to $5 bills in 1993. This thread is visible only when the bill is held up to a light and cannot be duplicated in photocopiers or printers.

The inks consist of dry color pigments blended with oils and extenders to produce especially thick printing inks. Black ink is used to print the front of the bills, and green ink is used on the backs (thus giving rise to the term greenbacks for paper money). The colored seals and serial numbers on the

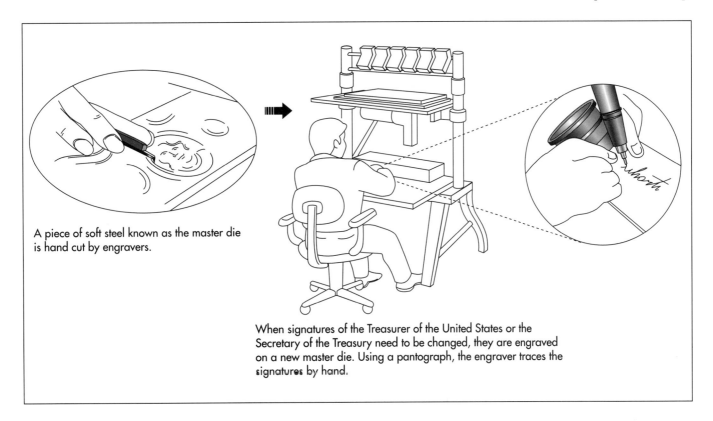

A piece of soft steel known as the master die is hand cut by engravers.

When signatures of the Treasurer of the United States or the Secretary of the Treasury need to be changed, they are engraved on a new master die. Using a pantograph, the engraver traces the signatures by hand.

front of the bill are printed separately using regular printing inks.

Design

The design of the front and back of each denomination bill is hand tooled by engravers working from a drawing or photograph. Each engraver is responsible for a single portion of the design — one doing the portrait, another the numerals, and so on.

The portrait on the face of each bill varies by the denomination. George Washington appears on the $1 bill, Abraham Lincoln on the $5, up to Benjamin Franklin on the $100 bill. These persons were selected because of their importance in history and the fact that their images are generally well known to the public. By law, no portrait of a living person may appear on paper money.

In 1955, Congress passed a law requiring that the words "In God We Trust" appear on all U.S. currency and coins. The first bills with this inscription were printed in 1957, and it now appears on the back of all paper money.

Starting in 1990, very small printing, called microprinting, was added around the outside of the portrait. This printing, which measures only 0.006-0.007 inches (0.15-0.18 mm) high, repeats the words "The United States of America." It appears on all paper money except the $1 bill.

The Manufacturing Process

In the United States, all paper money is engraved and printed by the Bureau of Engraving and Printing, which is part of the Department of the Treasury of the federal government. The Bureau also prints postage stamps, savings bonds, treasury notes, and many other items. The main production facility is located in Washington, D.C., and there is a smaller facility in Fort Worth, Texas. Every day, the Bureau prints approximately 38 million pieces of paper money. About 45% of this production are $1 bills and 25% are $20 bills. The rest of the production is divided between $5, $10, $50, and $100 bills. Although the $2 bill is still in circulation, it is rarely used, and therefore is rarely printed. Each bill, regardless of its denomination, costs the government about 3.8 cents to produce.

There are 65 separate operations in the production of paper money. Here are the major steps:

Engraving the master die

1 Engravers hand cut the design into a piece of soft steel, known as the master die, using very fine engraving tools and a magnifying glass. The portrait and images consist of numerous lines, dots, and dashes which are cut in various sizes and shapes. The fine crosshatched lines in the background of the portrait are produced by a ruling machine, and the scrollwork in the borders are cut using a geometric lathe.

2 Every time a new Treasurer of the United States or a new Secretary of the Treasury is appointed, their signatures must be engraved on a new master die for each denomination bill. First the signatures are photographically enlarged. An engraver then traces the signatures by hand with one end of a device known as a pantograph. This motion is mechanically reduced through a set of linkages, causing several diamond-tipped needles on the other end of the pantograph to cut the signatures into the master dies.

Making the master printing plate

3 Once the master die has been inspected, it is heated and a thin plastic sheet is pressed into it to form a raised impression of the design. Thirty-two of these raised plastic impressions are bonded together in a configuration of four across and eight down to form what is known as an alto. The master die is then placed in storage.

4 The plastic alto is placed in an electrolytic plating tank and is plated with copper. The plastic is stripped away leaving a thin plate of metal, known as a basso, with 32 recessed impressions of the design. The metal basso is then cleaned, polished, and inspected. If it passes inspection, it is plated with chromium to make the surface hard, and it becomes a master printing plate.

Printing the front and back of the bills

5 The principal printing process is known as intaglio printing. This process is used because of its ability to produce extremely fine detail that remains legible under repeated handling and is difficult to counterfeit. A stack of 10,000 sheets of paper is loaded into a high-speed, rotary intaglio printing press. Each sheet is sized to allow 32 individual bills to be printed on the same sheet. The paper is inspected to ensure that it contains the proper security thread for the denomination to be printed. A master printing plate of the proper denomination is secured around the master plate cylinder in the press.

6 The rotating master printing plate is coated with ink. A wiper removes the ink from the surface of the plate, leaving only the ink that is trapped in the engraved recesses of the design. A sheet of paper is fed into the press where it passes between the master plate cylinder and a hard, smooth impression cylinder under pressures reaching 15,000 psi (1,034 bar). The impression cylinder forces the paper into the fine, engraved lines of the printing plate to pick up the ink, leaving a raised image about 0.0008 in (0.02 mm) above the paper. This process is repeated at a rate of about 10,000 sheets per hour.

7 The printed sheets are then stacked on top of each other. The backs are printed with green ink first and are allowed to dry for 24-48 hours before the fronts are printed with black ink.

Printing the colored Treasury seal and serial numbers

8 After the intaglio printing process, the stacks are cut into two stacks of 10,000 sheets and are visually examined for defects. Each sheet is fed into a letterpress which prints the colored Treasury seal and serial numbers on the face of the bills. Sixteen serial numbers are printed at the same time. The press then automatically advances the numbers before the next sheet of sixteen is printed. The numbers on any sheet are separated by 20,000 between adjacent bills. Thus, the bill in the upper left-hand corner of the first sheet would be serial number 0000001 and the one below it on the same sheet would be 0020001, and so on. On the second sheet, all the numbers would advance by one giving 0000002 in the upper left, 0020002 below it, etc. In this manner, when the sheets are cut into separate stacks, the bills within each stack will have sequential serial numbers.

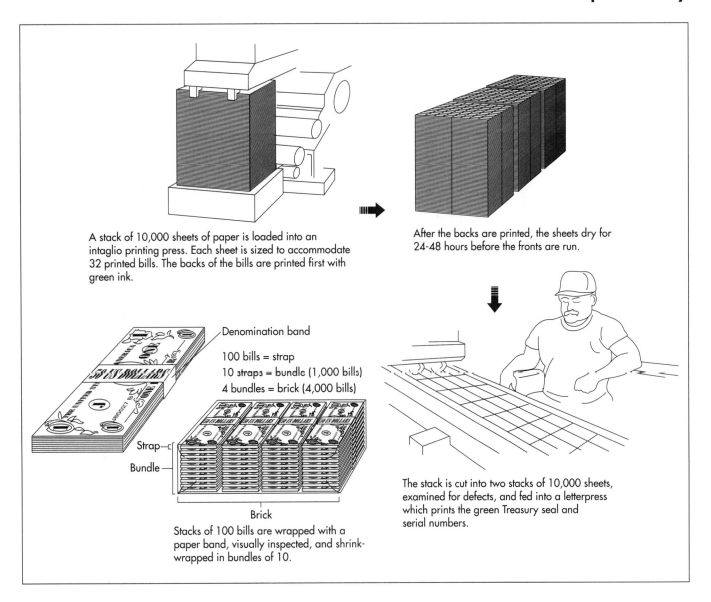

A stack of 10,000 sheets of paper is loaded into an intaglio printing press. Each sheet is sized to accommodate 32 printed bills. The backs of the bills are printed first with green ink.

After the backs are printed, the sheets dry for 24-48 hours before the fronts are run.

Denomination band

100 bills = strap
10 straps = bundle (1,000 bills)
4 bundles = brick (4,000 bills)

Strap

Bundle

Brick

Stacks of 100 bills are wrapped with a paper band, visually inspected, and shrink-wrapped in bundles of 10.

The stack is cut into two stacks of 10,000 sheets, examined for defects, and fed into a letterpress which prints the green Treasury seal and serial numbers.

9 The finished sheets are inspected with machine sensors, and any printing errors, folded paper, inclusion of foreign objects, or other defects are identified. Any bills which are found to be defective are marked for later removal. Such bills are replaced with star notes which are numbered in a different sequence and have a star printed after the serial number.

Cutting and wrapping the bills

10 The sheets are gathered in stacks of 100 and cut into 16 individual stacks of 100 bills each with a vertical guillotine knife. Any bills which have been identified as defective are replaced with star notes at this time. The stacks of 100 bills are then wrapped with a paper band. The banded stacks are given a final visual inspection and

are shrink-wrapped with plastic in bundles of 10 stacks. Four of these 10-stack bundles are then wrapped together to form a "brick" before they are shipped to the various federal reserve banks and other agencies.

Quality Control

Anything as important as money requires strict quality control standards. Flawed money is bad money and cannot be placed into circulation. In addition to the many inspections that occur during the printing process, the raw materials are also subject to strict inspections before they are used. The inks are tested for color, viscosity (thickness), and other properties. The paper is produced by a single manufacturer in a secret, tightly controlled process. The paper is tested for chemical composition, thickness, and other properties. It is illegal for anyone else to manufacture or possess this specific paper.

The finished bills are also tested periodically for durability. Some bills are put through a washing machine to determine the colorfastness of the inks, while others are repeatedly rolled into a cylinder and crushed on end to determine their resistance to handling. It is estimated that a bill can be folded and crumpled up to 4,000 times before it has to be replaced.

Destruction of Paper Money

Despite the use of high-quality paper and inks, the average life of a $1 bill in circulation is only about 18 months. Other denominations last somewhat longer. When a bill has been defaced, torn, or worn to the point where it is no longer identifiable or useable, it is taken out of circulation and returned to the federal reserve banks for destruction by shredding. Some of this shredded money is recycled to make roofing shingles or insulation. Money that is damaged or otherwise flawed during the printing process is shredded at the Bureau of Engraving and Printing plants.

The Future

For U.S. paper money, the future has arrived. Starting in 1996, the Department of the Treasury began issuing $100 bills with a new front and back design and several new security features designed to make counterfeiting more difficult. Other new bills in descending denominations will be printed at the rate of one new denomination per year.

The new bills use the same paper and are the same size and color as today's bills. Multicolor images, such as are commonly found on European currency, were not used because they were too easy to duplicate with color photocopies and printers. The image of Benjamin Franklin still appears on the front of the new $100 bill, although his portrait is larger and is shifted to the left. A watermark, formed by reducing the thickness of the paper during manufacture, has been placed to the right of the portrait and

shows a second image of Franklin when the bill is held up to the light. The imbedded security thread is also still there, although now it has been treated to glow red under ultraviolet light. The position of the thread varies depending on the denomination of the bill to prevent the counterfeiting practice of bleaching the ink off lower denomination bills and reprinting them as higher denominations.

Other new features include concentric fine lines behind Franklin's head on the front and behind the image of Independence Hall on the back. These lines are so fine that they are extremely difficult for copiers or printers to duplicate without blurring them into a solid background. Microprinting, which was introduced in 1990, is used in two places on the new $100 bill; the words "USA 100" appear within the lower left-hand numeral 100, and the words "United States of America" run down the lapel of Franklin's coat.

Perhaps the most high-tech feature is a special color-shifting ink which is used to print the numeral in the lower right-hand corner. When viewed from head on, this ink appears green, but changes to black when viewed from the side.

Lower denomination bills will have many, but not all, of the security features present in the new $100 bills. The highest level of protection was given to the $100 bill because it is the largest denomination being printed. It is also the most common bill in circulation outside the United States, and hence, is frequently counterfeited in other countries.

Some of the security features originally proposed for the new money—such as holograms, plastic films, and coded fiber optics—were not used for this latest change because they represented too great a departure from the current money or because of potential technical problems.

Looking further into the future, paper money may eventually be replaced by electronic money that is downloaded onto plastic "stored value" cards from an ATM or computer. Each card would have a computer chip memory, and the money would be electronically transferred through a card reader to make purchases.

Where to Learn More

Books

Friedberg, Robert. *Paper Money of the United States, 14th Edition.* The Coin and Currency Institute, Inc., 1995.

Krause, Chester L. and Robert F. Lemke. *Standard Catalog of U.S. Paper Money.* Krause Publications, 1990.

Periodicals

Freeman, David. "Change For a Hundred." *Popular Mechanics,* January 1996, pp. 72-73.

Geschickter, J. "Making Money." *National Geographic World,* November 1996, pp.30-33.

Hirschkorn, Phil. "The Buck May Stop Here." *George,* April/May 1996, pp. 92+.

Lipkin, Richard. "New Greenbacks." *Science News,* January 27, 1996, pp. 58-60.

Schafrik, Robert E. and Sara E. Church. "Protecting the Greenback." *Scientific American,* July 1995, pp. 40-46.

Other

"Engravers." The Department of the Treasury. http://www.ustreas.gov/treasury/bureaus/bep/proc/new/engrav.html

"The Money Factory." The Department of the Treasury. http://www.ustreas.gov/treasury/bureaus/bep/proc/new/hq.html

"Your Money Matters." The Department of the Treasury. http://www.ustreas.gov/treasury/whatsnew/newcurr/home.html

—*Chris Cavette*

Piano

Today, there are approximately 15 piano manufacturers in the United States, and Japan is the world's largest producer of pianos.

Background

The piano may be the best known and loved of all musical instruments. It also has the broadest range of any instrument, so music for all other instruments can be composed on it. It can be played solo, but most other instruments, including the voice, use the piano for accompaniment. Technically, the piano may also be the most complicated musical instrument with over 2,500 parts.

The piano is a stringed instrument. Its many parts are organized into five general structural and mechanical areas of either grand or vertical pianos. These are: the case of the wing-shaped grand piano (or the cabinet of the vertical or upright piano); the soundboard and the ribs and bridges that are its components; the cast iron plate; the strings; and, collectively, the keys, hammers, and piano action or mechanism. The case has many structural parts for attaching legs and tuning pins, but perhaps the rim and the keybed or shelf where the keys and piano action will be installed are most important. The soundboard amplifies the vibrations of the strings, which are transmitted through bridges.

The cast iron plate is installed over the soundboard and pinblock (part of the case), and it provides the strength to anchor the strings under tension. Nose bolts and perimeter bolts anchor the plate to the braces and inner rim of the case. The 220 to 240 strings of the piano are attached to hitch pins along the curved edge of the cast iron plate and to tuning pins across the front of the piano, roughly parallel to the keyboard. The piano action is still more complicated and includes the keys, hammers, and mechanism or action.

Names for pianos usually indicate their sizes. Grand (wing-shaped) pianos range in length from 4 ft 7 in-9 ft 6 in (1.4-2.9 m) from the front of the keyboard to end of the bend. The "baby" grand is 5 ft-5 ft 2 in (1.52-1.57 m) in length; smaller grand pianos are called "apartment size." The larger sizes are the medium grand and concert grand. Modern vertical piano design has changed little since 1935. Verticals range in height from 36-52 in (91-132 cm) with small variations in width and depth. The five standard sizes from smallest to tallest are the spinet, consolette, console, studio, and professional pianos. Pianos are frequently chosen for appearance, and cabinets are available in most furniture styles and finishes.

History

The piano's ancestors are the first stringed instruments. Plucking, striking, and bowing of strings was known among all ancient civilizations; the harp is mentioned in the Book of Genesis in the Bible. The psaltery was an ancient box-type instrument with strings that were plucked with a pick. Keys were added to stringed instruments to make the family of instruments led by the harpsichord, but keys are used to pluck strings in the harpsichord, the most popular instrument of the seventeenth century. A parallel development was the dulcimer, another stringed box with strings that are struck. Keys and strings were paired in a striking instrument in the clavichord, which led directly to the invention of the pianoforte or fortepiano.

Bartolomeo de Francesco Cristofori made harpsichords in Padua, Italy. He is credited with having invented the piano in 1700. Cristofori's piano had hammers that struck the strings by falling by momentum, after having been moved by the action parts linking the hammers to the keys. The hammers were caught by back checks or hammer checks to keep them from bouncing up and down on the strings after the initial strike. This method allowed the strings to continue to vibrate and make sound and for them to be struck loudly or softly, unlike the harpsichord. Johann Andreas Silbermann of Strasbourg, France, continued Cristofori's interest in the pianoforte, and the instrument became popular in Germany after Frederick the Great purchased several. Johann Sebastian Bach approved of it in 1747.

The piano had replaced the harpsichord in importance by the end of the eighteenth century. Cabinetmakers built beautiful cases for them. The square piano was built mid-century, and more musicians began writing music specific to the piano, rather than borrowing harpsichord tunes. Piano building began in America in 1775, and changes to the design of the hammers and to the playing mechanism or action improved the sound and responsiveness of the instrument. Jean Henri Pape of Paris patented 137 improvements for the piano during his life (1789-1875). In England, John Broadwood developed machines to manufacture pianos and reduce their cost.

Improvements continued from 1825 to 1851 with over 1,000 patents in Europe and the United States for stronger, more deft pianos with greater control and repetitive motion. By the mid-nineteenth century, the modern piano had emerged based on the development of the cast iron plate for structural strength and cross-stringing by fanning bass strings over trebles. By 1870, Steinway & Sons had developed this fanning method called the over-strung scale, so that the strings crossed most closely in the center of the soundboard where the best sound is produced.

C. F. Theodore Steinway also developed the continuous bent rim for the case, which enhanced sound transmission by using the acoustic properties of long wood fibers. These improvements were adapted to all styles of pianos including grand, upright, and square pianos. By 1911, there were 301 piano builders in the United States. Produc-

Advertisement for a Beckwith player piano from the 1915 Sears Roebuck catalog. (From the collections of Henry Ford Museum & Greenfield Village.)

In the early twentieth century, the player piano achieved great popularity, allowing people to feel artistic and produce music in their homes without having to invest endless hours in practice. The pianos, equipped with a built-in player mechanism, were activated by foot pedals or electricity and used perforated paper rolls to play a variety of music.

Manufacturers advertised their player pianos as good family entertainment and a source of cultural enrichment. An eager public responded with enthusiasm, purchasing over two million pianos by the end of the 1920s. Parents hoped that the pianos would interest their children in attaining musical skills—although they often had the opposite effect, since player pianos offered, as one manufacturer described it, "perfection without practice."

Dealers offered music rolls for a broad range of age groups, musical tastes, and interests. Young adults sang along with the latest tunes, while musical versions of nursery rhymes enchanted toddlers. Classical music enthusiasts listened to sonatas or operatic melodies. Many Greek, Italian, and Polish-Americans purchased song rolls with words printed in their native language.

Coin-operated player pianos were popular among hotel, dance hall, and restaurant owners, who purchased them to serenade customers and increase profits. Fitted with rolls that played several tunes, these pianos poured forth music at the drop of a coin. Customers glided across dance floors to waltzes and fox trots, dined in restaurants to popular melodies, or drank in speakeasies to uptempo tunes.

The enthusiasm for player pianos began to wane in the late 1920s, however, as phonographs and radio provided keen competition for leisure time and entertainment dollars.

Jeanine Head Miller

tion peaked in the 1920s and declined greatly because of the Great Depression in the 1930s. Today, there are approximately 15 piano manufacturers in the United States, and Japan is the world's largest producer of pianos.

The design of the piano has not changed appreciably since the late 1800s, although manufacturers may use different materials or approaches to the manufacturing process. The manufacturing process for the grand piano is described below; there are some differences in manufacturing the vertical or upright piano and in operation methods, particularly the angle at which the hammers strike the strings.

Raw Materials

Pianos are made of the finest materials, not only for appearances but for excellent sound production. The long fibers of maple wood are strong and supple for construction of the rim, but long fibers of spruce are needed for the strength of the braces. Wood is also needed for making patterns of other parts. Metal is used for a variety of parts, including the cast iron plate. Sand is needed for casting molds. The character of the sand is modified by using additives and binders such as bentonite (a type of clay) and coal dust. Molten iron for the casting is made of pig iron with some steel and scrap iron to add strength. Strings are made of high tensile steel wire that is manufactured at specialized piano string mills.

Design

Pianos are designed by specially trained and educated engineers called scale engineers. Scale engineers choose the materials, create the designs and specifications, and develop the interactions of the parts of the piano. Perhaps the most important aspect of design relates to the structural strength of the piano. About 160-200 lb per sq in (11.2-14 kg per sq cm) of tension is exerted on each of the 220 or more strings in the piano. The piano must perform well, but it also must remain stable over time as changing conditions affect the many materials in the piano differently.

The cast iron plate must support the tension of the stringing scale, covering the sound-board very little; it must have maximum mass for strength, but minimum mass for sound quality. Its shape is unique to the design of the piano because it conforms to the string layout, the placement of the bridges on the soundboard, and the paths of the strings. Because the material is brittle, it must be supported in places where the strings apply tension. Holes are designed in the curved side to prevent the plate from cracking due to thermal stress after it is poured and cooled, and this design allows sound to rise from the soundboard too. The scale engineer first sketches a proposed plate, draws it to scale, and makes a wood pattern; this design is later used for manufacture.

The Manufacturing Process

Bending the rim of the case

1 Steinway's method of rim bending is still used and is the first step in assembling the grand piano. Layers of long-fiber maple wood are glued together and bent in a metal press to form a continuous rim; both the inner and outer rim are made this way. Up to 22 layers form each piano rim, and the layers may be up to 25 ft (7.62 m) long. Resin glue is applied by machine, then the layers are carried to the press where they are shaped. The rims are stored in braces to keep them from changing shape. They are seasoned in controlled temperature and humidity conditions until the wood meets a specific moisture content where it will hold its contour. The bent inner rim is then fitted with other wood components, including the cross block, the pinblock, the cross braces, the keybed, and the backbottom. These are glued and doweled in place.

2 The cabinet is finished to improve sound properties as well as for appearance. The cabinet is sanded so stain is absorbed properly, wood is bleached to equalize appearance of the veneer, prestaining and staining are done next, wood fillers (sometimes with a washcoat) are added, and a first coat of sealer or lacquer is applied. The surface is sanded again, special glazes (for antiquing or other effects) are added followed by two more coats of lacquer, sanding is done again, special trims are added, and two final

coats of lacquer are used. The cabinet is dried for up to 21 days before it is hand-rubbed to its final finish.

Making the structural components

3 The wood components of the piano (collectively called the framework)—the pinblock and the cast iron plate—are the parts of the piano that support the tension of the strings. Braces are made of select spruce, and the pinblock or wrestplank is constructed of bonded layers of rock maple. The pinblock is quarter-sawn or rotary cut to maximize the grain structure's grip on the tuning pins. The laminated layers are also glued at different angles to each other so that the pins are surrounded with end grain wood. The pinblock has one hole per string, or up to 240 holes, drilled in it.

4 The cast iron plate is made in a piano plate foundry. Match-plates are made of metal from the wood pattern designed by the engineer with top and bottom pieces to match. Sand molds are made from the match-plates, and these are used to cast the plate. Molten iron is poured through the molds and allowed to harden during the founding process (a controlled cooling process) to produce a plate weighing about 600 lb (272.4 kg). After the plate is cooled and removed from the molds, sand is blasted off the plate with steel grit. The plate is transported on overhead conveyors to a drill room where holes are drilled for the tuning pins, nosebolts, bolts to the frame, and hitch pins. The hitch pins are inserted next; then the casting imperfections are removed from the plate by grinding and drilling. Oils are also removed. The plate is hand-sanded and rubbed, primed, and painted.

5 The cast iron plate is suspended above its piano during the process of fitting. The plate will be lowered and raised in and out of the piano several times as the pinblock, seal against the rim, and the soundboard and bridges are fitted.

Creating the soundboard

6 The soundboard is a thin panel of spruce that underlies the strings and the cast iron plate and rests on the rim braces. Its parts are the board itself, supporting ribs on the underside of the board, and the two bridges over which the strings are stretched.

The soundboard is made of spruce that is 0.25-0.375 in (0.635-0.95 cm) thick; it acts as a natural resonator, is strong for its weight, and can be vibrated by the strings because of its lightness. Spruce is air dried then kiln dried to a specific moisture content. It is then cut into strips that are 2-5 in (5.08-12.7 cm) wide, the edges are glued, and the strips are pressed together and dried. A pattern is superimposed, and the soundboard is trimmed to grand piano size.

7 The soundboard is curved to produce the right sound. The curve is called a crown that arches upward toward the strings. The arch is made by fitting ribs of lightweight spruce or sugar pine wood to the underside of the board. The ribs are carefully cut from patterns, then fitted and glued to the soundboard using a rib press that accurately positions the ribs, then forces the board into the proper curvature. The ribs are cut along the wood's lengthwise grain and fitted at right angles to the lengthwise grain of the soundboard, so that vibrations are evenly transmitted. The ends of the ribs are feathered, then fitted into notches in the framework of the piano that will exactly support the arch of the crown; the pianomakers use special patterns to guide these cuts in the frame.

8 The two soundboard bridges transfer the vibrations of the strings along their lengths to the soundboard. The long bridge is crossed by treble strings, and the bass strings that fan across the trebles cross the short bridge. The bridges are complicated because they must parallel the grain of the soundboard closely, curve with the crown, and support the strings, which exert a downbearing pressure on the bridges and therefore on the soundboard. This pressure must be supported by the strength of the bridges and the arch of the crown, or the tone of the strings will drop. The bridges are made of solid blocks of wood or of laminated wood. Hard maple is used in American-made pianos, and falcon wood (beech) is used in Europe. Laminated bridges must be placed with laminations perpendicular to the soundboard or the glue layers have a damping effect. The bridges are glued to the soundboard and also fastened to it with wood screws capped by soundboard buttons made of wood that act like washers and keep the screws from grinding into the board. The bridges are notched on both

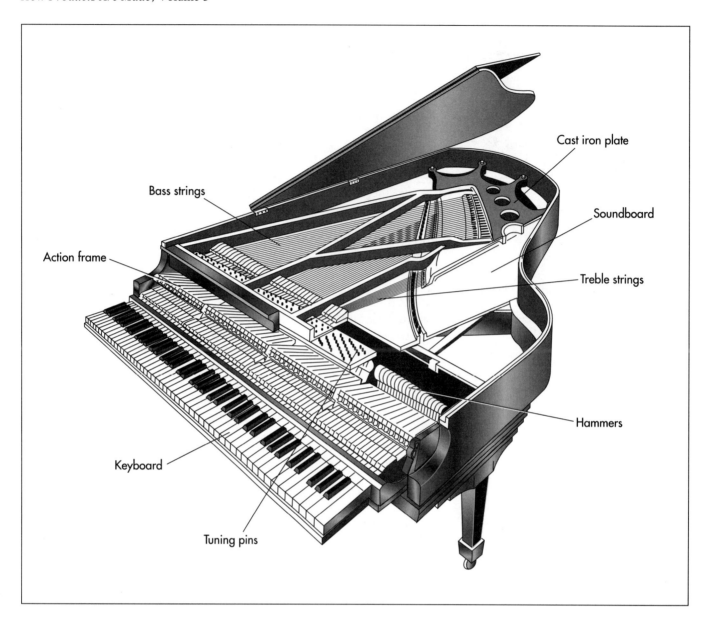

Cast iron plate

Bass strings

Soundboard

Action frame

Treble strings

Keyboard

Hammers

Tuning pins

Pianos have the greatest range of any instrument and over 2,500 parts. They are considered to belong to both the string family of instruments, because a piano's strings produce its sounds, as well as the percussion family, because the sound is produced when a hammer strikes a string.

sides where each string crosses, so the string strikes a small part of the bridge and can vibrate easily. Pins are inserted in the bridge, and strings are threaded between the pins.

Stringing and tuning

9 Piano string is made in specialized mills and consists of carbon steel wire. The bass strings are also wrapped with copper windings in a process called loading the strings. The windings add weight and thickness to the steel core strings so they vibrate more slowly and can be made to lengths that fit a piano of practical size; without loading, bass strings would have to be 30 ft (9.14 m) long to produce their sounds. Treble strings are short, are not wound with

copper, but are grouped in threes to make one tone. Scale sticks are used as standards for each string, acting as a gauge of each kind of wire and determining how many sizes of string are needed; up to 17 different diameters of wire may be used to string one piano. Piano strings require special care and handling because they lie straight after they are formed, cut, and loaded and are never wound on rolls. After the strings are strung, they are held in place near the tuning pins by metal bars and special brass studs called agraffes. Other bars position the strings properly near the hitch pins.

10 Tuning pins are made from steel wire. The wire is cut to the proper length, the ends are shaped with a die, and the pins

are loaded in a tumbler where rough edges are smoothed away. The tumbler empties them into a press where swags that fit tuning hammers are formed at the tops of the pins. Holes for the strings are drilled into the swagged ends of the pins, the pins are cleaned of metal chips and oil, and nickel-plating is applied to the pins to keep them from rusting. The pins are threaded to turn easily during tuning, then they are subjected to controlled heating called blueing, which oxidizes the outer surface of the threads of the pins (where the nickel plating was removed during threading) so the pins will grip the wood in the pinblock. Special machines insert several pins at a time through the holes in the cast iron plate and into the pinblock where they are fitted in place by hand.

Constructing the keyboard and action

11 Keyboards, key and action frames, and actions are made by specialty manufacturers. The keys balance and pivot on a set of either two or three rails that are covered with felt to prevent noise. Guide pins for each key are inserted in the front or head rail and the middle or balance rail. The keys themselves are made of lightweight wood that is cut to size and dried in kilns. The keys are covered with black or white plastic, although in the past ivory and ebony were used. The plastic key covers are molded to cover a group of keys that are later cut individually. Holes are drilled on the undersides to fit the guide pins. Capstan screws are mounted on the back edges of the keyboard extending inside the piano; the action will be seated on these. The keys are now cut into 88 individuals, which are sanded and polished on the sides. The black keys are also stained black before the black caps are glued on. The keys are rematched to the keyframe, punchings resembling washers are placed over the guide pins, and the keys are placed on them.

12 The voice of the piano depends on the quality of the hammers. Many materials from elkskin to rubber have been used over the history of the piano, but today, hammers are covered with premium wool felt of precision-graduated density. The felt is made by specialists who begin with select wool that is carded, combed, folded, and compressed into felt in tapered strips. The thinnest felt is used for the treble hammers, while thick felt is used for the bass. The core of each hammerhead is a wood molding, and an underfelt and top felt are bonded in place with resin to cover the molding. The hammerheads are made in long strips of the same size then sliced into individual hammerheads by hand or automation. The complete set of hammers is installed in the piano. The sound of the piano is adjusted by a specially trained tuner called a voicer. The key actions must respond with the same resistance. The felt hammers are modified with a sticker or needler that retextures the hammerheads and changes the sound.

13 The final parts are added, including the pedals and their trapwork, the fallboard or key cover, the music rack, the hinges and top lid, the topstick that supports the raised lid, and many other details. All parts are carefully made so they fit tightly and do not rattle or otherwise affect the sound of the instrument.

Quality Control

Pianos would not exist without quality control in all aspects of production because the instruments are too sensitive and dependent on the interaction of many parts and materials. For example, quality begins with the scale engineer's design. Metallurgists check the metal content of the iron plate; chemical analyses are made of the other contents, including carbon, sulfur, phosphorus, and manganese. Temperature is also critical; the molten iron is 2,750°F (1,510°C), and founding or hardening temperatures are also carefully monitored. String is similarly controlled and tested during manufacture for elasticity, resiliency, and tensile strength.

The Future

The process of piano manufacturing has remained essentially the same for a century, but scale engineers are always seeking new methods. Vacuum casting has recently been used to produce cast iron plates with smooth finishes requiring no grinding.

Where to Learn More

Books

Ardley, Neil. *Music: An Illustrated Encyclopedia.* Facts On File., Inc., 1986.

Bielefeldt, Catherine C.; Weil, Alfred R., ed. *The Wonders of the Piano: The Anatomy of the Instrument.* Belwin-Mills Publishing Corp., 1984.

Dolge, Alfred. *Pianos and their Makers: A Comprehensive History of the Development of the Piano.* Dover Publications, Inc., 1972.

Ehrlich, Cyril. *The Piano: A History.* Clarendon Press,1990.

Ford, Charles, ed. *Making Musical Instruments: Strings and Keyboard.* Pantheon Books, 1979.

—*Gillian S. Holmes*

Portable Toilet

Background

The portable toilet is a lightweight, transportable, efficient and more sanitary variation of a common facility for the elimination of human waste that existed before the advent of indoor plumbing—the outhouse. Before indoor plumbing allowed for the development of a system for transporting human waste from a receptacle to a sewer system through a series of pipes and other plumbing apparatus, humans often attended to their need to eliminate waste in an isolated stall located outside of living and working quarters, if they elected to use any type of structure at all. Typically, this stall contained a bench with a large hole cut into it. The waste was deposited through the hole directly to the ground below. A more refined version of the outhouse was the water closet, an indoor facility with a water tank and flushing system that deposited the waste in a cesspit below.

The advent of indoor plumbing led to the development of the first modern toilet in 1843, although toilets hooked up to sewage systems did not come into general use until the Victorian era when modern sewage systems began to be constructed. Still, this innovation was not useful for those who worked or otherwise congregated in outdoor areas with no access to such a facility. Some such venues, such as many roadway rest areas, camping facilities, and children's summer camps, still utilize rustic outhouses. Since at least the 1960s, many other venues, especially those where populations congregate only for temporary periods, have featured lightweight, sanitary portable toilet facilities—easily transportable, private, individual plastic stalls containing toilets, each with its own independent sanitary system consisting of rudimentary plumbing, a holding tank, and sanitizing chemicals.

Portable toilets are most commonly used at construction sites, outdoor parking lots, and other work environments where indoor plumbing is inaccessible, and at large outdoor gatherings such as concerts, fairs, and recreational events.

History

The earliest known toilet facilities date back to the third millennium B.C. Rudimentary lavatory facilities have been discovered in the form of recesses in stone walls of houses in Scotland dating back to around 2,800 B.C. Around the same time, it appears that Western-style lavatory facilities were being constructed from bricks with wooden seats in Pakistan. Sewage fell through a chute to a drain or cesspit. In Egypt, toilets have been found in the bathrooms of tombs, presumably for use in the afterlife. The first portable toilet dates back to the mid-fourteenth century B.C. In Egypt, a wooden stool with a large slot in the middle for use with a pottery vessel beneath it was discovered in the tomb of Kha, the senior official of the Thebes workmen's community.

Until the eighteenth century, the portable chamber pot, a slightly more modern variation of this first portable toilet, was the most commonly used lavatory facility. The notion of a flushable toilet was developed by the Elizabethan poet Sir John Harington, who designed an indoor water closet containing a toilet facility that could dilute sewage with water contained in a cistern. This invention was signifi-

The notion of a flushable toilet was developed by the Elizabethan poet Sir John Harington, who designed an indoor water closet containing a toilet facility that could dilute sewage with water contained in a cistern.

The main component of the portable toilet is lightweight sheet plastic, such as polyethylene, which forms the actual toilet unit as well as the cabana in which it is contained. A pump and holding tank form the portable sewage system. The facility is also equipped with a chemical supply container and inlet tube.

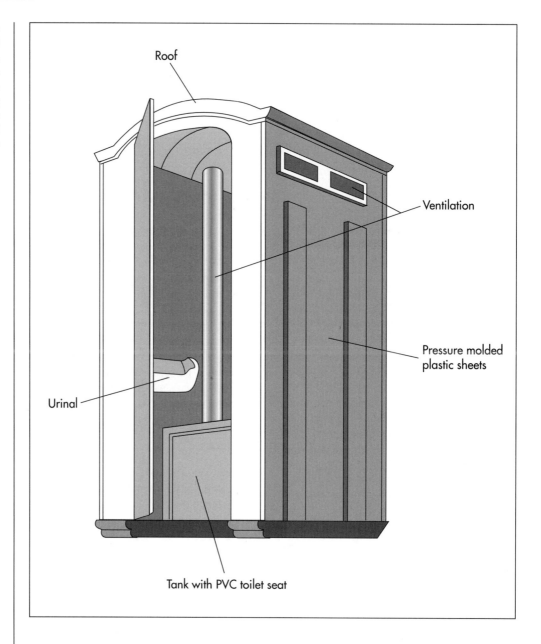

Roof

Ventilation

Pressure molded plastic sheets

Urinal

Tank with PVC toilet seat

cantly enhanced by Thomas Crapper, who in 1886 created the first flushable toilet featuring a water tank placed high above the toilet bowl to actually flush out the contents of the bowl, rather than merely dilute them. As public sewage systems became more developed, Crapper's invention became common.

In the twentieth century, inventors have combined the portability of the more ancient lavatories with the sanitary benefits of the modern toilet to create a contained system that is also compact, lightweight, and transportable. The portable toilet, commonly called the Porta-John after one prominent manufacturer of this product, is now a com-

mon feature at work sites and events that do not have access to sewer systems.

Raw Materials

Portable toilet assembly is relatively simple and few materials are needed. The main component of the facility is lightweight sheet plastic, such as polyethylene, which forms the actual toilet unit as well as the cabana in which it is contained. A pump and holding tank form the portable sewage system. These items are fastened with an assortment of screws, nails, rivets, bolts, and hinges. The facility is also equipped with a chemical supply container and inlet tube.

The Manufacturing Process

The toilet unit

1 The toilet unit is comprised of rigid, lightweight sheet plastic, which is formed into a box-like structure and secured with nuts, bolts, and rivets. The top sheet contains an opening for placement of the toilet tank. The top sheet may not be secured with these permanent fixtures, allowing for its easy removal to clean the tank. A lock may be placed over the top sheet to prevent its unauthorized removal.

2 The actual toilet tank, which is placed in this unit, is made of the same material and shaped with a flat, corrugated front wall and rounded rear wall. The upper edge of the toilet tank is formed as a peripheral flange that extends outward and downward.

3 The toilet tank is fitted with a cover formed of two flat semi-circular plastic sheets. The lower sheet has a peripheral edge lip that extends downward, the upper sheet has a front lip that extends downward, and the rear lip extends upward and outward to latch onto the peripheral flange of the toilet bowl. Both sheets are fitted with a central toilet opening.

4 A conventional toilet seat made of plastic is placed over the toilet bowl and connected to the assembly with hinges.

5 The seat is fitted with a pin, which pushes upward against a metal wear plate, which is secured to the bottom surface of the seat. The pin extends downward through the cover and a bracket. Under the bracket, a coil spring is placed around the pin. The upper end of the coil engages a washer fastened to the pin so that the seat maintains an upright position when not in use. (Note: Not all portable toilets are flushable. Those that are not do not contain this or the following two steps in the manufacturing process, but merely contain chemicals in the holding tank.)

6 A piston is placed underneath the lower end of the pin, and a mechanical, bellows-type pump is placed beneath the piston. The pump contains a spray opening and is connected to an inlet tube which is, in turn, connected to a chemical supply container. When the seat is raised, the piston will activate the pump.

7 The toilet opening is fitted with a pair of flat, plastic doors secured by hinges to bosses fastened to the bottom of the tank. These doors are connected to the toilet seat with metal links so that they are activated when the seat is lowered and raised.

The cabana

8 A cabana is formed with two lightweight plastic side panels, a similar back panel, and a front panel with a door opening. All parts are secured with nails, screws, bolts, and rivets. A variation of this model uses interfitting joints, so that the facility can be disassembled and reassembled for easier transportation.

9 A rounded top and a flat bottom containing a depression for the toilet tank, and a drain to release moisture are secured to the structure with rivets.

10 A vent pipe is placed through a small opening in the roof and another small opening in the toilet holding tank.

Additional features

11 A plastic door is fitted with an inset handle and a sliding lock that activates a "vacant-not vacant" sign on the reverse side of the door.

12 The door is attached to the front panel with hinges.

Byproducts/Waste

The contents of portable toilet holding tanks must be disposed of in accordance with state and federal environmental regulations. Typically, the chemicals used to sanitize the portable toilet facility are biodegradable and, thus, are not subject to any special disposal requirements. The waste in the holding tanks must be disposed of like any other form of sewage which is subject to local, state, and federal regulation. For example, in Michigan, portable toilet waste may be discharged to a publicly-owned treatment works (POTW) facility through the local sewage system or land-applied on farming property. Both of these activities require permits. The waste may also be transported to the POTW, which does not require a permit but does require approval of the POTW.

natives to chemical sanitizing solutions are also in the works. At least one company already manufacturers an organic solution that deodorizes and sanitizes in the same way as the commonly used chemical solutions.

The Future

Portable toilets will most likely be necessary as long as humans continue to congregate in outdoor areas and other sites without indoor plumbing. While this product is fairly simple, it is subject to innovations, especially with regard to design. For instance, at least one company manufactures a portable toilet facility constructed with removable joints, which allows for easy disassembly of the unit and enhances its transportability. Alter-

Where to Learn More

Other

Organica, Inc. http://www.organicain.com July 14, 1997).

Sanipages. 1997. http://sanipages.com (July 14, 1997).

—Kristin Palm

Potato Chip

Background

Potato chips are thin slices of potato, fried quickly in oil and then salted.

According to snack food folklore, the potato chip was invented in 1853 by a chef named George Crum at a restaurant called Moon's Lake House in Saratoga Spring, New York. Angered when a customer, some sources say it was none other than Cornelius Vanderbilt, returned his french fried potatoes to the kitchen for being too thick, Crum sarcastically shaved them paper thin and sent the plate back out. The customer, whoever he was, and others around him, loved the thin potatoes. Crum soon opened his own restaurant across the lake and his policy of not taking reservations did not keep the customers from standing in line to taste his potato chips.

The popularity of potato chips quickly spread across the country, particularly in speakeasies, spawning a flurry of home-based companies. Van de Camp's Saratoga Chips opened in Los Angeles on January 6, 1915. In 1921, Earl Wise, a grocer, was stuck with an overstock of potatoes. He peeled them, sliced them with a cabbage cutter and then fried them according to his mother's recipe and packaged them in brown paper bags. Leonard Japp and George Gavora started Jays Foods in the early 1920s, selling potato chips, nuts, and pretzels to speakeasies from the back of a dilapidated truck.

The chips were commonly prepared in someone's kitchen and then delivered immediately to stores and restaurants, or sold on the street. Shelf-life was virtually nil. Two innovations paved the way for mass production.

In 1925, the automatic potato-peeling machine was invented. A year later, several employees at Laura Scudder's potato chip company ironed sheets of waxed paper into bags. The chips were hand-packed into the bags, which were then ironed shut.

Potato chips received a further boost when the U.S. government declared them an essential food in 1942, allowing factories to remain open during World War II. In many cases, potato chips were the only ready-to-eat vegetables available. After the war, it was commonplace to serve chips with dips; French onion soup mix stirred into sour cream was a perennial favorite. Television also contributed to the chip's popularity as Americans brought snacks with them when they settled before their television sets each night.

In 1969, General Mills and Proctor & Gamble introduced fabricated potato chips, Chipos and Pringles®, respectively. They were made from potatoes that had been cooked, mashed, dehydrated, reconstituted into dough, and cut into uniform pieces. They further differed from previous chips in that they were packaged into breakproof, oxygen-free canisters. The Potato Chip Institute (now the Snack Food Association) filed suit to prevent General Mills and Proctor & Gamble from calling their products chips. Although the suit was dismissed, the USDA did stipulate that the new variety must be labeled as "potato chips made from dried potatoes." Although still on the market, fabricated chips have never achieved the popularity of the original.

Today, potato chips are the most popular snack in the United States. According to the Snack Food Association, potato chips consti-

Potato chips received a boost when the U.S. government declared them an essential food in 1942, allowing factories to remain open during World War II. In many cases, potato chips were the only ready-to-eat vegetables available.

tute 40% of snack food consumption, beating out pretzels and popcorn in spite of the fact that hardly anyone thinks potato chips are nutritious. Nonetheless, the major challenge faced by manufacturers in the 1990s was to develop a tasty low-fat potato chip.

Raw Materials

Even though Earl Wise started his business with old potatoes, today's product is made from farm-fresh potatoes delivered daily to manufacturing plants. The sources vary from season to season. In April and May, potatoes come from Florida; June, July and August bring potatoes from North Carolina and Virginia; in the autumn months, the Dakotas supply the majority of potatoes; during the winter, potato chip manufacturers depend on their stored supplies of potatoes. Stored potatoes are kept at a constant temperature, between 40-45°F (4.4-7.2°C), until several weeks before they are to be used. They are then moved to a reconditioning room that is heated to 70-75°F (21.1-23.9°C). Size and type are important in potato selection. White potatoes that are larger than a golf ball, but smaller than a baseball, are the best. It takes 100 lb (45.4 kg) of raw potatoes to produce 25 lb (11.3 kg) of chips.

The potatoes are fried in either corn oil, cottonseed oil, or a blend of vegetable oils. An antioxidizing agent is added to the oil to prevent rancidity. To further insure purification, the oil is passed through a filtration system daily. Salt and other flavoring ingredients, such as powdered sour cream and onion and barbecue flavor, are purchased from outside sources. Flake salt is used rather than crystal salt. Some manufacturers treat the potatoes with chemicals such as phosphoric acid, citric acid, hydrochloric acid, or calcium chloride to reduce the sugar level, and thus improve the product's color. The bags are designed and printed by the individual potato chip manufacturer. They are stored on rolls and brought to the assembly line as necessary.

The Manufacturing Process

1 When the potatoes arrive at the plant, they are examined and tasted for quality.

A half dozen or so buckets are randomly filled. Some are punched with holes in their cores so that they can be tracked through the cooking process. The potatoes are examined for green edges and blemishes. The pile of defective potatoes is weighed; if the weight exceeds a company's preset allowance, the entire truckload can be rejected.

2 The potatoes move along a conveyer belt to the various stages of manufacturing. The conveyer belts are powered by gentle vibrations to keep breakage to a minimum.

Destoning and peeling

3 The potatoes are loaded into a vertical helical screw conveyer which allows stones to fall to the bottom and pushes the potatoes up to a conveyer belt to the automatic peeling machine. After they have been peeled, the potatoes are washed with cold water.

Slicing

4 The potatoes pass through a revolving impaler/presser that cuts them into paper-thin slices, between 0.066-0.072 in (1.7-1.85 mm) in thickness. Straight blades produce regular chips while rippled blades produce ridged potato chips.

5 The slices fall into a second cold-water wash that removes the starch released when the potatoes are cut. Some manufacturers, who market their chips as natural, do not wash the starch off the potatoes.

Color treatment

6 If the potatoes need to be chemically treated to enhance their color, it is done at this stage. The potato slices are immersed in a solution that has been adjusted for pH, hardness, and mineral content.

Frying and salting

7 The slices pass under air jets that remove excess water as they flow into 40-75 ft (12.2-23 m) troughs filled with oil. The oil temperature is kept at 350-375°F (176.6-190.5°C). Paddles gently push the slices along. As the slices tumble, salt is sprinkled from receptacles positioned above the trough at the rate of about 1.75 lb (0.79 kg) of salt to each 100 lb (45.4 kg) of chips.

Opposite page:
Potatoes arrive daily at manufacturing plants. After they are checked for quality, they are stored at a constant temperature until they are processed into potato chips. Some manufacturers treat the potatoes with chemicals to improve the color of the final product. To make the chips, potatoes are fried in either corn oil, cottonseed oil, or a blend of vegetable oils. Flake salt rather than crystal salt is used to season the chips.

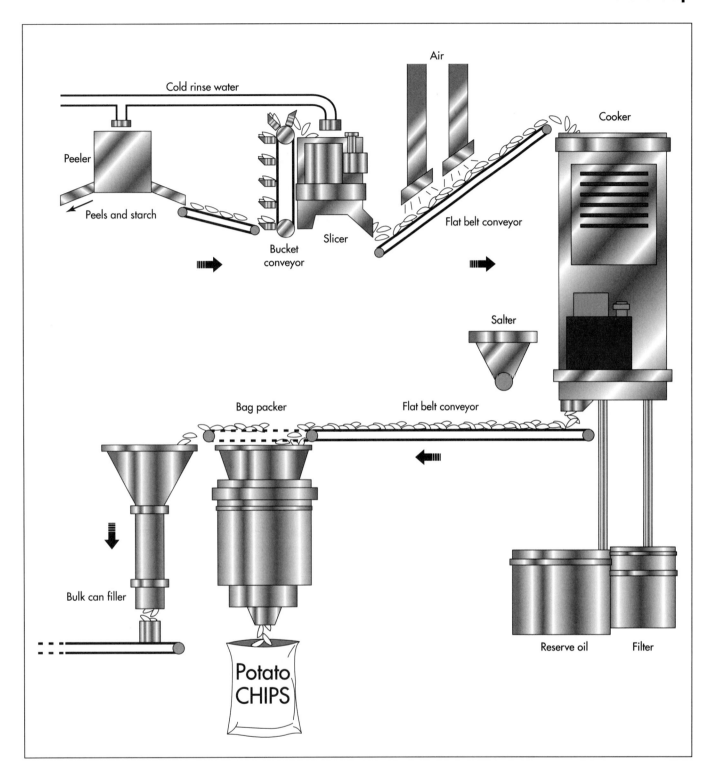

Cold rinse water

Air

Cooker

Peeler

Peels and starch

Bucket conveyer

Slicer

Flat belt conveyer

Salter

Bag packer

Flat belt conveyer

Bulk can filler

Potato CHIPS

Reserve oil

Filter

8 Potato chips that are to be flavored pass through a drum filled with the desired powdered seasonings.

Cooling and sorting

9 At the end of the trough, a wire mesh belt pulls out the hot chips. As the chips move along the mesh conveyer belt, excess

oil is drained off and the chips begin to cool. They then move under an optical sorter that picks out any burnt slices and removes them with puffs of air.

Packaging

10 The chips are conveyed to a packaging machine with a scale. As the preset weight of chips is measured, a metal detector checks the chips once more for any foreign matter such as metal pieces that could have come with the potatoes or been picked up in the frying process.

11 The bags flow down from a roll. A central processing unit (CPU) code on the bag tells the machine how many chips should be released into the bag. As the bag forms, (heat seals the top of the filled bag and seals the bottom of the next bag simultaneously) gates open and allow the proper amount of chips to fall into the bag.

12 The filling process must be accomplished without letting an overabundance of air into the bag, while also preventing the chips from breaking. Many manufacturers use nitrogen to fill the space in the bags. The sealed bags are conveyed to a collator and hand-packed into cartons.

13 Some companies pack potato chips in cans of various sizes. The chips flow down a chute into the cans. Workers weigh each can, make any necessary adjustments, and attach a top to the can.

Quality Control

Taste samples are made from each batch throughout the manufacturing process, usually at a rate of once per hour. The tasters check the chips for salt, seasoning, moisture, color, and overall flavor. Color is compared to charts that show acceptable chip colors.

Preventing breakage is a primary goal for potato chip manufacturers. Companies have installed safeguards at various points in the manufacturing process to decrease the chances for breakage. The heights that chips fall from conveyer belts to fryers have been decreased. Plastic conveyer belts have been replaced with wide mesh stainless steel belts. These allow only the larger chips to travel to the fryers and the smaller potato slivers to fall through the mesh.

Byproducts/Waste

Rejected potatoes and peelings are sent to farms to be used as animal feed. The starch that is removed in the rinsing process is sold to a starch processor.

The Future

Potato chips show no sign of declining in popularity. However, the public's increased demand for low-fat foods has put manufacturers on a fast track to produce a reduced-calorie chip that pleases the palate as well. In the late 1990s, Proctor and Gamble introduced olestra, a fat substitute that was being test-marketed in a variety of products, including potato chips.

Food technicians are using computer programs to design a crunchier chip. Upper- and lower-wave forms are fed into the computer at varying amplitudes, frequencies, and phases. The computer then spits out the corresponding models. Researchers are also working on genetically engineered potatoes with less sugar content since it is the sugar that produces brown spots on chips.

Where to Learn More

Books

Trager, James. *The Food Chronology.* Henry Holt, 1995.

Other

"Potato Chips," Jays Foods, Chicago, IL 60628. 773/731-8400.

—Mary F. McNulty

Propane

Background

Propane is a naturally occurring gas composed of three carbon atoms and eight hydrogen atoms. It is created along with a variety of other hydrocarbons (such as crude oil, butane, and gasoline) by the decomposition and reaction of organic matter over long periods of time. After it is released from oil fields deep within Earth, propane is separated from other petrochemicals and refined for commercial use. Propane belongs to a class of materials known as liquefied petroleum gases (LPGs), which are known for their ability to be converted to liquid under relatively low pressures. As a liquid, propane is 270 times more compact than it is as a gas, which allows it to be easily transported and stored as a liquid until ready for use. Approximately 15 billion gal (57 billion L) of propane are consumed annually in the United States as a fuel gas. The greatest consumers are the chemical and manufacturing industries, which use propane as chemical intermediates and aerosol propellants, followed by residential homes and commercial establishments, who use propane for heating and in dryers and portable grills.

The value of petroleum products has long been recognized by the civilized world, with documented examples of their use stretching back more than 5,000 years. The ancient Mesopotamians used petroleum-derived tar-like compounds for many applications, including caulking for masonry and bricks and adhesives for jewelry. About 2,000 years ago Arabian scientists learned one of the basic tenets of petroleum chemistry—that it can be distilled or separated into different parts, or fractions, based on their boiling points, and that each fraction has its own distinctive properties.

The modern era of refining is considered to have begun in 1859, when petroleum was found in Pennsylvania and the Sennaca Oil Company drilled the first oil well there. From a depth of 70 ft (21.2 m) the world's first oil well produced nearly 300 tons (305 metric tons) of oil in its first year, and thus an entire industry was born. Propane was first recognized as an important component of petroleum in 1910, when a Pittsburgh motor car owner asked chemist Dr. Walter Snelling why the gallon of gasoline he had purchased was half gone by the time he got home. The car owner thought the government should investigate why consumers were being cheated, because the gasoline was evaporating at a rapid and expensive rate. Snelling discovered a large part of liquid gasoline was actually composed of propane, butane, and other hydrocarbons. Using coils from an old hot water heater and other miscellaneous pieces of laboratory equipment, Snelling built a still that could separate the gasoline into its liquid and gaseous components. Since the days of Snelling, chemists have made tremendous advances in techniques for processing propane and other LPGs. Today, the manufacture of propane gas is an $8 billion industry in the United States.

Raw Materials

Because propane has natural origins, it is not "made" of other raw materials; instead, it is "found" in petroleum chemical mixtures deep within the earth. These petroleum mixtures are literally rock oil, combinations of various hydrocarbon-rich fluids

As a liquid, propane is 270 times more compact than it is as a gas, which allows it to be easily transported and stored as a liquid until ready for use.

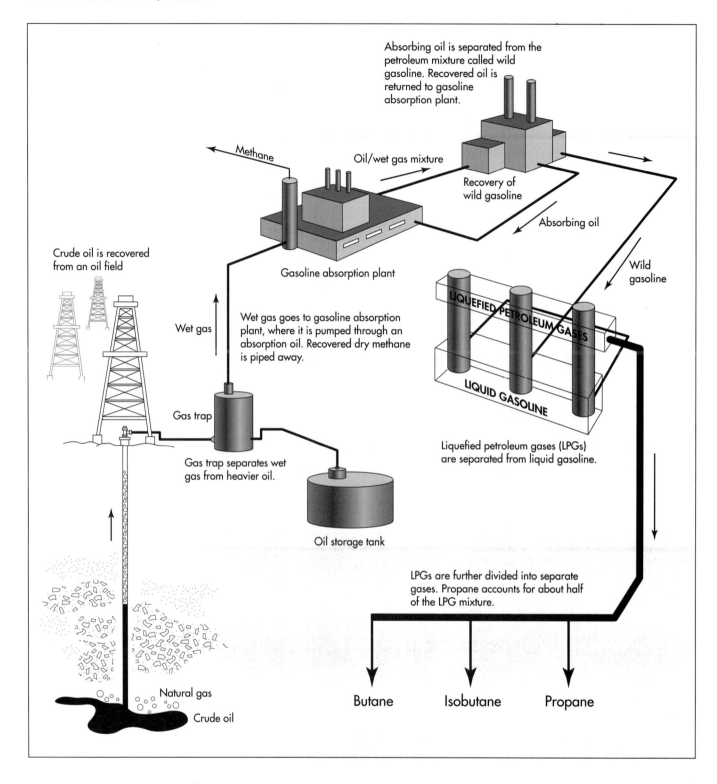

Absorbing oil is separated from the petroleum mixture called wild gasoline. Recovered oil is returned to gasoline absorption plant.

Methane

Oil/wet gas mixture

Recovery of wild gasoline

Absorbing oil

Crude oil is recovered from an oil field

Wild gasoline

Gasoline absorption plant

LIQUEFIED PETROLEUM GASES

Wet gas

Wet gas goes to gasoline absorption plant, where it is pumped through an absorption oil. Recovered dry methane is piped away.

LIQUID GASOLINE

Gas trap

Liquefied petroleum gases (LPGs) are separated from liquid gasoline.

Gas trap separates wet gas from heavier oil.

Oil storage tank

LPGs are further divided into separate gases. Propane accounts for about half of the LPG mixture.

Natural gas

Crude oil

Butane   Isobutane   Propane

which accumulate in subterranean reservoirs made of porous layers of sandstone and carbonate rock. Petroleum is derived from various living organisms buried with sediments of early geological eras. The organisms were trapped between rock layers without oxygen and could not break down, or oxidize, completely. Instead, over tens of millions of years, the residual organic mate-

rial was converted to propane-rich petroleum via two primary processes, diagenesis and catagenesis. Diagenesis occurs below 122°F (50°C) when the organic "soup" undergoes microbial action (and some chemical reactions) which result in dehydration, condensation, cyclization, and polymerization. Catagenesis, on the other hand, occurs under high temperatures of 122-424°F (50-

200°C) and causes the organic materials to react via thermocatalytic cracking, decarboxylation, and hydrogen disproportionation. These complex reactions form petroleum in the sedimentary rocks.

The Manufacturing Process

Propane manufacture involves separation and collection of the gas from its petroleum sources. Propane and other LPGs are isolated from petrochemical mixtures in one of two ways—by separation from the natural gas phase of petroleum and by refinement of crude oil.

1 Both processes begin when underground oil fields are tapped by drilling oil wells. The gas/oil hydrocarbon mixture is piped out of the well and into a gas trap, which separates the stream into crude oil and "wet" gas, which contains natural gasoline, liquefied petroleum gases, and natural gas.

2 Crude oil is heavier and sinks to the bottom of the trap; it is then pumped into an oil storage tank for later refinement. (Although propane is most easily isolated from the "wet gas" mixture, it can be produced from crude oil. Crude oil undergoes a variety of complex chemical processes, including catalytic cracking, crude distillation, and others. While the amount of propane produced by refinery processing is small compared to the amount separated from natural gas, it is still important because propane produced in this manner is commonly used as a fuel for refineries or to make LPG or ethylene.)

3 The "wet" gas comes off the top of the trap and is piped to a gasoline absorption plant, where it is cooled and pumped through an absorption oil to remove the natural gasoline and liquefied petroleum gases. The remaining dry gas, about 90% methane, comes off the top of the trap and is piped to towns and cities for distribution by gas utility companies.

4 The absorbing oil, saturated with hydrocarbons, is piped to a still where the hydrocarbons are boiled off. This petroleum mixture is known as "wild gasoline." The clean absorbing oil is then returned to the absorber, where it repeats the process.

5 The "wild gasoline" is pumped to stabilizer towers, where the natural liquid gasoline is removed from the bottom and a mixture of liquefied petroleum gases is drawn off the top.

6 This mixture of LP gases, which is about 10% of total gas mixture, can be used as a mixture or further separated into its three parts—butane, isobutane, and propane (about 5% of the total gas mixture).

Quality Control

As described above, propane must be carefully isolated from a complex mixture of petrochemicals which includes methane, ethane, ethene, propene, isobutane, isobutene, butadiene, pentane, and pentene, to name a few. If such impurities are not removed, the propane or propane and butane mixture will not liquefy properly. Liquefaction at appropriate temperature and pressure is critical for the gas to be economically useful. The liquefied gas industry has established standardized specifications that LPG mixtures must conform to in order to be considered acceptable for use as fuel gas. Standardized test methodologies for evaluating these specifications are approved and published by the American Society for Testing and Materials (ASTM). For example, the LPG known as "commercial propane" must have a maximum vapor pressure of 200 psig at 100°F (38°C) and can have no more than 0.0017 ounces (0.05 ml) of residual matter. Furthermore, the allowed amount of volatile residue is strictly limited, and the gas must meet established guidelines for corrosivity to copper, volatile sulfur content, and moisture. Other mixtures of propane and butane are commercially available which have slightly different target values.

These tightly held quality standards make propane an environmentally attractive fuel. In fact, to meet pipeline standards, nearly all pollutants are removed from propane before it is allowed to enter pipelines. When used in properly adjusted and maintained burners, propane's emissions easily meet the standards for clean air set by the Environmental Protection Agency (EPA). Their testing has proven that propane is environmentally safer than other hydrocarbon energy sources, and that properly processed propane can be used as a motor fuel which is significantly cleaner than gasoline. Studies have shown that, compared to gasoline, propane engines have as much as 45% less ozone-forming potential. Results of another recent EPA study show propane reduces total hydrocarbon emissions by 29% according to the new Federal Clean Air Standards. Furthermore, carbon monoxide emissions are 93% below the standard, hydrocarbon emissions are 73% below the standard, and nitrogen oxide emissions are 57% below the standard.

Byproducts/Waste

As detailed above, the manufacture of propane produces a variety of byproducts that are economically useful. Actually, it is more accurate to think of these not as byproducts but as co-products, since they are produced along with propane as part of petroleum refinement. These co-products may be in the form of solids, gases, or liquids. Solids (or semisolids) include bitumes, hydrogen sulfide, and carbon dioxide and are sold for fuel purposes. The liquid fractions include crude oil, which is further refined to give a variety of products. These oils vary dramatically in appearance and physical properties like boiling point, density, odor, and viscosity. The different fractions of crude oil are referred to as "light" or "heavy" depending on their density. Light crude is rich in low-boiling and paraffinic hydrocarbons; heavy crudes are higher-boiling and more viscous. They yield a variety of asphalt-like molecules. Many of the co-products of propane pro-

duction, such as propylene and butylene, are useful in gasoline refining, synthetic rubber manufacture, and the production of petrochemicals.

The Future

As the field of petroleum chemistry evolves, propane chemistry will continue to advance. Improvements will be made in the way propane is separated from petroleum. One area that offers opportunity for advancement is in the area of oil well production. Much natural gas is burned at remote oil wells because the extensive piping system required to transport it is prohibitively expensive. There are efforts underway to convert more of this wasted gas to condensable gases, which could be easily stored and transported. It is also important to note that propane is likely to become increasingly popular as a fuel gas based on economic factors and environmental concerns. In fact, in the Clean Air Act of 1990 Congress named LPGs as one of the clean-burning alternative fuels designated to take national air quality into the twenty-first century.

Where to Learn More

Books

Clark, William, ed. *Handbook of Butane/Propane Gases.* Butane-Propane News, Inc., 1972.

Other

National Propane Gas Association. http://www.propanegas.com/npga/ (July 14, 1997).

—*Randy Schueller*

Rammed Earth Construction

Background

Rammed earth is essentially manmade sedimentary rock. Rather than being compressed for thousands of years under deep layers of soil, it is formed in minutes by mechanically compacting properly prepared dirt. The compaction may be done manually with a hammer-like device, mechanically with a lever-operated brick-making press, or pneumatically with an air-driven tamping tool. Dynamic compaction using manual or power tampers not only compresses the soil, but it also vibrates the individual dirt particles, shifting them into the most tightly packed arrangement possible. When finished, rammed earth is about as strong as concrete.

Houses built of rammed earth have several advantages over wood-frame construction. The walls are fireproof, rot resistant, and impervious to termites. The solid, 18-24 in (45.72-60.96 cm) thick walls are nearly soundproof. The massive walls help maintain a comfortable temperature within the house, damping temperature swings that normally occur on hot summer days or cold winter nights. When designed and oriented to take the best advantage of solar energy, a rammed earth house can be comfortable with 80% less energy consumption than a wood-frame house. On the other hand, initial construction is about 5% more expensive than wood-frame construction because it is very labor intensive.

History

Humankind has used the earth itself to build homes and other structures for thousands of years. Jericho, the earliest city recorded in history, was built of earth. Temples, mosques, and churches were built of mud bricks and rammed earth throughout the ancient Middle East. Egyptian pharaohs ruled cities constructed of rammed earth. In the Far East, the technique was used not just for houses, but even for building ancient forerunners of the Great Wall of China. Romans and Phoenicians brought the technology to Europe, where it was used for more than 2,000 years.

In the United States, rammed earth construction enjoyed a period of popularity from 1780 until about 1850, when mass-produced fired bricks and sawed lumber became readily available. Houses could be built more quickly and easily with bricks and lumber, which were considered more modern and elegant materials than dirt. The supply shortages experienced after World War I and during the Great Depression brought rammed earth construction back into favor for two decades. Frank Lloyd Wright designed houses to be made of rammed earth.

When World War II ended, the country faced a large demand for housing, and wartime factories turned to manufacturing building materials that could be used for quicker types of construction. Rammed earth was brushed aside until it was repopularized during the environmentally conscious 1970s. A modified version of the technique, invented by Michael Reynolds, uses building blocks of discarded automobile tires rammed full of earth. These houses not only keep used tires out of landfills, but they can be built by inexperienced, first-time builders. When a homeowner uses the unpaid labor of himself, relatives,

When designed and oriented to take the best advantage of solar energy, a rammed earth house can be comfortable with 80% less energy consumption than a wood-frame house.

and friends, and when he can obtain many of the building materials for free, the construction cost can be held to less than half that of a wood-frame house.

For thousands of years, rammed earth construction was taught personally by one generation of builders to the next. In early twentieth-century America, such a network of experienced builders did not exist. The U.S. Department of Agriculture developed and published a manual titled *Rammed Earth Walls for Buildings* that showed people how to build their own homes. Research projects designed to improve the methods and quality of rammed earth construction were reported in academic journals, and more than 100 articles on rammed earth appeared in trade and popular magazines from 1926 to 1950.

Several more recent innovations increase the speed and ease of construction and enhance the structural integrity of the final product. For example, pneumatic tampers can be used to compact soil much more quickly than the traditional manual method. Easy-to-assemble forms allow walls to be built as solid panels rather than building them up as successive layers a foot or two at a time. Using time-honored manual methods, a four-person crew can erect 40-50 sq ft (12.19-15.24 sq m) of rammed earth wall per day; with power tools, the same crew can construct 300 sq ft (91.44 sq m) per day. David Easton, founder of Rammed Earth Works, has developed several earthquake-resistant designs to reinforce and structurally integrate the walls; the choice of design alternative depends on several considerations, including the distance to the nearest seismic fault.

Raw Materials

As the name implies, the primary material used in rammed earth construction is the earth itself. There are five basic types of soil (gravel, sand, silt, clay, and organic), and the dirt in a given location is generally some combination of all or most of these types. Historically, the longest lasting rammed earth walls were made of soil that was 70% sand and 30% clay. The soil from a new building site is tested to determine its suitability. Organic material must be re-

moved from the soil and, if necessary, a different type of soil can be trucked in and mixed with the existing dirt to create a blend that will work. Cement may be added to the soil to increase both its strength and its resistance to moisture—usually at about one-fourth the ratio that would be used to make concrete.

Steel reinforcing bars are placed in the foundations and sometimes in the walls. Plywood is used to make the removable forms for standard rammed earth construction. Sheets of three-quarter-inch (1.9 cm) plywood are thick enough. High-density-overlay (HDO) panels, which have a thin, plastic coating on one side, work especially well because they release more easily from the wall after construction. This not only leaves a clean finish on the just-completed wall, but it leaves the form boards in good condition to be used on future projects.

Rammed-earth tire construction uses discarded automobile tires, aluminum cans, and cardboard in addition to compacted soil. About 1,000 tires are used to build the walls of a 2,000 sq ft (609.6 sq m) house.

Design

Rammed earth houses are custom designed to make the most energy-efficient use of the site. They can be successfully designed for many climate regions, including humid areas with cold winters. The size and placement of windows is an important factor in taking advantage of solar heating in the winter and cooling breezes in the summer. The house can be positioned to take advantage of hills that offer protection from storms. Shade trees or trellised vines offer relief from summer heat but admit warm sunlight in winter.

The Manufacturing Process

Rammed earth houses can be built in one of three basic ways. Individual, rammed earth bricks can be formed and used with standard building techniques; in fact, such bricks may be used to form the floors in a rammed earth house built with other techniques. Standard rammed earth construction involves erecting wood forms and compact-

Standard rammed earth construction involves erecting wood forms and compacting the prepared soil into these molds, which are removed after the walls are completed.

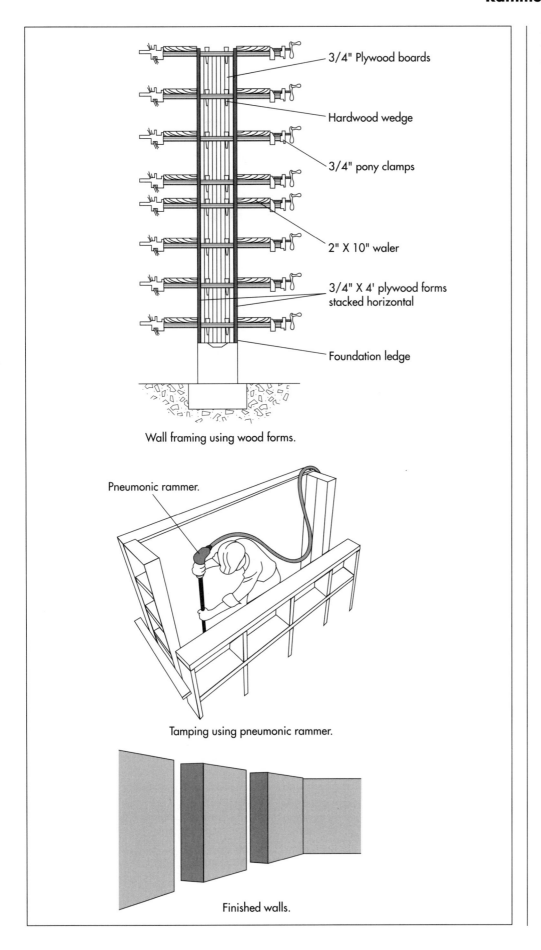

3/4" Plywood boards

Hardwood wedge

3/4" pony clamps

2" X 10" waler

3/4" X 4' plywood forms stacked horizontal

Foundation ledge

Wall framing using wood forms.

Pneumonic rammer.

Tamping using pneumonic rammer.

Finished walls.

In the tire method, a row of used automobile tires is simply laid atop the concrete footing, perhaps centered around steel reinforcing bars that extend out of the footing. The tires are then filled with soil. About 1,000 tires are used to build the walls of a 2,000 sq ft (609.6 sq m) house.

TIRE METHOD

Wall framing using automobile tires.

Soil is then shoveled into the tire and compacted by repeated blows with a sledge hammer. Once packed with dirt, the tire walls bulge, interlocking with the tire row below.

The walls of the rammed-earth tire house are constructed by stacking tires on top of each other in a running bond type pattern (each row off 1/2 tire from the one above and below).

ing the prepared soil into these molds, which are removed after the walls are completed. The rammed-earth tire method is a commonly used alternative. The descriptions that follow are overviews of the standard and tire methods.

Preparing the site

1 An inch or two (2.5-5 cm) of topsoil is removed from the building site and stored so it can be replaced around the com-

pleted structure. Organic matter such as weeds and roots are removed and may be composted for use in post-construction landscaping. After the site is cleared, the outline of the house is staked out. The soil is excavated to a depth that guarantees a level surface; the excavation includes the floor area of the building as well as a 3 ft (1 m) surrounding buffer zone. A trench may be dug so that the walls will be anchored into the ground to a depth below the winter freezing line.

Laying the foundation

2 The foundation, which is made of reinforced concrete, consists of a footing that may be as narrow as the thickness of the wall or up to three times that thickness, depending on the strength of the underlying soil. The footing is extended above ground level to form a short "stem wall" that will connect the rammed earth walls to the footing. Depending on the architectural design, a slab floor may also be poured.

Analyzing the soil

3 A variety of tests are conducted to determine the suitability of the local soil for construction material. For example, a particle determination test reveals the relative proportions of sand and silt in the sample. A compaction test is performed by forming a ball of mud and dropping it from a height of 3 ft (1 m); the degree to which the ball disintegrates on impact reveals its usefulness for building. Other, more precise, tests can be performed at a geotechnical laboratory. If the native soil is unsuitable or inadequate for building, it can be blended with or replaced by soil from another source. Soil may be purchased from a quarry, or it might be available as refuse from a nearby construction site, in which case it could be delivered free or at a minimal cost.

Framing the walls

4 Traditionally, wood forms were used to build up walls 2 ft (0.6 m) at a time. After the mold was filled with fully compacted soil, it would be removed and reset to form the next section of wall. More efficient methods now allow forms to be constructed for the entire height of the wall (even more than one story). Horizontally, the framework may form the complete length of wall, or it may form shorter panels [e.g., 8 ft (2.44 m) long] separated by 6 in (15 cm) gaps that can be filled with reinforced concrete for enhanced structural strength. Framing is a major component of the construction process, in terms of both importance and time; it usually takes less time to fill and compact the soil within the forms than it does to set, align, and remove the framework.

In the case of rammed-earth tire construction, wood forms are not used. A row of used automobile tires is simply laid atop the concrete footing, perhaps centered around steel reinforcing bars that extend out of the footing. After each layer of tires has been filled and compacted, another layer will be added, offset by half the tire diameter from the layer below.

Tamping the soil

5 Traditional tampers are made of a heavy wooden block with a handle extending upward through its center. A more compact version can be made from a 4 in (10 cm) square steel plate welded to a section of 1 in (2.5 cm) pipe. A 4-6 in (10-15 cm) layer of moistened soil is placed inside the form, and a worker drops the tamper from a height of 12-18 in (30-46 cm). In fact, most of the work is now done quickly with pneumatic tampers, and manual devices are used only in tight spaces around electrical boxes or plumbing pipes. After many repetitions with the tamper over the entire surface of the layer, the noise made by the impacting tamper changes from a dull thud to a ringing sound. This happens when the soil has been compacted to about half of its original volume. At this point, another layer of prepared soil is added, and the tamping process is repeated. When the tamping is finished, the wood forms are removed.

The tamping process is different when tires are used as the framework. In this case, a sheet of cardboard is placed across the bottom of the hole in the tire, and moistened soil is shoveled into the tire. The dirt is packed by hand into the interior of the tire, and then it is compacted by repeated blows with a sledge hammer. Using this technique, about three wheelbarrow loads (350 lb or 158.9 kg) of soil can be packed into each tire. Pounding the dirt causes the tire's walls to bulge, interlocking the tire to the row below. As additional layers are added and the wall becomes taller, scaffolding must be constructed so workers have a place to stand while filling and pounding the tires.

Finishing the walls

6 Interior faces of walls are often finished with plaster. If such a coating is not ap-

plied, the wall should be treated with a clear, penetrating sealant to prevent dust from sloughing off. Because stone (even when manmade from rammed earth) is somewhat porous, it may be necessary to apply sealant to weatherproof the exterior faces of the walls in certain climate areas.

Rammed earth tire walls are finished by inserting aluminum cans into gaps between the tires and filling remaining voids with adobe (straw-reinforced mud). Earth is packed against the exterior face of the wall, creating a flat surface that completely conceals the tires. Wall interiors are finished with 2-4 in (5-10 cm) of plaster or stucco.

Byproducts/Waste

Rammed earth structures use natural resources efficiently. Those made of packed tires even make productive use of some of society's trash. Because the tires are sealed within 3 ft (0.9 m) thick walls, neither oxygen nor the sun's ultraviolet rays can react with them. This means they cannot catch fire and they do not release toxic chemicals. The structures qualify for better fire ratings than wood-frame buildings, and they do not smell of rubber.

The Future

During the late 1700s, a French builder named Francois Cointeraux founded a school in Paris to study and publicize rammed earth construction, which he called *pisé de terre* (puddled clay of earth). Today, David Easton has developed a new version of rammed earth construction he calls PISÉ (Pneumatically Impacted Stabilized Earth). It involves spraying the prepared soil under high pressure against a one-sided form. This technique can produce 1,200 sq ft (365.76 sq m) of 18 in (45.72 cm) thick wall per day, which is four times faster than a typical, four-person crew can fill box-like forms and compact earth with power tampers.

Where to Learn More

Books

Easton, David. *The Rammed Earth House.* Chelsea Green Publishing Company, 1996.

McHenry, Paul Graham. *Adobe and Rammed Earth Buildings: Design and Construction.* University of Arizona Press, 1989.

Periodicals

"Rammed Earth Construction." *Countryside & Small Stock Journal,* March/April 1992, pp. 32-33.

Other

"Rammed-Earth Tire Homes" February 9, 1997. http://monticello.avenue.gen.va.us/ Community/Environment/YellowMtn/men u.html (May 22,1997).

—*Loretta Hall*

Recliner

Background

A reclining chair is an upholstered chair with a metal mechanism activated by the user so that the back is pushed out and a foot rest rises up to accommodate the user's lower legs. Most recliners are armchairs, meaning they include arms. Upholstered backs and seats make the chair a truly comfortable piece of furniture. Reclining chairs or sofas are called "action" or "motion" furniture in that they move or change shape. Some recliners are activated by the sitter pulling on a lever; others are activated by the sitter pushing back in the chair with some force.

History

One of the earliest designs for a reclining chair was published in a British periodical, *Ackermann's Repository of Arts*, in 1813. The "Reclining Patent Chair" was a prototype for hundreds of such chairs that were made throughout the century. Browne and Ash, upholsterers and cabinetmakers in New York, advertised "Improved Patent Self Acting Reclining, Elevating, and Revolving, Recumbent Chairs and Sofas" in 1855. Cabinetmaker George J. Henkels featured a reclining chair, which could be manipulated by the sitter without complicated machinery.

The late nineteenth century recliner known as the "Morris Chair" was not devised by artist and designer William Morris but by William Watt in 1883 and manufactured by the decorative arts manufacturing firm Morris and Company (founded by William Morris)—hence its name. Mechanisms that transformed this chair to a recliner varied; however, most were quite simple, consisting of a rod or series of pegs that permit the

back of the armchair to recline and rest against the pegs or metal. The frames were of wood (often quite massive) with upholstered cushions for back and seat. Such Morris Chairs were advertised as comfortable seating furniture, as are recliners today.

Throughout the late nineteenth century and early twentieth century, there were many companies that produced innovative, comfortable "easy chairs" that reclined. Some included built-in magazine racks and practical, washable cushions. La-Z-Boy Incorporated, founded by Edwin J. Shoemaker and Edward M. Knabusch in 1929, designed a patentable reclining chair mechanism that they felt was unbeatable. The pair guaranteed the mechanism for the lifetime of the recliner, an ingenious marketing tool for the company. Today, La-Z-Boy is one of the few companies that extensively tests the activating mechanism, as well as the upholstery and frame, to ensure that their customers are getting the quality they are guaranteed.

Raw Materials

Primary components of the recliner include the frame, the metal activating mechanism, the foam or upholstery padding, and the upholstery fabric. While materials of these components vary according to manufacturer and specific style charateristics, generally the materials are as described.

The frame is made out of wood, often hardwoods, since the chair receives much motion or weight shifting as it moves from a conventional chair to a recliner. The frame is reinforced with metal nuts, bolts, and steel corners or supports on some styles. Some **fiberboard** may be used in minor frame construction or back support.

La-Z-Boy Incorporated, founded by Edwin J. Shoemaker and Edward M. Knabusch in 1929, designed a patentable reclining chair mechanism that they felt was unbeatable. The pair guaranteed the mechanism for the lifetime of the recliner, an ingenious marketing tool for the company.

The metal activating mechanism, which converts the chair into a recliner, is of stamped steel. Metal is also used for the springs in the upholstered seat.

Padding of upholstered backs is a foam that may be sculpted, such as polyurethane foam. All hard corners of foam blocks are padded and softened by a material such as polyester batting sheets. Some companies have abandoned foam that contain chlorofluorocarbons (CFCs) in order to be more ecologically responsible.

Exposed wood on the arms, feet, or back may be more expensive hardwoods such as maple or oak that is stained and/or varnished.

Upholstery fabric varies dramatically from piece to piece even within one company. All must be fairly durable as the chairs are pushed and pulled as pieces of action furniture. Recliners are upholstered in leather, cotton, natural-synthetic fiber combinations, or solely of synthetic fibers that are easy to keep clean and fire resistant.

Design

The recliner manufacturing process generally includes phases that range from concept, to prototyping, to full production. These phases may be divided as follows:

Initial design

A residential furniture designer may bounce an idea off of a merchandising manager or marketer for an idea for a recliner; often, new products are the result of a design sketch completed by an inhouse designer. Every piece is targeted to sell for a specific price at this point in the design process; the market for the piece is identified; and the new product's fit in the line of products available is determined. Manufacturers of recliners often produce up to 100,000 recliners in a single style, so these companies must be confident that consumers will be drawn to the product.

The designers sketch and submit a prototype for final approval. Full-scale drawings of new products are required by some companies, and this, along with marketing information, is handed off to the product development team so that the new recliner may be devised for full-scale production.

Prototyping

The product development team takes the recliner from sketches, to working prototype, to manufacturing instructions for full-scale production. The team consists of at least one of each of the following: upholstery designer, sewing technician, pattern layout technician, a wood technician, and a Computer-Aided Design (CAD) engineer. The team physically builds a prototype; and once the prototype is approved, the preproduction phase begins.

Preproduction phase

A manufacturing plant is chosen and the prototype is test produced on the machinery by staff to see if it can be successfully created.

Pilot production

Once the prototype has been successfully produced at one factory, the specifications and materials are electronically shipped to all factories for another test run. The computer-based specifications assist in programming stamping machines, cutting machines, etc.

After the small run of the new recliner is successfully executed, full production is initiated, and each plant produces several thousand recliners per day.

The Manufacturing Process

Recliners, like many pieces of upholstered furniture, are assembled in stages with components put together in sections. All components are then pulled together in final assembly.

1 The frame, including the back, are constructed of planed, measured, shaped, and assembled wood. Rigid frames are put together with nuts, bolts, nails, and/or staples. The framing is laborious work, and individuals often work alone to hand-construct this part of the recliner. The frame is sent to a holding area, where it will await further assembly.

2 Die-making machines and electrical discharge machines make the design cuts for the metal activating mechanisms for recliners; parts for the finished metal mecha-

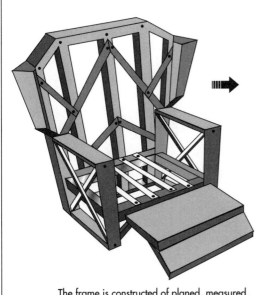

The frame is constructed of planed, measured, shaped, and assembled wood.

Metal die cutters and stamping machines cut steel pieces for the activating mechanism. Assembled mechanisms are attached to the frame.

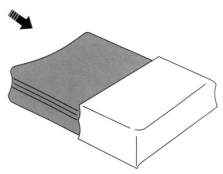

Rigid foam is sculpted for the back and seat cushions.

nism are cut from steel in stamping machines. The metal mechanisms are then attached to the chair frame.

3 Foam bodies for the upholstered parts (back and seat) are sculpted from a rigid foam, such as polyurethane, by cutting, slicing, or shaping with saws to conform with a desired profile.

4 Patterns or templates for cutting the upholstery fabric are sent electronically from the product development team to the cutting departments of the various manufacturing facilities. This computer data is programmed into the computer-driven cutting machine, which cuts approximately 40 layers of fabric at a time. The cut fabric is removed from the machine, sorted, marked, and stored for delivery to sewers.

5 The cut pieces of upholstery are delivered to the sewers, who work on industrial grade **sewing machines**, to stitch all pieces of the covers for the foam seats together. The empty cushion covers are then sent to the cataloguer, who stores the cushions for the upholsterer.

6 Next is upholstery assembly, in which polyurethane foam, polyester fiberfill, die-cut fiberboard, and covers or seats are sent to the upholsterer. Edges are softened by gluing padding of some sort, often polyester fiberfill, to the hard edges of the foam. The air bubbles are sucked out of the foam bodies, thus compressing them. The foam bodies are then pushed into the sewn cushion in order to fill it out neatly. The upholstery covers are nailed and stapled to the seats and backs.

7 During final assembly, the backs and seats are assembled to the seat frame, the activating mechanism is attached to the frame, the decorative skirt is added (depending on the style of the recliner), and final inspection occurs.

Quality Control

Furniture constructed for residential use is not required to undergo consumer testing, unlike furniture manufactured specifically for use in businesses, public buildings, or schools. However, furniture manufacturers consistently monitor quality by visual inspection; many have inspectors who examine each piece to ensure that it looks right (no protruding nails, puckers, etc.), feels right, and performs correctly. Once the piece passes visual inspection, the recliner is tagged and boxed for shipment out of the plant. Fabrics may be tested by the company to ensure that they conform to specifica-

The frame is made out of wood, often hardwoods, since the chair receives much motion or weight shifting as it moves from a conventional chair to a recliner. Padding of upholstered backs is a foam that may be sculpted, such as polyurethane foam. All hard corners of foam blocks are padded and softened by a material such as polyester batting sheets.

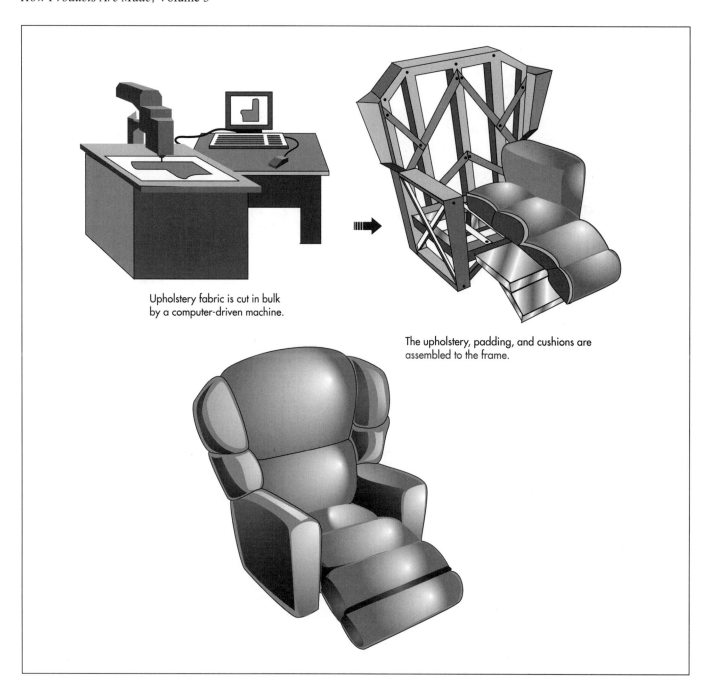

Upholstery fabric is cut in bulk by a computer-driven machine.

The upholstery, padding, and cushions are assembled to the frame.

Recliners are upholstered in leather, cotton, natural-synthetic fiber combinations, or solely of synthetic fibers that are easy to keep clean and fire resistant. During final assembly, the backs and seats are assembled to the seat frame, the activating mechanism is attached to the frame, the decorative skirt is added (depending on the style of the recliner), and final inspection occurs.

tions such as fabric stretch, seam slippage, colorfastness, and flammability. Some companies conduct much more extensive testing on their residential-use motion furniture, so that they may assure their clients that the product will perform consistently.

Where to Learn More

Books

Ensign, Robert. *Always a Leader*. La-Z-Boy Chair Company, 1987.

Hanks, David. *Innovative Furniture in America from 1800-Present*. Horizon Press, 1981.

Periodicals

Powell, James D., ed. "New La-Z-Boy Chair Plant Sets Stage for Growth." *Furniture Methods and Materials*, March 1976, pp. 10.

—*Nancy EV Bryk*

Ribbon

Background

Ribbons are useful and decorative fabrics that are almost infinite in their variety, texture, and color. Modern ribbons are manufactured from every kind of fabric, from velvet and satin to synthetics like nylon and rayon. They are patterned, printed, woven, braided, adorned with embroidery, decorated with pearls or sequins, shaped like ric-rac, skillfully made like lace, edged with metal so they can be molded and shaped, and crafted like motif ribbons. Ribbon is classified by the textile industry as a narrow fabric, and it ranges from 1/8 in-1 ft (0.32-30 cm) in width. Its uses may most often be thought of as decorative, but ribbons are also materials for making larger fabrics by weaving, crocheting, or knitting them together.

History

Ribbons appeared when civilizations began crafting fabrics. They are among the oldest decorative or adorning materials. People have always looked for ways to personalize their clothing and household goods. When all textiles were handmade, items with the finest threads were the most expensive. But the simplest, most coarse textiles in plain colors could be made more elegant and individual with a bit of ribbon as decoration. In the Middle Ages, peddlers traveled throughout Europe selling exotic ribbons; the tales of Geoffrey Chaucer mention "ribbands" used to adorn garments. Medieval and Renaissance patrons bought ribbons woven with gold and silver thread and made from silk and other rare fabrics from the Orient. The modern ribbon with selvedges (finished edges) came into being by 1500. Ribbons were so identified with luxury that, during the sixteenth century, the English Parliament tried to make the wearing of ribbons a right of only the nobility. They were also identified with certain orders of merit; the Knights of the Garter wear broad blue sashes to this day, and the Knights of Bath wear red.

By the seventeenth century, ribbons stormed the fashion world. Both men's and women's clothing of this period were extravagant, and every accessory from gloves to bonnets was festooned with ribbons in many forms. A length of ribbon could be given as a gift to decorate clothing, for use in braiding and curling hair, for ornamenting baskets and furniture, or for brightening linens. Ornately patterned household fabrics were further bedecked with ruchings (gathered ribbons), frills, and rosettes. The huge demand for more elaborate ribbons prompted a manufacturing revolution in which Coventry, England, and Lyons, France, became hubs of ribbon design and generation.

This ribbon industry sprang from the silk trade. Merchants who traveled the "Silk Road" to and from Asia sold raw silk to middlemen in Europe who boiled, cleaned, and dyed the ribbon yarn and sold it in "twists" to weavers. The weavers used specially scaled looms and scores of laborers to weave ribbons on hand-operated looms. The products were sold in the major cities and exported for trade. The enormous demand for ribbon was one of the sparks of the Industrial Revolution. In the 1770s, the Dutch engine loom was developed, and six types of ribbon could be produced simultaneously under the watchful eye of one operator. This development came just in time to decorate the towering wigs in fashion in the courts of Europe. Curiously, in the fledg-

Ribbon is classified by the textile industry as a narrow fabric, and it ranges from 1/8 in-1 ft (0.32-30 cm) in width.

ling colonies in the Americas, ribbons were seldom worn at this time, perhaps due to religious convictions or in opposition to the extravagances of European rulers.

Peasant costumes of many lands are often distinguished by single or braided ribbons that are dyed bright colors, decorated with lace or beads, or patterned. Unique designs came to characterize cultures. During the Napoleonic Wars early in the nineteenth century, the ribbon industry suffered a major decline because skilled weavers from England and Western Europe were recruited for military service. With the supply restricted, the demand for ribbon was even greater, and ribbons were a popular cargo for smugglers. The next ribbon "boom" occurred in 1813, when picot-edged ribbon (with tiny scallops along the sides) became a fashion must. Ribbon-weavers reaped the benefits for the two years picot-edged ribbon topped the fashion charts. Ribbons often followed fashion trends. Deaths at the courts of Europe stimulated the demand for black ribbon; military tapes, jacquards, and medal ribbons became symbols of military regiments and the highest awards nations could bestow.

The Victorian Era was the last to see a ribbon boom when the dresses, underclothes, coats and cloaks, and hats of Victorian ladies used yards of ribbon. Trade agreements between European countries killed the English manufacture of ribbon because cheap labor and ever-larger looms could not produce competitively priced products. These manufacturers survived by diversifying and producing braids, cords, fringes, silk pictures, and bookmarks. The development of synthetics and paper fibers for use in making gift wrap quickly extended to the ribbon world in our times, and ribbon became as adaptable to modern living as other fabrics. Many types of ribbon today are colorfast, shrink resistant, and able to be washed or dry cleaned.

Raw Materials

Ribbon can be manufactured from a wide range of materials, and their manufacture is classified by type and texture. The three principle categories of manufacture are cut-edge, woven-edge, and wire-edge ribbons. Woven-edge ribbons are most common to the textile industry; they are narrow pieces of fabric with two "selvedges" or woven edges that can be straight or shaped. These ribbons are usually washable because the woven edges prevent them from fraying. Wire-edge ribbons can be cut from broader strips of cloth with their edges wrapped over thin wires, or the wire can be woven into the fabric along the edges or down the middle. Wire mesh can also be woven to make ribbon with or without the addition of yarns or silks for color. Wire-edge ribbon is versatile because the wire allows it to hold a definite shape, but the material can not be washed. Cut-edge or craft ribbon is the type most often used for gift wrap. The fabric is patterned, printed, or decorated with designs transferred by heat then cut to the needed width. The product is then treated with a stiffener that prevents the edges from unraveling. High quality cut-edge ribbon is made of acetate, a thermoplastic, which is cut by a hot knife that fuses the edge instantly.

Ribbon used for decorating fabrics is typically made of fabric. Rayon, velvet, silk, and satin ribbon may be the most common types of fabric ribbon; but cotton, wool, and other synthetics can be processed in ribbon form. Various surface treatments can also be used to change the appearance of cloth ribbon or modify its performance characteristics. The six broad categories of ribbon textures include organdies, satins, velvets, grosgrains, metallics, and natural fibers. Organdies are delicate products made of very fine woven yarns, and they often have metal edges to provide shape. Satins are popular because of their shiny finish (either single- or double-face), their bright and bold colors, and their variety of edges and surface patterns. Velvet ribbon has soft pile, usually on one face only, and can be printed, flocked, or backed with satin. Grosgrains are woven, and the weave usually shows clearly in ribs. Grosgrains are made of cotton, polyester, or fiber blends, and they are very durable. Traditionally, grosgrains were used to decorate ladies' bonnets, but modern techniques give them a range of finishes, including patterns and pleats. Metallics are woven from lurex or other metallic yarns and are favored for their sparkle. Natural fibers include the whole range of paper ribbons, cotton tapes, jute, and linen. Jacquards are a specialized type of ribbon developed in France and

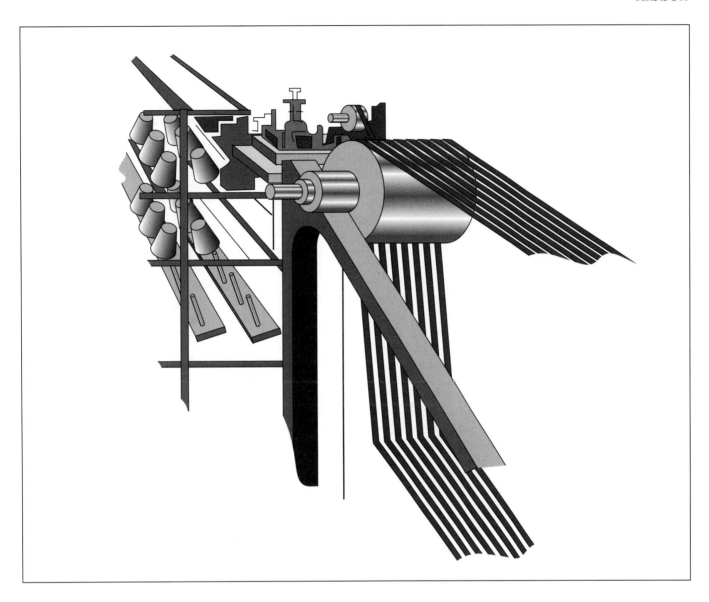

overlapping several textural types. Jacquards are prized for elaborate design woven into the ribbon, and they are very expensive to manufacture.

The desired behavior of the ribbon often dictates the material and any surface treatments used. Curling ribbon, for example, is bathed in glue that is pressed thin by rollers and dried. The glue gives the ribbon its curling properties. Other raw materials include ink for printing on finished ribbon, and paper and plastics if the ribbon manufacturers also make their own spools and packaging.

Design

Ribbons are designed in much the same way as fabrics. Colors are chosen depend-

ing on fashion trends, seasons, and intended uses. Materials are selected based on use, wearability, cleaning requirements, and fabric trends that the ribbons must match. Sales records are also considered because ribbons go in and out of fashion and are sometimes discontinued.

The width and pattern of the ribbon must also be designed. As narrow fabrics, ribbons are 1/8 in-1 ft (0.32-30 cm) wide, although the ribbon industry has adopted the French "ligne" as its unit of measure. The ligne is about 1/11 inch (0.67 mm) wide. Many patterns and designs can be woven into the ribbon, and ribbon can be printed or ornamented by virtually any type of printing method so the pattern or trim, such as sequins, appears on one side.

After the particular thread for ribbon has been spun, dyed, and treated, it is rolled on bobbins. The bobbins are placed on a ribbon loom that consists of a series of miniature looms, each with its own shuttle and warp sized to produce the desired width of ribbon. The woven product emerges on rollers that carry it forward for further processing such as adding glues, stiffeners, or fabric treatments. A winder then places the ribbon on spools for packaging and sale.

The Manufacturing Process

1 The process of manufacturing a particular kind of thread can vary widely, from the spinning of silk to the carding and processing of wool. After the particular thread for ribbon has been spun, dyed, and treated, it is rolled on bobbins. The bobbins are placed on a ribbon loom that consists of a series of miniature looms, each with its own shuttle and warp (lengths of yarn) sized to produce the desired width of ribbon. The ribbon loom may weave as many as 144 pieces of fabric simultaneously. Today's ribbon looms can be very elaborate and computerized to produce detailed designs like jacquards less expensively than the looms or weavers of the past. The threads leading from the bobbins are guided by a series of eye hooks that hold the position of each thread and raise and lower it as the fabric is woven. The bobbins (also called cheeses) control the warp and are a major difference between ribbon loom and a fabric loom, which uses a warp beam to raise and lower the warp and cloth. The bobbins may also be curved to save space on the machine. The tension of the warp thread on a ribbon loom is maintained by a series of pulleys. A rack and pinion mechanism is used to adjust the lay (flatness or slope) of the loom.

2 To produce fancy effects, ingenious devices, selection of fabrics, and weaving techniques are used. Threads of different colors or multiple fibers can be woven together. Odd color effects can be achieved because the fibers may take dye differently. The woven product emerges on rollers that carry it forward for further processing such as adding glues, stiffeners, or fabric treatments. Machines equipped with pairs of rollers press and dry the treated ribbon, and large reels are used to collect the treated product.

3 As the ribbon is wound onto spools, the tension is maintained by a governor so the ribbon does not fall slack on the spools. If the ribbon is to be printed or embossed, it is then processed through a calendar that smoothes the surface to be printed and through a printing or stamping machine. A winder then places the ribbon on spools for packaging and sale.

Quality Control

The machines used to process one type of ribbon, but perhaps multiple varieties or colors of it, are arranged in a series and in a layout so that one operator can monitor one ribbon loom producing many ribbons in a series. Careful attention is paid to the detail in the ribbon, and the operators control the quality of the product as well as maintain the machines.

Byproducts/Waste

Ribbon mills produce some fabric waste at the start and end of each ribbon production, and this is disposed. Ribbon mills usually produce a range of other ornamental products as well, such as braid, cord, and ric-rac.

The Future

Ribbon manufacturers seem to have guaranteed the future of their product by the variety and ingenuity of their output. While fashion trends may cause particular types of ribbon to fade in and out of favor, the outcasts are quickly replaced by new products. Computer techniques have enhanced both design and manufacturing processes. They allow infinite combinations to be generated on screen, and intricate procedures that were previously cost-prohibitive may be possible with computer-controlled manufacturing.

Where to Learn More

Books

Collier, Ann M. *A Handbook of Textiles.* Pergamon Press, 1974.

Corbman, Bernard P. *Textiles: Fiber to Fabric.* Gregg Division, McGraw-Hill Book Company, 1983.

Evans, Hilary. *Ribbonwork.* Bobbs-Merrill, 1976.

Hall, A. J. *The Standard Handbook of Textiles.* Heywood Books, 1969.

Kerridge, Eric. *Textile Manufactures in Early Modern England.* Manchester University Press, 1985.

Lewis, Annabel. *The Ultimate Ribbon Book*. Trafalgar Square Publishing, 1995.

Linton, George E. *The Modern Textile Dictionary*. Duell, Sloan and Pearce, 1954.

Miller, Edward. *Textiles: Properties and Behaviour*. Theatre Arts Books, 1970.

Periodicals

Allen, Frederick. "The Ribbon Factory." *Invention & Technology,* Spring 1995.

"Make silk ribbon embroidery." *Woman's Day,* July 16, 1996, p. 23.

"Stylist secrets." *Redbook*, January 1995, p. 25.

—*Gillian S. Holmes*

Sand

Background

Sand is a loose, fragmented, naturally-occurring material consisting of very small particles of decomposed rocks, corals, or shells. Sand is used to provide bulk, strength, and other properties to construction materials like asphalt and concrete. It is also used as a decorative material in landscaping. Specific types of sand are used in the manufacture of glass and as a molding material for metal casting. Other sand is used as an abrasive in sandblasting and to make sandpaper.

Sand was used as early as 6000 B.C. to grind and polish stones to make sharpened tools and other objects. The stones were rubbed on a piece of wetted sandstone to hone the cutting edge. In some cases, loose sand was scattered on a flat rock, and objects were rubbed against the sandy surface to smooth them. The first beads with a glass glaze appeared in Egypt in about 3,500-3,000 B.C. The glass was made by melting sand, although naturally-occurring glass formed by volcanic activity was probably known long before that time.

In the United States, sand was used to produce glass as early as 1607 with the founding of the short-lived Jamestown colony in Virginia. The first sustained glass-making venture was formed in 1739 in Wistarburgh, New Jersey, by Caspar Wistar. The production of sand for construction purposes grew significantly with the push for paved roads during World War I and through the 1920s. The housing boom of the late 1940s and early 1950s, coupled with the increased use of concrete for building construction, provided another boost in production.

Today, the processing of sand is a multi-billion dollar business with operations ranging from very small plants supplying sand and gravel to a few local building contractors to very large, highly automated plants supplying hundreds of truckloads of sand per day to a wide variety of customers over a large area.

Raw Materials

The most common sand is composed of particles of quartz and feldspar. Quartz sand particles are colorless or slightly pink, while feldspar sand has a pink or amber color. Black sands, such as those found in Hawaii, are composed of particles of obsidian formed by volcanic activity. Other black sands include materials such as magnetite and hornblende. Coral sands are white or gray, and sands composed of broken shell fragments are usually light brown. The white sands on the Gulf of Mexico are made of smooth particles of limestone known as oolite, derived from the Greek word meaning egg stone. The white sands of White Sands, New Mexico, are made of gypsum crystals. Ordinarily, gypsum is dissolved by rain water, but the area around White Sands is so arid that the crystals survive to form undulating dunes.

Quartz sands, which are high in silica content, are used to make glass. When quartz sands are crushed they produce particles with sharp, angular edges that are sometimes used to make sandpaper for smoothing wood. Some quartz sand is found in the form of sandstone. Sandstone is a sedimentary, rock-like material formed under pressure and composed of sand particles held together by a cementing material such as calcium carbonate. A few sandstones are composed of almost pure quartz particles and are the source of the silicon used to make semiconductor silicon chips for microprocessors.

Molding sands, or foundry sands, are used for metal casting. They are composed of about 80%-92% silica, up to 15% alumina, and 2% iron oxide. The alumina content gives the molding sand the proper binding properties required to hold the shape of the mold cavity.

Sand that is scooped up from the bank of a river and is not washed or sorted in any way is known as bank-run sand. It is used in general construction and landscaping.

The definition of the size of sand particles varies, but in general sand contains particles measuring about 0.0025-0.08 in (0.063-2.0 mm) in diameter. Particles smaller than this are classified as silt. Larger particles are either granules or gravel, depending on their size. In the construction business, all aggregate materials with particles smaller than 0.25 in (6.4 mm) are classified as fine aggregates. This includes sand. Materials with particles from 0.25 in (6.4 mm) up to about 6.0 in (15.2 cm) are classified as coarse aggregates.

Sand has a density of 2,600-3,100 lb per cubic yard (1,538-1,842 kg per cubic meter). The trapped water content between the sand particles can cause the density to vary substantially.

The Manufacturing Process

The preparation of sand consists of five basic processes: natural decomposition, extraction, sorting, washing, and in some cases crushing. The first process, natural decomposition, usually takes millions of years. The other processes take considerably less time.

The processing plant is located in the immediate vicinity of the natural deposit of material to minimize the costs of transportation. If the plant is located next to a sand dune or beach, the plant may process only sand. If it is located next to a riverbed, it will usually process both sand and gravel because the two materials are often intermixed. Most plants are stationary and may operate in the same location for decades. Some plants are mobile and can be broken into separate components to be towed to the quarry site. Mobile plants are used for remote construction projects, where there are not any stationary plants nearby.

The capacity of the processing plant is measured in tons per hour output of finished product. Stationary plants can produce several thousand tons per hour. Mobile plants are smaller and their output is usually in the range of 50-500 tons (50.8-508 metric tons) per hour.

In many locations, an asphalt production plant or a ready mixed concrete plant operates on the same site as the sand and gravel plant. In those cases, much of the sand and gravel output is conveyed directly into stockpiles for the asphalt and concrete plants.

The following steps are commonly used to process sand and gravel for construction purposes.

Natural decomposition

1 Solid rock is broken down into chunks by natural mechanical forces such as the movement of glaciers, the expansion of water in cracks during freezing, and the impacts of rocks falling on each other.

2 The chunks of rock are further broken down into grains by the chemical action of vegetation and rain combined with mechanical impacts as the progressively smaller particles are carried and worn by wind and water.

3 As the grains of rock are carried into waterways, some are deposited along the bank, while others eventually reach the sea, where they may join with fragments of coral or shells to form beaches. Wind-borne sand may form dunes.

Extraction

4 Extraction of sand can be as simple as scooping it up from the riverbank with a rubber-tired vehicle called a front loader. Some sand is excavated from under water using floating dredges. These dredges have a long boom with a rotating cutter head to loosen the sand deposits and a suction pipe to suck up the sand.

5 If the sand is extracted with a front loader, it is then dumped into a truck or train, or placed onto a conveyor belt for transportation to the nearby processing plant. If the sand is extracted from underwater with a dredge, the slurry of sand and water is pumped through a pipeline to the plant.

Sorting

6 In the processing plant, the incoming material is first mixed with water, if it is not already mixed as part of a slurry, and is discharged through a large perforated screen in the feeder to separate out rocks, lumps of clay, sticks, and other foreign material. If the material is heavily bound together with clay or soil, it may then pass through a blade mill which breaks it up into smaller chunks.

7 The material then pass through several perforated screens or plates with different hole diameters or openings to separate the particles according to size. The screens or plates measure up to 10 ft (3.1 m) wide by up to 28 ft (8.5 m) long and are tilted at an angle of about 20-45 degrees from the horizontal. They are vibrated to allow the trapped material on each level to work its way off the end of the screen and onto separate conveyor belts. The coarsest screen, with the largest holes, is on top, and the screens underneath have progressively smaller holes.

Washing

8 The material that comes off the coarsest screen is washed in a log washer before it is further screened. The name for this piece of equipment comes from the early practice of putting short lengths of wood logs inside a rotating drum filled with sand and gravel to add to the scrubbing action. A modern log washer consists of a slightly inclined horizontal trough with slowly rotating blades attached to a shaft that runs down the axis of the trough. The blades churn through the material as it passes through the trough to strip away any remaining clay or soft soil. The larger gravel particles are separated out and screened into different sizes, while any smaller sand particles that had been attached to the gravel may be carried back and added to the flow of incoming material.

9 The material that comes off the intermediate screen(s) may be stored and blended with either the coarser gravel or the finer sand to make various aggregate mixes.

10 The water and material that pass through the finest screen is pumped into a horizontal sand classifying tank. As the mixture flows from one end of the tank to the other, the sand sinks to the bottom where it is trapped in a series of bins. The larger, heavier sand particles drop out first, followed by the progressively smaller sand particles, while the lighter silt particles are carried off in the flow of water. The water and silt are then pumped out of the classifying tank and through a clarifier where the silt settles to the bottom and is removed. The clear water is recirculated to the feeder to be used again.

11 The sand is removed from the bins in the bottom of the classifying tank with rotating dewatering screws that slowly move the sand up the inside of an inclined cylinder. The differently sized sands are then washed again to remove any remaining silt and are transported by conveyor belts to stockpiles for storage.

Crushing

12 Some sand is crushed to produce a specific size or shape that is not available naturally. The crusher may be a rotating cone type in which the sand falls between an upper rotating cone and a lower fixed cone that are separated by a very small distance. Any particles larger than this separation distance are crushed between the heavy metal cones, and the resulting particles fall out the bottom.

Quality Control

Most large aggregate processing plants use a computer to control the flow of materials. The feed rate of incoming material, the vibration rate of the sorting screens, and the flow rate of the water through the sand classifying tank all determine the proportions of the finished products and must be monitored and controlled. Many specifications for asphalt and concrete mixes require a certain distribution of aggregate sizes and shapes, and the aggregate producer must

Opposite page:

The preparation of sand consists of five basic processes: natural decomposition, extraction, sorting, washing, and in some cases crushing. Sand is extracted from the location at which it occurs by either a floating dredge or front loader. The dredge delivers a slurry of sand and water to the processing plant via pipeline, while the front loader simply scoops the sand up and into trucks or onto conveyor belts for transportation to the plant. The sand is sorted through a series of screens that separate differently sized particles. The sand is washed, and the smaller particles are sent to the sand classifying tank, where the particles are further separated. Some particles may be crushed if smaller particles are needed.

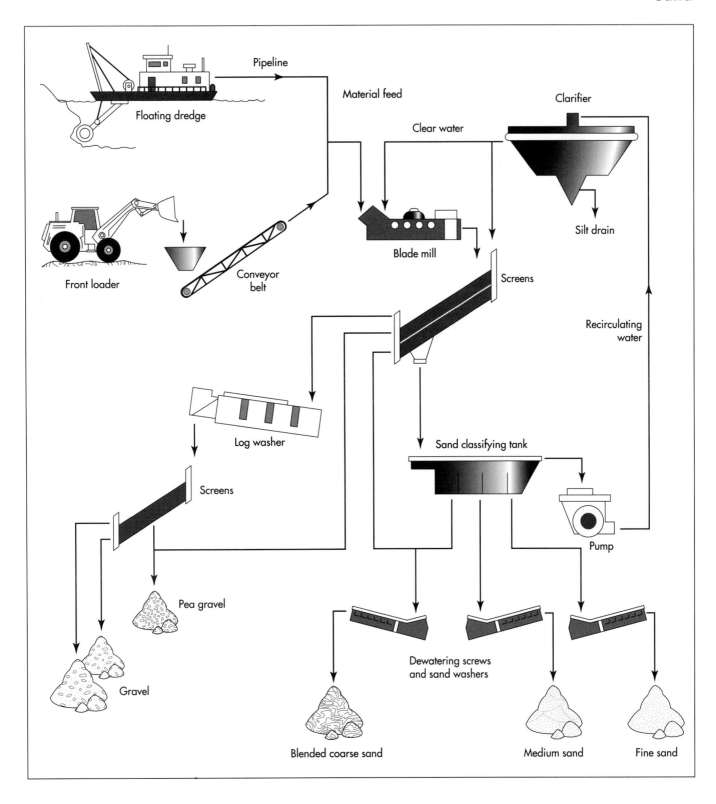

Floating dredge

Pipeline

Material feed

Clear water

Clarifier

Silt drain

Front loader

Conveyor belt

Blade mill

Screens

Recirculating water

Log washer

Screens

Pea gravel

Gravel

Sand classifying tank

Pump

Dewatering screws and sand washers

Blended coarse sand

Medium sand

Fine sand

ensure that the sand and gravel meets those specifications.

The Future

The production of sand and gravel in many areas has come under increasingly stringent restrictions. The United States Army Corps of Engineers, operating under the Federal Clean Water Act, has required permits for sand extraction from rivers, streams, and other waterways. The cost of the special studies required to obtain these permits is often too expensive to allow smaller companies to continue operation. In other cases, residential development in the vicinity of existing aggregate processing plants has led to restrictions regarding noise, dust, and truck traffic. The overall result of these restrictions in certain areas is that sand and gravel used for construction will have to be transported from outside the area at a significantly increased cost in the future.

Where to Learn More

Books

Brady, George S. and Henry R. Clauser. *Materials Handbook, 12th Edition.* McGraw-Hill, 1986.

Hornbostel, Caleb. *Construction Materials, 2nd Edition.* John Wiley and Sons, Inc., 1991.

Siever, Raymond. *Sand.* W.H. Freeman and Company, 1988.

Periodicals

Grover, Jennifer E., Bob Drake, and Steven Prokopy. "100 Years of Rock Products, History of an Industry: 1896-1996." *Rock Products,* July 1996, pp. 29+.

Mack, Walter N. and Elizabeth A. Leistikow. "Sands of the World." *Scientific American,* August 1996, pp. 62-67.

Miller, Russell V. "Changes in Construction Aggregate Availability in Major Urban Areas of California Between the Early 1980s and the Early 1990s." *California Geology,* January/February 1997, pp. 3-17.

—*Chris Cavette*

Saw

Background

A saw is a hand tool with a toothed blade used to cut hard materials such as wood or bone. They are among the oldest known tools. Innovations made over thousands of years are still present in modern, mass-produced examples.

The first flint saws appeared during the early Paleolithic Era, between 60,000 and 10,000 B.C. Stone saws and composite saws made of stone bladelets or "microliths" set into a bone handle also were made during this time. The first metal blades were made possible by the discovery of copper about 4,000 years ago.

As the Iron Age began, the weaker copper and bronze were discarded and raked teeth were finally made possible. Eventually it became apparent that increasing the number of teeth in a saw increased the efficiency of its use. Small saws were used for carpentry, with the Asian style of pull-saws being specifically used by the Ancient Egyptians. Hieroglyphics discovered in Egyptian monuments record the Egyptians' use of the saw in their methods of furniture making. Adjustments in saw design were made according to a saw's intended application. For example, spaced teeth allowed the saw to double as a rake after the cutting stroke, removing sawdust from the developing "kern" or cut.

Saws continued to be improved as innovations in metallurgy were developed. Leonardo da Vinci invented a marble saw during the fifteenth century, and many developers in Europe and abroad took advantage of improvements in steel to create a better cutting edge. Throughout the seventeenth century, the strongest blades were

still the narrowest. The bow saw—named for its structural similarity to the bow and arrow—continued to be popular because of this limitation. The popularity of the wooden frame saw among the early European settlers in America has been attributed to the scarcity of metal in the colonies at that time, as well as to the lack of wide-rolled steel.

With the advent of the Industrial Revolution, stronger, more durable saws were produced. For example, various forms of the circular saw were being made during the early eighteenth century, though the first patent in the United States was granted to Benjamin Cummins of New York in 1814. Today, a wide variety of manual and power saws are produced for consumer as well as commercial use.

Raw Materials

Tempered, high-grade tool steel, alloyed with certain other metals, is the main material used to manufacture the saw blade. Handles used to be made solely of wood, but modern tools can also be made with molded plastic.

Design

There are three major types of hand-held saws: the hacksaw, the bucksaw, and the iconically familiar crosscut or ripsaw.

The crosscut saw cuts across the grain, while the ripsaw cuts along the grain. The teeth of a saw are formulated differently to fulfill different needs. If the angle is too extreme, the teeth will catch on the wood. If the angle is too shallow, the teeth will be unable to cut at all. The teeth of a crosscut saw are angled more obtusely than those of a ripsaw, to slice

Eventually it became apparent that increasing the number of teeth in a saw increased the efficiency of its use.

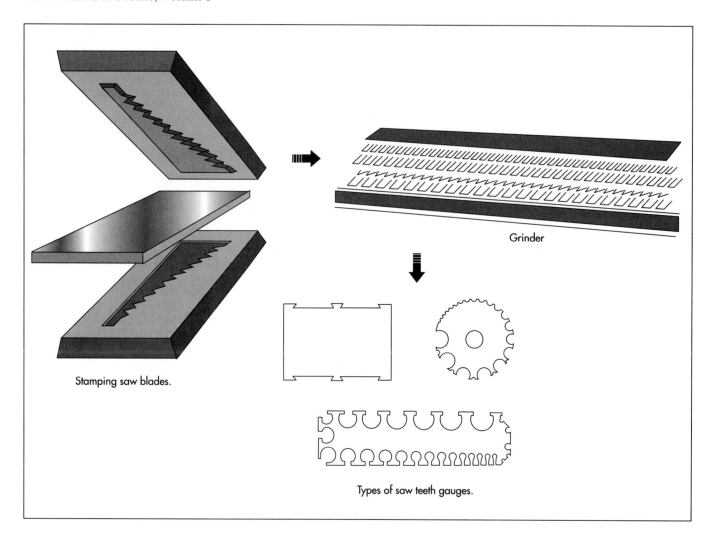

Stamping saw blades.

Grinder

Types of saw teeth gauges.

A special type of steel alloyed with tungsten is produced and rolled into strips. The blades are then stamped out of the alloyed steel using stamping machines. The saw is ground to specifications. Gauges are used to measure the angle of the saw's teeth, which determine its effectiveness in cutting. The blade is then hardened, and the handle is attached.

into the wood grain without chiseling it. Conversely, the cutting edge of the ripsaw is set at right angles to the actual blade, so the teeth act like little chisels. A hardwood saw's teeth are optimally angled at 60 degrees, while softer woods must be cut with teeth set at a more acute angle, generally 45 degrees. Seasoned and green woods also call for differently shaped blades. A coarse saw has about five teeth per inch (two per cm), which is best when cutting green or soft wood. A fine saw, with at least eight teeth per inch (two per cm), can make smooth cuts in seasoned hardwood intended for show, or for more intricate constructions like dovetailing.

The Manufacturing Process

1 A special type of steel alloyed with tungsten is produced and rolled into strips. The thickness of the strip sets the thickness

of the finished blade and is gauged by the same instruments used to measure wires.

2 The blades are then stamped out of the alloyed steel using stamping machines. The overall shape of the handsaw blade narrows from handle to tip. The best saws have a "crown" or curved cutting edge, rather than a straight one, so fewer teeth are in contact with the surface of the wood at any given time while the saw is in motion. Most inexpensive handsaws are of a uniform thickness.

3 The blade is then processed according to standards for optimal use. Depending on the type of saw, different techniques may be applied. The crosscut saw, for instance, is bevel filed. The back of a handsaw is ground thinner than the toothed edge to reduce friction during use. Handsaws are generally taper ground.

4 The "set" or adjustment of the blade's teeth is crucial to the saw's effective-

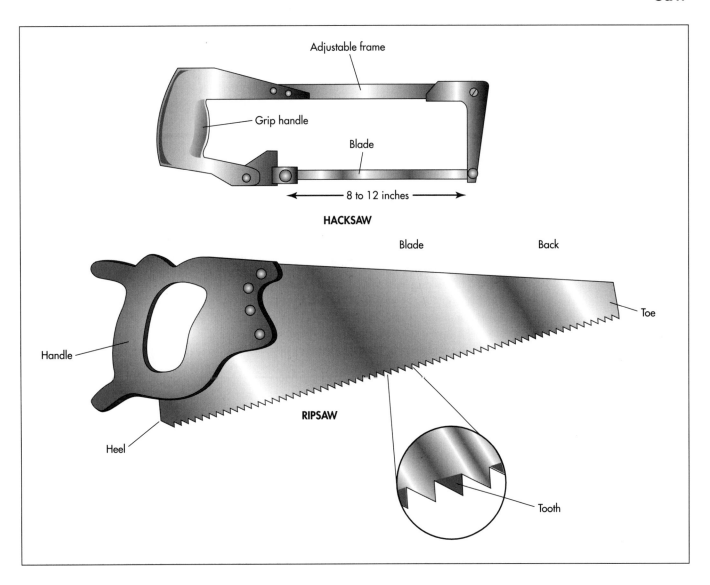

ness, so teeth are measured with a gauge made of plate steel that analyses three teeth at a time. Then, they are cut and bent in alternating directions. Too much angling away from each other, however, results in a saw that does not cut properly.

5 Hardening comes next. The classic technique used for centuries includes hammering the blade to render it "tensioned," so it displays the best combination of stiffness and flexibility. Many variations are possible. Today, professional hacksaws are hardened throughout, while those intended for home use have only their teeth hardened.

6 A coating of rust protection is sprayed on the hardened saw blade.

7 Finally, the blades are fastened to separately made, injection-molded handles.

Quality Control

The American National Standard is intended to regulate the set of blades for safety as well as optimal use, since a badly made saw can be a hazard. Hacksaw blades with 24-32 teeth per inch (10-13 teeth per cm) must be set wavy. Other types of saws require a "raker set" in which every third tooth is left unset. The composite of the metal used is also regulated. A standard steel blade, in order to be called that, cannot be more than 1.25% alloy. For industrial and high-power saws, a high-speed steel blade must be able to withstand a temperature up to 1,000°F (537.78°C).

The Future

Thanks to such user-friendly optimization software as computer aided design (CAD), refinements are being experimented with that may retard the effects of repetitive motion on the handsaw user. "Cumulative Trauma Disorders," as they are known in the field, must be counteracted by ergonomic research. Goals include spreading the impact of using a saw over a larger area, reducing the need for sharp corrective movements, and improving the fit of the handle to avoid uncomfortable hand and arm positions. Volunteer test groups are used to gauge consumer needs and professional users test prototypes of products under development.

The same approach is being made towards improving both large and small scale power tools. Computer-aided manufacturing (CAM), computer-integrated manufacturing (CIM), and numerical control (NC) techniques allow saw manufacturers to cut waste and improve efficiency. Circle saws in the lumber industry have been reinvented with the help of finite element analysis. The Saw Paw Corporation of Pennsylvania holds patents on a recyclable, one-piece carbide saw shank and bit. The streamlined design is geared towards improving efficiency, while lengthening life expectancy, in order to lower the long-term cost of use.

Where to Learn More

Books

Armentrout, Patricia and David. *The Saw.* Rourke Publishers Group, 1995.

Disston, Henry. *The Saw In History.* H. Disston & Sons, Inc., 1922.

Goodman, W. L. *The History of Woodworking Tools.* G. Bell and Sons, Ltd., 1964.

Saws. Miller Freeman Publishers, 1990.

Periodicals

Christianson, Rich. "Computer Advances Drive Panel Saws Forward." *Wood & Wood Products,* October 1996, pp. 83-90.

—*Jennifer Swift Kramer*

Scissors

Background

Scissors are cutting instruments consisting of a pair of metal blades connected in such a way that the blades meet and cut materials placed between them when the handles are brought together. The word shears is used to describe larger instruments of the same kind. As a general rule, scissors have blades less than 6 in (15 cm) long and usually have handles with finger holes of the same size. Shears have blades longer than 6 in (15 cm) and often have one small handle with a hole that fits the thumb and one large handle with a hole that will fit two or more fingers.

Scissors and shears exist in a wide variety of forms depending on their intended uses. Children's scissors, used only on paper, have dull blades to ensure safety. Scissors used to cut hair or fabric must be much sharper. The largest shears are used to cut metal or to trim shrubs and must have very strong blades.

Specialized scissors include sewing scissors, which often have one sharp point and one blunt point for intricate cutting of fabric, and nail scissors, which have curved blades for cutting fingernails and toenails. Special kinds of shears include pinking shears, which have notched blades that cut cloth to give it a wavy edge, and thinning shears, which have teeth that thin hair rather than trim it.

The earliest scissors known to exist appeared in the Middle East about 3,000 or 4,000 years ago and were known as spring scissors. They consisted of two bronze blades connected at the handles by a thin, curved strip of bronze. This strip served to bring the blades together when squeezed and to pull them apart when released. Steel shears of a similar design are still used to cut wool from sheep.

Pivoted scissors of bronze or iron, in which the blades were connected at a point between the tips and the handles, were used in ancient Rome, China, Japan, and Korea. Despite the early invention of this design, still used in almost all modern scissors, spring scissors continued to be used in Europe until the sixteenth century.

During the Middle Ages and Renaissance, spring scissors were made by heating a bar of iron or steel, then flattening and shaping its ends into blades on an anvil. The center of the bar was heated, bent to form the spring, then cooled and reheated to make it flexible. Pivoted scissors were not manufactured in large numbers until 1761, when Robert Hinchliffe of Sheffield, England, began using cast steel to make them. Cast steel, recently invented at the time by Benjamin Huntsman, also of Sheffield, was made by melting steel in clay crucibles and pouring it into molds. This resulted in a more uniform steel with fewer impurities.

During the nineteenth century, scissors were hand-forged with elaborately decorated handles. They were made by hammering steel on indented surfaces known as bosses to form the blades. The rings in the handles, known as bows, were made by punching a hole in the steel and enlarging it with the pointed end of an anvil.

By the beginning of the twentieth century, scissors were simplified in design to accommodate mechanized production. In-

The earliest scissors known to exist appeared in the Middle East about 3,000 or 4,000 years ago and were known as spring scissors.

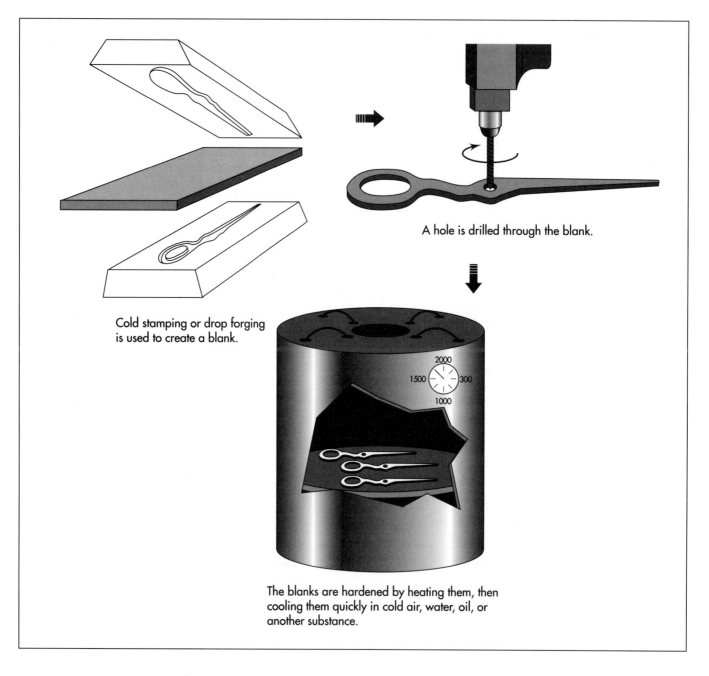

Cold stamping or drop forging is used to create a blank.

A hole is drilled through the blank.

The blanks are hardened by heating them, then cooling them quickly in cold air, water, oil, or another substance.

stead of being forged entirely by hand, blades and handles were now formed by using drop hammers. Powered by steam, these large, heavy devices used dies to shape the scissors from bars of steel. Modern versions of drop hammers are still used to manufacture scissors today.

Raw Materials

Scissors are usually made of steel. Some scissors used for special purposes are made from other metal alloys. Scissors used to cut cordite (an explosive substance resembling twine) must not produce sparks. Scissors used to cut magnetic tape must not interfere with magnetism.

Steel scissors exist in two basic forms. Carbon steel is used to make scissors in which the blade and the handle form one continuous piece. Carbon steel is manufactured from iron and about 1% carbon. It has the advantages of being strong and staying sharp. Scissors made from carbon steel are usually plated with nickel or chromium to prevent them from rusting.

Stainless steel is used to make scissors in which a plastic handle is fitted to the metal blade. Stainless steel is manufactured from

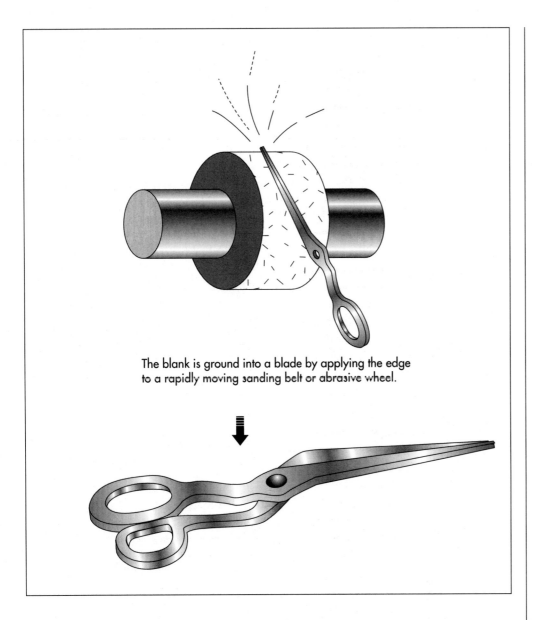

The blank is ground into a blade by applying the edge to a rapidly moving sanding belt or abrasive wheel.

iron, about 1% carbon, and at least 10% chromium. It has the advantages of being light and rustproof. The handles of stainless steel scissors are made from a strong, light substance such as ABS (acrylonitrile-buta-diene-styrene) plastic.

The Manufacturing Process

Making the blanks

1 Before they are sharpened and attached, the two halves of a pair of scissors are known as blanks. A blank may consist of a blade and a handle in one piece or it may consist of only the blade. In the latter case, a metal handle will be welded to the blade or a plastic handle will be attached to it.

2 Inexpensive scissors may be made from blanks formed by cold stamping. In this process, a sharp die in the shape of the blank is stamped into a sheet of unheated steel. The die cuts through the steel to form the blank.

3 Blanks may also be made by molding. Molten steel is poured into a mold in the shape of the blank. The steel cools back into a solid and the blank is removed.

4 Most quality scissors are made from blanks formed by drop forging. Like cold stamping, this process involves shaping the blanks with a die. This die, known as a drop hammer, pounds into a bar of red-hot steel to form the blank. The pressure of the drop hammer also strengthens the steel.

Processing the blanks

5 The blanks are trimmed to the proper shape by cutting away excess metal. A hole is drilled through the blank. This hole will later allow two completed blades to be attached to each other.

6 The trimmed blanks are hardened by heating them, then cooling them quickly in cold air, water, oil, or another substance. The temperature to which they are heated and the medium in which they are cooled varies depending on the type of steel from which they are made and the desired characteristics of the blade.

7 The hardened blanks are heated again and allowed to cool slowly in air. This second heating, known as tempering, gives the blank a uniform hardness. If the blades of a pair of scissors did not have uniform hardness, the harder places on one blade would soon wear out the softer places on the other blade.

8 The repeated heating and cooling causes the blanks to warp. They are straightened by being placed on an anvil and lightly tapped with a hammer. This process is known as peening.

Grinding and polishing

9 The blank is ground into a blade by applying the edge to a rapidly moving sanding belt or abrasive wheel. The surface of the belt or wheel is covered with small particles of an abrasive substance and works in the same way as sandpaper. The hard abrasive grinds away enough steel to form a sharp edge. During this process, the blade is cooled with water or various liquids known as cutting fluids to prevent it from heating and warping. The sharpened blade is then polished in a similar manner using belts or wheels, containing much smaller particles of abrasive.

Making the handles

10 For many scissors, the handles are formed from the start as part of the blank. If not, they may be made of a metal alloy or from plastic. If they are metal, they are made in the same way as the blanks and then welded to them. If they are plastic, they are made by injection molding. In this process, molten plastic is forced under pressure into a mold in the shape of the handles. It is allowed to cool and the mold is opened to remove the handles. The handles contain hollow slots into which the end of the blanks can be inserted. A strong adhesive is used to keep the handle firmly attached.

Assembling the scissors

11 Two polished blades are attached to each other by a rivet or screw through the previously drilled holes. Rivets, which cannot be adjusted by the consumer, are used to make less expensive scissors. Adjustable screws are used in more costly scissors.

12 The scissors are adjusted to ensure that the two blades work together correctly. They may be painted or plated with nickel or chrome to protect them from rust. The scissors are inspected for flaws, the screw or rivet is lubricated, and the scissors are wrapped for shipping to consumers.

Quality Control

The most important aspect of quality control for scissors is the proper alignment of the two blades. In order for scissors to cut smoothly, the blades must meet at two points only. These two points are the swivel (the point where the rivet or screw connects the blades) and the cutting point. The cutting point moves from just beyond the swivel to the tip as the scissors are closed. The blades are prevented from meeting at any other points by giving them a slight horizontal and vertical curve away from each other during manufacture.

In order to ensure that the blades meet correctly, the holes must be drilled to within one ten-thousandth of an inch (about one four-hundredth of a millimeter) of the correct position. The position of the blades is inspected visually to see if the blades meet evenly. If not, a portion of one blade will overlap the other. This defect is known as a wing. The tips are also inspected to ensure that they meet evenly, without a gap between them or any overlap.

Because even dull scissors are able to cut paper adequately, quality scissors are tested on tough synthetic fabrics. Sharpness is tested by making sure the blades cut the fabric rather than tear it. Strength is tested by cutting through multiple layers of fabric. The blades should come together with a constant pressure during cutting.

The consumer is responsible for maintaining the quality of the scissors. Scissors should only be used to cut the materials for which they were designed. They should be oiled and sharpened regularly, and the screw should be adjusted as necessary. Scissors should be stored in a closed position. Setting down scissors in an open position is the most common cause of dull blades.

The Future

Although scissors have remained in a standard form for hundreds of years, recent innovations may change the look of this ordinary household tool. Scissors using round, rolling blades have been designed. Ceramics made from zirconium oxide have been used to manufacture scissors with blades which are extremely strong, rustproof, and which never need sharpening.

Where to Learn More

Periodicals

"Scissors and Shears." *Consumer Reports*, October 1992, pp. 672-677.

Werner, Karen Flake. "Cutting With Scissors: Three Steps to Easy Snipping." *Parents Magazine*, January 1996, pp. 137-138.

Other

Allison, John. "The Anatomy of Quality Scissors." Knife Connection. May 30, 1996. http://www.knife.com/news/scissor.htm (July 14, 1997).

—*Rose Secrest*

Scratch and Sniff

The pull-apart perfume strip was introduced in 1981, and has since become the prevalent form of sampling new perfume.

Background

Scratch 'N Sniff™ is the trade name for a special kind of perfume or scent saturated printing in which the scent is enclosed in minute capsules, which can be broken open by friction. Individual beads of scented oil too small to be seen with the naked eye are encapsulated in plastic or gelatin, and with specialized printing techniques, the beads are printed on paper. The scent does not leak out until the beads are deliberately broken. Because scratch and sniff patches keep a scent localized—it can be smelled only when some one deliberately scratches and sniffs—scents can be used in printing without overwhelming the surroundings.

Scratch and sniff is popular in children's books, where the reader can scratch and sniff a picture of a cookie to smell the fresh-baked scent, for example. It is also widely used in advertising, where it may capture the scent of a new car or rubber tire, lumber, burnt match, flowers, the smell of a particular detergent or medicine, mushrooms, ham, ketchup, butter, mildew, or a host of other scents. Micro-encapsulated scent is most prevalent in perfume advertising. A strip of paper printed with micro-encapsulated perfume oil and tacked shut at the border of the advertisement is generally used instead of the open patch of scratch and sniff. The consumer tears open the strip, thus breaking the capsules in two and releasing the scent.

History

The micro-encapsulation technology that makes scratch and sniff possible was discovered by scientists endeavoring to make carbonless paper. Before the era of the word processor and photocopy machine, typists inserted carbon paper between second and third sheets of white paper to make multiple copies of documents. This could be a messy and aggravating process. In the early 1960s, an organic chemist, Gale Matson, working for 3M (Minnesota Mining and Manufacturing Company) patented a micro-encapsulation process that could be used to make ink copies without carbon paper. The Matson process used a particular plastic called polyoxymethylene urea (PMU). A researcher at National Cash Register came up with a similar micro-encapsulation process using gelatin. Both scientists were thinking only of carbonless paper, but the marketing department at 3M was given the task of finding alternate uses for the technology Matson had patented. It soon became clear that micro-encapsulation could be used for scented oil, and Scratch 'N Sniff™ debuted in 1965. The pull-apart perfume strip was introduced in 1981, and has since become the prevalent form of sampling new perfume.

Raw Materials

The basic ingredients of scratch and sniff or perfumed strips are water, oil, scent, and either gelatin or a water soluble polymer, usually polyoxymethylene urea. A certain chemical catalyst is used to bring about the reaction. A water-soluble adhesive is needed to affix the material to the paper during printing.

The Manufacturing Process

Reacting

1 The micro-encapsulation process is done in a large vat or kettle called a reactor. First, scented oil is added to a solution of

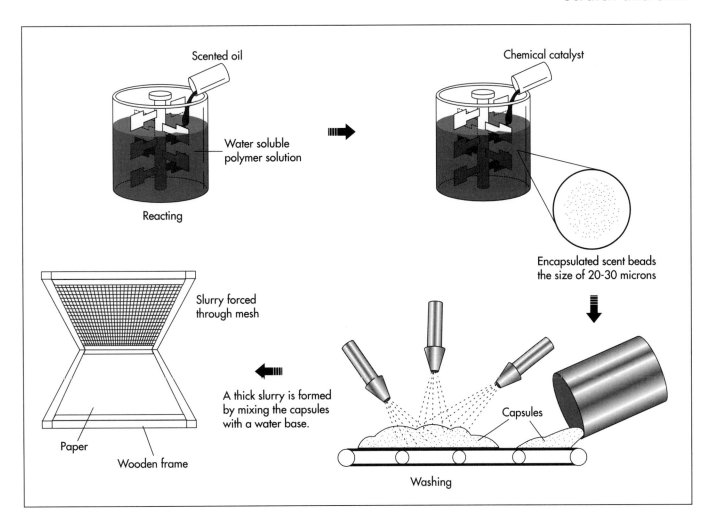

Scented oil

Water soluble
polymer solution

Reacting

Chemical catalyst

Encapsulated scent beads
the size of 20-30 microns

Slurry forced
through mesh

A thick slurry is formed
by mixing the capsules
with a water base.

Paper

Wooden frame

Capsules

Washing

water-soluble polymer in the reactor. At this stage, the oil and water do not mix, and are separate like oil and vinegar in a salad dressing. Then, the mixture is blended at high speed by means of a rotary blade. This part of the process is called high shear agitation. As the rotor blade mixes the oil and water, the oil breaks down into very small beads or droplets. After about 12 hours of agitation, the beads reach the size of from 20 to 30 microns. At this size, the individual beads are not visible to the naked eye, and their width is less than the diameter of a human hair.

Adding the catalyst

2 When the beads have reached the proper size, the agitation is stopped, and a chemical catalyst is added. The catalyst causes the molecular weight of the polymer to increase, and the polymer becomes water insoluble. The polymer then precipitates out of the water and "rains down" on the oil droplets. The precipitate forms a plastic (or gelatin) shell around each individual oil

bead. The oil beads have been encapsulated. At this point, the reaction is stopped.

Washing

3 The next step is to remove the capsules from the reactor and wash them. The capsules are removed to a belt, which moves them through a spray of water. The scratch and sniff capsules are washed only once, whereas perfume capsules are washed twice.

Making the slurry

4 The washed capsules are then loaded into a tank and mixed with a water base. A thick slurry is formed. For scratch and sniff, adhesives are added to the slurry. At this point, the slurry can be applied to paper using a variety of printing processes.

Printing

5 There are four basic methods of printing used with the micro-encapsulated slurry. It can be silk-screened onto paper or ap-

First, scented oil is added to a solution of water-soluble polymer in a reactor. Then, the mixture is blended at high speed by means of a rotary blade until the beads of oil are very small. A chemical catalyst is added, and the polymer surrounds each bead of oil, forming a shell. The newly formed capsules are washed once for scratch and sniff, twice for fragrance strips. Micro-encapsulated scents may be applied to paper in different ways. To make scratch and sniff papers, the slurry of capsules and water can be silk-screened.

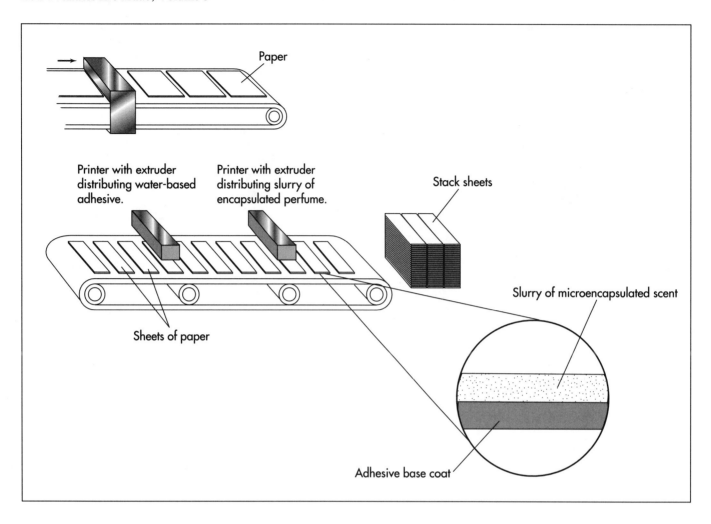

Paper

Printer with extruder distributing water-based adhesive.

Printer with extruder distributing slurry of encapsulated perfume.

Stack sheets

Slurry of microencapsulated scent

Sheets of paper

Adhesive base coat

In order to make fragrance strips, precut papers are pretreated with a base coat of special water-based adhesive which is extruded through pumps on the printer. Then, the base coat must be dried in an oven. After drying, the slurry of micro-encapsulated perfume is extruded onto the strip.

plied to a pattern gluer for web offset printing. Scratch and sniff is often used for stickers, and these are usually produced through a type of printing known as flexographic printing. The strips of fragrance commonly used in advertisements are produced by extrusion. The fragrance is extruded onto paper in a process similar to the way adhesive is extruded onto tape. The extrusion process is complex and requires state-of-the-art printing equipment. Usually presses capable of handling perfume strip printing have advanced computer controls. To produce the fragrance strip, the paper has to be pre-treated with a base coat of special water-based adhesive which is extruded through pumps on the printer. Then, the base coat must be dried in an oven. After drying, the slurry of micro-encapsulated perfume is extruded onto the strip.

Quality Control

The micro-capsules are subjected to many laboratory tests to determine their strength

and longevity under stressful conditions. They may be frozen or subjected to steam, and then examined under magnification. Finished printing for a customer is also checked to make sure the scent released is the correct scent and the correct strength. In the case of perfumed strips, if the scent is too weak or too strong, the printers may adjust the width of the strip, and the adhesion of the strip may be altered by adding or subtracting adhesive to the base coat.

The Future

The biggest growth in micro-encapsulating technology has been in perfume strips, possibly because of advancing printing capabilities. Fast, large-scale computer-operated printers with specialized extruder heads are necessary for accurate printing of perfume strips. There have been several new developments in perfume strips in the 1990s, such as pressure sensitive labels and strips that open to reveal tiny pearls of perfume powder that can be applied directly to skin. Future advances seem dependent on the coordination of perfumers, printing technology, and micro-encapsulation technology, in order to make commercially viable products.

Where to Learn More

Periodicals

Charbonneau, Jack and Keith Relyea. "The Technology Behind On-Page Fragrance Sampling." *Drug and Cosmetic Industry*, February 1997, pp. 48-52.

—*Angela Woodward*

Screw

Screws are part of a family of threaded fasteners that includes bolts and studs as well as specialized screws like carpenter's wood screws and the automotive cap screw.

Background

Screws are part of a family of threaded fasteners that includes bolts and studs as well as specialized screws like carpenter's wood screws and the automotive cap screw. The threads (or grooves) can run right handed or left, tapered, straight, or parallel. There are two types of screws, machine and wood screws. Both are made of metal, however the machine screw has a constant diameter and joins with nuts while the wood screw is tapered and grips to the actual wood surface.

History

Even though the concept of the screw dates back to around 200 B.C., the actual metal screw that is known today was not developed until the Renaissance. Early screws had to be handmade, so no two screws were ever alike. The time consuming process of hand filing the threads into the screw form made mass production and use virtually impossible. In 1586, the introduction of the first screw-cutting machine by Jacques Besson, court engineer for Charles IX of France, paved the way for more innovations.

Inspired by earlier designers and makers of scientific instruments like microscopes, clockmakers and gunsmiths led the way in screw-cutting machine design. In 1760, Job and William Wyatt, two English brothers, filed a patent for the first automatic screw-cutting device. Their machine could cut 10 screws per minute and was considered one of the precursors to mass production machinery.

During the early nineteenth century, Englishman Henry Maudslay produced the method of screw manufacture still in use today. His machine was the first power-driven, screw-cutting lathe. In the United States, at the same time, David Wilkinson also built a screw-cutting lathe and was awarded the first American screw patent. New innovations followed soon after. In 1845, Stephen Finch developed a turret lathe, and soon after the Civil War, Christopher Walker invented a fully automatic lathe.

The first screw factory, Aborn and Jackson, was opened in Rhode Island in 1810. By 1895 screw makers in America were forming unions and demanding a minimum wage of $1.75 per ten-hour day for a member and $1.25 for an apprentice. Smaller scale innovations continued to be made to improve efficiency. John E. Sweet devised the angular thread-cutting method to cut an entire thread from one side.

Today, machining of screws has been superseded by thread rolling. In 1836, American William Keane developed the thread rolling process, but at the time it had little success. The iron metal that was used to create the thread-rolled screws was too low grade and had the tendency to split during the die-cutting process. The eventual need to mass produce screws at a fraction of the cost of machining led to the reevaluation and establishment of the thread-rolling manufacture of screws.

Raw Materials

Screws are generally made from low to medium carbon steel wire, but other tough and inexpensive metals may be substituted, such as stainless steel, brass, nickel alloys, or aluminum alloy. Quality of the metal used is of utmost importance in order to avoid

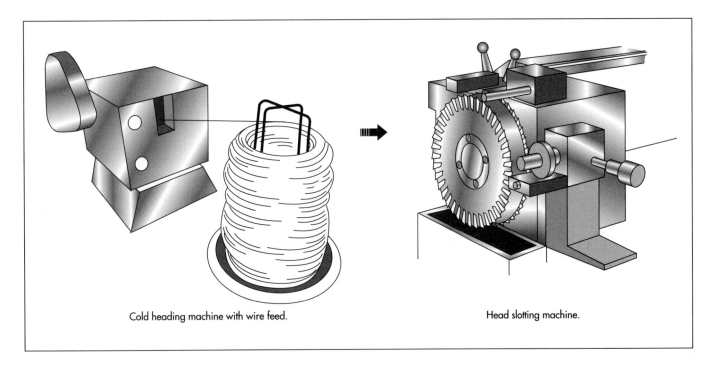

Cold heading machine with wire feed.

Head slotting machine.

cracking. If a finish is applied to the screw, it must be of a compatible makeup. Steel may be coated or plated with zinc, cadmium, nickel, or chromium for extra protection.

Design

On a single thread screw, the lead and pitch are identical, lead is twice the pitch on a double thread model, and three times as much on a triple thread. The pitch of a screw is the distance between two threads (or grooves) from the same point on each thread. It is also more commonly known as the number of threads per inch or centimeter. The lead of the screw measures how far it is driven in for each revolution.

The Manufacturing Process

Machining is only used on unique designs or with screws too small to be made any other way. The machining process is exact, but too time consuming, wasteful, and expensive. The bulk of all screws are mass manufactured using the thread rolling method, and that is the procedure described in further detail.

Cold heading

1 Wire is fed from a mechanical coil through a prestraightening machine. The straightened wire flows directly into a machine that automatically cuts the wire at a designated length and die cuts the head of the screw blank into a preprogrammed shape. The heading machine utilizes either an open or closed die that either requires one punch or two punches to create the screw head. The closed (or solid) die creates a more accurate screw blank. On average, the cold heading machine produces 100 to 550 screw blanks per minute.

Thread rolling

2 Once cold headed, the screw blanks are automatically fed to the thread-cutting dies from a vibrating hopper. The hopper guides the screw blanks down a chute to the dies, while making sure they are in the correct feed position.

3 The blank is then cut using one of three techniques. In the reciprocating die, two flat dies are used to cut the screw thread. One die is stationary, while the other moves in a reciprocating manner, and the screw blank is rolled between the two. When a centerless cylindrical die is used, the screw blank is rolled between two to three round dies in order to create the finished thread. The final method of thread rolling is the planetary rotary die process. It holds the screw blank stationary, while several die-cutting machines roll around the blank.

The cold heading machine cuts a length of wire and makes two blows on the end, forming a head. In the head slotting machine, the screw blanks are clamped in the grooves around the perimeter of the wheel. A circular cutter slots the screws as the wheel revolves.

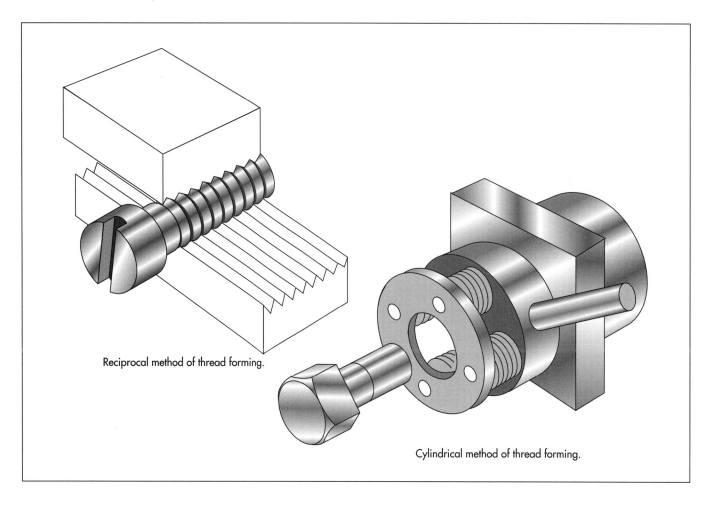

Reciprocal method of thread forming.

Cylindrical method of thread forming.

Threads can be cut into the blank by several methods. In the reciprocal method, the screw blank is rolled between two dies. In the cylindrical method, it is turned in the center of several rollers.

All three methods create higher quality screws than the machine-cut variety. This is because the thread is not literally cut into the blank during the thread-rolling process, rather it is impressed into the blank. Thus, no metal material is lost, and weakness in the metal is avoided. The threads are also more precisely positioned. The more productive of the thread-rolling techniques is by far the planetary rotary die, which creates screws at a speed of 60 to 2,000 parts per minute.

Quality Control

The National Screw Thread Commission established a standard for screw threads in 1928 for interchangeability. This was followed by an international Declaration of Accord in 1948, adopting a Unified Screw Thread system. The standards focus on three main elements: the number of threads per inch, the designated pitch and shape of the thread, and designated diameter sizes. In 1966, the International Standards Organization (ISO) suggested a universal restriction on threads to ISO metric and inch size ranges with coarse and fine pitches. Compliance with the ISO suggested standards has been global.

Where to Learn More

Books

Brittania Company. *Screws and Screw-Making.* James H. Wood, 1892.

Camm, F. J. *Screw Cutting.* Cassell and Company Ltd., 1920.

Glover, David. *Screws.* Rigby Education, 1997.

Periodicals

Koepfer, Chris. "Technology Gamble Pays Off." *Modern Machine Shop,* February 1995, pp. 94-104.

—Jennifer Swift Kramer

Sewing Machine

Background

Before 1900, women spent many of their daylight hours sewing clothes for themselves and their families by hand. Women also formed the majority of the labor force that sewed clothes in factories and wove fabrics in mills. The invention and proliferation of the sewing machine freed women of this chore, liberated workers from poorly paid long hours in factories, and produced a wide variety of less expensive clothing. The industrial sewing machine made a range of products possible and affordable. The home and portable sewing machines also introduced amateur seamstresses to the delights of sewing as a craft.

History

The pioneers in the development of the sewing machine were hard at work at the end of the eighteenth century in England, France, and the United States. The English cabinetmaker Thomas Saint garnered the first patent for a sewing machine in 1790. Leather and canvas could be stitched by this heavy machine, which used a notched needle and awl to create a chain stitch. Like many early machines, it copied the motions of hand sewing. In 1807, a critical innovation was patented by William and Edward Chapman in England. Their sewing machine used a needle with an eye in the point of the needle instead of at the top.

In France, Barthelémy Thimmonier's machine patented in 1830 literally caused a riot. A French tailor, Thimmonier developed a machine that stitched fabric together by chain stitching with a curved needle. His factory produced uniforms for the French Army and had 80 machines at work by 1841. A mob of tailors displaced by the factory rioted, destroyed the machines, and nearly killed Thimmonier.

Across the Atlantic, Walter Hunt made a machine with an eye-pointed needle that created a locked stitch with a second thread from underneath. Hunt's machine, devised in 1834, was never patented. Elias Howe, credited as the inventor of the sewing machine, designed and patented his creation in 1846. Howe was employed at a machine shop in Boston and was trying to support his family. A friend helped him financially while he perfected his invention, which also produced a lock stitch by using an eye-pointed needle and a bobbin that carried the second thread. Howe tried to market his machine in England, but, while he was overseas, others copied his invention. When he returned in 1849, he was again backed financially while he sued the other companies for patent infringement. By 1854, he had won the suits, thus also establishing the sewing machine as a landmark device in the evolution of patent law.

Chief among Howe's competitors was Isaac M. Singer, an inventor, actor, and mechanic who modified a poor design developed by others and obtained his own patent in 1851. His design featured an overhanging arm that positioned the needle over a flat table so the cloth could be worked under the bar in any direction. So many patents for assorted features of sewing machines had been issued by the early 1850s that a "patent pool" was established by four manufacturers so the rights of the pooled patents could be purchased. Howe benefited from this by earning royalties on his patents; Singer, in

Elias Howe tried to market his sewing machine in England, but, while he was overseas, others copied his invention. When he returned in 1849, he sued the other companies for patent infringement. By 1854, he had won the suits, thus also establishing the sewing machine as a landmark device in the evolution of patent law.

partnership with Edward Clark, merged the best of the pooled inventions and became the largest producer of sewing machines in the world by 1860. Massive orders for Civil War uniforms created a huge demand for the machines in the 1860s, and the patent pool made Howe and Singer the first millionaire inventors in the world.

Improvements to the sewing machine continued into the 1850s. Allen B. Wilson, an American cabinetmaker, devised two significant features, the rotary hook shuttle and four-motion (up, down, back, and forward) feed of fabric through the machine. Singer modified his invention until his death in 1875 and obtained many other patents for improvements and new features. As Howe revolutionized the patent world, Singer made great strides in merchandising. Through installment purchase plans, credit, a repair service, and a trade-in policy, Singer introduced the sewing machine to many homes and established sales techniques that were adopted by salesmen from other industries.

The sewing machine changed the face of industry by creating the new field of ready-to-wear clothing. Improvements to the carpeting industry, bookbinding, the boot and shoe trade, hosiery manufacture, and upholstery and furniture making multiplied with the application of the industrial sewing machine. Industrial machines used the swing-needle or zigzag stitch before 1900, although it took many years for this stitch to be adapted to the home machine. Electric sewing machines were first introduced by Singer in 1889. Modern electronic devices use computer technology to create buttonholes, embroidery, overcast seams, blind stitching, and an array of decorative stitches.

Raw Materials

Industrial machine

Industrial sewing machines require cast iron for their frames and a variety of metals for their fittings. Steel, brass, and a number of alloys are needed to make specialized parts that are durable enough for long hours of use in factory conditions. Some manufacturers cast, machine, and tool their own metal parts; but vendors also supply these

parts as well as pneumatic, electric, and electronic elements.

Home sewing machine

Unlike the industrial machine, the home sewing machine is prized for its versatility, flexibility, and portability. Lightweight housings are important, and most home machines have casings made of plastics and polymers that are light, easy to mold, easy to clean, and resistant to chipping and cracking. The frame of the home machine is made of injection-molded aluminum, again for weight considerations. Other metals, such as copper, chrome, and nickel are used to plate specific parts.

The home machine also requires an electric motor, a variety of precision-machined metal parts including feed gears, cam mechanisms, hooks, needles, and the needle bar, presser feet, and the main drive shaft. Bobbins can be made of metal or plastic but must be precisely shaped to feed the second thread properly. Circuit boards are also required specific to the main controls of the machine, the pattern and stitch selections, and a range of other features. Motors, machined metal parts, and circuit boards can be supplied by vendors or made by the manufacturers.

Design

Industrial machine

After the automobile, the sewing machine is the most precisely made machine in the world. Industrial sewing machines are larger and heavier than home machines and are designed to perform only one function. Manufacturers of clothing, for example, use a series of machines with distinct functions that, in succession, create a finished garment. Industrial machines also tend to apply chain or zigzag stitch rather than lock stitch, but machines may be fitted for up to nine threads for strength.

Makers of industrial machines may supply a single-function machine to several hundred garment plants all over the world. Consequently, field-testing in the customer's factory is an important element in design. To develop a new machine or make changes in a current model, customers are surveyed, the competition is evaluated, and the nature of the desired improvements (such as faster or quieter machines) are identified. Designs are drawn, and a prototype is made and tested in the customer's plant. If the prototype is satisfactory, the manufacturing engineering section takes over the design to coordinate tolerance of parts, identify parts to be manufactured in-house and the raw materials needed, locate parts to be provided by vendors, and purchase those components. Tools for manufacture, holding fixtures for the assembly line, safety devices for both the machine and the assembly line, and other elements of the manufacturing process must also be designed along with the machine itself.

When the design is complete and all parts are available, a first production run is scheduled. The first manufactured lot is carefully checked. Often, changes are identified, the design is returned to development, and the process is repeated until the product is satisfactory. A pilot lot of 10 or 20 machines is then released to a customer to use in production for three to six months. Such field tests prove the device under real conditions, after which larger scale manufacture can begin.

Home sewing machine

Design of the home machine begins in the home. Consumer focus groups learn from sewers the types of new features that are most desired. The research and development (R&D) department of a manufacturer works, in conjunction with the marketing department, to develop specifications for a new machine that is then designed as a prototype. Software for manufacturing the machine is developed, and working models are made and tested by users. Meanwhile, R&D engineers test the working models for durability and establish useful life criteria. In the sewing laboratory, stitch quality is precisely evaluated, and other performance tests are conducted under controlled conditions.

When the new machine is approved for production, product engineers develop manufacturing methods for the production of machine parts. They also identify the raw materials needed and the parts that are to be ordered from outside sources. Parts made in the factory are put into production as soon as the materials and plans are available.

A 1899 trade card for Singer sewing machines. (From the collections of Henry Ford Museum & Greenfield Village.)

Isaac Merritt Singer did not invent the sewing machine. He was not even a master mechanic, but an actor by trade. So, what was Singer's contribution that caused his name to become synonymous with sewing machines?

Singer's genius was in his vigorous marketing campaign, directed from the beginning at women and intended to combat the attitude that women did not and could not use machines. When Singer introduced his first home sewing machines in 1856, he confronted resistance from American families for both financial and psychological reasons. It was actually Singer's business partner, Edward Clark, who devised the innovative "hire/purchase plan" to alleviate initial reluctance on financial grounds. This plan allowed families who could not afford the $125 investment for a new sewing machine (the average family income only equaled about $500) to purchase the machine by paying in three- to five-dollar monthly installments.

Psychological impediments proved more difficult to overcome. Labor-saving devices in the home were a new concept in the 1850s. Why would women need these machines? What would they do with the time saved? Wasn't work done by hand of better quality? Weren't machines too taxing on women's minds and bodies, and weren't they just too closely associated with man's work and man's world outside the home? Singer tirelessly devised strategies to combat these attitudes, including advertising directly to women. He set up lavish showrooms that simulated elegant domestic parlors; he employed women to demonstrate and teach machine operations; and he used advertising to describe how women's increased free time could be seen as a positive virtue.

Donna R. Braden

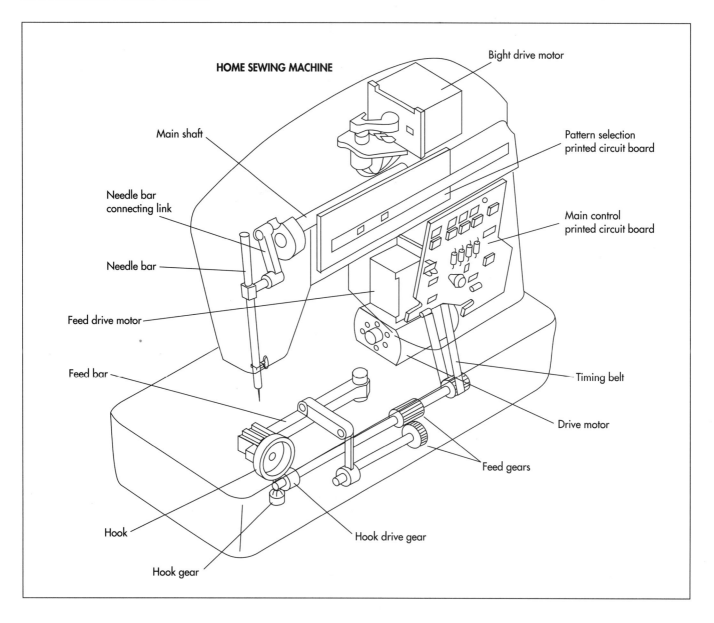

HOME SEWING MACHINE

Bight drive motor

Main shaft

Pattern selection printed circuit board

Needle bar connecting link

Main control printed circuit board

Needle bar

Feed drive motor

Feed bar

Timing belt

Drive motor

Feed gears

Hook

Hook drive gear

Hook gear

Unlike the industrial machine, the home sewing machine is prized for its versatility, flexibility, and portability. Lightweight housings are important, and most home machines have casings made of plastics and polymers that are light, easy to mold, easy to clean, and resistant to chipping and cracking.

The Manufacturing Process

Industrial machine

1 The basic part of the industrial machine is called the "bit" or frame and is the housing that characterizes the machine. The bit is made of cast iron on a computer numerical control (CNC) machine that creates the casting with the appropriate holes for inserting components. Manufacture of the bit requires steel castings, forging using bar steel, heat-treating, grinding, and polishing to finish the frame to the specifications needed to house the components.

2 Motors are usually not supplied by the manufacturer but are added by a suppli-

er. International differences in voltage and other mechanical and electrical standards make this approach more practical.

3 Pneumatic or electronic components may be produced by the manufacturer or supplied by vendors. For industrial machines, these are typically made of metal rather than plastic parts. Electronic components are not necessary in most industrial machines because of their single, specialized functions.

Home sewing machine

Parts production in the factory may include a number of precisely made components of the sewing machine.

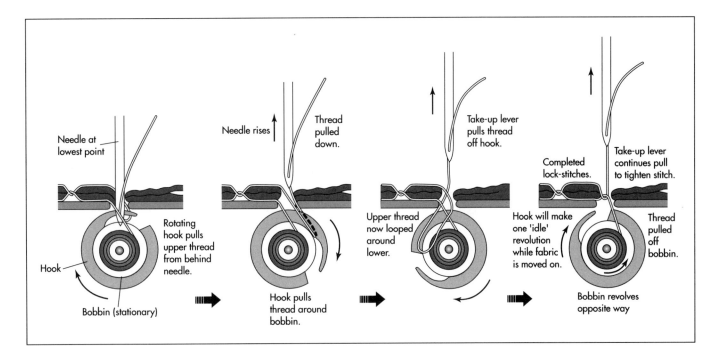

Needle at lowest point

Rotating hook pulls upper thread from behind needle.

Hook

Bobbin (stationary)

Needle rises

Thread pulled down.

Hook pulls thread around bobbin.

Take-up lever pulls thread off hook.

Upper thread now looped around lower.

Completed lock-stitches.

Hook will make one 'idle' revolution while fabric is moved on.

Take-up lever continues pull to tighten stitch.

Thread pulled off bobbin.

Bobbin revolves opposite way

How a sewing machine works.

4 Gears are made of injection-molded synthetics or may be specially tooled to suit the machine.

5 Drive shafts made of metal are hardened, ground, and tested for accuracy; some parts are plated with metals and alloys for specific uses or to provide suitable surfaces.

6 The presser feet are made for particular sewing applications and can be interchangeable on the machine. Appropriate grooves, bevels, and holes are machined into the feet for their application. The finished presser foot is hand polished and plated with nickel.

7 The frame for the home sewing machine is made of injection-molded aluminum. High-speed cutting tools equipped with ceramic, carbide, or diamond-edged blades are used to drill holes and to mill cuts and recesses to house features of the machine.

8 Covers for the machines are manufactured from high-impact synthetics. They are also precision-molded to fit around and protect the machine's components. Small, single parts are preassembled into modules, whenever possible.

9 The electronic circuit boards that control the machine's many operations are produced by high-speed robotics; they are then subjected to a burn-in period that is several hours long and are tested individually before being assembled in the machines.

10 All of the parts that are preassembled join a main assembly line. Robots move the frames from operation to operation, and teams of assemblers fit the modules and components into the machine until it is complete. The assembly teams take pride in their product and are responsible for purchasing the components, assembling them, and making quality control checks until the machines are completed. As a final quality check, every machine is tested for safety and various sewing procedures.

11 The home sewing machines are sent to packing where they are separately assembled by power control units that are foot-operated. A variety of accessories and instruction manuals are packed with the individual machines. The packaged products are shipped to local distribution centers.

Quality Control

The quality control department inspects all raw materials and all components furnished by suppliers when they arrive at the factory. These items are matched with plans and specifications. The parts are again checked along every step of manufacture by the makers, receivers, or persons who add the components along the assembly line. Independent quality control inspectors examine the product at various stages of assembly and when it is finished.

Byproducts/Waste

No byproducts result from sewing machine manufacture, although a number of specialized machines or models may be produced at one plant. Waste is also minimized. Steel, brass, and other metals are salvaged and melted down for precision castings whenever possible. Remaining metal waste is sold to a salvage dealer.

The Future

The merging of the capabilities of the electronic sewing machine and the software industry is creating an ever-widening range of creative features for this versatile machine. Efforts have been made to develop threadless machines that inject thermal fluids that harden with heat to finish seams, but these may fall outside the definition of "sewing." Large embroideries can be machine-produced based on designs developed onscreen using AUTOCAD or other design software. The software allows the designer to shrink, enlarge, rotate, mirror designs, and select colors and types of stitches that can then be embroidered on materials ranging from satin to leather to make products like baseball caps and jackets. The speed of the process lets products celebrating today's victories hit the street by tomorrow's business day. Because such features are add-ons, the home sewer can buy a basic home sewing machine and enhance it over the years with only those features most frequently used or of interest. Sewing machines become individual crafting devices and, therefore, seem to have a future as promising as the imagination of the operator.

Where to Learn More

Books

Finniston, Monty, ed. *Oxford Illustrated Encyclopedia of Invention and Technology.* Oxford University Press, 1992.

Travers, Bridget, ed. *World of Invention.* Gale Research, 1994.

Periodicals

Allen, O. "The power of patents." *American Heritage,* September/October 1990, p. 46.

Foote, Timothy. "1846." *Smithsonian,* April. 1996, p. 38.

Schwarz, Frederic D. "1846." *American Heritage,* September 1996, p. 101

—*Gillian S. Holmes*

Shampoo

Shampoos are cleaning formulations used for a wide range of applications, including personal care, pet use, and carpets. Most are manufactured in roughly the same manner. They are composed primarily of chemicals called surfactants that have the special ability to surround oily materials on surfaces and allow them to be rinsed away by water. Most commonly, shampoos are used for personal care, especially for washing the hair.

History

Before the advent of shampoos, people typically used soap for personal care. However, soap had the distinct disadvantages of being irritating to the eyes and incompatible with hard water, which made it leave a dull-looking film on the hair. In the early 1930s, the first synthetic detergent shampoo was introduced, although it still had some disadvantages. The 1960s brought the detergent technology we use today.

Over the years, many improvements have been made to shampoo formulations. New detergents are less irritating to the eyes and skin and have improved health and environmental qualities. Also, materials technology has advanced, enabling the incorporation of thousands of beneficial ingredients in shampoos, leaving hair feeling cleaner and better conditioned.

Raw Materials

New shampoos are initially created by cosmetic chemists in the laboratory. These scientists begin by determining what characteristics the shampoo formula will have. They must decide on aesthetic features such as how thick it should be, what color it will

be, and what it will smell like. They also consider performance attributes, such as how well it cleans, what the foam looks like, and how irritating it will be. Consumer testing often helps determine what these characteristics should be.

Once the features of the shampoo are identified, a formula is created in the laboratory. These initial batches are made in small beakers using various ingredients. In the personal care industry, nearly all of the ingredients that can be used are classified by the Cosmetic, Toiletry, and Fragrance Association (CTFA) in the governmentally approved collection known as the International Nomenclature of Cosmetic Ingredients (INCI). The more important ingredients in shampoo formulations are water, detergents, foam boosters, thickeners, conditioning agents, preservatives, modifiers, and special additives.

Water

The primary ingredient in all shampoos is water, typically making up about 70-80% of the entire formula. Deionized water, which is specially treated to remove various particles and ions, is used in shampoos. The source of the water can be underground wells, lakes, or rivers.

Detergents

The next most abundant ingredients in shampoos are the primary detergents. These materials, also known as surfactants, are the cleansing ingredients in shampoos. Surfactants are surface active ingredients, meaning they can interact with a surface. The chemical nature of a surfactant allows it to

Over the years, many improvements have been made to shampoo formulations. New detergents are less irritating to the eyes and skin and have improved health and environmental qualities.

surround and trap oily materials from surfaces. One portion of the molecule is oil compatible (soluble) while the other is water soluble. When a shampoo is applied to hair or textiles, the oil soluble portion aligns with the oily materials while the water soluble portion aligns in the water layer. When a number of surfactant molecules line up like this, they form a structure known as a micelle. This micelle has oil trapped in the middle and can be washed away with water, thus giving the shampoo its cleansing power.

Surfactants are derived from compounds known as fatty acids. Fatty acids are naturally occurring materials which are found in various plant and animal sources. The materials used most often to make the surfactants used in shampoos are extracted from coconut oil, palm kernel oil, and soy bean oil. Some common primary detergents used in shampoos are ammonium lauryl sulfate, sodium lauryl sulfate, and sodium lauryl ether sulfate.

Foam boosters

In addition to cleansing surfactants, other types of surfactants are added to shampoos to improve the foaming characteristics of the formulation. These materials, called alkanolamides, help increase the amount of foam and the size of the bubbles. Like primary detergents, they are also derived from fatty acids and have both water soluble and oil soluble characteristics. Typical materials include lauramide DEA or cocamide DEA.

Thickeners

To some extent, the alkanolamides that make shampoos foam also make the formulations thicker. However, other materials are also used to increase the viscosity. For example, methylcellulose, derived from plant cellulose, is included in shampoos to make them thicker. Sodium chloride (salt) also can be used to increase shampoo thickness.

Conditioning agents

Some materials are also added to shampoos to offset the sometimes harsh effect of surfactants on hair and fabrics. Typical conditioning agents include polymers, silicones, and quaternary agents. Each of these compounds deposit on the surface of the hair and improve its feel, softness, and comba-

bility, while reducing static charge. Shampoos that specifically feature conditioning as a benefit are called 2-in-1 shampoos because they clean and condition hair in the same step. Examples of conditioning agents include guar hydroxypropyltrimonium chloride which is a polymer, dimethicone which is a silicone, and quaternium 80, a quaternary agent.

Preservatives

Since shampoos are made from water and organic compounds, contamination from bacteria and other microbes is possible. Preservatives are added to prevent such growth. Two of the most common preservatives used in shampoos are DMDM hydantoin and methylparaben.

Modifiers

Other ingredients are added to shampoo formulas to modify specific characteristics. Opacifiers are added to make the formula opaque and give it a pearly look. Materials known as sequestering agents are added to offset the dulling effects of hard water. Acids or bases such as citric acid or sodium hydroxide are added to adjust the pH of a shampoo so the detergents will provide optimal cleaning.

Special additives

One of the primary factors that influence the purchase of a shampoo is its color and odor. To modify these characteristics, manufacturers add fragrance oils and governmentally approved and certified FD&C dyes. Other special additives can also have a similar effect. Natural materials such as botanical extracts, natural oils, proteins, and vitamins all impart special qualities and help sell shampoos. Additives such as zinc pyrithione are included to address the problem of dandruff. Other additives are dyes which can color the hair.

The Manufacturing Process

After a shampoo formula is developed, it is tested to ensure that its qualities will minimally change over time. This type of testing, called stability testing, is primarily used to detect physical changes in such things as color, odor, and thickness. It can

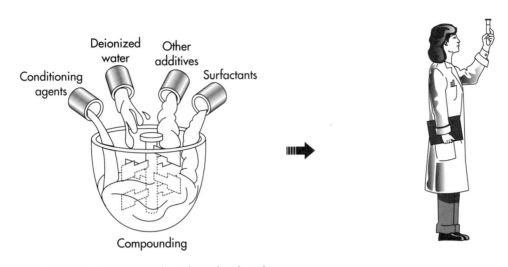

Conditioning agents

Deionized water

Other additives

Surfactants

Compounding

Raw materials are poured into large batch tank and thoroughly mixed. The mixture may be heated or cooled to facilitate blending.

A sample from the batch is sent to the Quality Control testing lab to ensure it meets product specifications.

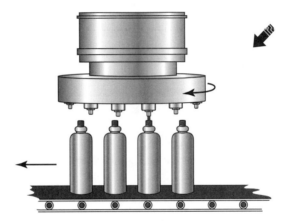

After the batch is approved, the shampoo goes to the filler, which delivers a measured amount into empty bottles as they move down a conveyor belt.

also provide information about other changes, like microbial contamination and performance differences. This testing is done to ensure that the bottle of shampoo that is on the store shelves will perform just like the bottle created in the laboratory.

The manufacturing process can be broken down into two steps. First a large batch of shampoo is made, and then the batch is packaged in individual bottles.

Compounding

1 Large batches of shampoo are made in a designated area of the manufacturing plant. Here workers, known as compounders, follow the formula instructions to make batches that can be 3,000 gal (11,000 l) or more. Raw materials, which are typically provided in drums as large as 55 gal (200 l) or in 50-lb (23-kg) bags, are delivered to the compounding area via forklift trucks. They are poured into the batch tank and thoroughly mixed.

2 Depending on the formula, these batches can be heated and cooled as necessary to help the raw materials combine more quickly. Some raw materials such as water or the primary detergents are pumped and metered directly into the batch tank.

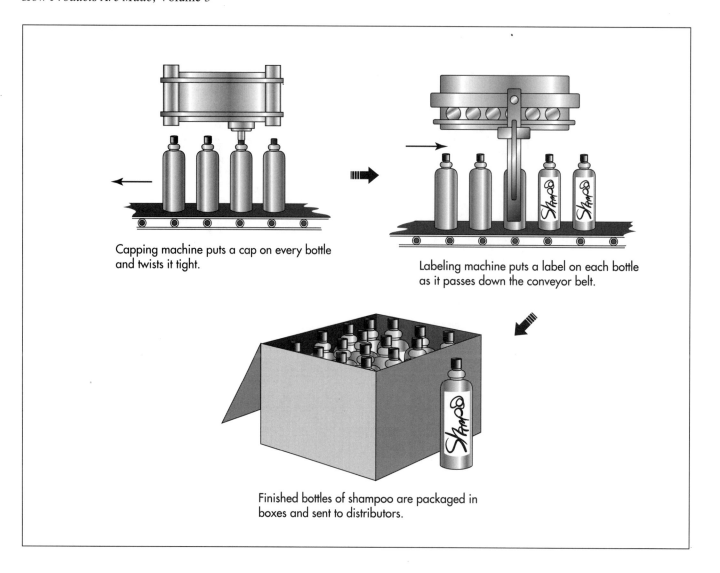

Capping machine puts a cap on every bottle and twists it tight.

Labeling machine puts a label on each bottle as it passes down the conveyor belt.

Finished bottles of shampoo are packaged in boxes and sent to distributors.

These materials are added simply by pressing a button on computerized controls. These controls also regulate the mixing speeds and the heating and cooling rates. Depending on the size and type of shampoo, making a 3,000-gal (11,000-l) batch can take anywhere from one to four hours.

Quality control check

3 After all the ingredients are added to the batch, a sample is taken to the Quality Control (QC) lab for testing. Physical characteristics are checked to make sure the batch adheres to the specifications outlined in the formula instructions. The QC group runs tests such as pH determination, viscosity checks, and appearance and odor evaluations. They can also check the amount of detergent that is in the formula and whether there is enough preservative. If the batch is found to be "out of spec," adjustments can

be made. For instance, acids or bases can be added to adjust the pH, or salt can be added to modify the viscosity. Colors can also be adjusted by adding more dye.

4 After a batch is approved by QC, it is pumped out of the main batch tank into a holding tank where it can be stored until the filling lines are ready. From the holding tank it gets pumped into the filler, which is made up of a carousel of piston filling heads.

Filling

5 At the start of the filling line, empty bottles are put in a large bin called a hopper. Here, the bottles are physically manipulated until they are correctly oriented and standing upright. They are then moved along a conveyor belt to the filling carousel, which holds the shampoo.

6 The filling carousel is made up of a series of piston filling heads that are calibrated to deliver exactly the correct amount of shampoo into the bottles. As the bottles move through this section of the filling line, they are filled with shampoo.

7 From here the bottles move to the capping machine. Much like the bin that holds the empty bottles, the caps are also put in a hopper and then correctly aligned. As the bottles move by the caps are put on and twisted tight.

8 After the caps are put on, the bottles move to the labeling machines (if necessary). Depending on the type of labels, they can either be stuck on using adhesives or heat pressed. Labels are stuck to the bottles as they pass by.

9 From the labeling area, the bottles move to the boxing area, where they are put into boxes, typically a dozen at a time. These boxes are then stacked onto pallets and hauled away in large trucks to distributors. Production lines like this can move at speeds of about 200 bottles a minute or more.

Quality Control

In addition to the initial checks to make sure the product meets specifications, other quality control checks are made. For example, line inspectors watch the bottles at specific points on the filling line to make sure everything looks right. They notice things like fill levels, label placement, and whether the cap is on correctly. The product is also routinely checked to see if there has been any microbial contamination. This is done by taking a bottle off the filling line and sending it to the QC lab. Here, a small amount of the shampoo product is smeared onto a plate and inoculated with bacteria and other organisms to see if they grow. Additionally, the packaging is also checked to see if it meets specifications. Things such as bottle thickness, appearance, and bottle weight are all checked.

The Future

Consumer product corporations will continue to manufacture new types of shampoos. These new formulas will be driven by ever-changing consumer desires and developing chemical technology. Currently, consumers like multi-functional shampoos, such as 2-in-1 shampoos, which provide cleansing and conditioning in one step, or shampoos that aid in styling. New shampoos will likely provide improved conditioning, styling, and coloring while cleaning the hair.

Shampoo technology will also improve as new ingredients are developed by raw material suppliers. Some important advances are being made in the development of compounds such as polymers, silicones, and surfactants. These materials will be less irritating, less expensive, more environmentally friendly, and also provide greater functionality and performance.

Where to Learn More

Books

Knowlton, John and Steven Pearce. *The Handbook of Cosmetic Science and Technology.* Elsevier Science Publishers, 1993.

Umbach, Wilfried. *Cosmetics and Toiletries Development, Production, and Use.* Ellis Horwood, 1991.

Periodicals

Kintish, Lisa. "Shampoos Get Specific." *Soap/Cosmetic/Chemical Specialties,* October 1995, pp. 20-30.

—Perry Romanowski

Shingle

Before being used in the manufacture of shingles, asphalt must be oxidized by a process called blowing. This is done by bubbling air through heated asphalt to which appropriate catalysts have been added, causing a chemical reaction.

Background

Roofing shingles are made from several types of materials. Wood shingles are sawed from red cedar or pine. Modern shingles are cut from new growth trees and must be treated with chemical preservatives to make them last as well as earlier versions that were cut from old growth trees. They must also be chemically treated to achieve a fire resistance rating comparable to other types of shingles; in fact, the highest rating can be attained only by installing them over a special subsurface layer. Aluminum shingles have a long life span, although they are comparatively expensive. Asphalt shingles cover about 80% of the homes in the United States. Their popularity is due to their relatively light weight, comparatively low cost, ease of installation, and low maintenance requirements.

A typical asphalt shingle is a rectangle about 12-18 in (30-46 cm) wide and 36-40 in (91-102 cm) long. Popular styles have several cutouts along one edge to form tabs that simulate smaller, individual shingles. Three tabs are common, but the number may range from two to five. Some styles are made to interlock with adjacent shingles during installation, creating a more wind-resistant surface.

History

Asphalt has been used as a building material for thousands of years. Ancient Babylonians used it as mortar between clay bricks and as a waterproofing liner in canals. Roll roofing, consisting of long strips of asphalt-coated felt with a finishing layer of finely crushed stone, has been manufactured in the United States since 1893. In 1903, Henry M. Reynolds began marketing asphalt shingles he cut from sheets of roll roofing. By the 1920s, this roofing material had become so popular it was sold through mail-order catalogs. By the 1950s, the typical asphalt shingle looked much as it does today, including the tab-forming cutouts.

Since the late 1950s, manufacturers have sought to develop inorganic base materials as alternatives to the traditional organic felt. Inorganic bases are desirable because they are more fire resistant than an organic base; furthermore, they absorb less asphalt during the manufacturing process, so the resulting shingles weigh less. Asbestos was used in shingle mats until its related health risks became well known. Improvements in fiberglass matting have made them the most popular asphalt shingle base material in the industry since the late 1970s.

Raw Materials

Asphalt shingles are sometimes called composite shingles. Their foundation is a base of either organic felt or fiberglass. Organic felt mats are made of cellulose fibers obtained from recycled waste paper or wood. These fibers are reduced to a water-based pulp, formed into sheets, dried, cut into strips, and wound onto rolls. Thinner, lighter shingles with a higher resistance to fire are made on a base of fiberglass. In a typical process, the fiberglass membrane is made by chopping fine, glass filaments and mixing them with water to form a pulp, which is formed into a sheet. The water is then vacuumed out of the pulp, and a binder is applied to the mat. After curing, the mat is sliced to appropriate widths and rolled.

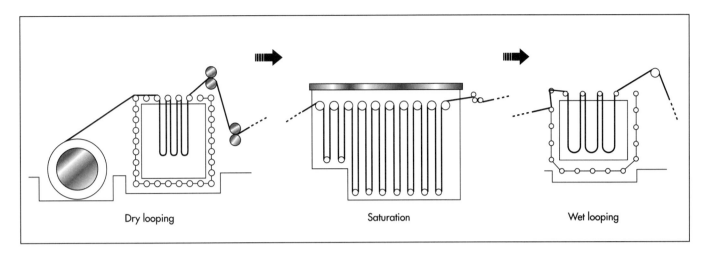

Dry looping   Saturation   Wet looping

Asphalt, a very thick hydrocarbon substance, can be obtained either from naturally occurring deposits or, more commonly, as a byproduct of crude oil refining. Before being used in the manufacture of shingles, asphalt must be oxidized by a process called blowing. This is done by bubbling air through heated asphalt to which appropriate catalysts have been added, causing a chemical reaction. The resulting form of asphalt softens the right amount at the right temperatures to make good shingles. To further process the blown asphalt into a proper coating material, a mineral stabilizer such as fly ash or finely ground limestone is added. This makes the material more durable and more resistant to fire and weather.

Various colors of ceramic-coated mineral granules are used as a top coat on shingles to protect them from the sun's ultraviolet rays, increase their resistance to fire, and add an attractive finish. The granules may be small rocks or particles of slag (a byproduct of ore smelting). Shingles designed for use in humid locations may include some copper-containing granules in the top coat to inhibit the growth of algae on the roof. The back surface of the shingles is coated with sand, talc, or fine particles of mica to keep the shingles from sticking together during storage.

Strips or spots of a thermoplastic adhesive are applied to most shingles during the manufacturing process. Once installed on a roof, the shingles are heated by the sun, and this adhesive is activated to bond overlapping shingles together for increased wind resistance.

The Manufacturing Process

Asphalt shingles are produced by passing the base material through a machine that successively adds the other components. The same machine can be used to make either shingles or roll roofing.

Dry looping

1 A jumbo roll [6 ft (1.83 m) in diameter] of either organic felt or fiberglass mat is mounted and fed into the roofing machine. The base material first passes through a dry looper. Matting is accumulated accordion-style in this reservoir, so that the machine can continue to operate when the supply roll is exhausted and a new one is mounted.

Saturation

2 The base material passes through a pre-saturation chamber, where it is sprayed on one side with hot asphalt to drive out any moisture that may be present. It then goes into a saturator tank filled with hot asphalt. Soaking in the asphalt coats the fibers within the mat and fills the voids between them.

Wet looping

3 The matting is again formed into accordion-like folds. While the asphalt coating on the mat cools, it is drawn into the felt, creating an even greater degree of saturation.

Coating

4 Coating asphalt, which has been stabilized with powdered minerals, is applied to both surfaces of the mat. The mat passes

To make shingles, a roll of organic felt or fiberglass mat is mounted and fed into a dry looper. The material passes through a presaturation chamber, then goes into a saturator tank filled with hot asphalt, which coats the fibers. If needed the material passes through the wet looping machine.

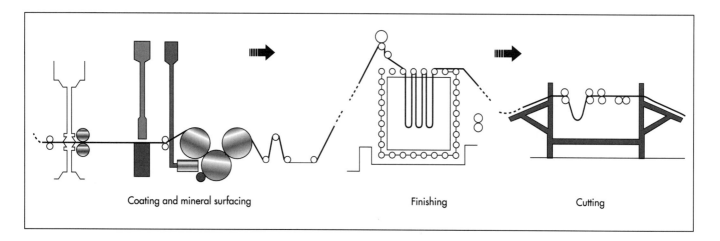

Coating and mineral surfacing   Finishing   Cutting

The mat passes between a pair of asphalt coating rolls. After, a coating of fine mineral particles is applied to the back surface of the mat, which then passes through a series of rollers that embed the coating particles in the asphalt and cool the material. The shingles are finished by cooling and cutting them to size.

between a pair of coating rolls, which are separated by an appropriate distance to ensure that the desired amount of coating asphalt is applied to the mat. This step may be sufficient to coat fiberglass filaments and fill voids between them; in this case, the saturation and wet looping steps can be bypassed.

Mineral surfacing

5 Granules of ceramic-coated minerals of the desired color are applied to the top surface of the asphalt-coated mat. A coating of fine particles of a mineral, such as talc or mica, is applied to the back surface of the mat. The sheet of treated mat then passes through a series of rollers that embed the coating particles in the asphalt and cool the material.

Finishing

6 The strip of roofing material is accumulated, accordion-style, on a cooling looper to finish cooling.

Cutting

7 The sheet of finished shingle material then passes into a cutting machine, where it is cut from the back side into the desired size and shape. The machine separates the shingles and stacks them in bundles. Bundles generally contain enough shingles to roof 25-35 sq ft (7.62-10.67 sq m).

Packaging

8 The bundles of shingles are transferred to equipment that wraps them and affixes labels.

Quality Control

Quality control begins during the manufacture of the base material. Not only must the material perform well in the final product, but it must also have enough tensile (pulling) strength and tear resistance to withstand the shingle-making process. In addition to these strength characteristics, organic mats are continuously monitored to ensure proper moisture content and absorbency. Fiberglass mats are monitored for proper fiber distribution and uniform weight.

During the manufacturing of the shingles themselves, numerous characteristics are monitored, including uniformity and thickness of the asphalt coating. Application of the mineral coatings is monitored for uniform distribution and proper embedding of the particles. Inspections of the finished shingles verify proper weight, count, size, and color of the product.

In addition to their own quality control measures, shingle manufacturers may also invite third-party, independent testing laboratories to their plants and send them product samples. These testing laboratories verify the product's compliance with ratings defined by Underwriters Laboratories (UL) and the American Society for Testing Materials (ASTM). Such ratings indicate that the product meets specified standards for fire resistance and other performance characteristics like strength and wind resistance.

Byproducts/Waste

Asphalt-coated waste from the manufacturing process includes shingle fragments and waste generated by cutting tabs on the shingles. In some cases, this waste is sold for use in making asphalt pavement for roads. For example, the state of North Carolina has allowed such material to be used in asphalt pavement since 1995 and has not encountered any problems with making or placing the paving material. Because of shipping costs, however, this option may not be practical unless a paving asphalt processing plant is located relatively close to the shingle plant.

Where to Learn More

Books

Gibson, Barbara G., ed. Sunset Books. *Roofing & Siding*. Lane Publishing Co., 1981.

Herbert, R.D., III. *Roofing: Design Criteria, Options, Selection*. R.S. Means Company, 1989.

Residential Asphalt Roofing Manual. Asphalt Roofing Manufacturers Association, 1993.

Other

"Frequently Asked Questions." *Certain-Teed Corporation*. http://www.certainteed.com/roofing/ctroof/faq.html (7 May 1997).

—*Loretta Hall*

Silver

Silver is the whitest metallic element. It is rare, strong, corrosion resistant, and unaffected by moisture, vegetable acids, or alkalis. Silver is also resonant, moldable, malleable, and possesses the highest thermal and electric conductivity of any substance. The chemical symbol for silver is Ag, from the Latin argentum, which means white and shining.

Background

Silver was one of the earliest metals known to humans, and it has been considered a precious metal since ancient times. Silver has been used as a form of currency by more people throughout history than any other metal, even gold. Although it is usually found in ores with less rare metals, such as copper, lead, and zinc, silver was apparently discovered in nugget form, called native silver, about 4000 B.C. Silver utensils and ornaments have been found in ancient tombs of Chaldea, Mesopotamia, Egypt, China, Persia, and Greece. In more recent times, the principal uses for silver were coinage and silverware.

In 1993, worldwide production of silver from mines totaled 548.2 million ounces (15.5 billion grams). During that year, Mexico was the world's largest producer of silver, with a total production of 75.7 million ounces (2.1 billion grams). The United States was the second leading producer, followed by Canada, Australia, Spain, Peru, and Russia. The vast majority of the world's silver is used in industrial applications, and the United States is the leading consumer. Other top consumers include Japan, India, and eastern European countries.

Silver mining in North America dates back to the eighteenth century. Around 1800, production began in the United States on the east coast and then moved west. The mining of silver was instrumental in the settlement of the state of Nevada. In 1994, Nevada was the largest producer of silver in the United States; Nevada mines produced 22.8 million troy ounces (709 million grams) of silver. Arizona, California, and Nevada are known for large-tonnage, low-grade silver deposits.

Physical Characteristics and Uses of Silver

Silver is the whitest metallic element. It is rare, strong, corrosion resistant, and unaffected by moisture, vegetable acids, or alkalis. Silver is also resonant, moldable, malleable, and possesses the highest thermal and electric conductivity of any substance. The chemical symbol for silver is Ag, from the Latin *argentum*, which means white and shining. Although silver does not react to many chemicals, it does react with sulfur, which is always present in the air, even in trace amounts. The reaction causes silver to tarnish, therefore, it must be polished periodically to retain its luster.

Silver possesses many special physical characteristics and qualities that make it useful in a variety of industries. The photography industry is the biggest user of silver compounds. Silver forms the most light-sensitive salts, or halides, which are essential to developing high-quality photography. Silver has the highest electrical conductivity per unit volume of any metal, including copper, so it is used extensively in electronics. Specialized uses include switch and relay contacts for automobile controls and accessories, automotive window heating, and in electrodes for electrocardiograms.

Silver is one of the strongest oxidants, making it an essential catalyst for the chemical process industry. It is used in the production of adhesives, dinnerware, mylar recording tape, and many other products. Silver is the most reflective of all metals,

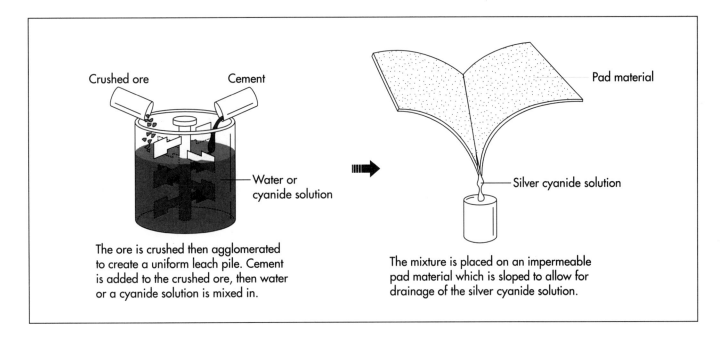

Crushed ore   Cement

Water or
cyanide solution

The ore is crushed then agglomerated
to create a uniform leach pile. Cement
is added to the crushed ore, then water
or a cyanide solution is mixed in.

Pad material

Silver cyanide solution

The mixture is placed on an impermeable
pad material which is sloped to allow for
drainage of the silver cyanide solution.

and is used to coat glass in mirrors. It is also used in x-ray vacuum tubes and as material for bearings. With the highest level of thermal conductivity among metals and resistance to combustion and sparks, silver is a valuable material for a range of other industrial processes. The most common consumer application of silver is its use in jewelry. Pure silver, which would be too soft to be durable, is mixed with 5-20% copper in an alloy known as sterling silver.

Today, a very small percentage of the world's silver is used in coinage, though silver coins were a popular form of currency until the recent past. As industrialized nations began to produce large numbers of silver coins in the twentieth century, silver became less available, and therefore more expensive. The United States Treasury, which until then had been minting 90% silver coins, changed their minting by a 1965 act of Congress. The Johnson Silver Coinage Act completely demonetized silver, and with the exception of bicentennial coins, all newly-minted United States coins are now made of an alloy of copper and nickel.

The Manufacturing Process

Silver was first obtained in sixteenth-century Mexico by a method called the patio process. It involved mixing silver ore, salt, copper sulphide, and water. The resultant silver chloride was then picked up by adding mercury. This inefficient method was superseded by the von Patera process. In this process, ore was heated with rock salt, producing silver chloride, which was leached out with sodium hyposulfite. Today, there are several processes used to extract silver from ores.

A method called the cyanide, or heap leach, process has gained acceptance within the mining industry because it is a low-cost way of processing lower-grade silver ores. However, the ores used in this method must have certain characteristics: the silver particles must be small; the silver must react with cyanide solutions; the silver ores must be relatively free of other mineral contaminants and/or foreign substances that might interfere with the cyanidation process; and the silver must be free from sulfide minerals. The idea for cyanidation actually dates back to the eighteenth century, when Spanish miners percolated acid solutions through large heaps of copper oxide ore. The process developed into its present form during the late nineteenth century. The cyanide process is described here.

Preparing the ore

1 Silver ore is crushed into pieces, usually with 1-1.5 in (2.5-3.75 cm) diameters, to make the material porous. Approximately 3-5 lb (1.4-2.3 kg) of lime per ton of silver ore is added to create an alkaline environ-

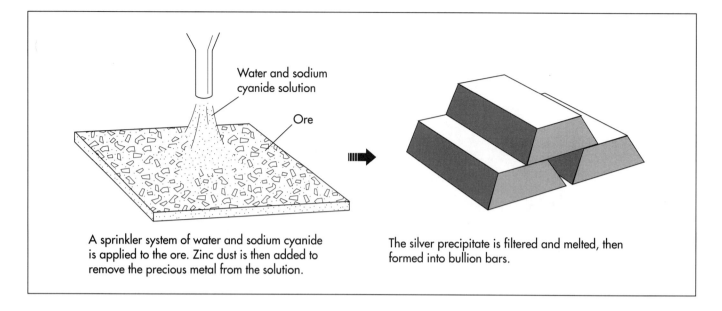

A sprinkler system of water and sodium cyanide is applied to the ore. Zinc dust is then added to remove the precious metal from the solution.

The silver precipitate is filtered and melted, then formed into bullion bars.

ment. The ore must be completely oxidized so the precious metal is not confined in sulfide minerals. Where fines or clays exist, the ore is agglomerated to create a uniform leach pile. This process consists of crushing the ore, adding cement, mixing, adding water or a cyanide solution, and curing in dry air for 24-48 hours.

2 Broken or crushed ore is stacked on impermeable pads to eliminate the loss of the silver cyanide solution. Pad material may be asphalt, plastic, rubber sheeting, and/or clays. These pads are sloped in two directions to facilitate drainage and the collection of the solutions.

Adding the cyanide solution and curing

3 A solution of water and sodium cyanide is added to the ore. Solutions are delivered to the heaps by sprinkler systems or methods of ponding, including ditches, injection, or seepage from capillaries.

Recovering the silver

4 Silver is recovered from heap leach solutions in one of several ways. Most common is Merrill-Crowe precipitation, which uses fine zinc dust to precipitate the precious metal from the solution. The silver precipitate is then filtered off, melted, and made into bullion bars.

5 Other methods of recovery are activated carbon absorption, where solutions are pumped through tanks or towers containing activated carbon, and the addition of a sodium sulfide solution, which forms a silver precipitate. In another method, the solution is passed through charged resin materials which attract the silver. The recovery method is generally decided based on economic factors.

Silver is rarely found alone, but mostly in ores which also contain lead, copper, gold, and other metals which may be commercially valuable. Silver emerges as a byproduct of processing these metals. To recover silver from zinc-bearing ores, the Parkes process is used. In this method, the ore is heated until it becomes molten. As the mixture of metals is allowed to cool, a crust of zinc and silver forms on the surface. The crust is removed, and the metals undergo a distillation process to remove the zinc from the silver.

To extract silver from copper-containing ores, an electrolytic refining process is used. The ore is placed in an electrolytic cell, which contains a positive electrode, or anode, and a negative electrode, or cathode, in an electrolyte solution. When electricity is passed through the solution, silver, with other metals, accumulates as a slime at the anode while copper is deposited on the cathode. The slimes are collected, then roasted, leached, and smelted to remove impurities. The metals are formed into blocks which are used as anodes in another round of electrolysis. As electricity is sent through a solution of silver nitrate, pure silver is deposited onto the cathode.

The Future

How much silver will be produced in the future depends on many factors, including the rate of production of other metals and future uses of silver. Industrial demand for silver appears to be steady overall. Because silver naturally occurs with other metals, future production is linked to the production of copper, lead, gold, and zinc.

In the future, silver will likely continue to be used for special industrial applications, as well as for consumer items, such as jewelry and silverware. In addition to these traditional uses, the value of silver will also depend on new uses for the metal. For example, using silver as a sanitizing agent is currently under development. Manufacturers have hustled in response to studies by the Atlanta-based Center for Disease Control that many viruses, including those linked to Acquired Immunodeficiency Syndrome (AIDS), will survive briefly outside an individual in fluids deposited on surfaces of plastic products, such as telephones. Matsushita Electric Industrial Co., Ltd. in Osaka, Japan, completed a project at the Research Institute for Microbial Diseases, Osaka University, to produce a surface treatment that provides long-lasting sanitization for its plastic products. Research revealed the most effective system to be a compound based on silver thiosulfate.

Currently marketed under the name Amenitop, the system consists of silica gel micros-pheres that contain a silver thiosulfate complex. The silica gel coating allows a gradual release of the silver compound onto the surface, which provides long-lasting sanitization. Studies suggest that Amenitop kills bacteria and viruses by destroying the cell's membranes.

Where to Learn More

Books

Coombs, Charles. *Gold and Other Precious Metals*. William Morrow and Company, 1981.

Robbins, Peter and Douglass Lee. *Guide to Precious Metals and Their Markets*. Nichols Publishing, 1979.

Smith, Jerome F. and Barbara Kelly Smith. *What's Behind the New Boom In Silver and How to Maximize Your Profits*. Griffin Publishing Company, 1983.

Periodicals

"A Listing of the End Users of Silver By Property." *The Silver Institute*, January 13, 1994, pp. 1-14.

"A Nevada Leader." *News from Las Vegas (Las Vegas News Bureau)*, March 1995, p. 1.

"New Silver Compound To Fight Spread of Viruses." *The Silver Institute Letter*, December 1994-January 1995, pp. 1, 2, 4.

Thorstad, Linda E. "How Heap Leaching Changed The West." *World Investment News*, February 1987, pp. 31, 33.

—*Susan Bard Hall*

Slinky Toy

Consumers have found ingenious uses for slinky toys, employing them in pecan-picking machines and using them as drapery holders, antennas, light fixtures, window decorations, gutter protectors, pigeon repellers, birdhouse protectors, therapeutic devices, wave motion coils, table decorations, and mail holders.

Background

The Slinky toy is a coil of wire or plastic that has the ability to "walk" on its own, usually down a flight of stairs.

History

Toys of infinite varieties have amused children, and some adults, of every civilization. Before the advent of the toy industry, playthings were made at home or by accomplished crafters. The industry grew from the area around Nuremburg, Germany, beginning in the Middle Ages. Toys were finely crafted of gold, silver, brass, iron, tin, wood, silk, and leather. With the Industrial Revolution came new machinery and production methods. Wood-pulp paper, aniline dyes, and chromolithography were used to mass produce storybooks, card and board games, and paper dolls. Advanced metal-metalworking techniques were exhibited in the plethora of steel-plated toys such as trains, boats, and animals.

By the late nineteenth century, three major toy companies in the United States, Milton Bradley, Selchow & Righter, and Parker Brothers, had been established. The intrusion of World War I and II severely curtailed European trade, and contributed to the rise of the toy industry in the United States.

During the latter period of that era the Slinky toy was invented, quite by accident. Richard James was a marine engineer employed at the Cramp Shipyard outside Philadelphia, Pennsylvania, during World War II. One day in 1943, as James was devising a spring that would hold shipboard marine torsion meters steady, one of the springs fell off his desk, springing end over end across the floor. Impressed and intrigued by the simplicity of it, James went home and told his wife he had a great idea for a new toy. He refined the coil of steel until he was satisfied with the end-over-end movement. The finished prototype was a 2.5 in (6.35 cm) stack of 98 coils. Betty James pored over the dictionary to find an appropriate name for the gadget, eventually deciding on the word slinky. Two years later, the Jameses borrowed $500 to manufacture a small inventory of slinky toys. Their initial attempts to sell the springy toy through small retail outlets were unsuccessful, and it seemed that they would have to abandon the project. Then, at the height of the Christmas holiday season, Gimbels department store agreed to sell 400 of the slinky toys. Richard James appeared in the store's toy department one night to demonstrate the toy and within an hour and a half, the entire inventory was sold out.

By the following year, Richard James had quit his job at the shipyard and the couple opened a factory to mass produce the slinky toys. At the 1946 American Toy Fair, slinky toys were a popular item. In the 1950s, business soared and the Jameses were monetarily quite successful. Richard James also became somewhat of an eccentric, joining a religious cult and tithing much of his money to it. In 1960, he left his wife Betty, their six children, and a very successful business to live in Bolivia.

Betty moved the factory to her hometown of Hollidaysburg, a small town outside Altoona, Pennsylvania, where the company remains today. The original machinery that Richard designed and engineered to twist the steel into a Slinky is still used. The orig-

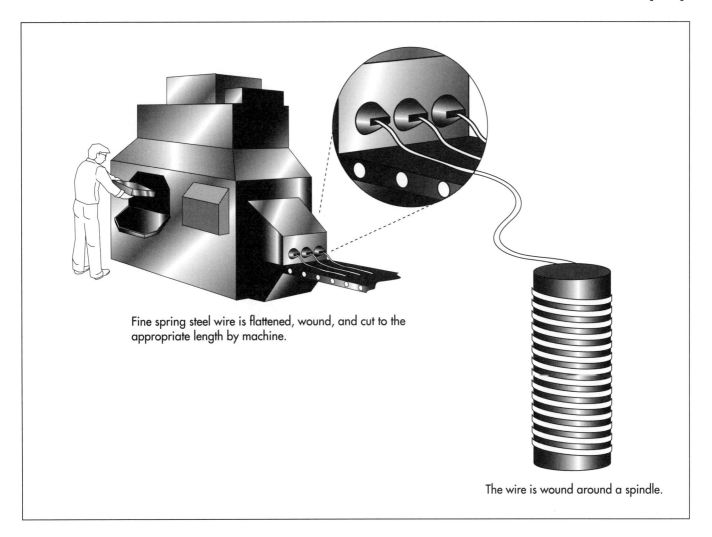

Fine spring steel wire is flattened, wound, and cut to the appropriate length by machine.

The wire is wound around a spindle.

inal metal Slinky was made from 80 ft (24.4 m) of wire. At the end of the twentieth century, the company was making two sizes in steel and two in plastic of various colors.

Consumers have found ingenious uses for slinky toys, employing them in pecan-picking machines and using them as drapery holders, antennas, light fixtures, window decorations, gutter protectors, pigeon repellers, birdhouse protectors, therapeutic devices, wave motion coils, table decorations, and mail holders. For physics teachers and students, it is the tool of choice to demonstrate the properties of wave motion. The Slinky is on display at the Smithsonian Institute in Washington, DC, and in the Metropolitan Museum of Art in New York, New York.

Raw Materials

Fine spring steel wire, 0.0575 in (0.146 cm) in diameter, is purchased from an outside source. It is composed of high carbon steel

and coated for durability. The plastic is also purchased from a supplier and then forced through an extruder to form long thin strands. Some of the slinky toys have animal heads and tails attached to either end. The animal parts are created from molds by an outside supplier.

The Manufacturing Process

Wire slinky toy

1 The wire is fed into the machinery by a factory worker. The machine flattens the wire and winds it on its end. While it is being wound, the machine automatically cuts it to the preset length. The nature of spring wire is such that once it has been wound, it never loses that shape.

2 The curled wire is removed from the machinery by another worker, who

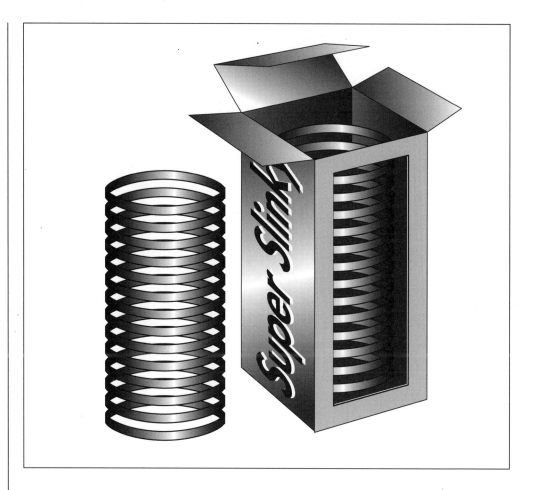

crimps one end to the coil next to it. The worker then turns the wire around and crimps the other end in the same fashion. If the slinky toy is to have animal attachments, it is placed on an additional conveyer belt, traveling to a work station where the heads and tails are applied manually.

Plastic slinky toy

3 When the plastic emerges from the extruder, it resembles a piece of warm taffy. It is then mechanically wrapped and cast around a mounted spindle called a mandrel. As the plastic hardens, it takes on the curled shape of the mandrel. The coiled plastic is dipped into a warm water bath to further seal it into its curled shape.

Packaging

4 The slinky toy is sent on a conveyer belt to a worker who inserts the toy into a box that has been opened by an automatic boxing machine. The boxed slinky toy moves back into the boxing machine, which then closes the box.

5 Further along on the conveyer belt, the box enters a shrink-wrap machine, which encases the box in plastic wrap. The boxes are then loaded into display or shipping cases.

The Future

It would seem that in an age of electronic gadgetry, the slinky toy would fade into obscurity, however it has enjoyed consistent success over the years primarily because of its simplicity. Although the package includes instructions on how to use the toy, something that has inspired various spoofs on the Internet, anyone who picks one up knows what to do with it. The appearance of slinky toys in the 1990s movies "Toy Story" and "Ace Ventura, Pet Detective" has only served to increase demand. In a 1993 *Philadelphia Inquirer Magazine* article, reprinted in the November 1993 issue of *Reader's Digest*, writer Jeanne Marie Laskas summed it up this way: "You don't have to be smart, athletic, rich or clever to appreciate the Slinky. It's a toy for regular people."

Where to Learn More

Periodicals

Laskas, Jeanne Marie. "The Lady Behind the Slinky." *Reader's Digest*, November 1993, p. 77.

Witchel, Alex. "Persevering for Family and Slinky." *The New York Times*, February 21, 1994, pp. C1, C7.

Other

Robinson, Kelly and Sherry Miller. Slinky Homepage. http://www.messiah.edu/hpages/facstaff/barrett/slinky/home.htm (July 9, 1997).

"The Slinky Story." James Industries. Hollidaysburg, Pennsylvania.

—Mary F. McNulty

Sofa

The style of a sofa is generally set by its arms, which double as artistic statements and rests.

Background

Sometimes called a couch or a davenport, a sofa is a long upholstered seat with both arms and a back. Today, it is a common luxury that indicates humans' progression away from the nomadic "pack and evacuate" lifestyle of our recent past.

History

Upholstery technically dates back to ancient Egypt, where pharoahs' tombs were furnished with comfortable appointments preserved to last a millennia. Ancient Egyptians and their Roman contemporaries reserved such items for royalty and other social elites. In the West, upholstery as we know it today developed slowly as building architecture improved. Prior to the 1500s, woven artifacts known as tapestries were the main source of insulation, protecting inhabitants from the damp and cold, that seeped in through their walls. Seating for two or more people was usually supplied by a hard bench.

Once the need for protection from the elements decreased, fabrics could be used for decoration and on individual pieces of furniture. Contributions to interior design were made from all major European centers. Germans introduced the use of horsehair padding, still a central feature of properly upholstered furniture. The English preferred dried sea moss. Italians introduced backrests and arms during the Renaissance. Upholstered chairs had been invented already, but were not popularized until this time. The sofa with a down cushion was an extension of the upholstered chair. Minor adjustments were made to stuffing methods, such as using buttons to secure padding rather than the practice of "tufting" (sewing raised loops or cut pile into the fabric).

The eighteenth century "upholder" was a combination designer and decorator who completed an architect's vision of a room. Cabinet makers like George Hepplewhite, Matthias Lock, Henry Copland, and the far more reknowned Thomas Chippendale extended their woodworking enterprises into this new and exciting field of upholstery. A rash of what were called "pattern books" by these and other practitioners, with such names as *The Cabinet-Maker and Upholsterer's Guide*, set the pace. They contained sofa designs as well as new ideas for other practical and decorative pieces.

During the nineteenth century, the advent of industrial technology had a major impact on modern methods of upholstery. In 1850, coil springs were invented. A modern sofa typically, though not always, contains springs to even out weight distribution. The **sewing machine** was also developed during this period, speeding up the upholstery process. New improvements such as modern welting would not be possible without the sewing machine.

Raw Materials

The frame of a sofa is made most often wood, though newer options include steel, plastic, and laminated boards or a combination of the above. Kiln-dried maple wood deemed free of knots, bark, and compromising defects is used under the upholstery. The show wood of the legs, arms, and back can also be maple, but sometimes mahogany, walnut, or fruitwoods are used for carved legs or moldings.

Padding is primarily made from animal hair, particularly hog or horse. Other paddings used in mass production are foam and polyester fiberfill wrap. Some preprocessing may be necessary, as with the pre-matted rubberized hair, where animal hair is arranged and bonded into shape with glue.

Cushions are fashioned from polyurethane foam, polyester fiber, down, cotton, latex, or cotton-wrapped springs.

A sofa may be covered with any choice of synthetic, natural, or blended fabric. Wool and nylon are the best choices in their respective categories of natural and synthetic fibers, but cotton, acetate, rayon, and polyester have their own functional properties. Exterior fabric may be finished with a protective anti-stain coating.

When used, springs are made of tempered steel. A typical sofa calls for 15 yd (13.71m) of burlap and at least 10 yd (9.14 m) of muslin for the interior. All materials are fastened with approximately 1,000 or more tacks, over 200 yd (182.8 m) of twine, and hundreds of yards of machine sewing thread.

Design

Sofas come in three major sizes. The full sofa is 84 in (2.13 m) wide. Smaller versions like the two-seater and love seat range between 60-80 in (1.52-2.03 m). Variations on the standard sofa include modular items and sofas with special uses such as daybeds or convertible sofa beds. Ornamental designs are not necessarily less durable, but they do not invite casual use. The design of a sofa can be adjusted to the use that will be made of it, and the average size of the people who will use it most. A deep seat, for instance, is good for taller people but does not easily accommodate shorter individuals. The style of a sofa is generally set by its arms, which double as artistic statements and rests. Some styles of seating furniture are known by the names of these arm designs. The overstuffed sofa is called that in the trade in order to indicate the use of more than one layer of muslin in the foundation.

The Manufacturing Process

A single sofa takes up 300 to 600 hours of skilled labor to make. Even small compa-

A traditional Victorian sofa purchased as part of a parlor suite by Mary Todd Lincoln after her husband's assassination. (From the collections of Henry Ford Museum & Greenfield Village.)

After President Abraham Lincoln's assassination, his wife, Mary Todd Lincoln, purchased an expensive parlor suite for use in her new life as a widow. The suite included a sofa, table, two arm chairs, and several side chairs, and was probably manufactured by J & J.W. Meeks of New York.

Epitomizing the Victorian era, the sofa represents the ultimate in mid-nineteenth century comfort and decoration. The technology of the time allowed for the use of coil springs, giving the seat a cushion-like softness that returns to its taut shape as soon as pressure is removed. New technologies also gave way to the lavishly carved show wood on the back. Ordinarily, the almost three dimensional fruit and flowers design would split the wood, however thin layers of rosewood were cross-directionally glued together to form a stronger wood laminate able to withstand the carving. Once glued together, the entire lamination was steamed and forced to curve with the back of the sofa.

The bent plywood system would be utilized again (100 years later) by Charles Eames in order to create his famous chairs. They would be the stylistic antithesis of Mrs. Lincoln's ornate Victorian sofa, but just as chip and crack resistant.

Henry Prebys

nies and individuals avail themselves of power **saws** and other motorized machinery, yet specialized hand tools are still applied to detail work. These include the regulator for stuffing, the "ripping tool," and a type of pliers called diagonal cutters.

The nitty-gritty of assembly is where the differences between traditional craft and

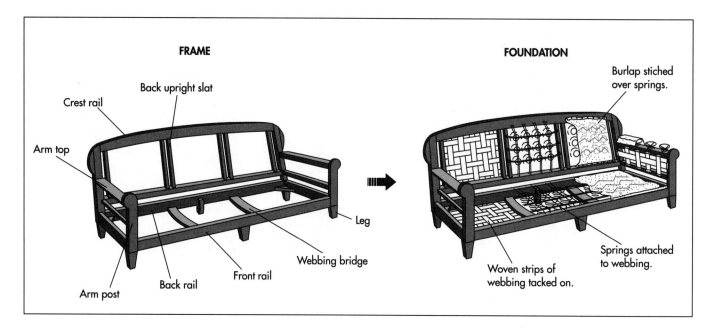

FRAME

Crest rail

Back upright slat

Arm top

Arm post

Back rail

Front rail

Webbing bridge

Leg

FOUNDATION

Burlap stiched over springs.

Woven strips of webbing tacked on.

Springs attached to webbing.

factory production become most obvious. Classic hand-tied springs, for instance, may give way to a mechanically attached spring grid if recent factory practice becomes the norm. Such industrial reforms have sparked controversy. The following breakdown is modeled after the handmade process that still defines the industry.

Framing

1 First the frame is constructed from wood that has been found clear of any defects. The thickness of the wood should allow for the heavy tension webbing to follow. If the frame is not sufficiently strong, it will not bear the weight redistributed into it by the webbing whenever someone sits down. Arms, back, or back sections, seat, and legs are attached. The preferred method is with clean-cut, fitted double doweled glue joints reinforced with corner braces, glued and also screwed into place. Each major part of the sofa will have to have springs attached separately, and also be padded separately. Consequently, they are "framed out" with reinforcing slats, arranged around the seat section.

Webbing and springs

2 The foundation is then set for padding. Jute, a kind of burlap made in India, is used as webbing. Strips of this material are interwoven, stretched across the frame, and tacked down. Flax twine is then used to strap the springs onto the webbing. Two lines of twine are tacked into position and

then tied around a spring back to front. Another pair of lines will run side to side on each row of springs after all the springs have been lashed into position individually. If heavy-gauge springs are used in the "front row," these are further tied down with a length of wire. This process is repeated for the back, with special attention to the springs at the base, which are treated like the front row of seat springs. If the back comes in sections (sometimes three for design purposes), then each part is separately tied off and the twine ends tacked onto the four-sided frame. The same is true for any sides and arms. Each part will be wrapped in its own sheet of burlap after being completely fitted with secured springs. The burlap is cut to size for each part, tacked into place initially, and then tightly lashed to the springs to minimize movement. This is to prevent the springs from wearing through the burlap over time.

Padding

3 Each part is separately padded as well, with layers of burlap and horsehair or chosen synthetic material. The padding is placed in a burlap envelope, arranged on the edge of the seat, pinned into place, and stitched down. As the stitching progresses, the pins can be removed one by one. This roll is then shaped according to design requirements and stitched with special needles and more twine. After this is secured yet still pliable, a layer of about 15 lb (6.81 kg)

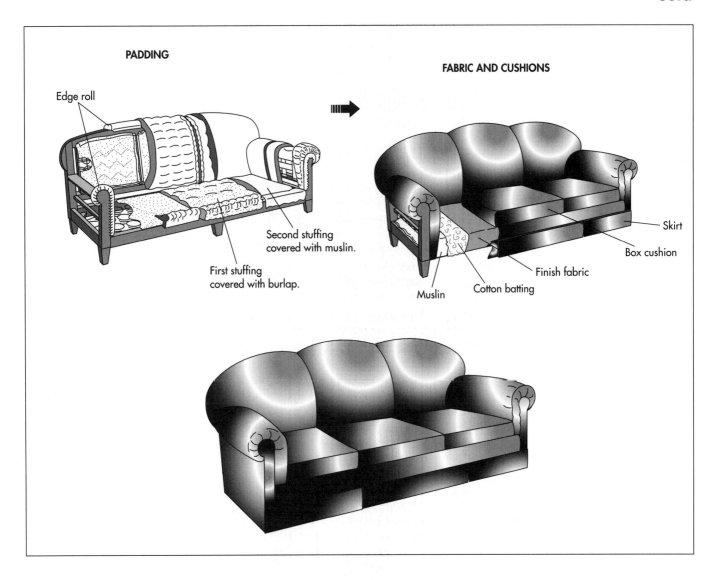

PADDING

Edge roll

FABRIC AND CUSHIONS

Second stuffing
covered with muslin.

First stuffing
covered with burlap.

Skirt

Box cushion

Finish fabric

Muslin

Cotton batting

of padding is distributed over the whole area of the seat, extending over the roll. The layer is basted into place with long, loose stitches and covered with lighter weight burlap. Tighter stitching divides the seat into two areas called the platform and the nose or front edge. This front part is re-shaped with hand stitching. After the shaping is completed, a final, thicker layer of padding is added to fill in dips left by stitching in the burlap, and basted like the previous layer. A muslin sheet of covering is applied, stitched into the break between the platform and nose, tightened across the front edge and back across the platform; its edges are tacked into place. Anomalies in the padding are addressed before proceeding.

4 The arms are done next in the same basic fashion. Layers of padding and burlap are fixed in succession and topped

with muslin. The arms also have a front edge of extra thick padding. Once the arms are properly shaped the back or back sections may be padded. If there is more than one part to the back, the center is padded first up until the second burlap layer. Then, the two flanking sections are padded up to that point, to match the center in size. The edge roll is formed around the top and back of the crest rail or uppermost part of the frame, or the corresponding area of each of the back parts, each of which must be kept parallel to the others. After inspecting and making any adjustment to the padding, the exposed wood parts can be stained and finished to taste or design specifications.

Fabric

5 Every piece and panel that will be fabric covered must be measured and recorded

in a cutting list. The fabric is purchased in one piece or lot. The panels are then plotted out in chalk so they match wherever their seams will meet when finally applied. If any of the panels and pieces need to be sewn together before being attached to the padded frame, this is taken care of first. The seat is covered with panels for the platform and nose and hand-stitched into place along the break between them over a layer of cotton batting. The nose is then covered first to check if the pattern continues along the front properly. The covering is fitted over the back or platform end and secured. The arms are covered next after being prepped with their own layers of cotton batting. A fan-pleated arm is a classic look. The fabric is folded into place around the front roll, in a series of pleats that look like an opened fan when finished. A series of strategically placed cuts may be made, so the fabric clears all obstructions presented by the frame. The top, bottom, back, and pleated front are operated on in succession. Temporary tacks are replaced one by one with permanent tacks.

6 Other parts to be covered, like the back or its sections, may require machine sewing and the attachment of pull tabs that will allow the fabric to be stretched between frame slats and secured. Cotton batting is layered on as well, and the appropriate panel of fabric laid down, basted, stretched fully into place, and fixed with tacks. The outside is the last part to be padded and covered, starting at the arms. The open area is covered with a layer of burlap, an outside cotton padding, and finally the finishing fabric. Covering is fabric-stitched on top and tacked into place on the bottom, front, and back. The largest panel left open is the outside back. If the webbing has left any gaps between frame slats, these must be stuffed. Padding should be basted over the gaps along the whole outside back. The fabric panel for this section may be welted, or edged with a decorative strip made of stuffing cord covered in matching fabric. The covering is basted, then sewn at the top and tacked at the bottom as with the other parts.

Finishing

7 After the sofa is flipped and covered at the base with a cambric (dust cover), fin-ishing touches are then applied. The sofa may be fitted with one of several choices of skirt. Arms may be supplied with welted panel covers. Cushions are made separately to cover the seat. These are constructed most often from a jacket of ticking, encasing two pads that in turn frame an inner core of foam. Each one is covered with finishing fabric panels supplied with a back zipper, so the case can be removed for dry cleaning.

Quality Control

Quality control is more a matter of individual or company standards than government regulations. Manufacturers' warranties range from five to 10 years to a lifetime.

The Future

Sofas continue to be made by individual craftsmen and small workshops as well as factories. There are different ways to learn how to make sofas and other upholstered items. North Carolina State University offers an industrial engineering bachelor's degree that specializes in furniture manufacturing. In addition to courses in product engineering and facilities design, they sponsor field trips to local factories and workshops in industry-specific computer applications.

Where to Learn More

Books

Beard, Geoffrey W. *Upholsterers and Their Work in England: 1530-1840.* Yale University Press, 1997.

Brumbaugh, James E. *Upholstering.* Audel Books, 1992.

Gheen, W. Lloyd. *Upholstery Techniques Illustrated.* Tab Books, 1994.

James, David. *Upholstery: A Complete Course.* Sterling Publishers, 1993.

Zimmerman, Fred W. *Upholstering Methods.* Goodheart-Willcox Co., 1992.

Periodicals

Colborn, Marge. "Is This A Good Sofa?" *The Detroit News Homestyle Section,* August 3, 1996.

UIU Journal/Upholsterers' Journal. Upholsterers' International Union of North America, 1922-present.

Upholstering Today/Furniture Today. Communications/Today Ltd., 1986-present.

Other

Joines, Jeff. "FMMC Sponsored Field Trips." March 1996. http://www.fmmcenter.ncsu.edu/education/furncomp.html (July 14, 1997).

Wright, Monte. "Heirloom Upholstery's Introduction to Upholstery Home Course" (video and companion booklet).

—Jennifer Swift Kramer

Solar Heating System

Background

In just one second, the Sun gives off 13 million times the energy that is generated by all the electricity consumed in one year in the United States. Only one millionth of the Sun's energy reaches Earth, but this scant amount would be more than sufficient to meet the energy requirements of our entire planet. The relative difficulty in extracting energy from the Sun, when compared to systems that derive energy from fossil fuels or nuclear power, has hindered its development as a widespread source of energy. On a smaller scale and in many experimental projects, however, solar energy has proven highly effective in producing both electricity and heat.

Solar energy was first explored for electrical purposes in the 1950s, when the need for continuous electric power generation on space satellites spawned the development of a solar cell in the Bell Telephone Laboratories of the United States. Even today, though, the best silicon solar cell converts sunlight to electric power with only 18% efficiency. Still, experiments have utilized sun-generated electricity with great success. One highly visible project focusing on solar power is the annual international solar-powered automobile race.

Solar energy has proven more effective and has been more widely utilized for both water and space heating and cooling systems. As a water heater, solar energy is most commonly used to heat swimming pools. For space heating, two main types of systems are used. A passive solar heating system admits solar energy directly into a building through large windows facing south (in the northern hemi-sphere) and directly heats the space within (this is known as direct gain and is also referred to as the greenhouse effect) or through a wall or roof that absorbs the solar radiation, stores the resulting heat, and transfers the heat into the building (this is known as indirect gain). A passive system may also utilize an absorber and storage components (such as a rock bed) that are not a part of the building, but are contained in their own separate chamber. This is known as an isolated-gain system.

Active solar heating systems use water or air to transport heat from collectors mounted on the south-facing side (again, in the northern hemisphere) of a building's roof to rock beds or water tanks. The stored heat may either be allowed into a room directly when utilizing rock beds and air as the transfer fluid, or through fan-coil units when the solar energy first heats water. The heat is then transferred through the coils to heat the air. Cooling systems are also divided into passive and active systems which utilize night air and condensation to cool the air in a building.

Different types of solar collector panels may be used and suited to the type of facility being heated or cooled. A flat-plate collector, a large, flat box with a glass top and a heat-absorbent black bottom containing pipes that run parallel to the top and bottom, is best suited to domestic use. A concentrating collector, made of reflecting materials and shaped like a trough or bowl, is best suited to industrial use.

History

The sun has served as a source of heat since the beginning of time, but the earliest docu-

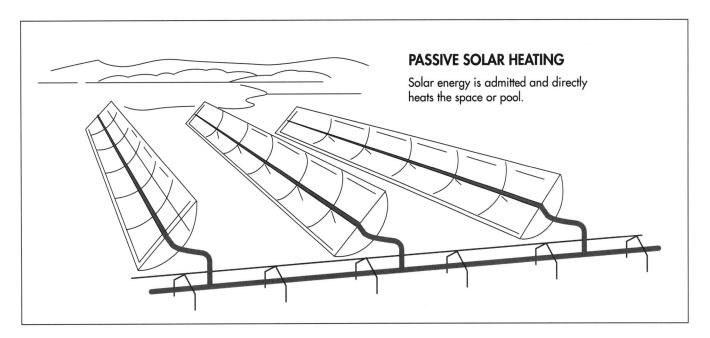

PASSIVE SOLAR HEATING

Solar energy is admitted and directly heats the space or pool.

mented use of a human-constructed solar collection system dates back to 1776 when Swiss scientist Horace de Saussure developed a rudimentary collector commonly referred to as a "hot box." Astronomer Sir John Herschel used hot boxes to cook food during an expedition to Africa in the 1830s, and the use of solar thermal energy for cooking and water distillation purposes subsequently became common in certain parts of Africa.

Early American pioneers of the late 1800s used black pots and pans to heat up water as they traveled during the day, thus popularizing the use of solar energy to heat water. The idea was put into widespread use in areas of the country which otherwise had to import fuel for water heating. For example, in 1897 nearly 30% of the houses in Pasadena, California, utilized solar water heaters.

Solar energy continued to be used moderately throughout the next century, but leapt in popularity during the energy and oil crisis of the 1970s. Use of solar energy tapered off as the energy crisis waned, but has risen again in popularity in the 1990s as populations become increasingly aware of the environmental and public health hazards caused by the burning of fossil fuels and the use of nuclear power.

Raw Materials

Because there are many types of solar heating systems, there is a wide variety of raw materials that may be used in their manufacture. This entry will focus on a basic residential passive system and a basic residential active system.

A passive solar heating system requires a black flat-plate collector panel made of a steel absorber plate covered with two sheets of glass and an insulating pad made of fiberglass insulation or polyurethane foam with an aluminum foil facing, which acts as a moisture barrier. The system is contained in a shallow box made of wood, galvanized steel, or aluminum. The system also typically requires a heat storage bin, containing dry pebbles or rocks. This unit stores heat when a building is sufficiently heated. In addition, the system requires a differential thermostat, an electronic system which allows manual control of the heat levels and, thus, room temperatures, and an air-handling module consisting of connecting ducts, air filters, a blower, and automatic dampers.

An active residential solar heating system utilizing water as the storage agent requires a flat-plate collector constructed of one or two sheets of glass or transparent plastic with black metal tubing and an insulation pad made of fiberglass board or a similar insulating material. The system also requires water pumps, a storage tank, heat exchanging coils, an auxiliary heater, a fan, filters, and a control valve. If rocks are used as the storage material, an insulated bin

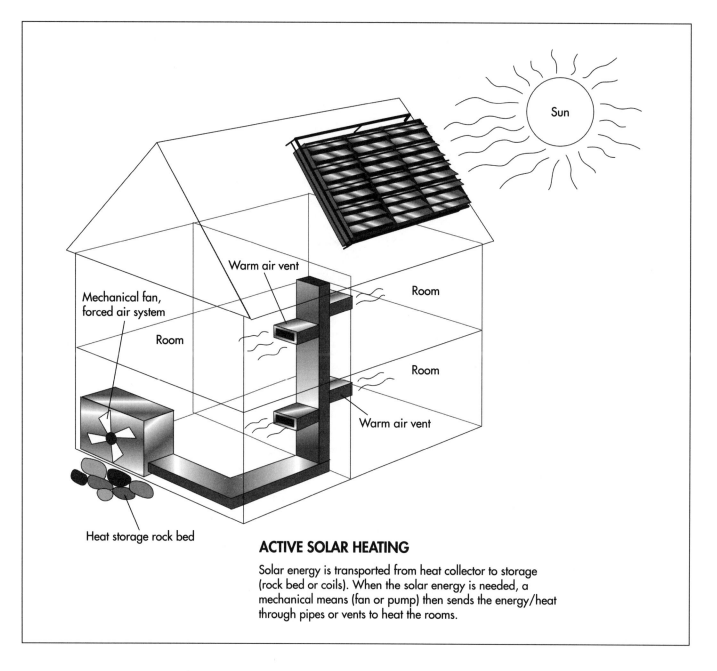

ACTIVE SOLAR HEATING

Solar energy is transported from heat collector to storage (rock bed or coils). When the solar energy is needed, a mechanical means (fan or pump) then sends the energy/heat through pipes or vents to heat the rooms.

must be used to contain them and, obviously, water pumps and coils are not necessary.

All solar heating systems utilize caulking, aluminum siding, and closure strips, most likely made of rubber, paint, and wood.

The Manufacturing Process

Collectors

1 Flat plate collectors are installed on either a pitched roof or vertical walls by installing shallow, sheet metal manifold pans, which resemble large baking pans, at

each end of the collectors and fastening the pans to the rafters of the roof or wall. Caulking is then applied liberally around the upper edge of the manifold pans. Sheathing is next attached to the roof or wall, and all seams are caulked. Manifold slots are cut in the sheathing, and blocking is fastened around the perimeter of the sheathing.

2 The absorber plate is installed over the sheathing, extending to the outer manifold blocking. The ends are sealed with end closure strips, typically made of rubber and caulking. The top and bottom edges are also caulked.

3 Battens are fastened to the absorber plate, and a glazing system consisting of a single layer of glass on a roof or a double layer of glass on a wall is attached.

4 Aluminum glazing bars are installed around the perimeter of the collector.

Air handling and control system

5 In a passive system, the air handling system consists solely of ductwork installed between the collector and the living space, as well as a fan and a manual control to activate the fan. An isolated gain system would also include a separate storage chamber and ductwork leading to and from the chamber.

6 In an active system, a storage bin holding rocks or water is required to store excess heat. Ductwork with on-off dampers are installed running from the collector to the house, the collector to the storage bin, and the storage bin to the house. Two fans are installed.

Storage

7 A rock storage system is constructed of an insulated bin filled with small rocks 1-5 in (2.5-12.7 cm) in width. Sufficient space is allowed between the rocks to facilitate the blowing of air through the duct connections at either end of the bin.

8 A water storage system is similar to the rock storage system, using water rather than rocks.

Quality Control

For a residential solar heating system, the fit of the end closures and adequacy of the caulking on the collectors may be tested using a smoke bomb to detect any leaks. The smoke test is conducted prior to glazing.

The Future

As more prototypes of solar-heated and solar-powered living and commercial units are developed, this source of energy shows serious signs of growing into a source of heat and electricity that is more than experimental. Environmental and natural resource groups, as well as and often in cooperation with the United States Department of Energy and several international organizations, continue to push for a more widespread reliance on solar heat and power as well as continually develop new and innovative uses for this source of energy. While the work inside the Department of Energy's Solar Energy Research Facility focuses on photovoltaic power, the building itself stands as a testament to the success of solar energy. The building's innovative, window-laden, stairstep-like design allows for direct sun lighting and heating and stands as an example of how successful the reliance on solar energy can be—and how prevalent such reliance could become in the future.

The American Solar Energy Society demonstrates the practicality of solar energy in its annual solar home tours. The organization also keeps its members posted on legislative developments regarding solar energy matters. In July 1996, for instance, the organization was active in lobbying for the passage of renewable energy funding legislation. While no such legislation has been enacted to date, introduction of such bills indicates an interest in solar energy issues among at least some members of Congress. The International Solar Energy Society is concerned with similar issues on a global level, and works for a greater reliance on solar energy worldwide.

Where to Learn More

Other

American Solar Energy Society. June 3, 1997. http://www.csn.net/solar/ (July 14, 1997).

Center for Renewable Energy and Sustainable Technology. http://solstice.crest.org (July 14, 1997).

International Solar Energy Society. http://www.ises.org (July 14, 1997).

Passive Solar Industries Council. http://www.psic.org (July 14, 1997).

—*Kristin Palm*

Soy Sauce

Background

Soy sauce is one of the world's oldest condiments and has been used in China for more than 2,500 years. It is made from fermenting a mixture of mashed soybeans, salt, and enzymes. It is also made artificially through a chemical process known as acid hydrolysis.

History

The prehistoric people of Asia preserved meat and fish by packing them in salt. The liquid byproducts that leeched from meat preserved in this way were commonly used as liquid seasonings for other foods. In the sixth century, as Buddhism became more widely practiced, new vegetarian dietary restrictions came into fashion. These restrictions lead to the replacement of meat seasonings with vegetarian alternatives. One such substitute was a salty paste of fermented grains, an early precursor of modern soy sauce. A Japanese Zen priest came across this seasoning while studying in China and brought the idea back to Japan, where he made his own improvements on the recipe. One major change the priest made was to make the paste from a blend of grains, specifically wheat and soy in equal parts. This change provided a more mellow flavor which enhanced the taste of other foods without overpowering them.

By the seventeenth century this recipe had evolved into something very similar to the soy sauce we know today. This evolution occurred primarily as a result of efforts by the wife of a warrior of one of Japan's premier warlords, Toyotomi Hideyori. In 1615 Hideyori's castle was overrun by rival troops. One of the warrior's wives, Maki Shige, survived the siege by fleeing the castle to the village of Noda. There she learned the soy brewing process and eventually opened the world's first commercial soy sauce brewery. News of the tasty sauce soon spread throughout the world, and it has since been used as a flavoring agent to give foods a rich, meaty flavor.

Today soy sauce is made by two methods: the traditional brewing method, or fermentation, and the non-brewed method, or chemical-hydrolyzation. The fermentation method takes up to six months to complete and results in a transparent, delicately colored broth with balanced flavor and aroma. The non-brewed sauces take only two days to make and are often opaque with a harsh flavor and chemical aroma. Soy sauce has been used to enhance the flavor profiles of many types of food, including chicken and beef entrees, soups, pasta, and vegetable entrees. Its sweet, sour, salty, and bitter tastes add interest to flat-tasting processed foods. The flavor enhancing properties, or *umami*, of the soy extract are recognized to help blend and balance taste. The condiment also has functional preservative aspects in that its acid, alcohol, and salt content help prevent the spoilage of foods.

Raw Materials

Soybeans

Soybeans (*Glycine max*) are also called soya beans, soja beans, Chinese peas, soy peas, and Manchurian beans. They have been referred to as the "King of Legumes" because of their valuable nutritive properties. Of all beans, soybeans are lowest in starch and have the most complete and best protein mix. They are also high in minerals, particularly calcium and magnesium, and in Vitamin B. They have been cultivated since the dawn of civilization in China and Japan and were introduced into the United States in the nineteenth century. In the 1920s and 1930s, soybeans gained popularity in the U.S. as a food crop.

Soybeans are short, hairy pods containing two or three seeds which may be small and round or larger and more elongated. Their color varies from yellow to brown, green, and black. The variety designated yellow #2 are most commonly used for food products. These soybeans get their name from the yellow hilum or seed scar which runs down the side of the pod. The grades of grain allowed for trading are established by the United States Grain Standards which are administered by the U.S. Department of Agriculture. Soybeans are unusual in that, unlike other grains, most are used in processing or exporting, and not much as direct animal feed. This is because soybeans contain "anti-nutritional" factors that must be removed from the beans before they can be of nutritional value to animals. The soybeans used in soy sauce are mashed prior to mixing them with other ingredients.

Wheat

In many traditional brewed recipes, wheat is blended in equal parts with the soybeans. Pulverized wheat is made part of the mash along with crushed soy beans. The non-brewed variety does not generally use wheat.

Salt

Salt, or sodium chloride, is added at the beginning of fermentation at approximately 12-18% of the finished product weight. The salt is not just added for flavor; it also helps establish the proper chemical environment for the lactic acid bacteria and yeast to ferment properly. The high salt concentration is also necessary to help protect the finished product from spoilage.

Fermenting agents

The wheat-soy mixture is exposed to specific strains of mold called *Aspergillus oryzae* or *Aspergillus soyae*, which break down the proteins in the mash. Further fermentation occurs through addition of spe-

Henry Ford demonstrates the durability of automobile components made from soybeans by striking the trunk of a car with an axe. (From the collections of Henry Ford Museum & Greenfield Village.)

American farmers produced surpluses of many agricultural commodities in 1930, but soybeans were not one of them. During the early years of the Great Depression, few farmers raised soybeans, but this changed in just 10 years. In 1929, American farmers produced less than 10 million bushels (352 million L) of soybeans. By 1939 production approached 100 million bushels (3.5 billion L), and in 1995, American farmers raised more than 2.1 billion (74 billion L) bushels of soybeans. No one surpassed Henry Ford as a promoter of soybean production in the 1930s.

In 1929, Henry Ford constructed a research laboratory in Greenfield Village and hired Robert Boyer to oversee experimentation related to farm crops. Ford hired additional scientists to investigate the industrial uses of many agricultural commodities, including vegetables such as carrots. The greatest success was in soybean experimentation. The researchers developed soy-based plastics and made parts for automobiles out of the products. The scientists manufactured ink made from soy oil, and produced soy-based whipped topping. Many of these processes and products remain in use.

Ford believed that farmers should have one foot on the soil and the other in industry. Ford promoted agricultural production of soybeans through an exhibit in a barn at the Chicago "Century of Progress" World Exposition in 1933. He hosted a meal which included a variety of soybean items and supported the publication of recipe booklets full of soybean-based recipes.

Henry Ford wished to see farmers to produce soybeans on their farms and process them for industrial purposes. Though his vision was not realized, the importance of soybeans in American agriculture came to fruition. Soybeans are one of most important crops raised in America, and provide American farmers millions of dollars in income.

Leo Landis

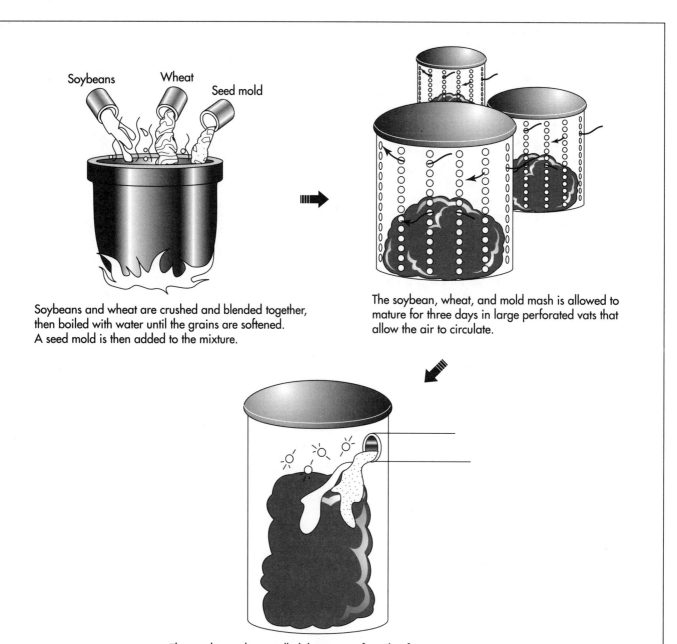

Soybeans and wheat are crushed and blended together, then boiled with water until the grains are softened. A seed mold is then added to the mixture.

The soybean, wheat, and mold mash is allowed to mature for three days in large perforated vats that allow the air to circulate.

The resulting culture, called *koji,* is transferred to fermentation tanks, where it is mixed with water and salt to produce a mash called *moromi.*

cific bacteria (*lactobaccillus*) and yeasts which enzymatically react with the protein residues to produce a number of amino acids and peptides, including glutamic and aspartic acid, lysine, alanine, glycine, and tryptophane. These protein derivatives all contribute flavor to the end product.

Preservatives and other additives

Sodium benzoate or benzoic acid is added to help inhibit microbial growth in finished soy sauce. The non-brewed process requires addition of extra color and flavor agents.

The Manufacturing Process

Traditional brewed method

Brewing, the traditional method of making soy sauce, consists of three steps: *koji*-making, brine fermentation, and refinement.

After the *moromi* is allowed to ferment for about six months, the soy sauce is separated from the wheat and soy residue by pressing it through a filtration cloth.

The raw sauce is pasteurized and bottled.

Koji-*making*

1 Carefully selected soybeans and wheat are crushed and blended together under controlled conditions. Water is added to the mixture, which is boiled until the grains are thoroughly cooked and softened. The mash, as it is known, is allowed to cool to about 80°F (27°C) before a proprietary seed mold (*Aspergillus*) is added. The mixture is allowed to mature for three days in large perforated vats through which air is circulated. This resulting culture of soy, wheat, and mold is known as *koji*.

Brine *fermentation*

2 The *koji* is transferred to fermentation tanks, where it is mixed with water and salt to produce a mash called *moromi*. Lactic acid bacteria and yeasts are then added to promote further fermentation. The *moromi* must ferment for several months, during which time the soy and wheat paste turns into a semi-liquid, reddish-brown "mature mash." This fermentation process creates over 200 different flavor compounds.

Refinement

3 After approximately six months of *moromi* fermentation, the raw soy sauce is separated from the cake of wheat and soy residue by pressing it through layers of filtration cloth. The liquid that emerges is then pasteurized. The pasteurization process serves two purposes. It helps prolong the

shelf life of the finished product, and it forms additional aromatic and flavor compounds. Finally, the liquid is bottled as soy sauce.

Non-brewed method (chemical hydrolysis)

Instead of fermenting, many modern manufactures artificially break down the soy proteins by a chemical process known as hydrolysis because it is much faster. (Hydrolysis takes a few days as compared to several months for brewing.)

1 In this method, soybeans are boiled in hydrochloric acid for 15-20 hours to remove the amino acids. When the maximum amount has been removed, the mixture is cooled to stop the hydrolytic reaction.

2 The amino acid liquid is neutralized with sodium carbonate, pressed through a filter, mixed with active carbon, and purified through filtration. This solution is known as hydrolyzed vegetable protein.

3 Caramel color, corn syrup, and salt are added to this protein mixture to obtain the appropriate color and flavor. The mixture is then refined and packaged.

Sauces produced by the chemical method are harsher and do not have as desirable a taste profile as those produced in the traditional brewed manner. The difference in taste occurs because the acid hydrolysis used in the non-brewed method tends to be more complete than its fermentation counterpart. This means that almost all the proteins in the non-brewed soy sauce are converted into amino acids, while in the brewed product more of the amino acids stay together as peptides, providing a different flavor. The brewed product also has alcohols, esters, and other compounds which contribute a different aroma and feel in the mouth.

In addition to the brewed method and the non-brewed method, there is also a semi-brewed method, in which hydrolyzed soy proteins are partially fermented with a wheat mixture. This method is said to produce higher quality sauces than can be produced from straight hydrolysis.

Quality Control

Numerous analytical tests are conducted to ensure the finished sauce meets minimum quality requirements. For example, in brewed sauces, there are several recommended specifications. Total salt should be 13-16% of the final product; the pH level should be 4.6-5.2; and the total sugar content should be 6%. For the non-brewed type, there is 42% minimum of hydrolyzed protein; corn syrup should be less than 10%; and carmel color 1-3%.

In the United States, the quality of the finished sauce is protected under federal specification EE-S-610G (established in 1978) which requires that fermented sauce must be made from fermented mash, salt brine, and preservatives (either sodium benzoate or benzoic acid). This specification also states that the final product should be a clear, reddish brown liquid which is essentially free from sediment. The non-fermented sauce is defined as a formulated product consisting of hydrolyzed vegetable protein, corn syrup, salt, caramel color, water, and a preservative. It should be a dark brown, clear liquid.

The Japanese, on the other hand, are more specific in grading the quality of their soy sauces. They have five types of soy sauce: *koikuchi-shoyu* (regular soy sauce), *usukuchi-shoyu* (light colored soy sauce), *tamari-shoyu, saishikomi-shoyu,* and *shiro-shoyu*. These types are classified into three grades, Special, Upper, and Standard, depending upon sensory characteristics such as taste, odor, and feel in the mouth, as well as analytical values for nitrogen content, alcohol level, and soluble solids.

Byproducts/Waste

The fermentation process produces many "byproducts" that are actually useful flavor compounds. For example, the various sugars are derived from the vegetable starches by action of the *moromi* enzymes. These help subdue the saltiness of the finished product. Also, alcohols are formed by yeast acting on sugars. Ethanol is the most common of these alcohols, and it imparts both flavor and odor. Acids are generated from the alcohols and sugars, which round out the flavor and provide tartness. Finally, aromatic esters (chemicals that contribute flavor and aroma) are formed when ethanol combines with organic acids.

Chemical hydrolyzation also leads to byproducts, but these are generally considered undesirable. The byproducts are a result of secondary reactions that create objectionable flavoring components such as furfural, dimethyl sulfide, hydrogen sulfide, levulinic acid, and formic acid. Some of these chemicals contribute off odors and flavors to the finished product.

The Future

The future of soy sauce is constantly evolving as advances are made in food technology. Improved processing techniques have already allowed development of specialized types of soy sauces, such as low-sodium and preservative-free varieties. In addition, dehydrated soy flavors have been prepared by spray drying liquid sauces. These powdered materials are used in coating mixes, soup bases, seasoning rubs, and other dry flavorant applications. In the future, it is conceivable that advances in biotechnology will lead to improved understanding of enzymatic reactions and lead to better fermentation methods. Technology may someday allow true brewed flavor to be reproduced through synthetic chemical processes.

Where to Learn More

Books

Farrel, Kenneth T. *Spices, Condiments and Seasonings.* Van Nostrand Reinhold Company Inc., 1985.

Periodicals

"A Tale of Two Soy Sauces." *Prepared Foods*, October 1996, p. 57.

The Soy Sauce Handbook, A Reference Manual for the Food Manufacturer. Kikkoman Corporation., 1996.

—*Randy Schueller*

Stetson Hat

Background

The Stetson hat, named after its inventor, John B. Stetson, is synonymous with the more generic cowboy hat. A symbol of Western pride and bravado, this modified sombrero, with its large crown and wide brim, has graced the heads of America's most treasured Western heroes, from old-time favorites like actor John Wayne, Clayton Moore as the Lone Ranger, and country singer Gene Autry, to modern-day popular artists like Garth Brooks and Larry Hagman as J.R. Ewing on the television series *Dallas*. (J.R.'s hat is now displayed in the Smithsonian National Museum of American History's contemporary Americana exhibit.) The Stetson hat is not just a male fashion statement, either. Prominent country singers from Dale Evans to Trisha Yearwood, spurred on by legendary female maverick Annie Oakley, have proven that females can carry off this most essential Western look, too.

The Stetson Hat Company was established in Philadelphia, Pennsylvania, in 1865 when John B. Stetson decided to mass-produce the modified sombrero he had fashioned for himself out of necessity during a lengthy Western expedition. Stetson's "Boss of the Plains" model, with its high, creased crown and wide, molded brim, became the prototype for all other cowboy hat designs. Now located in St. Joseph, Missouri, the Stetson hat factory there and its second factory in Galveston, Texas, continue to turn out the "Boss of the Plains," along with over 100 variations for men and women.

History

John B. Stetson was born in 1830 in Orange, New Jersey, the 12th of 13 children born to Stephen Stetson, a hatmaker. As a youth, the younger Stetson worked in the hatmaking business with his father until he was diagnosed with tuberculosis and his doctor predicted he would have only a short time to live. Given this dire prognosis, Stetson left the hatmaking business to explore the American West, afraid this would be his only chance to see it. He eventually settled in St. Joseph, Missouri, a trading post where expeditions to Pike's Peak and similar western destinations were outfitted.

In the 1860s, unable to enlist to serve in the Civil War due to his poor health, Stetson set off on a Pike's Peak expedition himself and found himself becoming healthier throughout the course of the journey. When his party was unable to find suitable shelter, Stetson and the others fashioned makeshift cover by sewing together the skins of the muskrat, rabbit, beaver, and coyote they had shot for food. Given their rustic environs, the men had no means of tanning the hides, however.

Then, Stetson suggested that he make cloth for tents using the felting process he had learned in his father's hatmaking business. He shaved the fur from some of the skins and fashioned a hunter's bow from a section of a hickory sapling and a throng from one of the skins. Stetson agitated the fur with the bow, keeping it in a small cloud in the air and eventually allowing it to fall to the ground and naturally distribute itself over a small area. As the fur fell to the ground, Stetson blew a fine spray of water from his mouth through the fur, creating a mat that could be lifted from the ground and rolled. Stetson then dipped the sheet of matted fur into a pot of boiling water. As the sheet began to shrink, he manipulated it with his hands, rapidly and repeatedly dipping it in hot water and eventually forming a small, soft blanket. By repeating this process, Stetson and the other members of his party created enough of this water-repellent material to construct a tent. This same method is employed in felt-making today, although the fur is raised to the air by a fan and the water is sprayed mechanically.

To shield himself from the daytime sun, wind, and rain, Stetson also fashioned a hat from the felt. The high, indented crown and wide brim were modeled after the Mexican sombrero. According to legend, the other members of Stetson's party ridiculed him for wearing the hat until a passing bullwhacker from Mexico one day offered Stetson a five dollar gold piece for his invention.

After mining gold at Pike's Peak for a year, Stetson traveled to Philadelphia in 1865 and set up a small hatmaking business with $100 for the purchase of tools and fur. After designing several unsuccessful models, Stetson again created his modified sombrero with a 4 in (10 cm) crown and a 4 in (10 cm) brim and, when he was unable to sell Easterners on the innovation, began to market his product, grandly dubbed the "Boss of the Plains," in the Southwest, where it took off almost immediately. By the time Stetson died in 1906, his business was a booming success, and the company that bears his name still turns out a wide variety of Western hats at its St. Joseph and Galveston factories. Today, the Western hat is nearly as popular in the eastern United States, not to mention internationally, as it is in the American West.

Raw Materials

The primary component of the Stetson hat is felt, which is fashioned from a variety of fur, preferably beaver, rabbit, and wild hare. Hot water is also integral to the felting process. Dyes are used to achieve a variety of felt colors (the original color of the felt depends on the color of the original fur). Powder is used to soften the felt. Leather is another component of the process, used to form the interior sweatband of the hat. Attaching the sweatband may require glue.

Two-ply or two-cord band is used to create the ribbon that encircles the outside of the crown where it meets the brim, and thread is used to stitch the ribbon. Small metal eyelets are also typically used for venting.

Design

For the style-conscious, three of the most important considerations in purchasing a Stetson hat are the slope of the crown, the roll of the brim, and the number and arrangement of the creases, or pinches, in the crown, which are viewed as giving each Stetson its distinctive character. The pinch can be prefabricated or chosen by the consumer and blocked by the hatter. Cowboy hats may also be adorned with feathers, embroidery, silver accessories, and the like. The appeal of the Stetson hat for many is that, when fitted in cowboy boots and a Stetson, the wearer appears at least 6 in (15 cm) taller.

The Manufacturing Process

Carroting, cutting, and sorting

1 Typically beaver, rabbit, and wild hare pelts are used. They are cleaned to remove grease and other impurities.

2 An acid solution is applied to the fur which prepares it for the felting process. This procedure is known as carroting.

3 The skins are then fed through a cutting machine, fur side down, and the skin is cut from the fur.

4 The fur is then sorted by feeding it through blowing and picking machines.

5 Next, the fur is bagged in 5 lb (2.27 kg) increments and baled for shipment to the rough body plant.

Felt mixing and initial shaping

6 First, the fur must be weighed out and the three types combined to the company's specifications for the appropriate mix.

7 The fur is cleaned and then placed in a giant bin and mixed into a fine blend.

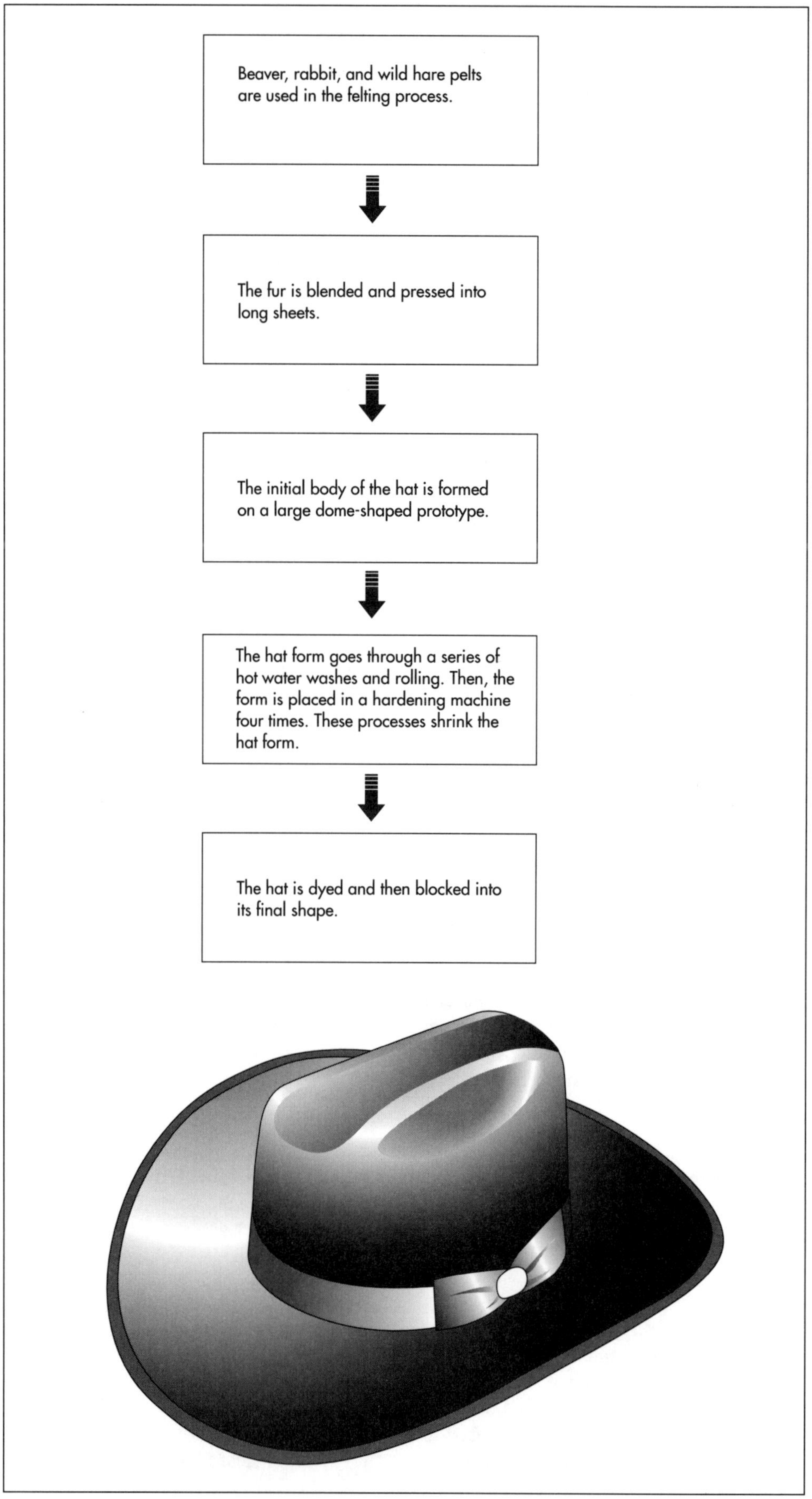

Beaver, rabbit, and wild hare pelts are used in the felting process.

↓

The fur is blended and pressed into long sheets.

↓

The initial body of the hat is formed on a large dome-shaped prototype.

↓

The hat form goes through a series of hot water washes and rolling. Then, the form is placed in a hardening machine four times. These processes shrink the hat form.

↓

The hat is dyed and then blocked into its final shape.

8 Next, the fur is forwarded into a feeder, where it is broken down into an even softer blend.

9 From the feeder, the fur is forwarded into a blowing machine and, upon exit from the machine, bad fibers are manually removed from the mix.

10 Next, the mixture is placed on a conveyer, which presses the mixture out into long sheets.

11 These sheets are then blown onto a large dome-shaped prototype, known as a former dome, which sets the initial body of the hat. The dome spins around rapidly for approximately 30 seconds, gathering the fur much in the same way a spinner wraps cotton candy.

12 The former dome is placed in hot water for 35 seconds to mold the fur fibers into place.

Felting and dyeing

13 Through a series of applications of hot water, pressure, and rolling, known as "starting" and "stumping," the dome-shaped figure is dipped in hot water and twisted.

14 The hat is then placed in a hardening machine four times. Over the course of these two steps, the hat begins to shrink from its original size of almost 2 ft (61 cm).

15 The hat bodies are then dyed with various pigments to create a range of colors.

Initial blocking and pouncing

16 Blocking, or shaping, is performed by a three-person team. First, a tipper stretches the crown of the hat. Next, a brimmer forms the band-line and brim. A blocker then sets the shape of the hat and applies a stiffening substance.

17 The hat is then turned inside out and hung to dry overnight.

18 The next day, the hat is sandpapered to remove any surface hair that was not eliminated in the blowing process. This procedure is known as pouncing and the overall operation is technically known as the back-shop.

19 The two-ply or two-cord band is stitched around the outer bottom of the crown of the hat, where it meets the brim.

Western blocking and finishing

20 The hat is steamed and an appropriately sized wooden block is pressed in the crown for shaping.

21 The crown of the hat is then ironed and the brim is plated.

22 The hat is then placed on a crown-pouncing machine, which uses very fine sand paper to remove any excess fur and finish the hat.

23 Powder is applied to make the hat softer and the color richer (this step is eliminated for dark-colored hats).

24 The hat is again placed in the pouncing machine.

25 The shape of the brim is set and dry heat is applied. The brim is then pressed to maintain its shape.

Sweatband

26 Excess plastic wire, known as polyreed, is cut off the leather sweatband to make it lay flat.

27 The band is hand-fitted inside the hat.

28 Any excess leather is cut off.

29 The band is stitched and then tacked or glued.

30 The hat is cleaned, brushed, and steamed.

31 The final cutting of the brim is executed.

Creasing and miscellaneous details

32 A bow is stitched to the outer two-ply or two-cord hatband.

33 The hat is hand-creased.

34 A satin name pad is pressed and steamed inside the crown.

35 If eyelets are required, they are placed in the crown at this point.

36 A cleaning instructions tag is glued to the crown.

Quality Control

Stetson hats are subject to many stages of inspection to check that the finish, shape, body, and feel are appropriate and to identify any flaws. One major factor in determining the price of a Stetson hat is the quality of the felt, demarcated by the number of "x"s that appear on the sweatband. Factors contributing to the quality of the felt include the fur and skin types, the living environment of the animal from which the fur was culled (wild, domestic, or season), the age of the fur, and its color.

The Future

Western hats are considered as stylish today as they were when they were invented. The Stetson Hat Company and many others are continually developing new variations on the style of this product and exploring different materials such as straw and leather.

Where to Learn More

Books

Reynolds, William and Ritch Rand. *The Cowboy Hat Book*. Gibbs Smith, 1995.

Other

Biltmore Hats. "The Story of the Cowboy Hat." http://www.in.on.ca/~biltmore/story.html (July 9, 1997).

—*Kristin Palm*

Sunglasses

Background

Sunglasses are eyewear designed to help protect the eyes from excessive sunlight. Eyes are extremely light sensitive and can be easily damaged by overexposure to radiation in the visible and nonvisible spectra. Bright sunlight can be merely a distracting annoyance, but extended exposure can cause soreness, headaches, or even permanent damage to the lens, retina, and cornea. Short term effects of sun overexposure include a temporary reduction in vision, known as snow blindness or welders' flash. Long-term effects include cataracts and loss of night vision. In both cases, the damage is caused by ultraviolet (UV) light, which literally burns the surface of the cornea.

Sunglasses were originally invented to reduce distracting glare and allow more comfortable viewing in bright light. Early sunglasses were simply tinted glass or plastic lenses that were primarily meant to reduce brightness. Darker lenses were considered to be better because they screened out more light. As our understanding of the damaging nature of sunlight evolved, the need for better eye protection was recognized, and technology was developed to help sunglasses better screen out the harmful rays of the sun, especially UV rays. From inexpensive models with plastic lens and frames to costly designer brands with ground glass lenses and custom-made frames, sunglasses are available in a staggering array of styles and prices. Unfortunately there is no way to tell from the color or darkness of the lens how well it will screen out UV light. Similarly, there is little relationship between price of glasses and their ability to block UV light.

Raw Materials

Sunglasses consist of a pair of light-filtering lenses and a frame to hold them in place. The vast majority of lenses are made of colorized plastic, such as polycarbonate. However, glass is still employed for high quality brands. The highest quality lenses are optically accurate and do not distort shapes and lines. These lenses, like camera lenses, are made from distortion-free ground and polished optical glass. The borosilicate glass used in these lenses is scratch resistant and is made impact resistant by tempering it with various chemical treatments.

Soluble organic dyes and metallic oxide pigments are added to the lens material to absorb or reflect light of certain frequencies. These additives must not distort colors excessively, however; for example, badly colored lenses may make it difficult to discern the correct color of traffic lights. Gray lenses produce the least distortion for most people, although amber and brown are good too. Blue and purple tend to distort too much color. The additives also should block at least part of the blue light which is part of the lower frequency UV rays. Brown or amber screen out blue light the best, but at the cost of some color distortion. Various chemical coatings which are added to the lens can enhance viewing by reducing reflection or screening out polarized light.

Sunglass frames are made from metal or plastic. Metal frames, particularly expensive ones, are often made of mixtures of nickel and other metals such as silver. These frames have precisely engineered features, such as sculpted and gimbaled nose-pads, durable hinges with self-locking screws, and flexible temples. Upscale man-

Bright sunlight can be merely a distracting annoyance, but extended exposure can cause soreness, headaches, or even permanent damage to the lens, retina, and cornea.

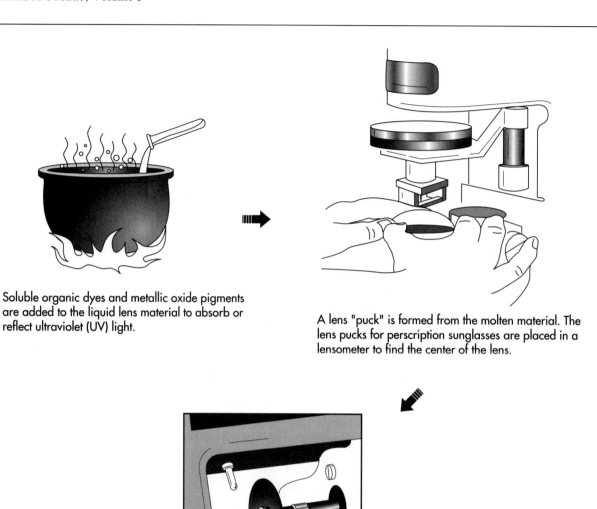

Soluble organic dyes and metallic oxide pigments are added to the liquid lens material to absorb or reflect ultraviolet (UV) light.

A lens "puck" is formed from the molten material. The lens pucks for perscription sunglasses are placed in a lensometer to find the center of the lens.

The lens is placed in a curve generator to grind out the lens according to specifications.

ufacturers use combinations of nickel, silver, stainless steel, graphite, and nylon in their leading-edge designs.

Design

There are two key elements to consider regarding sunglasses design, fashion and function. In the last few decades sunglasses have become a high fashion item, and the current design process reflects this status. Upscale clothing designers, fragrance marketers, and sporting goods vendors custom-design sunglasses to promote their own spe-cific image. By and large these design changes are not functional; they are intend-ed to increase the fashion appeal of the glasses. Stylized frames, uniquely shaped lenses, and embossed logos are all part of this designer mystique. While some designs are considered "classic" and timeless, oth-ers must be continually updated to satisfy the public's constantly changing tastes. The children's sunglasses market is another area which requires frequent redesign, since the style of the glasses changes from season to season based on merchandising tie-ins with popular cartoon or other characters.

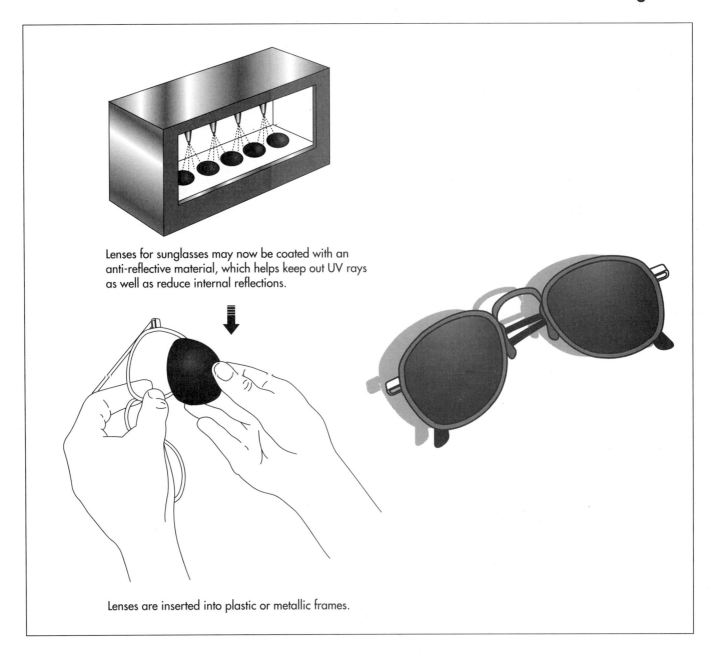

Lenses for sunglasses may now be coated with an anti-reflective material, which helps keep out UV rays as well as reduce internal reflections.

Lenses are inserted into plastic or metallic frames.

From a functional standpoint, sunglasses are designed specifically for a variety of outdoor activities. Sports enthusiasts have specific requirements that are reflected in sunglass design. For example, sunglasses designed for trap shooters are designed to provide maximum contrast to allow better viewing of their clay pigeon targets. On the other hand, sunglasses for skiers are designed to counter the light reflected of snow-covered surfaces. Lenses of the this types are known as blue blockers, because they filter out violet, blue, and some UV rays. Fisherman and boaters have their own special needs that must be addressed as well. Today there are cus-

tom-designed sunglasses for these activities and many more.

Sunglasses can protect the eyes in several ways. The glasses can either absorb or reflect certain frequencies of light, for both reduce the amount of light that enters the eyes. The absorbing types use various substances that are added to the lens material to selectively absorb light of specific frequencies. This range of frequencies can be controlled by changing the mixture of colorizing additives. The strength of the absorption is controlled by adjusting the amount of additive. Reflecting lenses have multilayer antireflective coatings, consisting primarily of metallic particles. These metallic coatings reflect all colors of light and UV radiation equally well. There are reflective types with non-metallic coatings, which create a coloring effect. By varying the type and amount of colorant additive or coating, a large variety of lenses can be produced. The color of the finished lens indicates which portion of the visible spectrum is being transmitted. For example, if the lenses are dark yellowish, they absorb violet, blue, and probably some UV rays.

A special type of absorbing lens filters out polarized light. Light actually consists of two waves, one propagated in the horizontal plane and one in the vertical plane. When light bounces off a flat surface, such as snow, roadways, or a shiny metallic objects, the horizontal component is seen as glare. Polarized lenses are made using a special optical filter which absorbs the horizontal component of light and transmits only the vertical component. As a result bright reflected light is eliminated and eye strain is reduced. However, polarized lenses do not block UV light, so they require additional coatings or coloring agents to provide complete eye protection.

Another type of lens, the photochromatic lens, contains silver salts like those used in photographic film. These lenses darken outdoors and lighten indoors. In this way the lenses change color in response to UV exposure. However, the range of lenses' color change is not broad enough to be effective against most light frequencies, and although they are widely used, photochromatic lenses are not universally accepted by optometrists.

The Manufacturing Process

Lenses

1 Colorant can be added to lenses in two primary ways, either by adding color to the molten lens material before the lens is formed, or by chemically post-coating the finished lens to achieve the desired hue. In the former method, the colorant additives are incorporated into the lens while the plastic or glass is at high temperatures and still liquefied. Soluble organic dyes or metallic oxide pigments are added to plas-

tic. Metallic oxide or metal particles are incorporated into glass.

2 After the appropriate additives have been blended in, the molten plastic or glass is then cast into the general lens shape, or "puck." Inexpensive lenses are simple pucks that placed into frames. Expensive lenses are prepared in a manner similar to the method by which prescription lenses are made. First the appropriate lens puck is placed in a lensometer, an instrument that is used to find the optical center of the blank.

3 The lens is then put in a curve generator, which grinds out the back of the lens according to the patient's prescription. An edge grinder then grinds the outer rim to its proper shape and puts a bevel on the edge, allowing the lens to fit properly into the frames.

4 Lenses may now be coated with an antireflective material. The post-coating method produces lenses that are more evenly coated regardless of the lens configuration. It also allows for the coating to be removed and recoated after the lens is made. Such coating may be applied through a vacuum coating method, used to deposit an antireflective layer on the lens surface to reduce internal reflections. This faint bluish coating is also commonly used on camera lenses and binoculars.

Frames

5 The finished lenses are now ready to be mounted in eyeglass frames. Frames are constructed to hold the lenses in place using either a tension mount or screw mount design. Tension mounting is typically used in plastic frames. In this type of frame, the dimensions of the lens opening on the front surface of the frame are somewhat smaller than the lens itself. In this way the lenses can be pressed into their respective openings through the front edge of the frame without falling out its rear edge. A groove is formed in the periphery of each of the openings, and these mate with a ridge formed on the periphery of the respective left and right lenses. The plastic frame material permits sufficient stretching or elongation to allow the lenses to snap into these grooves.

6 Metal frames use a screw mount design because metal tends to deform easily and cannot hold the lenses as well. The metal structure of the frames has thin extruded sections that are bent into desired shapes. The frame structure surrounding the lens openings forms an open loop into which the lens is inserted. After the lens is inserted, the loop is closed by attaching a screw to the two open ends.

Quality Control

Beyond the regulations that ensure the glass and plastic used in lenses is shatterproof, there is little governmental regulation of sunglasses. Labeling of the absorbance rates of both types of UV light, UVA and UVB, is voluntary, but the American National Standards Institute (ANSI) has established transmittance guidelines for general purpose and special lenses. According to these standards, general purpose cosmetic lenses must block 70% of UVB, general purpose must block 95% of UVB and most UVA light. Special purpose must block 99% of UVB.

To a large extent, the degree of quality control imposed on sunglasses manufacture is a function of the type of sunglasses. Inexpensive plastic models have little concern with optical perfection; they may contain flaws which will distort the wearer's vision. On the other hand, expensive glass lenses strive for high optical quality and are checked accordingly. There are a variety of instrumental methods used to evaluate distortion of the finished lens, but one simple test is to simply hold the glasses at arms' length and look at a straight line in the distance. Slowly move the lenses across the line. If the lens causes the line to sway or bend, the lenses are optically imperfect. For best results, look through the outer edges of the lens as well as the center.

Byproducts/Waste

There are no particular byproducts resulting from sunglasses manufacture. Waste materials include plastic, glass, and metallic scrap from grinding the lenses and making the frames.

The Future

Sunglasses manufacturing processes have become increasingly sophisticated in response to greater demand for high quality, stylish glasses. New coatings and colorants which deliver better protection against UV radiation continue to be developed. Improvements in the way frames are manufactured continue to be made. For example, U.S. patent 5,583,199 discloses a new way to make frames from a single piece of metal. New types of high performance sun-protective eyewear will be developed as advances are made in the fields of optics, surface chemistry, metallurgy, and others.

Where to Learn More

Books

Ahrens, Kathleen. *Opportunities in Eye Care Careers*. VGM Career Horizons, 1991.

Gourley, Paul and Gail Gourley. *Protect Your Life in the Sun*. High Light Publishing, 1993.

Zinn, Walter J. and Herbert Soloman. *The Complete Guide to Eyecare, Eyeglasses, and Contact Lenses*. Frederick Fell Publishers, 1986.

Periodicals

Berkeley Wellness Letter 9. The University of California, Berkeley, June 1993, p. 5.

—*Randy Schueller*

Surge Suppressor

Most surge suppressors commonly used today utilize spark-gap technology—a system whereby the suppressor breaks down, or shorts out, currents as the voltage applied to an electronic device exceeds the maximum tolerance, or rating, of the device.

Background

Power line disturbances occur on an average of four times a day, according to studies by the Institute of Electrical and Electronic Engineers (IEEE). These disturbances—increases in current that can damage electronic devices plugged into outlets—may be caused by lightning or other weather-related incidents; traffic accidents affecting power lines; the use of electrical products such as motors, compressors, and fluorescent lights; high-powered electrical equipment and voltage fluctuations initiated by a power company; and high-frequency noise. Such disturbances can interrupt or wipe out power service and damage electrical equipment. Both surges and spikes can be prevented by using a surge suppressor, an apparatus that serves to ground the interference or, in other words, run the interference into the ground rather than into the electrical equipment, and/or absorb the excess flow of current throughout an electrical system. Spikes and surges can damage electronic equipment even when it is not turned on.

History

One of the first surge suppressors was developed by the General Electric company in the 1950s. A similar device was being developed in Japan around the same time. Initial surge suppressors utilized selenium rectifiers (components used to convert direct current to alternating current) and, later, carbon piles (disk-shaped components used to carry current).

Most surge suppressors commonly used today utilize spark-gap technology—a system whereby the suppressor breaks down, or shorts out, currents as the voltage applied to an electronic device exceeds the maximum tolerance, or rating, of the device. There are three types of spark-gap surge suppressors, gas tube suppressors, metal oxide varistor (MOV) suppressors, and silicon avalanche suppressors (a specific type of transient voltage suppressor [TVS]). The type names refer to the materials that are the main component of each type of surge suppressor.

A TVS utilizes a process called "reverse bias voltage clamping." In this process, the excess current flowing through a TVS diode (a tubular semiconductor device that is similar to a resistor) breaks down and becomes a short circuit when the voltage applied to a circuit exceeds that circuit's rated avalanche, or maximum, level. A TVS is only used with direct current (DC) circuits, where the current only flows one way (such as in an automobile battery), but two TVSs placed back to back will protect an alternating current (AC) system, where current flows both ways.

A gas tube suppressor is commonly used with communications and voltage power supply lines. The gas tube suppressor shorts out a power surge and carries any associated high current to the ground, bypassing the circuit which would otherwise be affected, by providing a low-voltage path for the excess current through a ceramic or metal tube. Gas in the tube ionizes during a surge and creates a conducting state within the suppressor. The surge is transformed in this conducting atmosphere into a discharge arc, which shorts out the suppressor, and any high current is grounded. The gas subsequently deionizes and the suppressor is reset.

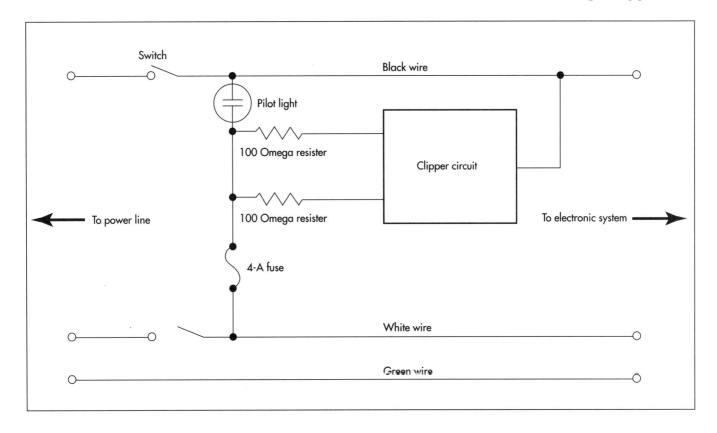

Switch

Black wire

Pilot light

100 Omega resister

Clipper circuit

To power line

100 Omega resister

To electronic system

4-A fuse

White wire

Green wire

A MOV uses variable resistors, or varistors, and is thus most functional with an AC system. An MOV suppressor absorbs extra voltage during both positive and negative swings in an AC system through a process known as clamping. When the voltage applied to the circuit exceeds the MOV rating — the load above which the MOV is supposed to be triggered — a short circuit is caused and the voltage is transferred into the absorbent body of the MOV suppressor, thereby bypassing the circuit that would have otherwise been disrupted.

Raw Materials

The actual assembly of a surge suppressor is relatively simple. The complexity of the instrument lies in the functions of the various components. The main component of a gas tube suppressor is a cylindrical ceramic or metal tube to contain the oxygen gas, which will carry the excess current. This type of suppressor also utilizes brazing washers made of copper or an alloy of copper and silver, thin-walled hollow electrodes made of a substance such as Kovar with radial flanges and absorbent getter material.

The main components of an MOV suppressor are the disc-shaped, ceramic MOVs,

which absorb the excess current. This type of suppressor also utilizes ring-shaped components surrounded by coils. These components are called balanced toroidal chokes, a type of inductor which is a component that stores electrical energy as an electromagnetic field. The inductor expands and collapses to maintain a constant flow of current throughout the coil. Flat rectangular or disk-shaped high-frequency capacitors and VHF capacitors may be used for filtering electromagnetic and radio frequency interference. A capacitor stores an electrostatic charge and increases or decreases the charge to maintain a constant flow of voltage across the component. The components are all mounted on a circuit board made of rigid, nonconductive material, such as fiberglass.

The Manufacturing Process

Gas tube supressor

1 First, the ends of the ceramic or metal tube must be fitted with conductive metal material such as copper. This is known as metallization.

Circuit diagram of a surge suppressor.

2 Next, brazing washers are attached to the metallized ends of the tube.

3 Thin-walled, hollow electrodes with radial flanges for connecting with the washers are then inserted into either end of the tube.

4 The assembly is then heated to braze the components, and the interior of the tube is evacuated.

5 The tube may also be lined with getter material to absorb any undesired gases in the interior of the tube.

MOV suppressor

6 First, the circuit board is metallized (a circuit is traced) with copper foil, copper deposition, or another highly conductive material.

7 Next, slots for the resistors and capacitors are etched into the metallized circuit board.

8 The MOVs, toroidal chokes, and capacitors are soldered to the circuit board.

9 Outlets and an on/off switch are attached, and the unit is screwed into a plastic or metal casing with openings for the outlets and the switch.

Quality Control

Electronic devices are subject to strict specification and quality control requirements. The IEEE, which is a standard-setting body accredited by the American National Standards Institute, sets standards that American electrical equipment must meet in order to be sold. Most, if not all, electrical equipment for sale in the United States is also tested by Underwriters Laboratories (UL), an independent company which provides product safety and performance testing. The UL seal on the packaging of electrical equipment indicates that those products withstood this testing. UL also assigns ratings to products based on safety and performance factors. Thus, not only does the seal indicate that a product passed UL's tests,

but the rating indicates how well the product performed in these tests.

The Future

Because surge suppressor design is fairly simple, new innovations center more on enhancing the original product rather than changing its makeup. While MOV surge suppressors remain the most common apparatus for residential and commercial surge protection, many models of MOV surge suppressors are available. Some models incorporate resistance to phone line interference and state-of-the-art methods of eliminating noise interference. Many changes are package-oriented. The outlet configuration is altered to allow for more appliances to be connected to the suppressor and to accomodate plug-in transformers; the casing is made of more durable materials, like aluminum; and diagnostic LED indicators (lights indicating the status of the line and whether there is any interference) are added. As residential and commercial electronics needs continue to grow at a quick pace, and protection of electronic equipment—especially computer and telecommunications equipment—becomes paramount, this small, simple electronic device is fast becoming a necessary component of any electronic setup.

Where to Learn More

Books

Miller, Mark and Rex Miller. *Electronics the Easy Way.* Barron's Educational Series, Inc., 1995.

Slone, G. Randy. *Understanding Electricity and Electronics.* McGraw-Hill Companies, Inc., 1996.

Periodicals

Breeze, Frank. "Applying Surge Suppressors." *Consulting-Specifying Engineer,* February 1988, pp. 82-85.

—*Kristin Palm*

Syringe

The hypodermic syringe, also known as the hypodermic needle, is a device used by medical professionals to transfer liquids into or out of the body. It is made up of a hollow needle, which is attached to a tube and a plunger. When the plunger handle is pulled back, fluids are drawn into the tube. The fluid is forced out through the needle when the handle is pushed down. The syringe was introduced in the mid 1800s and has steadily improved with the development of new materials and designs. Today, it has become such an important medical tool that it is nearly a symbol synonymous with the practicing physician.

History

Since the advent of pharmaceutical drugs, methods for administering those drugs have been sought. Various important developments needed to occur before injections through a hypodermic syringe could be conceived. Early nineteenth century physicians were not aware that drugs could be introduced into the body through the skin. One early experiment that demonstrated this idea, however, was performed by Francois Magendie in 1809. In his published work, he outlined a method for introducing strychnine into a dog by using a coated wooden barb. In 1825, A. J. Lesieur described another method for administering drugs through the skin, applying them directly to blisters on the skin. Expanding on results from these experiments, G. V. Lafargue developed a procedure for introducing morphine under the skin using a lancet. A drip needle was invented by F. Rynd in 1844 for the same purpose. However, he did not publish his method until 1861, eight years after the first hypodermic syringe was described.

The first true hypodermic syringe was created by Alexander Wood in 1853. He modified a regular syringe, which at that time was used for treating birthmarks, by adding a needle. He then used this new device for introducing morphine into the skin of patients who suffered from sleeping disorders. A few years later, he added a graduated scale on the barrel and a finer needle. These modifications were enough to attract the attention of the rest of the medical community, resulting in its more widespread use.

Over the years hypodermic syringes have undergone significant changes that have made them more efficient, more useful, and safer. One such improvement was the incorporation of a glass piston within the cylinder. This innovation prevented leaks and reduced the chances of infections, making the device more reliable. The technology for the mass production of hypodermic syringes was developed in the late nineteenth century. As plastics developed, they were incorporated into the design, reducing cost and further improving safety.

Background

The way in which a hypodermic needle works is simple. Fluid, such as a drug or blood, is drawn up through a hollow needle into the main tube when the plunger handle is pulled back. As long as the needle tip remains in the fluid while the plunger handle is pulled, air will not enter. The user can determine exactly how much material is in the tube by reading the measuring marks on the side of the tube. The liquid is dispensed out through the needle when the plunger handle is pushed back down.

The term hypodermic syringe comes from the Greek words *hypo*, meaning under, and *derma*, meaning skin. These terms are appropriate because they describe exactly how the device functions. The needle is used to pierce the top layer of the skin, and the material in the tube is injected in the layer below. In this subcutaneous layer, most injected materials will be readily accepted into the bloodstream and then circulated throughout the body.

A syringe is one of three primary methods for introducing a drug into the body. The others are transepidermal (through the skin) and oral. Using a hypodermic needle as the method of drug administration has some significant advantages over oral ingestion. First, the drugs are protected from the digestive system. This prevents them from being chemically altered or broken down before they can be effective. Second, since the active compounds are quickly absorbed into the bloodstream, they begin working faster. Finally, it is more difficult for the body to reject drugs that are administered by syringe. Transepidermal drug administration is a relatively new technology, and its effects are generally not as immediate as direct injection.

Design

There are many hypodermic syringe designs available. However, all of them have the same general features, including a barrel, plunger, needle, and cap. The barrel is the part of the hypodermic needle that contains the material that is injected or withdrawn. A movable plunger is contained within this tube. The width of the barrel is variable. Some manufacturers make short, wide tubes, and others make long, thin ones. The exact design will depend to some extent on how the device will be used. The end of the barrel to which the needle is attached is tapered. This ensures that only the desired amount of material will be dispensed through the needle. At the base of the barrel away from the needle attachment, two arms flare out. These pieces allow the needle user to press on the plunger with the thumb while holding the tube in place with two fingers. The other end of the barrel is tapered.

The plunger, which is responsible for creating the vacuum to draw up materials and then discharge them, is made of a long, straight piece with a handle at one end and a rubber plunger head on the other. The rubber head fits snugly against the walls of the barrel, making an airtight seal. In addition to ensuring an accurate amount of material is drawn in, the squeegee action of the plunger head keeps materials off the inner walls of the tube.

The needle is the part of the device that actually pierces the layers of the skin. Depending on how deep the injection or fluid extraction will be, the needle orifice can be thinner or wider, and its length varies. It can also be permanently affixed to the body of the syringe or interchangeable. For the latter type of system, a variety of needles would be available to use for different applications. To prevent accidental needle stick injuries, a protective cap is placed over the top of the needle when it is not in use.

Raw Materials

Since hypodermic syringes come in direct contact with the interior of the body, government regulations require that they be made from biocompatible materials which are pharmacologically inert. Additionally, they must be sterilizable and nontoxic. Many different types of materials are used to construct the wide variety of hypodermic needles available. The needles are generally made of a heat-treatable stainless steel or carbon steel. To prevent corrosion, many are nickel plated. Depending on the style of device used, the main body of the tube can be made of plastic, glass, or both. Plastics are also used to make the plunger handle and flexible synthetic rubber for the plunger head.

The Manufacturing Process

There are many manufacturers of hypodermic needles, and while each one uses a slightly different process for production, the basic steps remain the same, including needle formation, plastic component molding, piece assembly, packaging, labeling, and shipping.

Making the needle

1 The needle is produced from steel, which is first heated until it is molten and then

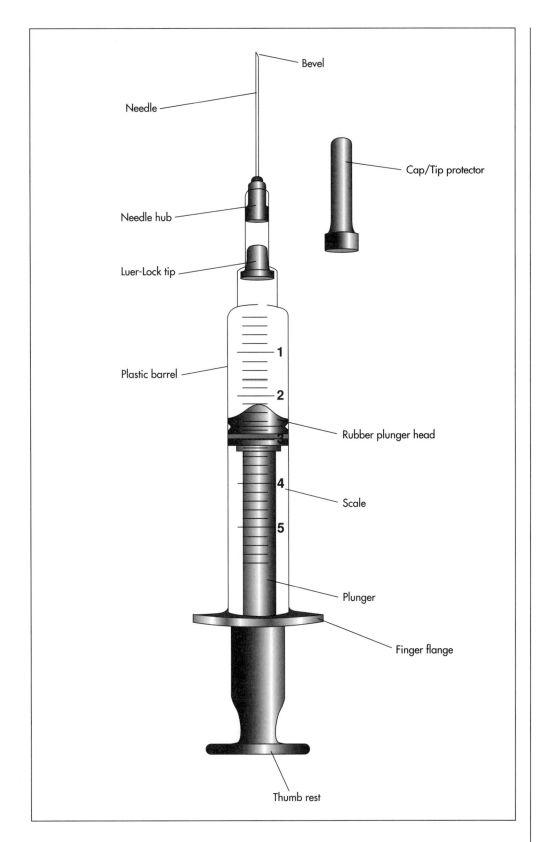

Bevel

Needle

Cap/Tip protector

Needle hub

Luer-Lock tip

Plastic barrel

1

2

Rubber plunger head

3

4

Scale

5

Plunger

Finger flange

Thumb rest

Diagram of a hypodermic syringe. Retraction of the plunger creates the vacuum to draw up materials, which can then be discharged by pushing on the plunger. Its rubber head makes an airtight seal against the walls of the barrel.

drawn through a die designed to meet the size requirements of the needle. As it moves along the production line, the steel is further formed and rolled into a continuous, hollow wire. The wire is appropriately cut to form the needle. Some needles are significantly more complex and are produced directly from a die casting. Other metal components on the needle are also produced in this manner.

Making the barrel and plunger

2 There are various ways that the syringe tube can be fashioned, depending on the design needed and the raw materials used. One method of production is extrusion molding. The plastic or glass is supplied as granules or powder and is fed into a large hopper. The extrusion process involves a large spiral screw, which forces the material through a heated chamber and makes it a thick, flowing mass. It is then forced through a die, producing a continuous tube that is cooled and cut.

3 For pieces that have more complex shapes like the ends, the plunger, or the safety caps, injection molding is used. In this process the plastic is heated, converting it into a liquid. It is then forcibly injected into a mold that is the inverse of the desired shape. After it cools, it solidifies and maintains its shape after the die is opened. Although the head of the plunger is rubber, it can also be manufactured by injection molding. Later, the head of the plunger is attached to the plunger handle.

Assembly and packaging

4 When all of the component pieces are available, final assembly can occur. As the tubes travel down a conveyor, the plunger is inserted and held into place. The ends that cap the tube are affixed. Graduation markings may also be printed on the main tube body at this point in the manufacturing process. The machines that print these markings are specially calibrated to ensure they print measurements on accurately. Depending on the design, the needle can also be attached at this time, along with the safety cap.

5 After all of the components are in place and printing is complete, the hypodermic syringes are put into appropriate packaging. Since sterility of the device is imperative, steps are taken to ensure they are free from disease-causing agents. They are typically packaged individually in airtight plastic. Groups of syringes are packed into boxes, stacked on pallets, and shipped to distributors.

Quality Control

The quality of the components of these devices are checked during each phase of manufacture. Since thousands of parts are made daily, complete inspection is impossible. Consequently, line inspectors randomly check components at fixed time intervals to ensure they meet size, shape, and consistency specifications. These random samples give a good indication of the quality of the hypodermic syringe produced. Visual inspection is the primary test method. However, more rigorous measurements are also performed. Measuring equipment is used to check the length, width, and thickness of the component pieces. Typically, devices such as a vernier caliper, a micrometer, or a microscope are used. Each of these differ in accuracy and application. In addition to specific tests, line inspectors are stationed at various points of the production process and visually inspect the components as they are made. They check for things such as deformed needles or tubes, pieces that fit together incorrectly, or inappropriate packaging.

Hypodermic syringe production is strictly controlled by the United States government, specifically the Food and Drug Administration (FDA). They have compiled a list of specifications to which every manufacturer must comply. They perform inspections of each of these companies to ensure that they are following good manufacturing practices, handling complaints appropriately, and keeping adequate records related to design and production. Additionally, individual manufacturers have their own product requirements.

The Future

Since Alexander Wood introduced the first device, hypodermic syringe technology has greatly improved. Future research will focus on designing better devices that will be safer, more durable, more reliable, and less expensive to produce. Also, improvements in device manufacture will also continue. One example of this is the trend toward utilizing materials such as metals and plastics that have undergone a minimum of processing from their normal state. This should minimize waste, increase production speed, and reduce costs.

Where to Learn More

Books

Chicka, C. and Anthony Chimpa. *Diabetic's Jet Ejectors. Diabetic Gun for Personal Insulin Injection.* H.W. Parker, 1989.

Trissel, Lawrence. *Pocket Guide to Injectable Drugs: Companion to Handbook of Injectable Drugs.* American Society of Health-System Pharmacists, 1994.

—*Perry Romanowski*

Teddy Bear

A teddy bear flew on the Space Shuttle when, in February 1995, Magellan T. Bear from the Elk Creek Elementary School in Pine, Colorado, joined a NASA mission as the school's ambassador.

Background

Most people born in this century have probably encountered teddy bears during their lives, for the teddy bear was developed around the turn of the century. Toy bears developed out of admiration for real bears. About 110,000 years ago, Neanderthal hunters collected skulls of a large brown bear (now extinct) in a shrine where the Cult of the Bear worshiped for over 50,000 years. In modern times, the bear is still considered a symbol of strength, courage, and endurance. Bears share many characteristics with humans, including the abilities to stand upright and to hug, and they also fiercely protect their cubs. Bears are sometimes called the "clowns of the woods" because they dance, sit on their haunches, and roll head over hind paws.

In medieval stories, Bruin the bear was a popular character. In Russia, the bear of folklore evolved into a caricature named "Mishka." The rest of the world learned of Mishka during the 1980 Olympic Games when he became the mascot of the games and a collectible toy. Since the teddy bear's invention, Winnie-the-Pooh, Paddington Bear, Big Teddy and Little Teddy (characters in a set of stories by H. C. Craddock), Yogi and Boo-Boo Bear, Smokey, and Sesame Street's Fozzie bear have become much loved friends and toys from the bear kingdom. Psychologists explain our connection with the teddy bear as "transitional;" children rely on teddies as secretive confidants who help them move away from total dependence on their parents.

History

The teddy bear was born in two parts of the world at about the same time. In 1903 in Giengen, Germany, Margarete Steiff made toy animals out of felt in a small factory owned by her family. Her nephew, Richard Stieff, encouraged her to make a bear based on his sketches following a visit to the Stuttgart Zoo.

Margarete was afraid a toy bear would be too frightening, so she softened the bear's snout into a friendly, pert nose and gave him a slightly hunched back like a real bear. She cut a pattern out of brown mohair pile fabric and created a bear whose head, arms, and legs were articulated so they could move independently and so the bear could sit or stand. The toy was stuffed with excelsior (wood shavings used as packing material), and he had shoe-button eyes and an embroidered nose and other features. At a toy show in Leipzig, Germany, Richard displayed the bear, which caught the attention of an American toy buyer who ordered 3,000 bears. Steiff bears in many variations from Margarete's original have been made in the Steiff factory in Germany ever since, where thousands are now produced every day.

Meanwhile, in the United States, President Theodore Roosevelt was becoming known as a champion of the natural wonders and wildlife of America. While on a diplomatic mission to settle the disputed boundary between the states of Louisiana and Mississippi, he went hunting for the brown bear famous in the area, but the bears eluded him. His hosts did not want to disappoint the President, so they captured a bear for him. But the captive was only a cub, and the President would not hurt a creature who had not been fairly hunted. A political cartoonist named Clifford Berryman drew a characterization of the bespectacled President and the fluffy, sweet-faced bear he had refused to shoot, and the cartoon appeared in newspapers on November 16, 1902.

At a candy store in Brooklyn, New York, Morris Michtom read about the President and the cub. His wife, Rose, made toy ponies to sell in their shop. Mr. Michtom asked her to make a bear instead, and they began selling "Teddy's Bears" in honor of the President. Curiously, the Michtom's bears, later known as Morris bears, looked much like those of Margarete Steiff with button eyes, embroidered mouths and noses, articulated joints that allowed limbs and heads to move, cloth soles, and felt claws.

Michtom wrote to the President for permission to name the bears after him, and the President officially approved the teddy bear. The Steiffs claimed that some of first shipment of 3,000 bears were used to decorate tables at a wedding President Roosevelt attended. By 1907, almost one million teddy bears had been sold, and, since the early 1950s, bear sales have typically been on the order of one-quarter of a million bears per year.

Raw Materials

The original teddy bears were made (on both sides of the Atlantic) with mohair fabric "fur" that was commonly used for upholstery, black leather shoe-button eyes, and excelsior packing as stuffing. In the 1920s, glass eyes were used, but both the glass and button eyes pulled off easily. In 1948, Wendy Boston patented a screw-in eye made of molded nylon. These were supplanted in the 1950s by plastic eyes mounted on stems and fastened securely to the inside of the fabric with grommets or washers. These safety eyes became standard by the 1960s.

Changes have also occurred in the construction of articulated bears. Materials for the original designs included disks and cotter pins (twistable fasteners) that attached separately made arms, legs, and heads to a body that had to be firmly stuffed to support the pins. The early disks were made of wood with leather coverings to protect the outer fur. Companies using this process today have substituted plastic disks, but the manufacture is still largely

The Teddy Bear and the Doll at Christmas
How to Dress Them: By Ida Cleve Van Auken

3673—This Teddy-Bear Outfit Consists of a Pajama Suit, Shown on the Left, a Rough-Rider Suit, in the Centre, and a Play Suit, on the Right

Teddy bear photos from a Ladies' Home Journal article on how to dress them for Christmas, December 1907. (From the collections of Henry Ford Museum & Greenfield Village.)

The teddy bear craze reached its height in America between 1906 and 1908, coinciding with President Theodore "Teddy" Roosevelt's second term in office. Across the country, adults and children alike were going "teddy bear mad."

In addition to the huge variety of regular teddy bears produced by manufacturers both at home and abroad, many unusual teddy bears were also introduced at this time. For example, a "Laughing Roosevelt Bear" was designed to reproduce President Roosevelt's toothy grin. A self-whistling bear produced a whistling sound when it was turned upside down and back upright again. An "Electric Eye" bear had a mechanism in its stomach that, when pressed, activated lights in its eyes (unfortunately, these mechanisms quickly broke).

Also during these years, teddy bear images appeared on many other consumer goods, including automobile accessories, baby rattles, jigsaw puzzles, postcards and greeting cards, and even the cover for a hot-water bottle. And it was in 1907 that John W. Bratton composed "The Teddy Bear Two-Step," to become famous later as the tune for "The Teddy Bear's Picnic."

Some teddy bear lovers enjoyed humanizing their bears by dressing them up in doll-like clothes. The above photos feature a set of teddy-bear clothes that could be sewn for children as a Christmas gift. The pattern, which cost 15 cents, came in three sizes to fit 12-, 16-, and 20-inch bears. The article that accompanied these photos claimed that, "Even the crossest teddy bear would be pleased if he found this nice set of clothes in his Christmas stocking!"

Donna R. Braden

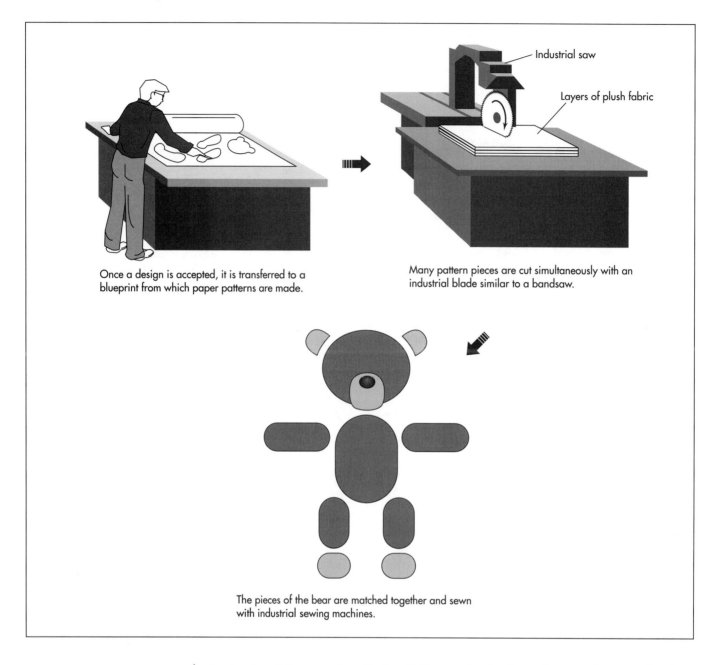

Once a design is accepted, it is transferred to a blueprint from which paper patterns are made.

Many pattern pieces are cut simultaneously with an industrial blade similar to a bandsaw.

The pieces of the bear are matched together and sewn with industrial sewing machines.

by hand and is expensive. In the 1940s, teddy bears were made with prestuffed arms that were sewn into the body seams and legs. The joints were stuffed loosely, so these bears could flex at the joints without being articulated.

Today's materials are most varied in fabric. Plushes made of many fibers are popular because they are fur-like. Early bears were made of mohair that consisted of Angora (goat's wool), sheep's wool, rayon, or silk. Today's plush may be wool, silk, rayon, nylon, other synthetics, or combinations of fibers. Velvet or velveteen (velvet made of cotton instead of silk or synthetics) is pop-ular for its softness and range of colors. Fake fur fabrics are classics for modern bears. Fake fur is different from plush be-cause it has a knitted instead of woven backing that is lightweight and flexible; the shaping of the fiber can eliminate sheen and closely resemble real fur with blended colors. Real fur can also be used for teddy bears, but fur bears are delicate and require special care.

Other fabrics like corduroy, denim, calico, terry cloth, and velour are also bear basics. They are selected for variety in producing durable bears of terry cloth for babies and cowboy bears of denim, for example.

Once the bear pieces are sewn together and pulled right side out, stuffing is blown into the bear through the small diameter tube of the stuffing machine.

Back of the bear is sewn.

Leather and suede (real or synthetic) produces handsome feet and paws. Felt can also be used for these and other features or for entire bears, but this fabric is not strong and tends to wear. Similarly, bears can be knitted or crocheted, but the resulting fabric stretches. Other raw materials include thread, embroidery floss for noses and other features, glue, Velcro™ for fastenings, and eye assemblies made of plastic and metal. Polyester stuffing has replaced the wood shavings used in the original teddy bears to produce products that are more durable and huggable. If the manufactured bears are clothed or decorated, a wide range of fabrics, ribbons, fasteners, and decorations (like eyeglasses for a Teddy Roosevelt bear) may be required.

Design

The design for a new model of teddy bear is first sketched by an artist experienced in toy design and the manufacturing process. Based on a sketch or conception of the planned bear, pieces of the bear are also drawn to be used in making a pattern. The pattern is cut out and assembled, and the prototype bear is examined for "character flaws." If the design prevents the bear from sitting properly, for example, or if the prototype is not appropriately cuddly or distinguished, the design is redrawn, shapes of pieces are changed, or different colors or fabrics may be used to make another prototype. Many trials may be required to perfect the design before it is ready for large-scale manufacture. Factors such as popular interests and headlines may be considered in planning new designs where a respectable volume of sales is needed to justify manufacture. For instance, a bear producer in California sells a stuffed bear resembling the bruin on the California state flag, but this item would not necessarily sell well elsewhere.

The Manufacturing Process

1 After the design is accepted, it is transferred to a blueprint from which paper patterns are made. The shapes are cut out and pinned to fabric. Many layers of plush may be stacked with the pattern pinned on the top, and a cutter with a blade much like a band saw is used to cut out many pieces at one time. Rows of workers sit behind industrial **sewing machine**s. Each is responsible for one style of bear. He or she will assemble the small pieces first, add the eyes and grommet attachments to the face, and then stitch the parts together. The bears are inside out with their seams exposed and openings down the long seam in the back. Finally, the bears are pulled right side out and dispatched to be stuffed. The volume a worker produces depends on the type of bear; one may produce 35 to 45 baby bears in one day or eight or nine larger species.

2 The stitched bear is then stuffed. The manufacturer purchases polyester fiber in 500 lb (227 kg) bales in which the fiber is tightly packed. The fiber is dual density with a very fine fiber and a thicker, slightly wavy fiber; together, these fibers have the most desirable packing quality for stuffing animals. Because the polyester comes so tightly packed, batches of it are tossed into a picker, which is a barrel lined with spikes that fluffs the stuffing. The fluffed material is blown by air into a stuffing machine. These machines have evolved little since World War II, when the machines were used to stuff flight jackets for the United States Navy and Army Air Force. Air pressure blows the stuffing into the bear through a small diameter tube. The operator can manipulate the bear to direct the stuffing to various parts of the toy. A pedal release on the stuffing machine controls the puffs of stuffing. Two pushes on the pedal, for example, may be needed to pump enough stuffing from the machine's nozzle into the nose of the bear, which needs to be firm. The head, feet, and paws also need to be firm, but the body should be squeezably soft. The operator can always apply the "hug test" to evaluate satisfactory cuddliness and provide quality control from bear to bear.

3 The stuffed bear is then passed to another worker, a "bear surgeon," who stitches up the opening in the back of the bear. The whole bear is then groomed. Because the plush fabric was stitched inside out, "fur" is caught in the seams and must be pulled out so the seams do not show. An electrically powered wire brush is used to fluff the seams, and the bear is then blown through an array of air jets to remove the loose fuzz and brush the fur. Final details like costumes, ribbons, and accessories are added before the bear is packaged for transport and sale.

Quality Control

Although teddy bears are mass produced, their design and production requires hands-on attention throughout the manufacturing process. Fabric cutters, assemblers, stuffing machine operators, and bear surgeons, groomers, and dressers all share great pride in knowing their product will give and receive much love over the years of its life. Hand-production ensures that the bears are inspected at every step in their manufacture, and that mistakes do not reach the packaging department.

Byproducts/Waste

There are no byproducts from the manufacture of teddy bears, although there are co-products consisting of other types of stuffed animals and dolls. Small stuffed creatures are sometimes designed with pattern pieces to fit between bear segments so fabric is not wasted. Fabric scraps and errant stuffing constitute most of the waste from bear production, and this material cannot be recycled and is disposed.

Safety concerns are important in the bear factory. The machinery is powered by electricity, requiring safety precautions; fabric cutters, sewing machines, wire brushes, and other machines are equipped with emergency shutoffs and other safety devices. Operators wear masks over their mouths and noses to prevent them from inhaling the airborne fluff. Safety glasses are also worn during some operations.

The Future

The future for teddy bears can only be a picnic. New interpretations of this much-loved creature follow trends in movies, television, and toy fashion from character bears to miniature, beanbag-like versions that are inexpensive and collectible. A teddy bear has even flown on the Space Shuttle when, in February 1995, Magellan T. Bear from the Elk Creek Elementary School in Pine, Colorado, joined a NASA mission as the school's ambassador. Teddy bears are popular with children, obviously, but also with adult collectors who build on their childhood friends and often invest in limited editions or bears made from prized designs and rare materials. In a world with increasing emphasis on technology, teddy bears remind us of our childhood and provide an unlimited supply of fuzzy hugs.

Where to Learn More

Books

Bialosky, Peggy and Alan. *The Teddy Bear Catalog*. Workman Publishing, 1980.

Börk, Christina. *Big Bear's Book by Himself*. Rabén & Sörgren, 1994.

Bull, Peter. *The Hug of Teddy Bears*. NAL-Dutton, 1984.

Cockrill, Pauline. *The Teddy Bear Encyclopedia*. Dorling Kindersley, 1993.

Hillier, Mary. *Teddy Bears: A Celebration*. Beaufort Books, 1985.

Hutchings, Margaret. *Teddy Bears and How to Make Them*. Dover Publications, 1964.

Kay, Helen. *The First Teddy Bear*. Stemmer House, 1985.

King, Constance Eileen. *The Encyclopedia of Toys*. Crown Publishers, Inc., 1978; pp. 81-84.

Morris, Ramona and Desmond. *Men and Pandas: Why People Like Pandas (and Teddy Bears) So Much*. McGraw-Hill, 1966.

Schoonmaker, Patricia N. *A Collector's History of the Teddy Bear*. Hobby House Press, 1981.

Vosburg Hall, Carolyn. *The Teddy Bear Craft Book*. Van Nostrand Reinhold Company, 1983.

Periodicals

Coy, Peter. "How the talking 'TV Teddy' tunes in." *Business Week*, October 25, 1993, p. 95.

Luscombe, Belinda. "Bearable Occasion." *Time,* October 7, 1996, p. 101.

Merkin, Daphne. "The Meaning of a Bear Market." *New Yorker,* December 26, 1994-January 2, 1995, p. 52.

"The Fur Flies." *People,* February 6, 1995, p. 54.

"The Ultimate Bear Market." *Economist,* September 29, 1990, p. 66.

—Gillian S. Holmes

Television

In the United States, more than 98% of households own at least one television set and 61% receive cable television.

Background

Among the technical developments that have come to dominate our lives, television is surely one of the top ten. In the United States, more than 98% of households own at least one television set and 61% receive cable television. The average household watches television for seven hours per day, which helps to explain why news, sports, and educational entities, as well as advertisers, value the device for communication.

The device we call the television is really a television receiver that is the end point of a broadcast system that starts with a television camera or transmitter and requires a complicated network of broadcast transmitters using ground-based towers, cables, and satellites to deliver the original picture to our living rooms. The U.S. television picture, whether black and white or color, consists of 525 horizontal lines that are projected onto screens with a four to three ratio of width to height. By electronic methods, 30 images per second, each broken into these horizontal lines, are scanned onto the screen.

History

The development of the television occurred over a number of years, in many countries, and using a wide application of sciences, including electricity, mechanical engineering, electromagnetism, sound technology, and electrochemistry. No single person invented the television; instead, it is a compilation of inventions perfected by fierce competition.

Chemicals that are conductors of electricity were among the first discoveries leading to the TV. Baron Öns Berzelius of Sweden isolated selenium in 1817, and Louis May of Great Britain discovered, in 1873, that the element is a strong electrical conductor. Sir William Crookes invented the cathode ray tube in 1878, but these discoveries took many years to merge into the common ground of television.

Paul Nipkow of Germany made the first crude television in 1884. His mechanical system used a scanning disk with small holes to pick up image fragments and imprint them on a light-sensitive selenium tube. A receiver reassembled the picture. In 1888, W. Hallwachs applied photoelectric cells in cameras; cathode rays were demonstrated as devices for reassembling the image at the receiver by Boris Rosing of Russia and A. A. Campbell-Swinton of Great Britain, both working independently in 1907. Countless radio pioneers including Thomas Edison invented methods of broadcasting television signals.

John Logie Baird of Scotland and Charles F. Jenkins of the United States constructed the first true television sets in the 1920s by combining Nipkow's mechanical scanning disk with vacuum-tube amplifiers and photoelectric cells. The 1920s were the critical decade in television development because a number of major corporations including General Electric (GE), the Radio Corporation of America (RCA), Westinghouse, and American Telephone & Telegraph (AT&T) began serious television research. By 1935, mechanical systems for transmitting black-and-white images were replaced completely by electronic methods that could generate hundreds of horizontal bands at 30 frames per second. Vladimir K. Zworykin, a Russian immigrant who first worked for Westing-house then RCA, patented an electronic camera tube based on the cathode tube. Philo T. Farnsworth and Allen B. Dumont, both Americans, developed a pickup tube that became the home television receiver by 1939.

The Columbia Broadcasting System (CBS) had entered the color TV fray and battled with RCA to perfect color television, initially with mechanical methods until an all-electronic color system could be developed. Rival broadcasts appeared throughout the 1940s although progress was slowed by both World War II and the Korean War. The first CBS color broadcast on June 25, 1951, featured Ed Sullivan and other stars of the network. Commercial color television broadcasts were underway in the United States by 1954.

Raw Materials

The television consists of four principle sets of parts, including the exterior or housing, the audio reception and speaker system, the picture tube, and a complicated mass of electronics including cable and antennae input and output devices, a built-in antenna in most sets, a remote control receiver, computer chips, and access buttons. The remote control or "clicker" may be considered a fifth set of parts.

The housing of the set is made of injection-molded plastic, although wood cabinets are still available for some models. Metals and plastics also comprise the audio system. The picture tube requires precision-made glass, fluorescent chemical coatings, and electronic attachments around and at the rear of the tube. The tube is supported inside the housing by brackets and braces molded into the housing. The antennae and most of the input-output connections are made of metal, and some are coated with special metals or plastic to improve the quality of the connection or insulate the device. The chips, of course, are made of metal, solder, and silicon.

Design

The design of the television requires input and teamwork on the part of a range of design engineers. Audio, video, plastics, fiber

Prismatic scanning disc mount made by C. Francis Jenkins in 1923. (From the collections of Henry Ford Museum & Greenfield Village.)

To the surprise of most people, television transmission began almost 25 years before the end of World War II. John Logie Baird, in England, and C. Francis Jenkins, in the United States, both made public demonstrations of television in 1925. Unlike post-war electronic televisions, these early systems used mechanical scanning methods.

Jenkins made significant contributions to optical transmission research during the 1920s. During 1922-23, he constructed mechanical prismatic disc scanners to transmit images. These scanners focused and refracted light through prisms ground into the edges of overlapping glass discs. As the discs rotated, a point of light scanned horizontally and vertically across a light-sensitive surface. This generated electrical signals necessary for transmission. In 1922 Jenkins sent facsimiles of photographs by telephone, and the following year transmitted images of President Harding and others by radio with an improved scanner. Unlike television, however, these first tests only sent still pictures.

Jenkins publicly broadcast moving images with his equipment in 1925. His first 10-minute broadcast showed in silhouette the motions of a small operating windmill. By 1931, he had experimental television stations operating in New York and Washington D.C. He sold receiver kits to those wishing to view his telecasts and encouraged amateur participation. With other companies, Jenkins contributed to a small, short-lived mechanical television "boom." By 1933, however, the poor image quality of mechanical scanning convinced larger manufacturers to pursue the possibilities of electronic technologies, and the mechanical television era ended.

Erik Manthey

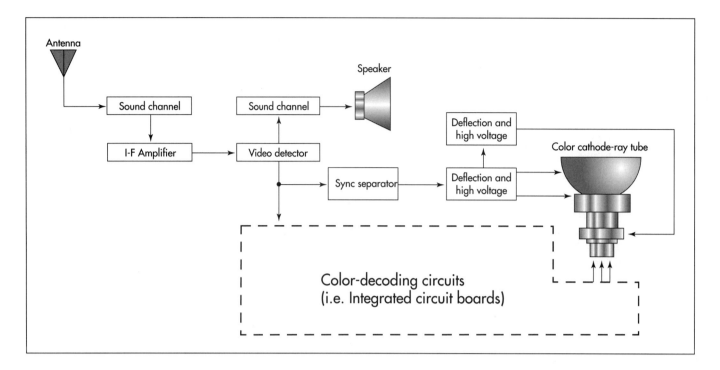

Diagram of a television receiver.

optics, and electronics engineers all participate in conceptualizing a new television design and the technical and sales features that will set it apart. A new design of television may have one or many new applications of technology as features. It may only be a different size of an existing model, or it may include an array of new features such as an improved sound system, a remote control that also controls other entertainment devices, and an improved screen or picture, such as the flat black screens that have entered the marketplace recently.

Conceptual plans for the new set are produced by the engineering team. The concept may change and be redrawn many times before the design is preliminarily approved for manufacture. The engineering specialists then select and design the components of the set, and a prototype is made to prove out the design. The prototype is essential, not only for confirming the design, appearance, and function of the set, but also for production engineers to determine the production processes, machining, tools, robots, and modifications to existing factory production lines that also have to be designed or modified to suit the proposed new design. When the prototype passes rigid reviews and is approved for manufacture by management, detailed plans and specifications for design and production of the

model are produced. Raw materials and components manufactured by others can then be ordered, the production line can be constructed and tested, and the first sets can begin their ride down the assembly line.

The Manufacturing Process

Housing

1 Almost all television housings are made of plastic by the process of injection molding, in which precision molds are made and liquid plastic is injected under high pressure to fill the molds. The pieces are released from the molds, trimmed, and cleaned. They are then assembled to complete the housing. The molds are designed so that brackets and supports for the various components are part of the housing.

Picture tube

2 The television picture tube, or cathode ray tube (CRT), is made of precision glass that is shaped to have a slightly curved plate at the front or screen. It may also have a dark tint added to the face plate glass, either during production of the glass or by application directly to the inside of the screen. Darker face plates produce improved picture contrast. When the tube is manufactured, a water suspension of phos-

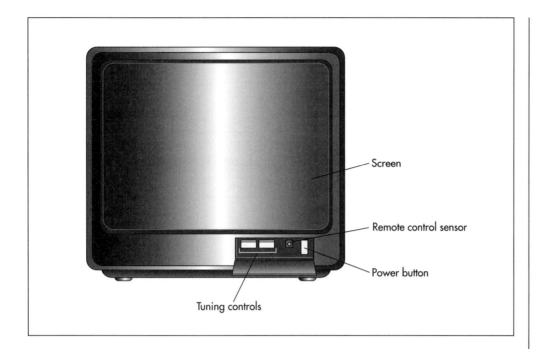

Screen

Remote control sensor

Power button

Tuning controls

phor chemicals is allowed to settle on the inside of the face plate, and this coating is then overlaid with a thin film of aluminum that lets electrons pass through. The aluminum serves as a mirror to prevent light from bouncing back into the tube.

Glass for picture tubes is supplied by a limited number of manufacturers in Japan and Germany. Quantities of the quality of glass needed for picture tubes are limited, and the emergence of large-screen sets has created a shortage in this portion of the industry. The large screens are also very heavy, so flat-panel displays using plasma-addressed liquid crystal (PALC) displays were developed in the 1980s. This gas plasma technology uses electrodes to excite layers of neon or magnesium oxide, so they release ultraviolet radiation that activates the phosphor on the back of the television screen. Because the gas is trapped in a thin layer, the screen can also be thin and lightweight. Projection TVs use digital micro mirror devices (DMDs) to project their pictures.

A shadow mask with 200,000 holes lies immediately behind the phosphor screen; the holes are precisely machined to align the colors emitted by three electron beams. Today's best picture tubes have shadow masks that are manufactured from a nickel-iron alloy called Invar; lesser quality sets have masks of iron. The alloy allows the tube to operate at a higher temperature

without distorting the picture, and higher temperatures allow brighter pictures. Rare-earth elements have also been added to the phosphor coating inside the tube to improve brightness.

The electrons are fired by three tubular, metal electron guns that are carefully seated in the neck, or narrow end, of the tube. After the electron guns are placed inside the tube, the picture tube is evacuated to a near vacuum so air does not interfere with the movement of the electrons. The small opening at the rear of the tube is sealed with a fitted electrical plug that will be positioned near the back of the set. A deflection yoke, consisting of several electromagnetic coils, is fitted around the outside of the neck of the picture tube. The coils cause pulses of high voltage to direct the scanning electron beams in the proper direction and speed.

Audio system

3 The housing also contains fittings for speakers, wiring, and other parts of the audio system. The speakers are usually made by a specialized manufacturer to the specifications of the television manufacturer, so they are assembled in the set as components or a subassembly. Electronic sound controls and integrated circuitry are assembled in panels in the set as it travels along the assembly line.

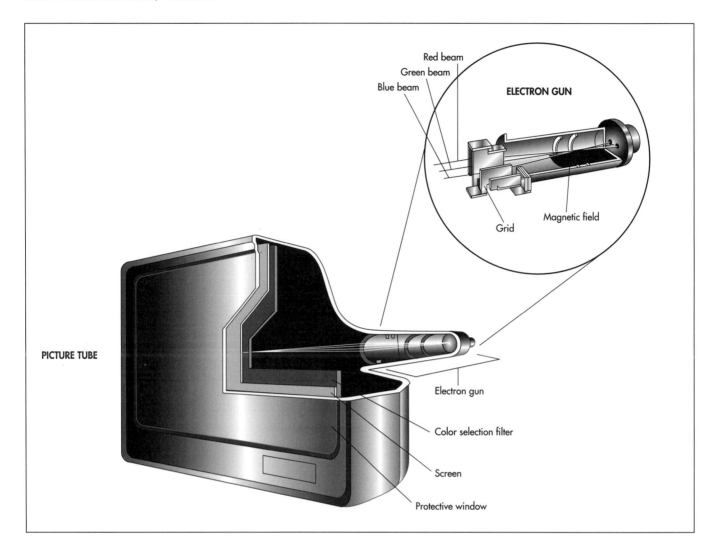

PICTURE TUBE

Red beam
Green beam
Blue beam

ELECTRON GUN

Grid

Magnetic field

Electron gun

Color selection filter

Screen

Protective window

The electrons are fired by three tubular, metal electron guns seated in the neck, or narrow end, of the picture tube. After the electron guns are placed inside the tube, the picture tube is evacuated to a near vacuum so air does not interfere with the movement of the electrons. A color selection filter with 200,000 holes lies immediately behind the television screen; the holes are precisely machined to align the colors emitted by three electron beams.

Electronic parts

4 When the picture tube and the audio speakers and attachments are assembled in the set, other electronic elements are added to the rear of the set. The antennae, cable jacks, other input and output jacks, the electronics for receiving remote control signals, and other devices are prepared by specialty contractors or as subassemblies elsewhere on the assembly line. They are then mounted in the set, and the housing is closed.

Quality Control

As with all precision devices, quality control for the manufacture of the television is a rigid process. Inspections, laboratory testing, and field testing are performed during the development of prototypes and throughout manufacture so the resulting television is not only technologically sound but safe for use in homes and businesses.

Byproducts/Waste

There are no byproducts from the manufacture of the television, although many other devices are a part of the television "family" and are often produced by the same manufacturer. These include the remote control, computer monitors, video recorders (VCRs), laser disc players, and a host of devices that may require compatible design and components. Specialized televisions are produced for some industries, including television studios and mobile broadcast facilities, hospitals, and for surveillance applications for public safety and use in inaccessible or dangerous locations.

Wastes may include metals, plastics, glass, and chemicals. Metals, plastics, and glass are isolated and recycled unless they have been specially treated or coated. Chemicals are carefully monitored and controlled; often, they can be purified and recycled, so disposal of hazardous wastes can be minimized. Hazardous waste plans are in effect in all stages of manufacture, both to minimize quantities of waste and to protect workers.

The Future

The future of television is now. High Definition Television (HDTV) was developed by the Japanese Broadcast Corporation and first demonstrated in 1982. This system produces a movie-quality picture by using a 1,125-line picture on a "letter-box" format screen with a 16 to nine width to height ratio. High-quality, flat screens suitable for HDTV are being perfected using synthetic diamond film to emit electrons in the first application of synthetic diamonds in electronic components. Other developments in the receiver include gold-plated jacks, an internal polarity switch on large screens that compensates for the effect of Earth's magnetic field on image reception, accessories to eliminate ghosts on the screen, the Invar shadow mask to improve brightness, and audio amplifiers. Liquid crystal display (LCD) technology is also advancing rapidly as an alternative to the cumbersome television screen. Assorted computer chips add functions like channel labeling, time and data displays, swap and freeze motions, parental channel control, touch screens, and a range of channel-surfing options.

Digital television of the future will allow the viewer to manipulate the angle of the camera, communicate with the sports commentator, and splice and edit movies on screen. Two-way TV will also be possible. Current screens may be used thanks to converter boxes that change the analog signal that presently energizes the phosphors on the back of your television screen to digital signals that are subject to less distortion—and are the language of computers. Computer technology will then allow a world of manipulation of the data as well as broadcast of six times as much data.

The future of television manufacture may be anywhere but in the United States. Thirty percent of all televisions manufactured by Japanese companies are made in factories in Mexico. The factories themselves will soon be producing hybrids in which the television, computer monitor, and telephone are a single unit, although this development will take further improvements in compatibility between machines that speak analog versus digital language and the creation of PC-to-video bridges. Proof of the possibility of this integrated future exists now in Internet access that is now available through television cable converters and the living room TV screen.

Where to Learn More

Periodicals

Barker, Dennis P. "High-tech tubes: today's technology delivers the best TV pictures ever." *Popular Mechanics*, April 1997, p. 60.

"Bell Atlantic puts on its producer's hat." *Business Week*, April 18, 1994, p. 116E.

Braithwaite, Lancelot. "Ghost busted: a first look at Magnavox's ghost canceler unearths new levels of image clarity." *Video Magazine*, November 1996, p. 56.

Doherty, Brian. "Made in America?" *Reason*, August/September 1993, p. 50.

Fisher, David E. and Marshall Jon Fisher. "The Color War." *Invention & Technology*, Winter 1997, pp. 8-18.

Goldberg, Ron. "Adding TV to the PC." *Popular Mechanics*, April 1993, p. 138.

Heald, Tom. "The next wave." *Video Magazine*, September 1996, p. 32.

Levine, Martin. "Dark tubes stake a claim." *Video Magazine*, November 1993, p. 64.

Lewyn, Mark. "Two-way TV isn't quite ready for Prime Time." *Business Week*, April 13, 1992, pp. 38-39.

Miller, Michael J. "Yet Another Dinosaur?" *PC Magazine*, September 14, 1993, p. 81.

"Mi TV es Su TV?" *Business Week*, November 1, 1993; p. 8.

"Romancing the Stone." *Video*, December 1993, p. 12.

"TV design receives gas assist." *Design News*, August 15, 1994, p. 28.

"TV does digital: in a world of bits and bytes, you control the camera angles and everything you see on TV." *Science World,* February 7, 1997, p. 18.

"Videotest: ProBono." *Video,* April 1996, p. 53.

—*Gillian S. Holmes*

Tennis Racket

Background

The game of tennis dates back officially to 1873, when the first book of rules was published by Major Walter Clopton Wingfield of north Wales. But tennis has antecedents in ball games played with the hand that evolved in Europe before the Renaissance. These games were played first with the bare hand, later with gloved hands, then with hands wrapped in rope. Later, a wooden bat was introduced, and the first rackets seem to have showed up during the fifteenth century. These early rackets were smaller than modern tennis rackets, and were strung in various patterns. When rules of tennis were standardized by Wingfield and others following him, the shape and size of the court was specified, and the kind of ball that could be used. There were, however, no rules governing the racket size, shape, or material makeup.

Until 1965, all professional tennis rackets were made of wood. A steel tennis racket was patented in 1965 by the French player Rene Lacoste, and in 1968 the Spalding company marketed the first aluminum rackets. These metal rackets caught on gradually. What the metal rackets made possible was a change in design to allow a broader head. Wooden rackets could not be made wider or longer in the head without causing problems with the stringing: if the head was too broad, string tension became too great, and the racket did not play well. But the greater strength of metal frames could accommodate greater string tension. An oversized aluminum racket developed by Howard Head in the mid 1970s was at first scoffed at by professionals, but amateurs quickly discovered that they could hit better with it. The prime hitting area, or so called

"sweet spot," was doubled in size in the the new, larger rackets, and so for most people, it was easier to use. The larger rackets became the standard at all levels of play by the early 1980s.

The International Tennis Federation finally adopted rules defining acceptable tennis rackets in 1981. The Federation had banned a racket introduced in 1977 that used an innovative stringing technique. Players using "spaghetti string" rackets scored huge upsets over high-ranked opponents, and after only five months, these rackets were not allowed in professional play. The first racket rules allowed the racket and strings to be made of any materials, and did not limit the size, weight, or shape. Strings were required to interlace or be bonded at cross points at least a quarter inch (0.64 cm) and not more than a half inch (1.3 cm) apart. No attachments were allowed that might alter the flight of the ball, and the weight distribution along the longitudinal axis of the racket must not change in play. Later the maximum length of rackets was limited to 32 in (81 cm). This was modified again in January 1997, bringing the length back down to 29 in (74 cm).

The average racket is now about 28 in (71 cm) long, and weighs from 10-14 oz (284-397 g). There have been many recent innovations in racket technology, not all of which have caught on with players. One maker markets a hexagonal racket, while others are making rackets with extra wide bodies. A racket made of a new material—graphite fiber-reinforced thermoplastic viscoelastic polymer—was designed to have variable flexibility, depending on how hard the ball is struck. A design to alleviate ten-

Rackets are now being designed by laboratory scientists who use mathematics to calculate the effects of weight, size, and material changes.

nis elbow employs small lead bearings enclosed in plastic chambers inside the head frame. The movement of the bearings as the racket connects with the ball is supposed to cushion the vibrations that might cause pain to the player's arm. But the most common rackets are now made of aluminum or of a composite of graphite, fiberglass, and other materials.

Raw Materials

Aluminum rackets are usually made of one of several alloys. One popular alloy contains 2% silicon, as well as traces of magnesium, copper, and chromium. Another widely used alloy contains 10% zinc, with magnesium, copper, and chromium. The zinc alloy is harder, though more brittle, and the silicon alloy is easier to work. Composite rackets may contain many different materials. They usually consist of a sandwich of different layers around a hollow core or a polyurethane foam core. The typical layers of a composite racket are fiberglass, graphite, and boron or kevlar. Other materials may be used as well, such as ceramic fibers for added strength.

Other materials found in tennis rackets are nylon, gut, or synthetic gut for the strings, and leather or synthetic material for the handle grip. Nylon is probably the most common string material, and only a few professionals still use gut, which is made from twisted cow or sheep intestine. Synthetic gut is made from nylon which has been twisted to achieve the same effect as natural gut. Old wooden rackets usually used a leather handle grip, but modern rackets generally use a leather-like replacement such as vinyl. Rackets may have plastic parts too, such as the yoke at the base of the head and the cap at the bottom of the handle.

The Manufacturing Process

Most rackets sold in the United States are mass produced at one of several large factories in Japan or elsewhere in Asia. So regardless of the brand, chances are the racket was made by one of the methods described below. Rackets with unusual features might be exceptions. Also, top of the line rackets are often sold unstrung, and the buyer has it strung to his or her specifications at a pro shop. So in this case, the stringing step at the factory would be skipped.

Aluminum racket

1 *Forming the frame.* There are two methods for forming aluminum rackets. The aluminum may be melted and forced through a die in the shape of the racket frame. Or the metal may first be melted and extruded into a tube, and then the tubing drawn through a die.

2 *Drilling and sanding.* The rough racket is then placed in a drilling machine, and holes are drilled for the yoke—the throat piece that holds the bottom of the strings—on the sides for the strings, and at the base of the stick. The drilling machine uses multiple spindles, each holding a drill bit in position for each string hole. The racket is held in place horizontally in the center of the machine. The drills are then activated, and all the holes are drilled simultaneously. The frames are then placed in a sander to smooth out sharp edges left from the drilling.

3 *Tempering.* At this stage, the rackets are tempered, that is, subjected to heat and rapid cooling. This process hardens the aluminum, giving the racket additional strength. The rackets are placed on a tray in an oven and heated to white-hot. Then the tray is removed from the oven and the rackets are immersed in water. After tempering, the rackets may also be anodized. They are immersed in a mild sulfuric acid solution, and an electric current is passed through the bath. This treatment changes the surface of the aluminum, and gives the rackets a shiny finish.

4 *Stringing.* A grommet strip is inlaid in the groove around the edge of the head. The flexible grommet strip, usually plastic, has been predrilled so that its holes fit over the string holes in the frame head. Then the yoke is fitted into the base of the racket head. Now the racket is ready for stringing. Each racket is strung individually, by a worker seated at a stringing machine. The worker first clamps the racket into the machine, which holds it horizontally. The worker forces the strings through the holes using a powerful threader mounted on a movable bar above the racket. The lengthwise strings are pulled through first, then

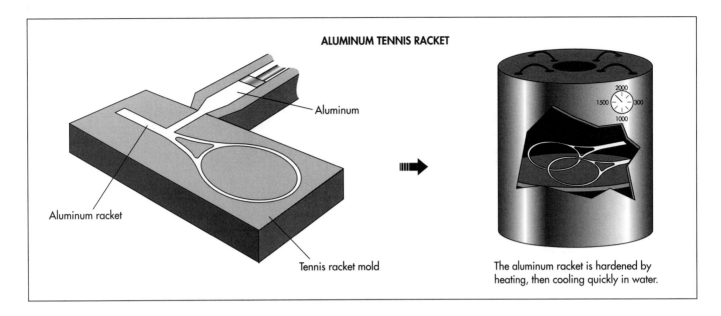

ALUMINUM TENNIS RACKET

Aluminum

Aluminum racket

Tennis racket mold

The aluminum racket is hardened by heating, then cooling quickly in water.

the cross strings are woven through, and the tension is adjusted.

5 *Finishing.* To finish the racket, a worker cuts the end of the handle and inserts a cap called the butt cap. Next the worker wraps strong double-stick tape around the handle, followed by vinyl grip tape. After this, the strings may be imprinted with a logo, and the frame may be stamped with a decal. Inspectors check the racket for nicks and mars, and make sure it conforms to size and weight specifications. Rackets may then go through a final cleaning stage. Then workers place them in protective covers, the rackets are packaged, and finally sent to a warehouse for distribution.

Composite racket

6 *Forming the frame.* Composite rackets are made out of layers of different materials, usually graphite and fiberglass, and perhaps other layers containing boron, kevlar, or a material similar to fiberglass that contains ceramic particles. The racket manufacturer begins by assembling the layers as a flat sandwich. The sandwich is then cut into strips, and the strips rolled around a hollow, flexible tube. The wrapped tube is then placed into a racket-shaped mold. The tube extends all the way through the racket, and is connected to a pump. Then the mold is heated, and air pumped into the tube. The pressure of the air in the tube, along with the heat, bond the layers of the sandwich. Alternately, the hollow tube may be filled with

polyurethene foam. The foam expands as the mold is heated, consolidating the materials.

7 *Drilling and sealing.* Workers release the rackets from the molds and carry them to an inspection area, where any defective ones are removed. The end of the frame is cut, and then the rackets are placed in a drilling machine and the string holes are drilled, as above. After drilling, the rackets are brushed with a polymer coating and placed in a dryer. This step is repeated several times, and then the rackets are sanded. Before the final coating, the brand name decal is applied.

8 *Stringing and finishing.* The next steps are the same as for the aluminum racket previously described. A grommet strip and yoke are fixed in the appropriate grooves, and workers string the rackets one at a time on stringing machines. A logo or brand name may be screen printed on the strings. Workers insert the butt cap, then wind double-stick tape and grip tape around the handle. Then the rackets are cleaned, inspected, packaged, and sent to a warehouse.

Quality Control

Inspectors check the rackets at many points in the manufacturing process. When the frames are first taken from the molds, they are inspected visually. Defective rackets are set apart, and passing rackets may be roughly graded for quality. Aluminum rackets are subjected to stress tests to determine

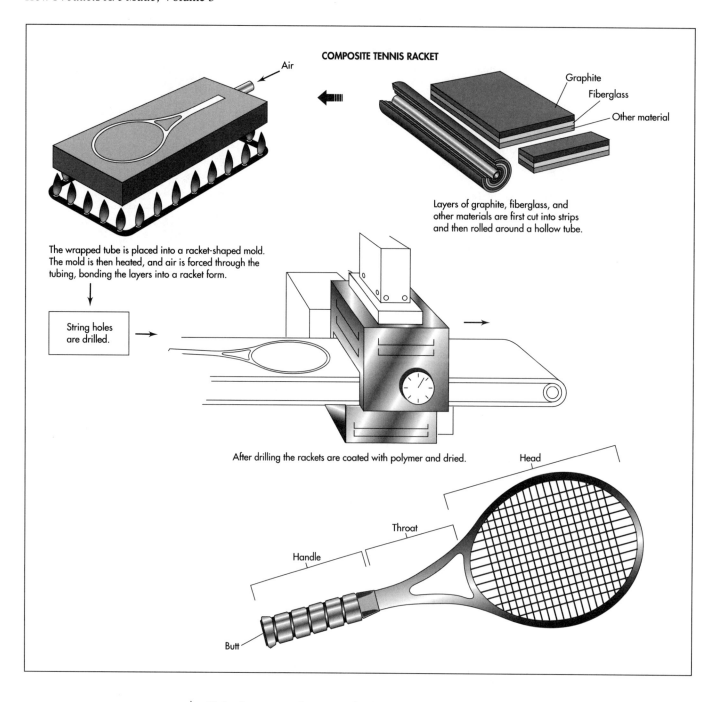

COMPOSITE TENNIS RACKET

Air

Graphite
Fiberglass
Other material

Layers of graphite, fiberglass, and other materials are first cut into strips and then rolled around a hollow tube.

The wrapped tube is placed into a racket-shaped mold. The mold is then heated, and air is forced through the tubing, bonding the layers into a racket form.

String holes are drilled.

After drilling the rackets are coated with polymer and dried.

Head

Throat

Handle

Butt

if the frames are the proper hardness. Composite rackets are also tested for stiffness. Inspectors weigh both types of racket, usually before and after stringing, to make sure they meet specifications. They also check the balance, as this is extremely important to how well the racket plays. It should not be too heavy at the head or in the handle, but balance close to the mid-point (though some models are designed to be deliberately head-heavy). The grommet holes are inspected. If these are not smooth or even, string tension is affected, and strings may break against rough edges. The finishing details are also subjected to visual inspection. The butt cap should fit snugly, and the printing on the frame and strings should be even and clear. The grip should be wound smoothly, and there should be no nicks or scratches. Some rackets may be play-tested, especially if it is a new design.

The Future

The science of tennis rackets is surprisingly complex—not the manufacturing process but the physics of string and frame vibration as the ball connects with the racket.

Rackets are now being designed by laboratory scientists who use mathematics to calculate the effects of weight, size, and material changes. Since the rules governing acceptable rackets are very broad, innovators have a lot of leeway. New rackets are also being made with computer-aided design (CAD) and computer-aided manufacturing (CAM), which allows precise calculation of material rigidity and center of gravity. As such advanced science is being lavished on the tennis racket, doubtless new models with eccentric features will continue to be developed. The trend today is toward lighter, bigger rackets, and these are viable because of advanced materials engineering.

Where to Learn More

Periodicals

Brody, Howard. "How Would a Physicist Design a Tennis Racket?" *Physics Today*, March 1995, pp. 26-31.

Fisher, Marshall Jon. "Racket Science." *The Sciences*, November/December 1996, pp. 10-11.

Gelberg, Nadine J. "The Big Technological Tennis Upset." *Invention & Technology*, Spring 1997, pp. 56-61.

Sparrow, David. "More Length, More Strength." *Sports Illustrated*, May 27, 1996, p. 16.

—*Angela Woodward*

Tissue with Lotion

Tissues of this type are made by a process in which the nonwoven fabric is made from a solution of cellulose fibers and water, formed into a sheet, then coated with softening agents. Finally, the coated fabric is cut into individual tissues, folded, and packaged for sale.

Background

Facial tissues belong to a class of paper products used extensively for personal hygiene in modern society. Other products of this type include paper towels, napkins, and sanitary (or toilet) tissue. These products are designed to be highly absorbent, soft, and flexible. These pleasant tactile properties are especially important for facial and bathroom tissues, considering their use. To optimize pleasant skin feel, tissues have been developed with softening agents or lotion-type ingredients to reduce any chafing effect on delicate parts of the body.

Tissues of this type are made by a process in which the nonwoven fabric is made from a solution of cellulose fibers and water, formed into a sheet, then coated with softening agents. Finally, the coated fabric is cut into individual tissues, folded, and packaged for sale.

Tissue softness is a tactile perception characterized by the sheet's physical properties, such as flexibility or stiffness, texture, and frictional properties. Historically it has been difficult to soften the tissue surface without interfering with other properties of the fabric. For example, softness can be increased by adding agents that interfere with the way the fibers within the tissue interact, making them less closely bonded to each other. These are known as debonding agents. However, these materials tend to decrease the tensile strength of the fabric and may irritate skin on contact. Enhanced softness can also be achieved by coating the fabric with oily materials. However, this limits the amount of moisture the tissue can absorb. In fact, the coating can also make the fabric so hydrophobic (water hating)

that it can not be processed properly in sewage plants. Another problem is that some coating materials may decrease the strength of the fabric to the point where the tissue is not usable. To overcome this problem, fabric strength may be increased by adding certain resins or by mechanical processes which ensure the fibers bond together better. However, increasing strength tends to make the fabric stiffer and harsher to the touch. Rising to these challenges, tissue manufacturers have designed methods that successfully balance softness with absorbency and strength to create a product that consumers find acceptable.

Raw Materials

Nonwoven tissue paper

Tissue paper is a nonwoven fabric made from cellulosic fiber pulp. Common fibers used in tissue paper pulp include wood (from either deciduous or coniferous trees), rayon, bagasse (a type of sugar cane stalk), and recycled paper. These fibers are macerated in a machine known as a hydropulper, which is a cylindrical tank with a rapidly revolving rotor at the bottom that breaks fiber bundles apart. In this process the fibers are mixed in a cooking liquor with water and either calcium, magnesium, ammonia, or sodium bisulfite. This mixture is cooked into a viscous slurry containing about 0.5% solids on the basis of weight. Bleaching agents are added to this mixture to whiten and brighten the pulp. Common bleaching agents include chlorine, peroxides, or hydrosulfites. The pulp is then washed and filtered multiple times until it is the fibers are completely free from contaminants. This mixture of pulp and water, known as a "furnish," is then ready for the papermaking process.

Lotion (Softening additives)

Softening agents are oily or waxy materials that are coated onto the tissue fabric to improve its tactile properties. These materials are too concentrated to coat directly on the paper, so they must be diluted with water first. However, these oils do not dissolve in water, they must be dispersed in water with the aid of chemicals known as surface active agents, or surfactants. A mixture of water, oils, and a surfactant is known as an emulsion. Mayonnaise is an example of a food product emulsion.

The oily materials used in lotions typically include vegetable and mineral oil, plant or animal derived waxes, fatty materials, and silicone-based oils. While theoretically all of these materials would be appropriate tissue paper softening agents, experience has shown that many of them do not function well because they interfere with other desirable properties of the paper, like its absorbency. The tissue industry has had to develop its own patented combinations of lotion materials which, when blended and applied in the correct ratio, provide appropriate softening without negatively affecting the tissue. These materials include polyhydroxy compounds with multiple oxygen-hydrogen groups that allow them to interact with water. Therefore, these compounds are able to soften the paper surface without blocking too much water. Examples of polyhhydroxy compounds include glycerine, propylene glycol, polyoxyethlyne glycol, and polyoxypropylene glycol. They are employed at concentrations between 0.1 and 1% on the basis of the dry tissue weight. Other useful agents include mixtures of petroleum- and silicone-based oils, which are judiciously added to further soften the paper. These oils must be used at low levels to avoid waterproofing the web and robbing it of absorbency. Surfactants are added to disperse the oils in water. A typical surfactant used in paper treatment emulsions is cetyl alcohol, a fatty material whose chemical structure allows it to combine oil and water.

The Manufacturing Process

Preparation of the nonwoven

A variety of specialized equipment is used to press the pulp mixture, or furnish, into a nonwoven sheet of fabric-like paper. Nonwoven fabrics are different from traditional fabrics because of the way they are made. Traditional fabrics are made by weaving fibers together to create an interlocking network of fiber loops. Nonwovens are assembled by mechanically, chemically, or thermally interlocking the fibers. There are two primary methods of assembling nonwovens, the wet laid process and the dry laid process. The wet laid process is employed for making the type of nonwoven used in tissue production.

1 The slurry flows into a device known as a headbox, which in turn spreads it on a moving wire mesh known as a Fourdrinier. The Fourdrinier is a continuous wire belt, approximately 50 feet (15 m) or more in length, which is stretched out like a table. As the fibers are travel down this belt much of the water drains through the holes in the wire mesh. The wet sheet of fibers is carried by a series of woolen blankets, called felts, between several sets of rolls, which further compress it and remove more water. At this point the sheet is strong enough to be transferred to a drying machine that is especially adapted for making tissue papers.

2 The tissue paper drier is called a Yankee Dryer and consists of a steam-heated, highly polished roller 10-12 feet (3-4 m) in diameter. The wet sheet is carried by a heavy canvas felt, which is threaded over and around the rollers. With each successive pass, the rollers remove more water until the paper is adequately dried. If desired, a pattern may be imprinted in the tissue by juxtaposing the web on an array of supports during the dewatering process. (Alternately, the web can be dewatered and transferred to a separate imprinting line.) The raised supports on the line create bumps and valleys on the fabric. These are regions of varying fiber density and are visible as tiny patterned "pillows" on the final sheet. If necessary, these high bulk areas can be densified even further by applying a vacuum to selected portions of the sheet.

After the fabric been compressed to the desired thickness, it is referred to as a "web." The web is now ready for additional processing. It may be coated or stored on large vertical rollers, known as calendar stacks, to await further operations.

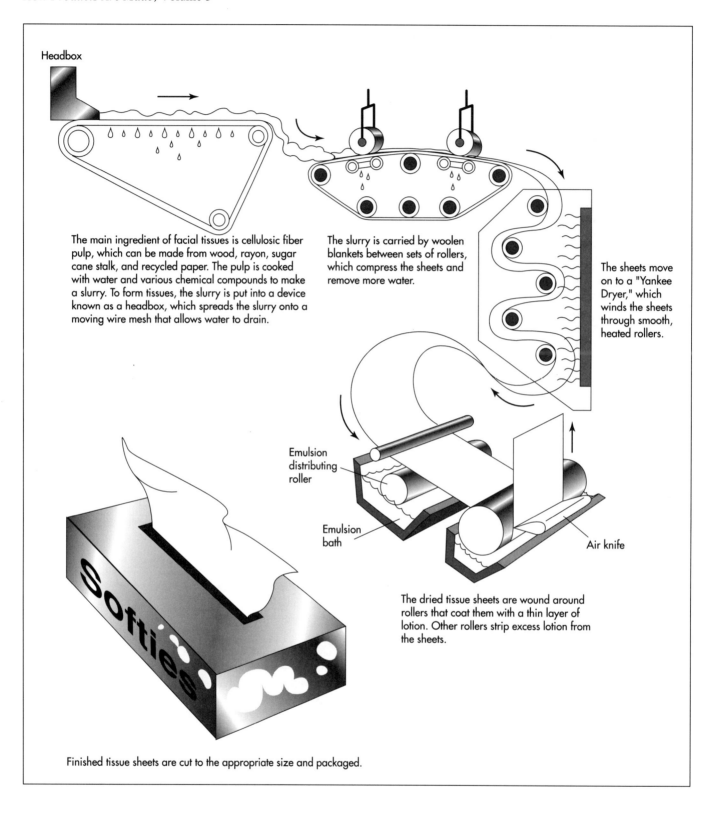

Headbox

The main ingredient of facial tissues is cellulosic fiber pulp, which can be made from wood, rayon, sugar cane stalk, and recycled paper. The pulp is cooked with water and various chemical compounds to make a slurry. To form tissues, the slurry is put into a device known as a headbox, which spreads the slurry onto a moving wire mesh that allows water to drain.

The slurry is carried by woolen blankets between sets of rollers, which compress the sheets and remove more water.

The sheets move on to a "Yankee Dryer," which winds the sheets through smooth, heated rollers.

Emulsion distributing roller

Emulsion bath

Air knife

The dried tissue sheets are wound around rollers that coat them with a thin layer of lotion. Other rollers strip excess lotion from the sheets.

Finished tissue sheets are cut to the appropriate size and packaged.

Lotion preparation and application

3 The lotion is prepared in steam-heated batch tanks equipped with high speed mixing blades. The oils and water may be preheated and are blended together with high shear to form an emulsion. The com- pleted lotion is ready to be applied to the paper surface and is pumped from the batch tanks to a holding vessel connected to the coating equipment.

4 The nonwoven web is fed onto a series of papermaking belts. As it travels over the belts the web comes into contact with an emulsion distributing roller, which pulls lotion out of the holding tank and coats a thin film onto the web. Ideally low amounts of lotion are applied (0.3% or less) to prevent the web from being overcoated. However, higher levels can be used if the coating is designed with additional surfactants that will act as wetting agents to help the tissue absorb moisture through its hydrophobic layer. After passing through the coating rollers, the web continues along the belts to other rollers, which strip excess lotion from the fabric. In between processes the belts are kept clean by belt-cleaning showers that remove paper fibers, adhesives, and other additives.

Forming operations and packaging

5 The web passes through a series of rotating knives that cut it to the desired width. The coated tissue is then sliced at tissue-sized intervals, folded, and packaged in boxes or cellophane wrap.

Byproducts/Waste

The tissue manufacturing and coating process generates considerable amounts of waste material, but much of this is reclaimable. Waste fibers from the pulping process can be washed and reused. The water used in the slurry and in subsequent operations can be recycled. Unfortunately, there is little or no recovery of the chemicals used in coating and other treatments, and the disposal of the various spent solutions is a problem for the industry.

Quality Control

There are many quality control measures used in the tissue paper industry. The ones related to lotion application include analytical testing and subjective panel evaluations. Since the amount of material deposited on the tissue is critical, the industry has established various tests to measure how much is actually present on the tissue surface. For example, the amount of polyhydroxy compounds present can be determined by stripping the compounds from a tissue sample using a method known as the Webul solvent extraction. The amount of compound is then measured on a spectroscope or chromatograph. The concentration of surfactants can be established in a similar manner.

While these analytical techniques can precisely determine the levels of specific chemicals, they can not evaluate fabric softness. This tactile property is assessed by subjective evaluation by trained panelists. Prior to these evaluations, the tissue fabric is equilibrated to a constant temperature of 72-111°F (22-44°C) and relative humidity of 10-35%. The fabric is then conditioned for another 24 hours at 50% humidity. Panelists are then asked to feel swatches and rate the degree of softness, flexibility, and smoothness. The evaluation is done by paired comparison, as described by the American Society for Testing Materials (ASTM). Subjects are presented with samples on a blind basis and required to choose one on the basis of tactile softness. The results are reported in Panel Score Units which grade the fabric on a scale of "Much Softer," "Slightly Softer," "Equally Soft," "Less Soft," etc.

Absorbency, the ability of the tissue to be wetted with water, is quantified by measuring the period of time required for dry tissue to become completely saturated with water. This measurement is known as wetting time. Once again, the fabric is equilibrated to a specified temperature and humidity. It is then cut into small squares, crumpled into a ball, and placed on surface of a 3-qt (3-l) beaker of water. A timer is started when the ball hits the water and the amount of time for the ball to be completely wetted by the water is measured. Five sets of five balls are tested to obtain an average measurement. Absorbency is measured on fresh tissue samples immediately after manufacture and on samples aged at least two weeks. This is important because absorbency will decrease over time, as the coating agents cure on the surface of the tissue.

The density of the tissue is also measured with a thickness tester to evaluate how thick cloth is, then its mass, volume, and area are calculated. Linting (the amount of loose lint which detaches from the tissue) is measured by abrading a sample against a piece of black wool by a motor driven device known as the Sutherland Rub Tester. Colormetric analysis can then be used to determine the quantity of lint transferred to the wool.

The Future

The increased environmental concern about waste chemicals may lead to improved lotion formulations employing biodegradable or recyclable raw materials in the future. The industry is continually researching ways to make the manufacturing process faster and more energy efficient. Finally, methods may be developed to improve the strength of nonwoven fabrics without sacrificing the pleasant tactile characteristics that make lotion-coated tissues so desirable.

Where to Learn More

Periodicals

Ingman, Lars C. "Accepting 'The New.' (new products and techniques in the paper industry)" *Pulp & Paper,* 62, no. 5 (May 1988):127(2).

"Nonwovens: making a better product," (International Textile Machinery Association) *Textile World* 141, no. 12 (Dec 1991):63(4).

"TAPPI brings librarians and paper industry together." (Technical Association for the Pulp and Paper Industry), *Library Journal* 113, no. 20 (Dec 1988):35(1).

"Pulping conference focus is energy conservation, process optimization." *Pulp & Paper* 55 (Dec 1981):108(3).

—*Randy Schueller*

Toothpaste

Background

Toothpaste has a history that stretches back nearly 4,000 years. Until the mid-nineteenth century, abrasives used to clean teeth did not resemble modern toothpastes. People were primarily concerned with cleaning stains from their teeth and used harsh, sometimes toxic ingredients to meet that goal. Ancient Egyptians used a mixture of green lead, verdigris (the green crust that forms on certain metals like copper or brass when exposed to salt water or air), and incense. Ground fish bones were used by the early Chinese.

In the Middle Ages, fine sand and pumice were the primary ingredients in teeth-cleaning formulas used by Arabs. Arabs realized that using such harsh abrasives harmed the enamel of the teeth. Concurrently, however, Europeans used strong acids to lift stains. In western cultures, similarly corrosive mixtures were widely used until the twentieth century. Table salt was also used to clean teeth.

In 1850, Dr. Washington Wentworth Sheffield, a dental surgeon and chemist, invented the first toothpaste. He was 23 years old and lived in New London, Connecticut. Dr. Sheffield had been using his invention, which he called Creme Dentifrice, in his private practice. The positive response of his patients encouraged him to market the paste. He constructed a laboratory to improve his invention and a small factory to manufacture it.

Modern toothpaste was invented to aid in the removal foreign particles and food substances, as well as clean the teeth. When originally marketed to consumers, toothpaste was packaged in jars. Chalk was commonly used as the abrasive in the early part of the twentieth century.

Sheffield Labs claims it was the first company to put toothpaste in tubes. Washington Wentworth Sheffield's son, Lucius, studied in Paris, France, in the late nineteenth century. Lucius noticed the collapsible metal tubes being used for paints. He thought putting the jar-packaged dentifrice in these tubes would be a good idea. Needless to say, it was adopted for toothpaste, as well as other pharmaceutical uses. The Colgate-Palmolive Company also asserts that it sold the first toothpaste in a collapsible tube in 1896. The product was called Colgate Ribbon Dental Creme. In 1934, in the United States, toothpaste standards were developed by the American Dental Association's Council on Dental Therapeutics. They rated products on the following scale: Accepted, Unaccepted, or Provisionally Accepted.

The next big milestone in toothpaste development happened in the mid-twentieth century (1940-60, depending on source). After studies proving fluoride aided in protection from tooth decay, many toothpastes were reformulated to include sodium fluoride. Fluoride's effectiveness was not universally accepted. Some consumers wanted fluoride-free toothpaste, as well as artificial sweetener-free toothpaste. The most commonly used artificial sweetener is saccharin. The amount of saccharin used in toothpaste is minuscule. Companies like Tom's of Maine responded to this demand by manufacturing both fluoridated and non-fluoridated toothpastes, and toothpastes without artificial sweetening.

In 1934, in the United States, toothpaste standards were developed by the American Dental Association's Council on Dental Therapeutics. They rated products on the following scale: Accepted, Unaccepted, or Provisionally Accepted.

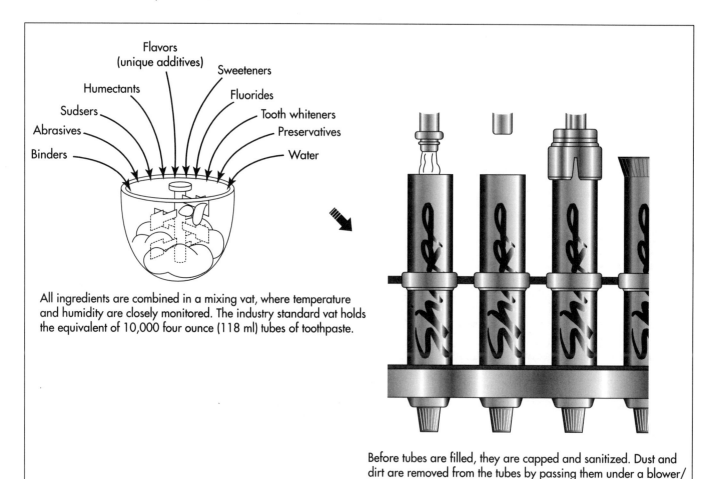

All ingredients are combined in a mixing vat, where temperature and humidity are closely monitored. The industry standard vat holds the equivalent of 10,000 four ounce (118 ml) tubes of toothpaste.

Before tubes are filled, they are capped and sanitized. Dust and dirt are removed from the tubes by passing them under a blower/vacuum mechanism.

Many of the innovations in toothpaste after the fluoride breakthrough involved the addition of ingredients with "special" abilities to toothpastes and toothpaste packaging. In the 1980s, tartar control became the buzz word in the dentifrice industry. Tarter control toothpastes claimed they could control tartar build-up around teeth. In the 1990s, toothpaste for sensitive teeth was introduced. Bicarbonate of soda and other ingredients were also added in the 1990s with claims of aiding in tartar removal and promoting healthy gums. Some of these benefits have been largely debated and have not been officially corroborated.

Packaging toothpaste in pumps and stand-up tubes was introduced during the 1980s and marketed as a neater alternative to the collapsible tube. In 1984, the Colgate pump was introduced nationally, and in the 1990s, stand-up tubes spread throughout the industry, though the collapsible tubes are still available.

Raw Materials

Every toothpaste contains the following ingredients: binders, abrasives, sudsers, humectants, flavors (unique additives), sweeteners, fluorides, tooth whiteners, a preservative, and water. Binders thicken toothpastes. They prevent separation of the solid and liquid components, especially during storage. They also affect the speed and volume of foam production, the rate of flavor release and product dispersal, the appearance of the toothpaste ribbon on the toothbrush, and the rinsibility from the toothbrush. Some binders are karaya gum, bentonite, sodium alginate, methylcellulose, carrageenan, and magnesium aluminum silicate.

Abrasives scrub the outside of the teeth to get rid of plaque and loosen particles on teeth. Abrasives also contribute to the degree of opacity of the paste or gel. Abrasives may affect the paste's consistency, cost, and taste. Some abrasives are more

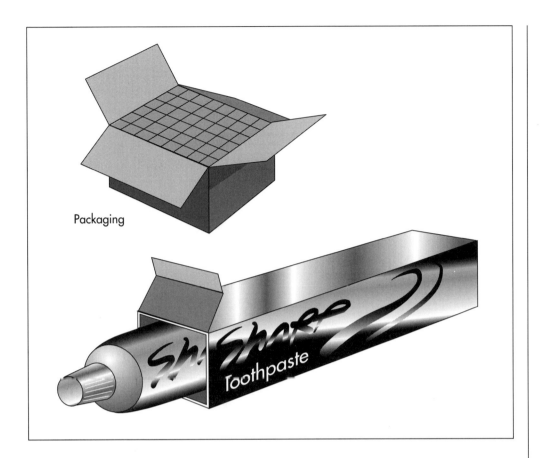

Packaging

harsh than others, sometimes resulting in unnecessary damage to the tooth enamel.

The most commonly used abrasives are hydrated silica (softened silica), calcium carbonate (also known as chalk), and sodium bicarbonate (baking soda). Other abrasives include dibasic calcium phosphate, calcium sulfate, tricalcium phosphate, and sodium metaphosphate hydrated alumina. Each abrasive also has slightly different cleaning properties, and a combination of them might be used in the final product.

Sudsers, also known as foaming agents, are surfactants. They lower the surface tension of water so that bubbles are formed. Multiple bubbles together make foam. Sudsers help in removing particles from teeth. Sudsers are usually a combination of an organic alcohol or a fatty acid with an alkali metal. Common sudsers are sodium lauryl sulfate, sodium lauryl sulfoacetate, dioctyl sodium sulfosuccinate, sulfolaurate, sodium lauryl sarcosinate, sodium stearyl fumarate, and sodium stearyl lactate.

Humectants retain water to maintain the paste in toothpaste. Humectants keep the solid and liquid phases of toothpaste together. They also can add a coolness and/or sweetness to the toothpaste; this makes toothpaste feel pleasant in the mouth when used. Most toothpastes use sorbitol or glycerin as humectants. Propylene glycol can also be used as a humecant.

Toothpastes have flavors to make them more palatable. Mint is the most common flavor used because it imparts a feeling of freshness. This feeling of freshness is the result of long term conditioning by the toothpaste industry. The American public associates mint with freshness. There may be a basis for this in fact; mint flavors contain oils that volatize in the mouth's warm environment. This volatizing action imparts a cooling sensation in the mouth. The most common toothpaste flavors are spearmint, peppermint, wintergreen, and cinnamon. Some of the more exotic toothpaste flavors include bourbon, rye, anise, clove, caraway, coriander, eucalyptus, nutmeg, and thyme.

In addition to flavors, toothpastes contain sweeteners to make it pleasant to the palate because of humecants. The most commonly used humectants (sorbitol and glycerin) have a sweetness level about 60% of table sugar. They require an artificial flavor to make the toothpaste palatable. Saccharin is the most common sweetener used, though some toothpastes contain ammoniated diglyzzherizins and/or aspartame.

Fluorides reduce decay by increasing the strength of teeth. Sodium fluoride is the most commonly used fluoride. Sodium perborate is used as a tooth whitening ingredient. Most toothpastes contain the preservative p-hydrozybenzoate. Water is also used for dilution purposes.

The Manufacturing Process

Weighing and mixing

1 After transporting the raw materials into the factory, the ingredients are both manually and mechanically weighed. This ensures accuracy in the ingredients' proportions. Then the ingredients are mixed together. Usually, the glycerin-water mixture is done first.

2 All the ingredients are mixed together in the mixing vat. The temperature and humidity of vat are watched closely. This is important to ensuring that the mix comes together correctly. A commonly used vat in the toothpaste industry mixes a batch that is the equivalent of 10,000 four-ounce (118 ml) tubes.

Filling the tubes

3 Before tubes are filled with toothpaste, the tube itself passes under a blower and a vacuum to ensure cleanliness. Dust and particles are blown out in this step. The tube is capped, and the opposite end is opened so the filling machine can load the paste.

4 After the ingredients are mixed together, the tubes are filled by the filling machine. To make sure the tube is aligned correctly, an optical device rotates the tube. Then the tube is filled by a descending pump. After it is filled, the end is sealed (or crimped) closed. The tube also gets a code stamped on it indicating where and when it was manufactured.

Packaging and shipment

5 After tubes are filled, they are inserted into open paperboard boxes. Some companies do this by hand.

6 The boxes are cased and shipped to warehouses and stores.

Quality Control

Each batch of ingredients is tested for quality as it is brought into the factory. The testing lab also checks samples of final product.

Where to Learn More

Books

Garfield, Sydney. *Teeth Teeth Teeth.* Simon and Schuster, 1969.

Other

Colgate-Palmolive. 1996. http://www.colgate.com/ (July 9, 1997).

Crest web site. 1996. http://www.pg.com/docYourhome/docCrest/directory_map.html (July 9, 1997).

—Annette Petrusso

Trampoline

Background

A trampoline is an elevated, essentially buoyant webbed bed or canvas sheet supported by springs or elastic shock cords. It is surrounded by a metal frame and used as a springboard for tumbling.

The trampoline is used in the sport of trampolining (sometimes called rebound tumbling). In the sport, the trampoline is used to rebound the athlete, so he or she can perform acrobatic movements in midair. The trampoline is also a training tool for gymnasts, divers, and pole vaulters.

History

Tumbling and acrobatic moves have existed for centuries. Humans have tried to get in the air in many ways. One of the first ways was the springboard. The springboard allowed the performer to leap high with little effort and do acrobatic stunts.

Another device used to get airborne was called "the leaps." It was made of a resilient, rather narrow wood plank supported at both ends by blocks that kept the plank off the floor. Court jesters in the Middle Ages jumped on it when they performed at court.

Circus lore has it that a Frenchman, named du Trampoline, helped develop the basics of the trampoline as we know it. For years, circus performers had used a net under the trapeze so they could rebound. By developing a system of spring suspensions, the Frenchman, a former trapeze artist, moved trampoline development forward. He adapted safety nets and experimented with spring suspension systems to make the earliest form of the trampoline.

The trampoline was not widely popular until the 1900s when circus performers made it a feature attraction. It became a modern sport in 1936 when the present-day trampoline was developed by American gymnast George Nissen. In the United States, trampoline was originally a trademark for the apparatus perfected by Nissen.

Surprisingly, trampolining became popular in the United States when World War II broke out. It was used for recreation and physical education purposes in the armed forces. It was especially important for pilot and air crew training because it helped to instruct trainees in body position and sensations associated with flight.

After the war, physical education teachers introduced trampolining in schools because of its physical benefits as well as its potential for fun. Its use spread to universities and places like the YMCA/YWCA as a competitive sport for students.

Unofficial trampolining contests in the United States first took place in 1947, with official competitions soon following in 1954. International trampolining events began in 1964. An international governing body was formed, the International Trampoline Association, to govern the sport worldwide. In a trampoline competition, each trampolinist performs two routines, one compulsory and one optional. During each routine, the performer can only make eight contacts with the trampoline. Scoring is based on difficulty, execution, and form. The winner of the compulsory and the optional events performs another optional routine to determine an overall champion.

Since the 1950s, trampolines have also been used by visual therapists and special educa-

It takes about 80 people working an eight-hour shift in the trampoline factory to make 500 to 600 trampolines.

tion teachers to improve vision, balance, and coordination in their students. In 1977, however, trampolines were discarded as part of the physical education curriculum in public schools because of a negative report issued by the American Academy of Pediatrics. It said that trampolines were dangerous and could cause injuries such as broken bones and quadriplegia.

Although in a 1995 Circuit Court of Appeals case, Dea Richter versus Limax International, Inc., the court decided that a trampoline manufacturer can be liable for injuries resulting from repetitive use of the apparatus, the trampoline has remained a popular source of exercise and fun at home. Today, trampolines are used for improving health through cardiovascular exercise. Using a mini-trampoline, called an aerobic rebounder, can improve stamina, strength, and coordination. Rebounders were first manufactured in 1975 and marketed as an indoor jogging aid.

Raw Materials

Trampolines are made of four basic components: the tubing, springs, jumping mat, and safety pads.

The tubing, used to make the frame and legs of the trampoline, is usually made of galvanized steel and is bought at a certain length and width from a supplier. Using galvanized steel protects the frame from rust and environmental conditions as many home trampolines are for outside use.

The springs, which give the trampoline its bounce, are commonly made to the specifications of the trampoline company by another supplier. Usually, the company making the springs specializes in spring manufacture.

The jumping mats are made of woven fibers. Today, mats are made of artificial fibers like polyethylene or nylon. The heavily woven fabric is UV-protected to prevent fading when used out of doors.

Safety pads go over the springs and frame and are made of foam. They have vinyl covers and pie straps to connect the pads to the frame. The manufacture of the foam core of the safety pads is also outsourced and made to manufacturer specifications.

The Manufacturing Process

(Most trampolines intended for home use are packed to be assembled by the consumer. Therefore, this section will describe the process of fabricating individual components.)

1 When the tubing arrives at the trampoline factory, it is bent into an arc on the bending jig. Holes are punched into the sides for the springs.

2 Sockets are welded so that the U-shaped legs can fit into the frame. The tubing often comes in four pieces. When connected, all four create a circular design. One end of each arc is swaged—squeezed down—so that the pieces can be put together.

3 Once delivered to the factory, the fabric for the jumping mat is cut to size. Using industrial **sewing machines**, the fabric is edged, and reinforced web strapping is sewn on. D-rings are also sewed into the fabric to hold the mat to the frame.

4 Around the foam core of the safety pads, a vinyl cover and pie straps are sewn to increase durability.

5 The springs do not require any additional manufacturing processes at the factory. They are packaged, along with the other parts, for final shipment.

It takes about 80 people working an eight-hour shift in the trampoline factory to make 500 to 600 trampolines.

Quality Control

The American Society for Testing and Materials has established safety and quality standards for trampolines. All materials used in trampoline manufacture are checked at regular intervals to see if they meet established guidelines. The steel frame's bend, gauge, and thickness of galvanization are inspected. Before the mat is sewn, the weaving is checked for flaws. The tension of the thread is subject to durability testing as well. The sewing on the mat is inspected after it is completed. Before the springs are packaged with the rest of the components, they are inspected for flaws.

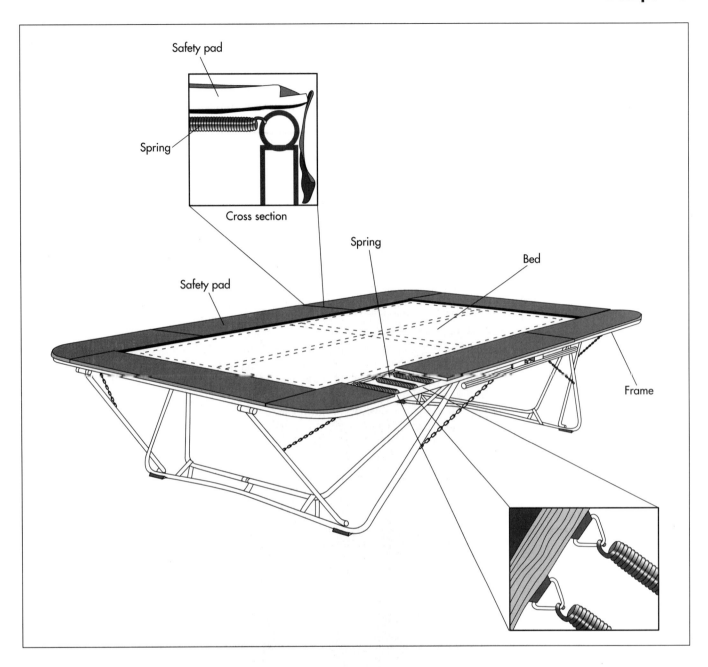

Safety pad

Spring

Cross section

Spring

Bed

Safety pad

Frame

The Future

Recently in Great Britain, the Inwood Ryan Company designed the "frameless" trampoline. It is supposed to be safer because it prevents children from hurting themselves on the metal frames. Presently, it is uncertain whether it will become a widespread innovation.

Where to Learn More

Books

Griswold, Larry and Glenn Wilson. *Trampoline Tumbling Today*. A.S. Barnes and Company, 1970.

Hennessy, Jeff T. *Trampolining*. William C. Brown Company Publishers, 1968.

Other

International Trampolining. 1996. http://www.worldsport.com/sports/trampoline/home.html (July 14, 1997).

United States Trampolining. http://www.geocites.com/Colosseum/9196

—*Annette Petrusso*

Trampolines are made of four basic components: the tubing, springs, jumping mat, and safety pads. The springs connect the jumping mat to the tubing and give the trampoline its bounce. Safety pads go over the springs and frame and are made of foam.

Vitamin

As new discoveries are made, and old claims either debunked or reinforced often, it is safest to say that more is understood about the consequences of lack of vitamins than what particular vitamins may do.

Background

Vitamins are organic compounds that are necessary in small amounts in animal and human diets to sustain life and health. The absence of certain vitamins can cause disease, poor growth, and a variety of syndromes. Thirteen vitamins have been identified as necessary for human health, and there are several more vitamin-like substances that may also contribute to good nutrition. Originally, it was thought that vitamins were particular chemical compounds called amines, but now it is known that the vitamins are unrelated chemically. Their actions are different, and though exhaustively studied, not everything is understood about how they work and what they do. The vitamins are named by letters—vitamin A, vitamin C, D, E, K, and the group of B vitamins. The eight B vitamins were originally thought to be one vitamin, and as more was learned about them, they were given numeral subscripts: vitamin B_1, B_2, etc. The B vitamins are now commonly called more aptly by chemically descriptive names: B_1 is thiamine, B_2 is riboflavin, B_6 and B_{12} retain their numeral names, and the other B vitamins are niacin, pantothenic acid, biotin, and folic acid. The vitamins are found in plant and animal food sources. They have also been chemically synthesized and so can be ingested in their pure form as nutritional supplements. It is not known precisely how much of each vitamin each person needs, but there are recommended daily allowances for 10 vitamins.

Some researchers have made extravagant claims about the benefits of large doses of specific vitamins as either preventatives or cures for diseases from acne to cancer. As new discoveries are made and old claims are either debunked or reinforced often, it is safest to say that more is understood about the consequences of lack of vitamins than what particular vitamins may do. For example, deficiency of vitamin A leads to breakdown of the photosensitive cells in the retina of the eye, causing night blindness. Absence of vitamin C in the diet leads to scurvy, a disease formerly the bane of sailors. Absence of vitamin D may lead to rickets, a bone disease.

History

Many researchers were responsible for piecing together the existence of vitamins as necessary components of the human and animal diets. One of the first people to study nutrition from a chemical standpoint was English physician William Prout. In 1827, he defined the three essentials of the human diet as the oily, the saccharin, and the albuminous, which in modern-day terms are fats and oils, carbohydrates, and proteins. In 1906, an English biochemist, Frederick Hopkins, discovered that mice fed on a pure diet of the three essentials could not survive unless they were given supplementary small amounts of milk and vegetables. A Polish scientist, Casimir Funk, coined the term vitamines in 1912 to describe the chemicals he believed were found in the supplementary food that helped the mice survive. Funk first believed that the vitamines were chemically related amines, thus *vita* (life) plus amines. As other vitamins were isolated that were not amines, the spelling of the word changed. Other researchers working on diseases such as scurvy and beriberi, which are caused by vitamin deficiency, contributed to the isolation of the different vitamins. Still, little was generally understood about vitamins at the beginning of the twentieth century. For instance, though the use of lime juice to prevent scurvy in sailors dates back to at least 1795, the physician who accompanied Scott's voyage to the South Pole in 1910 believed scurvy was caused by bacteria, and inadequate nutritional measures were taken to prevent the disease among the explorers. Between 1925 and 1955, the known vitamins were all isolated and synthesized. Research continues today on the function of the various vitamins.

Raw Materials

Vitamins can be derived from plant or animal products, or produced synthetically in a laboratory. Vitamin A, for example, can be derived from fish liver oil, and vitamin C from citrus fruits or rose hips. Most commercial vitamins are made from synthetic vitamins, which are cheaper and easier to produce than natural derivatives. So vitamin A may be synthesized from acetone, and vitamin C from keto acid. There is no chemical difference between the purified vitamins derived from plant or animal sources and those produced synthetically. Different laboratories may use different techniques to produce synthetic vitamins, as many can be derived from various chemical reactions.

Vitamin tablets or capsules usually contain additives that aid in the manufacturing process or in how the vitamin pill is accepted by the body. Microcrystalline cellulose, lactose, calcium, or malto-dextrin are added to many vitamins as a filler, to give the vitamin the proper bulk. Magnesium stearate or stearic acid is usually added to vitamin tablets as a lubricant, and silicon dioxide as a flow agent. These additives help the vitamin powder run smoothly through the tablet-making or encapsulating machine. Modified cellulose gum or starch is often added to vitamins as a disintegration agent. That is, it helps the vitamin compound break up once it is ingested. Vitamin tablets are also usually coated, to give the tablets a particular color or flavor, or to determine how the tablet is absorbed (in the stomach versus in the intestine, slowly versus all at once, etc.). Many coatings are made from a cellulose base. An additional coating of car-

nauba wax is often put on as well, to give the tablet a polished appearance.

Herbs of various kinds may be added to vitamin compounds, as well as minerals such as calcium, iron, and zinc. Typically, specialized laboratories produce purified vitamins and minerals. A distributor buys these from the laboratories and sells them to manufacturers, who put them together in different compounds such as multivitamin tablets or B-complex capsules.

The Manufacturing Process

Preliminary check

1 A vitamin manufacturer purchases raw vitamins and other ingredients from distributors. Raw vitamins from a reputable distributor arrive with a Certificate of Analysis, stating what the vitamins are and how potent they are. In many cases, the manufacturer will nevertheless test the raw materials or send samples to an independent laboratory for analysis. If herbs are to be an ingredient in the vitamin capsule, these must be tested for identity and potency, and for possible bacterial contamination as well.

Preblending

2 Often, the raw vitamins arrive at the manufacturer in a fine powder, and they need no preliminary processing. However, if the ingredients are not finely granulated, they will be run through a mill and ground. Some vitamins may be preblended with a filler ingredient such as microcrystalline cellulose or malto-dextrin, because this produces a more even granule which aids further processing steps. Laboratory technicians may run test batches when working with new ingredients and determine if preblending is necessary.

Wet granulation

3 For vitamin tablets, particle size is extremely important in determining how well the formula will run through the tabletting machine. In some cases, the raw vitamins arrive from the distributor milled to the appropriate size for tabletting. In other cases, a wet granulation step is necessary. In wet granulation, the fine vitamin powder is mixed with a variety of cellulose particles,

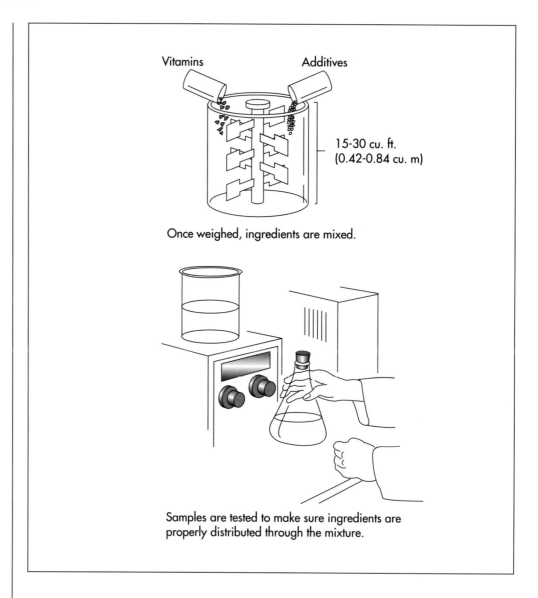

Vitamins   Additives

15-30 cu. ft.
(0.42-0.84 cu. m)

Once weighed, ingredients are mixed.

Samples are tested to make sure ingredients are
properly distributed through the mixture.

then wetted. The mixture is then dried in a dryer. After drying, the formula may be in chunks as large as a dime. These chunks are sized by being run through a mill. The mill forces the chunks through a small hole of the desired diameter of the granule. These granules can then be weighed and mixed.

Weighing and mixing

4 When all the vitamin ingredients are ready, a worker takes them to the weigh station and weighs them out on a scale. The required weights for each ingredient in the batch are listed on a formula batch record. After weighing, the worker dumps all the ingredients into a mixer. The volume of a typical mixer may be from 15-30 cu ft (0.42-0.84 cu m), though in a large manufacturing facility, it may be many times

that large. The ingredients spend from 15 to 30 minutes in the mixer. At this point, samples are taken from different sides of the mixer and checked in the laboratory. The lab technicians verify that all the ingredients are distributed in the same proportion throughout the mix. If the manufacturer is making a large batch, workers may check the first three or four lots in the mixer, and then only re-check periodically. After mixing is complete, workers take the vitamin formula to either an encapsulating or a tablet-making machine.

Encapsulating machine

5 If the lot in the mixer has been approved, workers tote the mixture to the encapsulating machine and dump it in a hopper. At the beginning of a batch, work-

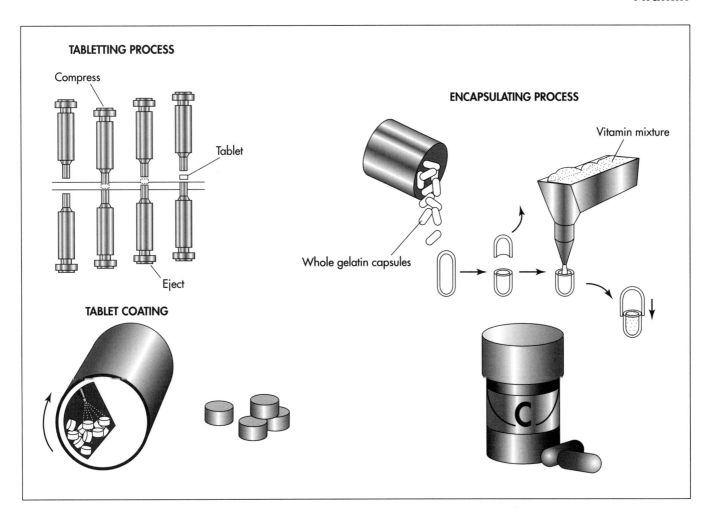

TABLETTING PROCESS

Compress

Tablet

Eject

TABLET COATING

ENCAPSULATING PROCESS

Vitamin mixture

Whole gelatin capsules

C

ers will test-run the encapsulating machine and check that the capsules are the proper and consistent weight. Workers also check the capsules visually to see if they seem to be splitting or dimpling. If the test batches run correctly, workers run the entire batch. The vitamin mixture flows through one hopper, and another hopper holds whole gelatin capsules. The capsules are broken into halves by the machine. The bottom half of the capsule falls through a funnel into a rotating dosing dish. Then the machine measures a precise amount of the powdered vitamin mixture into each open capsule half. Tamping pins push the powder down. Then the top halves of the capsules are pushed down onto the filled bottoms.

Polishing and inspection

6 The filled vitamin capsules are next run through a polishing machine. The vitamins are circulated on a belt through a series of soft brushes. Any excess dust or vitamin powder is removed from the exterior of the capsules by the brushes. The polished capsules are then poured onto an inspection table. The inspection table has a belt of rotating rods. The vitamins fall in the grooves between the rods, and the vitamins rotate as the rods turn. Thus, all sides of the vitamin are visible for the inspector to see. The inspector removes any capsules that are too long, split, dimpled, or otherwise imperfect. The vitamins that pass inspection are then taken over to the packaging area.

Tableting

7 Vitamin tablets are made in a tableting machine. After the vitamin blend has been mixed in the mixer, workers dump it into a hopper above the machine. The vitamin powder then flows through the hopper to a filling station beneath, and flows from there to a rotating table. The rotating table may be 2-4 ft (0.6-1.2m) in diameter, or even bigger, and is fitted with holes on its outside edge that hold dies in the shape of the desired tablet (oval, round, animal, etc.).

The finished vitamin mixture can be compressed into tablets, sometimes with a coating, or encapsulated in preformed gelatin capsules.

The dies are interchangeable, so the same table can produce whatever shape the manufacturer wishes, as long as the proper dies are installed. The vitamin powder flows from the filling station to fill the die. When the table rotates, the filled die moves into a punch press. When the upper and lower halves of the punch meet, 4-10 tons (3.6-9 metric tons) of pressure is exerted on the vitamin powder. The pressure compresses the vitamin powder into a compact tablet. The punch releases, and the lower punch lifts to eject the tablet. Some tableting machines may have two punches, one on each side, so two tablets are made simultaneously. The speed of the rotation of the table determines how many tablets are made per minute. The tablets eject onto a vibrating belt which vibrates any loose dust off the tablets. The tablets then are moved to the coating area.

Coating

8 Vitamin tablets are usually coated for a variety of reasons. The coating may make the tablet easier to swallow. It may mask an unpleasant taste, and it may give the tablet a pleasant color. A manufacturer may coat in two different colors tablets that are the same size and shape, for identification. Tablets may also be given an enteric coating—a pH sensitive chemical coating that resists gastric acid. Tablets with an enteric coating will not break open in the stomach, but move to the intestine before dissolving. Other coatings determine the timing of the tablet's dissolution, so the vitamins can be absorbed slowly, or all at once, depending on what is appropriate to that tablet.

Once the tablets are taken from the tableting area, they are placed in the coating pan. The coating pan is a large rotating pan surrounded by one to six spray guns operated by pumps. As the tablets revolve in the pan, the pumps spray coating over them. Many tablets also receive a second coating of carnauba wax. After air drying, the tablets are ready for packaging. The packaging step is the same for tablets as for capsules.

Packaging

9 Packaging the vitamins takes several steps, and different machines carry out these steps. So in the packaging area, the vitamins pass through a row of machines. Once the vitamins are dumped in the hopper of the first machine, no human touches them. The worker sets the machine to count out the required number of capsules or tablets per bottle, and the rest is done automatically. The capsules or tablets fall into a bottle, and the bottle is passed to the next machine to be sealed, capped, labelled, and shrink-wrapped. The finished bottles are then set in boxes and are ready for distribution.

Quality Control

Checks for quality are taken at many stages of vitamin manufacturing. All the ingredients of vitamin tablets or capsules are checked for identity and potency before they are used. Often this is tested both by the raw vitamin distributor and by the manufacturer. The mixed vitamin powder is checked before it is tableted or encapsulated, and the finished product is also thoroughly inspected. Federal regulations govern what substances can be used in vitamins and what claims manufacturers can make for their products. Vitamin ingredients must be proven safe before they can be made available to consumers.

The Future

Vitamin research is a volatile field, with new studies constantly suggesting new roles for vitamins in health and prevention of disease. Certain vitamins or vitamin-like substances go through fads of consumer popularity as some of this research surfaces. Nevertheless, the manufacturing process remains the same for new substances. The future of vitamins will likely change most conceptually, in how much we understand about how vitamins work.

Where to Learn More

Books

Bender, David A. *Nutritional Biochemistry of the Vitamins.* Cambridge University Press, 1992.

Hendler, Sheldon Saul. *The Doctor's Vitamin and Mineral Encyclopedia.* Simon and Schuster, 1991.

Lieberman, Shari and Nancy Bruning. *The Real Vitamin & Mineral Book.* Avery Publishing Group, 1990.

—*Angela Woodward*

Wallpaper

Background

Wallpaper is a nonwoven (paper) or woven (fabric) backing, decoratively printed for application to walls of a residence or business. Wallpaper is not considered essential to the decoration of a structure; however, it has become a primary method by which to impart style, atmosphere, or color into a room.

The wallpaper industry divides the manufacture of wallpaper into those used in residences and those hung in businesses or other public buildings. The two categories of paper differ in weight, serviceability, and quality standards. Residential-use wallpapers are made from various materials and can be purchased prepasted or unpasted. There are no mandated serviceability tests for residential papers. The commercial-grade wallpapers are divided into categories based on weight, backing composition, and laminate/coating thickness. All commercial-use wallpapers must have a vinyl surface and pass rigorous physical and visual tests as mandated by the Chemical Fabrics and Film Association.

There are four popular methods used to print wallpapers, and designers choose the printing technique based on cost as well as aesthetics.

History

The earliest wallpapers used in Europe as early as the thirteenth century were painted with images of popular religious icons. These "domino papers" were pasted within homes of the devout; however, they also enlivened the bleak homes of the poor. Within the next few centuries, papers were hand block-printed, but only remained popular with the poor.

By the sixteenth century, however, more expensive wallcovering, depicting tapestries hung in homes of nobility, became popular with the middle class. Small sheets either carried a repeating image, or several blocks produced a pattern spread across many sheets. Fashionable eighteenth century Americans puchased wallpapers from France and England; "paper stainers" were producing wallpapers in this country by the early nineteenth century if not before.

Two problems plagued wallpaper stainers until the mid-nineteenth century. One was the problem of producing long sheets of paper for printing, the other was printing attractive wallpaper inexpensively. Until the mid-1700s, rag-based paper was individually printed in sheets, then applied to walls. Then, wallpaper manufacturers were pasting the pieces together, ground coating them, then printing. In the late nineteenth century, the paper industry developed "endless" paper, or paper made in very long strips. By 1870, wood pulp had supplanted rag stock, resulting in a very cheap backing for wallcovering.

In the nineteenth century, printing costs were greatly reduced by abandoning labor-intensive block printing in favor of cylinder printing. Wood-block printers applied each color by hand using a separate block for each color in the pattern. Thus, each block had to be inked with the right color, pressed down on the paper, tapped to ensure a quality imprint, lifted up, and reinked as the printer moved down the paper roll—an expensive process. Wood blocks were supplanted by copper cylinders, which carried the design below the surface of the roll, each roll printing a single color. The cylinders were mounted within one machine and

By 1885, wood-pulp paper printed with cylinders so greatly reduced wallpaper costs that it was cheaper to wallpaper a house in the United States than to paint it.

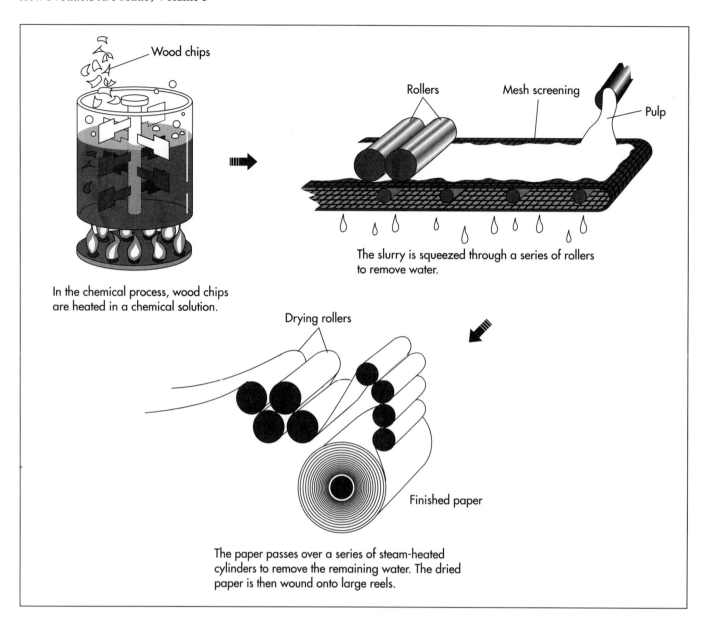

Wood chips

Rollers

Mesh screening

Pulp

The slurry is squeezed through a series of rollers to remove water.

In the chemical process, wood chips are heated in a chemical solution.

Drying rollers

Finished paper

The paper passes over a series of steam-heated cylinders to remove the remaining water. The dried paper is then wound onto large reels.

the paper was mechanically fed between cylinders until the paper was completely printed—no hand printing involved. Thus, by about 1885 wood pulp paper printed with cylinders so greatly reduced wallpaper costs that it was cheaper to wallpaper a house in the United States than to paint it.

More recent advances include development of additional printing methods, new inks and solvents, and use of latex and vinyl as coatings or laminates.

Raw Materials

Wallpaper consists of a backing, ground coat, applied ink, and sometimes paste on the backing used to adhere the paper to the wall.

Non-woven backings can be of ground wood, wood pulp, or wood pulp with synthetic material. Woven backings are those made of sturdy woven textiles such as drill (heavy woven cotton much like jean material). The woven backing is then coated and printed.

The ground coat is the background color laid on the surface, which receives the printed pattern. Coatings or laminates are made of latex or vinyl (polyvinyl chloride) and render the paper durable and strippable. Ground coats also include additives that enhance the ease of handling, opacity, and drapability of the paper.

The paper is printed with inks composed of pigment and a vehicle which ties the ink to the backing. Solvents can be acetone or water, for example. Printers choose inks carefully as the solvents they include affect the drying time and production time between color applications of the paper.

Pastes may or may not be applied to wallpapers. If they are, they are usually made up of cornstarch or wheat starch and are applied wet to the backing. Prepasted wallpapers must be rewetted for adhesion to the wall.

Design

New wallpaper designs are generally derived from sketches purchased from a staff designer or freelance wallpaper designer. The artist lays out the design on tracing paper and completes at least a partial pencil sketch. The marketing and design staff will then decide if the paper is the right "fit" for a specific look or line. If the design is accepted, the artist produces a full-scale color sketch in various colors and palettes.

After the printing process is chosen, the sketch is fine tuned to fit the requirements of the printing process and the pattern is sent to the engraver or screen-maker. Once the cylinder or screens are in place and a few pattern repeats are printed, a "strike off" (sample wallpaper) is printed to test the color and pattern. When okayed, the paper is commercially printed in large runs.

The Manufacturing Process

Making the paper

1 Ground wood sheets of paper are produced by using an entire tree, removing the bark, and pressing the tree against a revolving tread, which grinds the wood into slurry. The slurry is used to make a ground wood sheet—a relatively inexpensive wallpaper backing.

Wood pulp sheets are made by debarking a tree and chipping the tree into a slurry. The mixture is run through a pulp mill where chlorine dioxide and oxygen are added to separate the lignin (which cements the woody cell walls together) from the rest of the wood pulp and bleaches the pulp. Wood pulp sheets with fibers can have synthetic fibers added to give the paper additional texture.

A roll of paper from the paper mill is 65 in (1.65 m) wide, possibly as long as 22,000 ft (6,706 m), and weighs approximately one ton. Once sold to a printer, each paper roll is cut into six sub-rolls which are 21 in (53 cm) wide by 10,000 ft (3,048 m) long.

Coating

2 Before the pattern is printed, the backing must be coated with a ground color. Ground wood sheets are coated with colored vinyl (PVC), which varies in thickness depending on the durability and strippability of paper under production. Vinyl may also be laminated to backings for exceptional serviceability.

Wood pulp sheets are coated with one or all of the following: kaolin clay for drapability, titanium dioxide for opacity, and latex for ease in handling and color.

Printing

There are four possible types of printing techniques.

3 *Surface printing.* Metal rollers impregnated with a raised rubber pattern are mounted on a single machine. Ink is applied to the surface roller, and the ink lays in the hills or rubber pattern sitting above the surface of the roller. The ink is then pressed onto the paper.

4 *Gravure printing.* Each color of the pattern is printed with a single roller. Copper cylinders are laser-etched then chrome-plated for durability. Large gravure-printing machines hold a maximum of 12 cylinders that together create the whole pattern. The paper roll moves to one cylinder, a back roller picks up color and pushes it against an engraved roller. A steel doctor blade pushes against the engraved cylinder, forcing ink into the etched detail. A rubber roller then presses paper against the cylinder, enabling it to pick up the ink in the valleys of the engraving. Finally, rollers carry the paper away from the cylinder into a dryer, where the ink is set. Once the ink is dry, the process begins again with the next cylinder.

5 *Silk screen printing.* Stencils for each color present in the pattern are created from silk mesh screen, using a photographic

Patterns can be printed onto wall-paper by one of several printing methods, including gravure printing, rotary printing, and silk-screen printing.

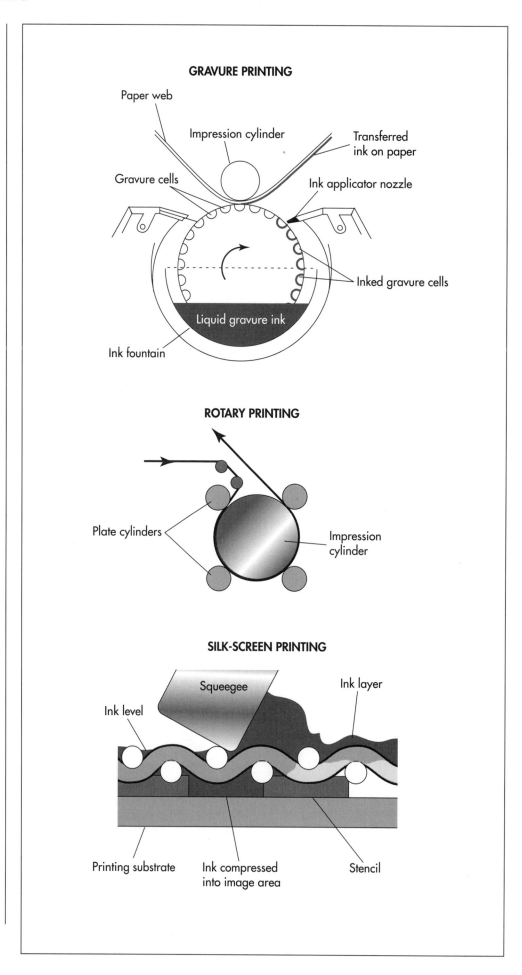

GRAVURE PRINTING

Paper web

Impression cylinder

Transferred ink on paper

Gravure cells

Ink applicator nozzle

Inked gravure cells

Liquid gravure ink

Ink fountain

ROTARY PRINTING

Plate cylinders

Impression cylinder

SILK-SCREEN PRINTING

Squeegee

Ink layer

Ink level

Printing substrate

Ink compressed into image area

Stencil

process. First, a photographic negative is made of the pattern. Then, a silk screen approximately 36 in (91 cm) long is stretched taut over a magnesium or wood frame. The screen is coated with a light sensitive emulsion, and the negative is placed on top of the screen. Once hit with bright light, the emulsion hardens in the areas not covered by the negative, forming a stencil.

Paper is set upon a long table, a screen stencil is placed on top, and ink is applied by a scraper or squeegee. Color is deposited on the paper where the screen permits the ink to pass through to the paper. The stencil is picked up, moved down the paper, and inked again along the entire length of the paper roll.

Before the next color is applied, the ink is thoroughly dried. Each screen is carefully put down with blocks, guides, etc. so that the pattern is aligned and repeats without breaks. Hand printing produces patterns with thick, evenly applied color. Theoretically, the number of colors used in the screening process is limitless; however, the high cost of hand printing necessarily limits the number of colors companies can include in the pattern.

6 *Rotary printing.* This type of printing process combines the mechanics of gravure printing with the precision of photographically produced stencils. Mesh stencils are wrapped around hollow tubes mounted within a machine. Ink continuously flows through the film-wrapped tubes and onto paper, imparting a tremendous amount of color (a maximum of 12 colors). This technique resembles the more expensive silk-screening, but it can print much more quickly—approximately 80 yd (73.12 m) of wallpaper per minute.

Prepasting

7 Printed wallpaper is rolled with a wet cornstarch or wheat starch-based coating and then dried.

Packaging

8 Residential-use wallpapers are cut down from 3,000 yd (2,742 m) rolls to 15 yd (13.71 m) rolls. Commercial-use rolls are generally packaged in 30, 45, and 60 yd (27, 41, and 55 m) rolls. A printed label, run number, and hanging instructions are placed against each roll and shrink wrapped together. Rolls are stored in a warehouse until final shipment.

Quality Control

The Chemical Fabric and Film Association (CFFA) has devised quality standards for commercial-use vinyl coated wallcoverings. The various categories of commercial-use papers have different physical test requirements specified in CFFA Quality Standard documents. All papers must undergo testing on such attributes as minimum coating weight, tensile strength, tear strength, coating adhesion, abrasion resistance, flame spread, smoke development, shrinkage, heat aging, stain resistance, etc.

Each wallpaper printing company conducts visual inspections in the form of spot checks or representative product samplings to ensure their product conforms to certain values established by the manufacturer. Generally, wood pulp and ground wood paper backings are given visual checks to see if there is foreign matter imbedded in the backing. When woven backings are received by printers, the printer checks thread count and physically tests the fabric for minimum requirements.

As the backing is printed, constant visual checks ensure proper adhesion of vinyl to backing, correct color, no streaking or unwanted shading, trimmed edges, etc. Representative samples are physically and visually examined before being cut into smaller rolls.

Where to Learn More

Books

Chemical Fabrics and Film Association, Inc. *CFFA Quality Standard for Vinyl Coated Wallcovering.* Cleveland, OH: CFFA, 1995.

Lynn, Catherine. *Wallpaper in America.* New York: W. W. Norton, 1980.

Swedlow, Robert M. *Step by Step Guide to Screen-Process Printing.* Englewood Cliffs, NJ: Prentice Hall, 1985.

Teynac, Francoise, et al. *Wallpaper: A History.* New York: Rizzoli, 1982.

—Nancy EV Bryk

Wig

Wigs of synthetic (e.g., acrylic, modacrylic, nylon, or polyester) hair are popular for several reasons. They are comparatively inexpensive (costing one-fifth to one-twentieth as much as a human hair wig). During the past decade, significant improvements in materials.

Background

Wigs are worn for either prosthetic, cosmetic, or convenience reasons. People who have lost all or part of their own hair due to illness or natural baldness can disguise the condition. For strictly cosmetic reasons (or perhaps to alter their appearance), people might wear a wig to quickly achieve a longer or fuller hairstyle or a different color. In an article in *Vogue* magazine, the wife of a prominent politician was described as using a wardrobe of wigs to avoid $8,400 and 160 or more hours spent with professional hairdressers each year, in addition to the complicated task of finding appropriate hair care while traveling.

History

Based on an ivory carving of a woman's head found in southwestern France, anthropologists speculate that wigs may have been used as long as 100,000 years ago. Wigs were quite popular among ancient Egyptians, who cut their hair short or shaved their heads in the interests of cleanliness and comfort (i.e., relief from the desert heat). While the poor wore felt caps to protect their heads from the sun, those who could afford them wore wigs of human hair, sheep's wool, or palm-leaf fiber mounted on a porous fabric. An Egyptian clay figure that dates to about 2500 B.C. wears a removable wig of black clay. The British Museum holds a beautifully made wig at least 3,000 years old that was found in the Temple of Isis at Thebes; its hundreds of tiny curls still retain their carefully arranged shape.

Wigs were popular in ancient Greece, both for personal use and in the theater (the color

and style of wigs disclosed the nature of individual characters). In Imperial Rome, fashionable women wore blond or red-haired wigs made from the heads of Germanic captives, and Caesar used a wig and a laurel wreath to hide his baldness. Both Hannibal and Nero wore wigs as disguises. A portrait bust of Plautilla (ca. 210 A.D.) was made without hair so wigs of current fashion could always adorn this image of Emperor Caracalla's wife.

During the reign of Stephen in the middle third of the twelfth century, wigs were introduced in England; they became increasingly common, and women began to wear them in the late sixteenth century. Italian wigs of that time were made of either human hair or silk thread. In 1630, embarrassed by his baldness, Louis XIII began wearing a wig made of hair sewn onto a linen foundation. Wigs became fashionable, increasing in popularity during the reign of Louis XIV, who not only wore them to hide his baldness but also to make himself seem taller by means of towering hair. During the Plague of 1665, hair was in such short supply that there were persistent rumors of the hair of disease victims being used to manufacture wigs. This shortage of hair was partially remedied by using wool or the hair of goats or horses to make lower grades of wigs (in fact, horsehair proved useful since it retained curls effectively). For several decades around 1700, men were warned to be watchful as they walked the streets of London, lest their wigs be snatched right off their heads by daring thieves.

The enormous popularity of wigs in England declined markedly during the reign of George III, except for individuals who continued to wear them as a symbols of their professions (e.g., judges, doctors, and clergymen). In fact, so many wigmakers were facing financial ruin that they marched through London in February 1765 to present George III with a petition for relief. Bystanders were infuriated, noticing that few of the wigmakers were wearing wigs although they wanted to protect their jobs by forcing others to wear them. A riot ensued, during which the wigmakers were forcibly shorn.

During the late eighteenth century, Louis XVI wore wigs to hide his baldness, and wigs were very fashionable throughout France. The modern technique of ventilating (attaching hairs to a net foundation) was invented in this environment. By 1784, springs were being sewn into French wigs to make them fit securely. In 1805, a Frenchman invented the flesh-colored hair net for use in wigmaking. A series of other improvements followed rapidly, including knotting techniques, fitting methods, and the use of silk net foundations. These matters were so important that a major lawsuit arose, and one inventor committed suicide after selling his patent cheaply and watching others become rich using his technique. One of the manufacturing processes that was tried at this time was based on the use of pig or sheep bladders to simulate bald heads on actors. In the mid 1800s, some wigs and toupees were made by implanting hairs in such bladders using an embroidery needle. In the late nineteenth century, children and apprentices of wigmakers amused themselves by playing the "wig game," in which each participant accumulated points by throwing an old wig up to touch the ceiling and catching it on the head as it fell.

Raw Materials

By the early 1900s, jute fibers were being used as imitation hair in theatrical wigs. Today, a favorite material for theatrical wigs, particularly those worn by clowns, is yak hair from Tibet. The hair of this ox species holds a set well, is easily dyed, and withstands food and shaving cream assaults.

Wigs of synthetic (e.g., acrylic, modacrylic, nylon, or polyester) hair are popular for several reasons. They are comparatively in-expensive (costing one-fifth to one-twentieth as much as a human hair wig). During the past decade, significant improvements in materials have made synthetic hair look and feel more like natural hair. In addition, synthetic wigs weigh noticeably less than human hair versions. They hold a style well—so well, in fact, that they can be difficult to restyle. On the other hand, synthetic fibers tend not to move as naturally as human hairs, and they tend to frizz from friction along collar lines. Synthetic hair is also sensitive to heat and can easily be damaged (e.g., from an open oven, a candle flame, or a cigarette glow).

Human hair remains a popular choice for wigs, particularly because it looks and feels natural. It is easily styled; unlike synthetic hair, it can be permed or colored. During periods of scarcity of cut human hair for wigs, manufacturers have used combings (hairs that fall out naturally at the end of their life cycle). However, actively growing hair that is cut for wigmaking is preferred. United States wigmakers import most of their hair. Italy is known as a prime source of hair with desirable characteristics; other colors and textures of hair are purchased in Spain, France, Germany, India, China, and Japan. Women contract with hair merchants to grow and sell their hair. After cutting, the hair is treated to strip the outer cuticle layer, making the hair more manageable. Wigmakers pay $80 or more per ounce for virgin hair, which has never been dyed or permed; a wig requires at least 4 oz (113.4 g) of hair.

Some manufacturers blend synthetic and human hair for wigs that have both the style-retaining qualities of synthetic hair and the natural movement of human hair. However, this can complicate maintenance, since the different types of hair require different kinds of care.

Types of Wigs

Ready-made wigs are available in stores and by mail order. They are one-size-fits-all models that adjust to individual heads by means of either a stretchy foundation or adjustable sections around the edge of the foundation. Ready-made wigs can be made of either synthetic or human hair and are available in either machine-made or hand-tied versions. Customers who are willing to pay more for a better fit can purchase semi-custom wigs that are hand-knotted on different sizes and shapes of stock foundations. The best fit, however, is achieved with a custom-made wig. Made to the customer's exact head measurements, these wigs are held in place by tension springs or adhesive strips, or can be clasped to existing growth hair. Silicone foundations can be molded to the exact head shape, so that they are held in place by a suction fit.

Machine-made wigs are fabricated by weaving hair into wefts (hair shafts that are woven together at one end into a long strip). These can be sewn in rows to a net foundation. When the hair is disturbed, by blowing wind for example, the foundation shows through the hair. Thus, such wigs are less desirable for people who have no growth hair under the wig. Hand-tied wigs, on the other hand, give a more natural look, particularly if slightly different shades of hair are blended before being applied to the foundation. Hand-tied wigs shed hair and must be repaired from time to time. With proper care, human-hair wigs generally last for two to six years.

The Manufacturing Process

The following description reflects the making of a full, custom-fitted, hand-tied, human-hair wig. Such a wig would take four to eight weeks to make, and would sell for approximately $2,000 to $4,000.

Preparing the hair

1 The wigmaker must first make sure that the individual hairs are lying in the same direction. This is done by holding a small bunch of hair in the hand, and rubbing the ends between the finger and thumb. The tips (uncut ends) turn back during the rubbing, while the cut ends (which were closer to the hair's root) lie straight. If the hairs in the bunch are running in both directions, they must be turned by sorting the "root down" hairs into one pile and the "root up" hairs into another before recombining them into a single, organized bunch.

2 Very short hairs less than 3 in (7.5 cm) long are separated out by drawing the bunch through a hackle (wire brush) that is clamped to the workbench. After hackling, the usable hairs are tied together into bundles of convenient size. Fine string is used to tie the bundles tight enough to hold them securely, but loose enough to allow the string to be shifted while washing the hair.

3 The hair is carefully inspected for nits (louse eggs). If any are found, they are removed by boiling the hair in an acetic acid solution and combing it through a steel-toothed nitting machine.

4 Each bundle of hair is gently, but thoroughly, hand washed in a bowl of hot, soapy water that contains a disinfectant. The hair is then rinsed several times in clear water. The bundles are carefully squeezed in a towel and allowed to dry either in open air or in an oven set at 176°-212°F(80°-100°C).

5 Bundles of clean, dry hair are again hackled to straighten them. They are then passed through a set of drawing brushes, so the wigmaker can sort them into bunches of equal length, which are tied near the root end.

6 If desired, hair can now be permanently curled or waved. After the hair is wound onto curlers, it is boiled in water for 15 to 60 minutes (depending on the tightness desired) and then dried in a warm oven for 24 hours or more.

7 Heads of growing hair are not uniform in color. The wigmaker may prepare hair for a specific wig by blending as many as five or more slightly different shades of hair together to produce a more natural look.

Preparing the pattern

8 In order to get the best possible fit, the foundation of a custom wig is made as close as possible to the shape of the client's head. This can be accomplished either by measuring several aspects of the head directly, or by making a plaster cast of the head and using it as a model.

9 Six basic measurements are taken of the customer's head. The circumference is measured a half inch (1.27 cm) above the hairline at the nape of the neck, above each ear, and across the front of the head a half

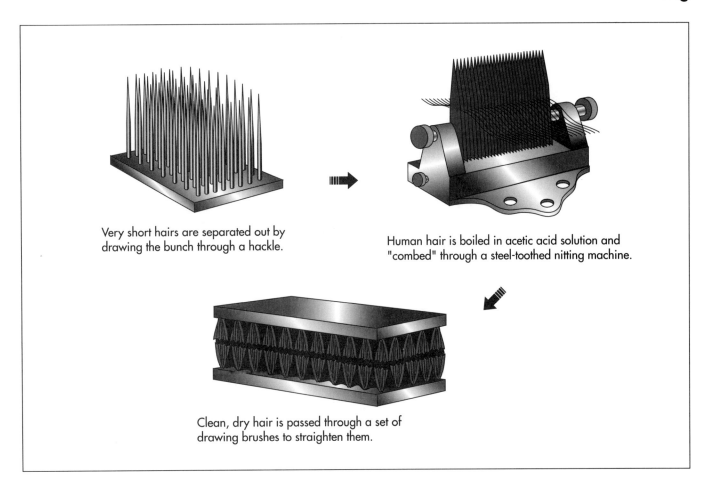

Very short hairs are separated out by drawing the bunch through a hackle.

Human hair is boiled in acetic acid solution and "combed" through a steel-toothed nitting machine.

Clean, dry hair is passed through a set of drawing brushes to straighten them.

inch (1.27 cm) above the hairline. For adults, this measurement ranges from 19-24 in (48-61 cm). The second measurement is from the hairline at the front of the head to the hairline at the nape of the neck. The third is taken between points just in front of each ear, along the hairline at the front of the head. The fourth goes across the crown of the head, from just above one ear to just above the other. The fifth runs straight across the back of the head, from one temple to the other. Finally, the sixth measurement traces the nape of the neck.

In addition, the wigmaker must note such information as any unusual shape of the head, the length and location of any desired parting of the hair, and the desired style of the hair on the finished wig.

10 A pattern is drawn and cut from light-colored paper.

Making the foundation

11 The edge of the wig's foundation is cut from fine-mesh silk netting that matches the desired hair color. This piece varies in width from two or more inches (5 cm) in the front to one inch (2.5 cm) in the back. The crown of the foundation is cut from a coarser net made of silk, cotton, or nylon. If a part is to be incorporated in the wig, a strip of very fine, silk net (white or flesh-colored) is cut and inserted in the appropriate location on the foundation.

12 The paper pattern is carefully positioned on an appropriate size block (a wooden, head-shaped form). With this paper lying under the net foundation, the mesh of the net is easier to see and knotting is facilitated. The pieces of the net foundation are joined together on the block by sewing them to pieces of galloon (fine, strong silk ribbon that matches the color of the foundation netting). The foundation is held in place on the block by cotton thread sewn through the galloon and laced through anchoring points (steel loops hammered into the block).

13 Springs sewn into the foundation at strategic locations will hold the com-

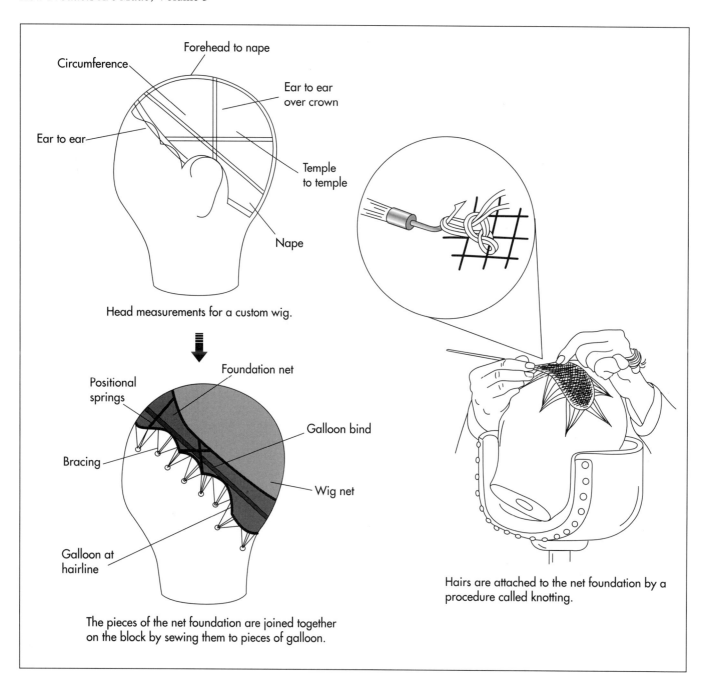

Head measurements for a custom wig.

Hairs are attached to the net foundation by a procedure called knotting.

The pieces of the net foundation are joined together on the block by sewing them to pieces of galloon.

pleted wig in place on the customer's head. These devices, 1.5-2 in (3.8-5.1 cm) long, are often made from steel watch springs or elastic bands and are encased in galloon.

Knotting

14 Hairs are attached to the net foundation by a procedure called knotting. Although various wigmakers use at least three types of knotting, the single, or "full V," knot is the most common. It is similar to the knot used in making a latch-hook rug. Using this knot, hairs 25 in (63.5 cm) long

must be used to make a wig with 12 in (30.5 cm) long hair. Different sizes of ventilating needles can be used, depending on the number of hairs that are to be tied together in one knot. Along parts and the front edges of the wig, knots are usually made with single hairs, while in the crown up to eight hairs may be knotted together. A full wig requires 30,000 to 40,000 knots, which take a total of about 40 hours of tying.

15 When the wig is fully ventilated (i.e., all the hair is attached), it is removed from the wooden block and mounted wrong-side-out on a soft block made of canvas stuffed with sawdust. A final row of knotting is done around the edge. The inside surface of the wig is pressed with a heated iron to secure the knots.

Styling

16 The completed wig is pinned to a soft block for styling. The hair is gently dampened by combing through with a wet comb. Curls are formed as pincurls or on rollers or cotton forms. The wig is covered with a net and dried in a warm oven. The curls are then unpinned, and the hair is combed and styled. A net is carefully placed over the finished style, and the wig is returned to the oven to set the style.

Where to Learn More

Books

Anderson, Jean. *Wigmaking: Step by Step.* JA Publications, 1992.

Botham, Mary, and L. Sharrad. *Manual of Wigmaking.* Funk & Wagnalls, 1964. Reprint, David & Charles, 1983.

Davenport, Millia. *The Book of Costume.* Crown Publishers, 1948.

Sutton, Alfred M. *Boardwork or The Art of Wigmaking, etc.* R. Hovenden & Sons, Ltd., 1921.

Periodicals

Cunliffe, Lesley. "With Today's Well-Made Wigs, You Can Have Your Hair and Cut It Too." *Vogue,* August 1989, pp. 149-50.

"Yak." *The New Yorker,* May 11, 1992, pp.30-31.

Other

Knight, Peggy. "Wig Buyer's Guide." *Alopecia Frequently Asked Questions.*http://npntserver.mcg.edu/html/alopecia/AlopeciaFAQ(appendixA).html#AlopeciaFAQ-appendixA.WigGuide (April 28,1997).

—*Loretta Hall*

Yarn

Researchers have developed genetically-altered cotton plants, whose fibers are especially good at retaining warmth. Each fiber is a blend of normal cotton and small amounts of a natural plastic called polyhydroxybutyrate.

Background

Yarn consists of several strands of material twisted together. Each strand is, in turn, made of fibers, all shorter than the piece of yarn that they form. These short fibers are spun into longer filaments to make the yarn. Long continuous strands may only require additional twisting to make them into yarns. Sometimes they are put through an additional process called texturing.

The characteristics of spun yarn depend, in part, on the amount of twist given to the fibers during spinning. A fairly high degree of twist produces strong yarn; a low twist produces softer, more lustrous yarn; and a very tight twist produces crepe yarn. Yarns are also classified by their number of parts. A single yarn is made from a group of filament or staple fibers twisted together. Ply yarns are made by twisting two or more single yarns. Cord yarns are made by twisting together two or more ply yarns.

Almost eight billion pounds (3.6 billion kg) of spun yarn was produced in the United States during 1995, with 40% being produced in North Carolina alone. Over 50% of spun yarn is made from cotton. Textured, crimped, or bulked yarn comprised one half of the total spun. Textured yarn has higher volume due to physical, chemical, or heat treatments. Crimped yarn is made of thermoplastic fibers of deformed shape. Bulked yarn is formed from fibers that are inherently bulky and cannot be closely packed.

Yarn is used to make textiles using a variety of processes, including weaving, knitting, and felting. Nearly four billion pounds (1.8 billion kg) of weaving yarn, three billion pounds (1.4 kg) of machine knitting yarn, and one billion pounds (450 million kg) of carpet and rug yarn was produced in the United States during in 1995. The U.S. textile industry employs over 600,000 workers and consumes around 16 billion pounds (7 billion kg) of mill fiber per year, with industry profits estimated at $2.1 billion in 1996. Exports represent more than 11% of industry sales, approaching $7 billion. The apparel industry employs another one million workers.

History

Natural fibers—cotton, flax, silk, and wool—represent the major fibers available to ancient civilizations. The earliest known samples of yarn and fabric of any kind were found near Robenhausen, Switzerland, where bundles of flax fibers and yarns and fragments of plain-weave linen fabric, were estimated to be about 7,000 years old.

Cotton has also been cultivated and used to make fabrics for at least 7,000 years. It may have existed in Egypt as early as 12,000 B.C. Fragments of cotton fabrics have been found by archeologists in Mexico (from 3500 B.C.)., in India (3000 B.C.), in Peru (2500 B.C.), and in the southwestern United States (500 B.C.). Cotton did not achieve commercial importance in Europe until after the colonization of the New World. Silk culture remained a specialty of the Chinese from its beginnings (2600 B.C.) until the sixth century, when silkworms were first raised in the Byzantine Empire.

Synthetic fibers did not appear until much later. The first synthetic, rayon, made from cotton or wood fibers, was developed in

1891, but not commercially produced until 1911. Almost a half a century later, nylon was invented, followed by the various forms of polyester. Synthetic fibers reduced the world demand for natural fibers and expanded applications.

Until about 1300, yarn was spun on the spindle and whorl. A spindle is a rounded stick with tapered ends to which the fibers are attached and twisted; a whorl is a weight attached to the spindle that acts as a flywheel to keep the spindle rotating. The fibers were pulled by hand from a bundle of carded fibers tied to a stick called a distaff. In hand carding, fibers are placed between two boards covered with leather, through which protrude fine wire hooks that catch the fibers as one board is pulled gently across the other.

The spindle, which hangs from the fibers, twists the fibers as it rotates downward, and spins a length of yarn as it pulls away from the fiber bundle. When the spindle reaches the floor, the spinner winds the yarn around the spindle to secure it and then starts the process again. This is continued until all of the fiber is spun or until the spindle is full.

A major improvement was the spinning wheel, invented in India between 500 and 1000 A.D. and first used in Europe during the Middle Ages. A horizontally mounted spindle is connected to a large, hand-driven wheel by a circular band. The distaff is mounted at one end of the spinning wheel and the fiber is fed by hand to the spindle, which turns as the wheel turns. A component called the flyer twists the thread just before it is wound on a bobbin. The spindle and bobbin are attached to the wheel by separate parts, so that the bobbin turns more slowly than does the spindle. Thus, thread can be twisted and wound at the same time. About 150 years later, the Saxon wheel was introduced. Operated by a foot pedal, the Saxon wheel allowed both hands the freedom to work the fibers.

A number of developments during the eighteenth century further mechanized the spinning process. In 1733, the flying shuttle was invented by John Kay, followed by Hargreaves' spinning jenny in 1766. The jenny featured a series of spindles set in a row, enabling one operator to produce large quantities of yarn. Several years later Richard Arkwright patented the spinning frame, a machine that used a series of rotating rollers to draw out the fibers. A decade later Samule Cromptons' mule machine was invented, which could spin any type of yarn in one continuous operation.

The ring frame was invented in 1828 by the American John Thorp and is still widely used today. This system involves hundreds of spindles mounted vertically inside a metal ring. Many natural fibers are now spun by the open-end system, where the fibers are drawn by air into a rapidly rotating cup and pulled out on the other side as a finished yarn.

Raw Materials

About 15 different types of fibers are used to make yarn. These fibers fall into two categories, natural and synthetic. Natural fibers are those that are obtained from a plant or an animal and are mainly used in weaving textiles. The most abundant and commonly used plant fiber is cotton, gathered from the cotton boil or seed pod when it is mature. In fact, cotton is the best-selling fiber in America, outselling all synthetic fibers combined.

Fibers taken from the plant leaf or stem are generally used for rope. Other plant fibers include acetate (made from wood pulp or cotton linters) and linen, made from flax, a vegetable fiber. Animal fibers include wool, made from sheep hair, and mohair, made from angora goats and rabbits. Silk is a protein extruded in long, continuous strands by the silkworm as it weaves its cocoon.

Synthetic fibers are made by forcing a thick solution of polymerized chemicals through spinneret nozzles and hardening the resulting filament in a chemical bath. These include acrylic, nylon, polyester, polyolefin, rayon, spandex, and triacetate. Some of these fibers have similar characteristics to the natural fibers without the shrinkage problems. Other fibers have special properties for specific applications. For instance, spandex can be stretched over 500% without breaking.

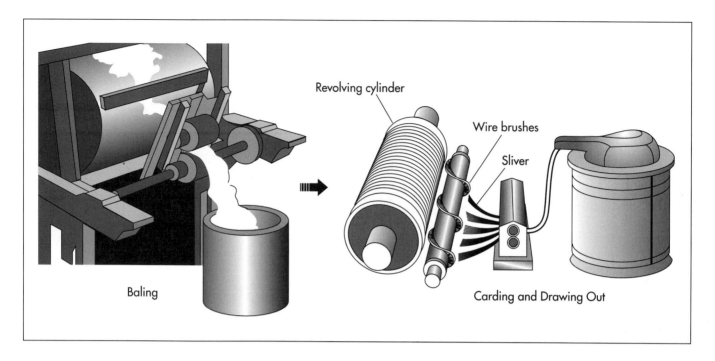

Baling

Revolving cylinder

Wire brushes

Sliver

Carding and Drawing Out

Fibers are shipped in bales, which are opened by hand or machine. The picker loosens and separates the lumps of fiber and also cleans the fiber if necessary. The carding machine separates the fibers and pulls them into somewhat parallel form. The thin web of fibers formed then passes through a funnel-shaped device that produces a ropelike strand of parallel fibers. Rollers elongate the strand, called a sliver, into a single more uniform strand that is given a small amount of twist and fed into large cans.

The Manufacturing Process

There are three major spinning processes: cotton, worsted or long-staple, or wool. Synthetic staple fibers can be made with any of these processes. Since more yarn is produced with the cotton process than the other two, its manufacture is described below.

Preparing the fibers

1 Fibers are shipped in bales, which are opened by hand or machine. Natural fibers may require cleaning, whereas synthetic fibers only require separating. The picker loosens and separates the lumps of fiber and also cleans the fiber if necessary. Blending of different staple fibers may be required for certain applications. Blending may be done during formation of the lap, during carding, or during drawing out. Quantities of each fiber are measured carefully and their proportions are consistently maintained.

Carding

2 The carding machine is set with hundreds of fine wires that separate the fibers and pull them into somewhat parallel form. A thin web of fiber is formed, and as it moves along, it passes through a funnel-shaped device that produces a ropelike strand of parallel fibers. Blending can take place by joining laps of different fibers.

Combing

3 When a smoother, finer yarn is required, fibers are subjected to a further paralleling method. A comblike device arranges fibers into parallel form, with short fibers falling out of the strand.

Drawing out

4 After carding or combing, the fiber mass is referred to as the sliver. Several slivers are combined before this process. A series of rollers rotating at different rates of speed elongate the sliver into a single more uniform strand that is given a small amount of twist and fed into large cans. Carded slivers are drawn twice after carding. Combed slivers are drawn once before combing and twice more after combing.

Twisting

5 The sliver is fed through a machine called the roving frame, where the strands of fiber are further elongated and given additional twist. These strands are called the roving.

Spinning

6 The predominant commercial systems of yarn formation are ring spinning and open-end spinning. In ring spinning, the roving is fed from the spool through rollers. These rollers elongate the roving, which passes through the eyelet, moving down

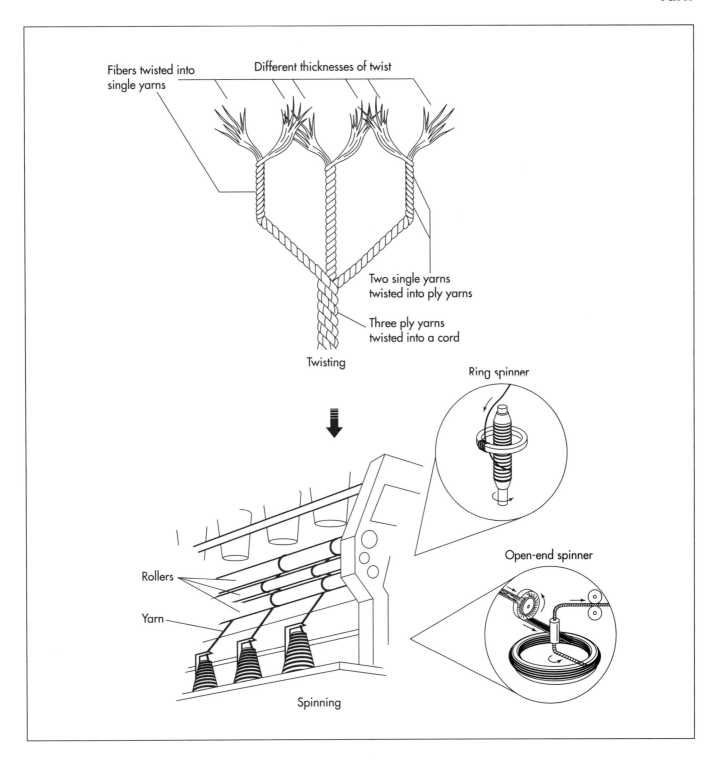

Fibers twisted into single yarns

Different thicknesses of twist

Two single yarns twisted into ply yarns

Three ply yarns twisted into a cord

Twisting

Ring spinner

Open-end spinner

Rollers

Yarn

Spinning

and through the traveler. The traveler moves freely around the stationary ring at 4,000 to 12,000 revolutions per minute. The spindle turns the bobbin at a constant speed. This turning of the bobbin and the movement of the traveler twists and winds the yarn in one operation.

7 Open-end spinning omits the roving step. Instead, a sliver of fibers is fed into the spinner by a stream of air. The sliver is delivered to a rotary beater that separates the fibers into a thin stream that is carried into the rotor by a current of air through a tube or duct and is deposited in a V-shaped groove along the sides of the rotor. As the rotor turns, twist is produced. A constant stream of new fibers enters the rotor, is distributed in the groove, and is removed at the end of the formed yarn.

The sliver is fed through a machine called the roving frame, where the strands of fiber are further elongated and given additional twist. The predominant commercial systems of yarn formation are ring spinning and open-end spinning. Open-end spinning omits the roving step.

Quality Control

Automation has made achieving quality easier, with electronics controlling operations, temperatures, speeds, twists, and efficiency. The American Society for Testing of Materials has also established standardized methods for determining such properties as drawforce, bulk, and shrinkage.

The Future

Spinning systems and yarn manufacturing machinery will continue to become more automated and will be integrated as part of a manufacturing unit rather than as a separate process. Spinning machines have already been developed that combine carding and drawing functions. Production rates will increase by orders of magnitude as machines become available with even more spindles. Robot-controlled equipment will become standard.

Domestic yarn producers will continue to be threatened by competition from Asian countries, as these countries continue to buy the latest textile machinery technology. Higher domestic material prices will not help, since the cost of the raw material can represent up to 73% of the total cost of producing the yarn. U.S. yarn producers will continue to form alliances with their customers and customers' customers to remain competitive. The textile industry is also forming unique partnerships. The American Textile Partnership is a collaborative research and development program among industry, government, and academia aimed at strengthening the competitiveness of the U.S. industry.

Another continuing challenge for the industry will be compliance with stricter environmental regulations. Recycling is already an issue and processes are under development to manufacture yarn from scrap material, including denim. Yarn producers will have to incorporate pollution prevention measures to meet the air and water quality restrictions. Equipment manufactures will continue to play an important role in this endeavor.

Genetic engineering will become more widely used for developing fibers with unique properties. Researchers have developed genetically-altered cotton plants, whose fibers are especially good at retaining warmth. Each fiber is a blend of normal cotton and small amounts of a natural plastic called polyhydroxybutyrate. It is predicted that dye-binding properties and greater stability will be possible with new fibers in the next generation.

New synthetic fibers will also be developed that combine the best qualities of two different polymers. Some of these fibers will be produced through a chemical process, whereas others will be generated biologically by using yeast, bacteria, or fungi.

Where to Learn More

Books

Needles, H. L. *Textile Fibers, Dyes, Finishes, and Processes.* 1986.

Periodicals

Clune, Ray. "AYSA head exhorts yarn spinners to take more proactive stance." *Daily News Record*, May 8, 1996, p. 9.

Isaacs, McAllister. "Texturing gets automation as TYAA celebrates 20." *Textile World*, May 1992, p. 54.

"Long-staple processing moves to cut costs." *Textile World*, September 1992, p. 42.

Weiss, Rick. "Molecular biologists grow gene-altered cotton plants." *Washington Post*, November 12, 1996.

Tortora, Phyllis G. "Making Fibers into Yarns." *Understanding Textiles.* Macmillan Publishing Company, 1987.

Other

American Textile Manufacturers Institute, http://www.atmi.org

Cotton Incorporated, 4605 Creedmoor Road, Raleigh, North Carolina 27612, tel: 919/782-6330, fax: 919/881-9874, http://www.cottoninc.com

Current Industrial Reports, MA22F—Yarn Production.1995. Department of Commerce, Bureau of Census. http://www.census.gov

—*Laurel M. Sheppard*

Yo-yo

Background

A yo-yo is a toy that has two disks connected together and sandwiching a long string. Traditionally made of wood, the disks are now commonly made of plastic. Attached to the center of the disks, the string winds, unwinds, and rewinds, while attached to a person's finger.

The specific origin of the yo-yo is uncertain. Early versions of the toy have been placed in China, Greece, and the Philippines. The National Museum of Athens houses several vases dating from around 500 B.C. depicting young Greeks playing with discs tethered to a cord. The word yo-yo means come come in the native Philippine language, Tagalog, and yo-yos have been hand-carved in that country for thousands of years.

History

The European introduction to the yo-yo occurred much more recently. The toy arrived in France during the eighteenth century, probably brought by missionaries returning from China, and it became a craze among the elite. The future king at that time, Louis XVII, was painted at age four with a yo-yo in hand. The French called the toy *l'emigrette*, after the aristocrats who popularized the toy and who were forced to emigrate to safer territories when the revolution began.

From France the yo-yo traveled to England where again it sparked a craze among the upper classes. The English dubbed the toy bandalore and also quiz, and illustrations from the period show soldiers, aristocrats, and even kings playing with it.

In 1927, a Filipino busboy named Pedro Flores began carving and selling a toy from his childhood to the guests at the Santa Monica, California, hotel where he worked. By 1929, the Flores Yo-yo Corporation had two factories in Los Angeles, California, feeding a craze for the toy that was sweeping across the United States. Flores' yo-yo utilized a unique innovation, the slip string. Previous designs had the string tied to the yo-yo's axle, and so the yo-yo would only go up and down. With Flores' design, a twisted length of string looped around the axle, allowing the yo-yo to spin or "sleep" so that various tricks could be performed.

Soon Flores' spinning top attracted the attention of marketing genius Donald Duncan. Duncan opened a Flores Yo-yo factory in Chicago, Illinois, and soon after, he bought the company. To promote his new purchase, Duncan staged yo-yo contests in cities across the country. The yo-yo became a nationwide craze; one 30-day campaign in 1930 sold three million of them.

A shortage of wood and labor put the yo-yo on hiatus during World War II. When production restarted in 1947, Duncan was not alone in the market. For years, the company was able to maintain an edge over the competition with a fierce defense of their trademark on the name yo-yo, forcing competitors to give their toys names such as return tops or Filipino twirlers.

In 1962, the yo-yo reached new heights in the United States, selling 45 million units in a country that only had 40 million children. Feeling that they were in a position to finally take back the market, Duncan sued its chief rival Royal Tops for trademark infringement. They lost. The court ruled that Duncan could not legally hold a trademark

In 1962, the yo-yo reached new heights in the United States, selling 45 million units in a country that only had 40 million children.

on the name yo-yo because it was, and always had been, simply the name of the toy, like kite or baseball. Three years later, deeply in debt from court battles, expensive television promotions, and the cost of retooling production lines from wooden to plastic yo-yos, Duncan was forced by its creditors into bankruptcy. In 1968, Flambeau Plastics Company bought the Duncan name and restarted production; the company still makes Duncan yo-yos today.

Raw Materials

The Greeks fashioned their discs from terra cotta. The French *emigrettes* were ivory and brass. Philippine yo-yos are carved from water buffalo horn or from wood. Until the switch to plastic in the late 1960s, American-made yo-yos were carved from solid blocks of maple. Most modern manufacturers of wooden yo-yos still use maple as it has the ideal density to give the yo-yo the proper weight at the required size. But 90% of the yo-yos sold today are plastic. Manufacturers use a plastic called K-resin. K-resin takes color well, is non-toxic, and is resilient enough to survive repeated abuse, but soft enough not to develop sharp edges.

The string is, and has been since 1927, pure Egyptian cotton. Recently, some manufacturers have introduced synthetic strings, but these do not maintain the proper friction against the inside of the yo-yo and against the axle and will not perform properly.

Traditionally, the axle was made out of the same wood as the sides of the yo-yo as the toy was carved out of a single, solid block. While wood provides ideal friction against the string, this friction eventually causes the string to cut the axle. And a broken axle cannot be repaired. So most modern axles are made from either aluminum or steel.

Design

A yo-yo functions on the two basic physical properties of friction and rotational inertia. Rotational inertia says that a spinning object will resist moving from the axis on which it is currently spinning, which is why a yo-yo will descend and return in a straight path rather than twisting and twirling on the end of the string. It also means that the ob-

ject will continue spinning until some other force —usually friction— stops it. The friction in a yo-yo comes from the contact of the string with the inside surfaces of the two halves and from the contact of the string with the axle. Decreasing the diameter of the axle reduces friction, allowing the yo-yo to spin longer, giving more time to perform tricks. But if the axle is too small, the yo-yo will not return properly. The distance between the two halves requires similar considerations; widening the gap decreases the friction against the string and allows the yo-yo to spin longer, but that friction of the string against the sides as the yo-yo descends is part of what makes the yo-yo spin. The width of the string has the same effect; a thicker or thinner string is essentially the same as a wider or narrower gap and will produce the same results. In addition, yo-yo string is designed to have a natural twist so that it will not unwind and let loose the yo-yo. But if the string is given too much twist, the end loop will wrap too tightly around the axle and stop the yo-yo from sleeping.

Rotational inertia increases as weight is distributed to the outside rim of the spinning object. A bicycle wheel, for example, has a lot of rotational inertia so it spins for a long time and is very stable while it spins, ideal for a yo-yo. Unfortunately, a yo-yo the size of a bicycle wheel is difficult to hold in one hand and is almost impossible to throw, so most yo-yos are made considerably smaller.

Considering all of these variables, each yo-yo manufacturer arrives at a different conclusion about specifics, usually varying by a sixteenth of an inch (0.16 cm) in any dimension and a sixteenth of an ounce (1.75 g) in overall weight. But they all end with the same basic conclusion, a yo-yo that weighs about one and three-quarters ounces (49 g) and is about two and five-eighths inches (6.67 cm) in diameter.

The Manufacturing Process

The first yo-yos manufactured in the United States were carved out of solid blocks of maple. This was a time-intensive but straight forward process. Lumber was first dried in huge kilns. This step is critical in

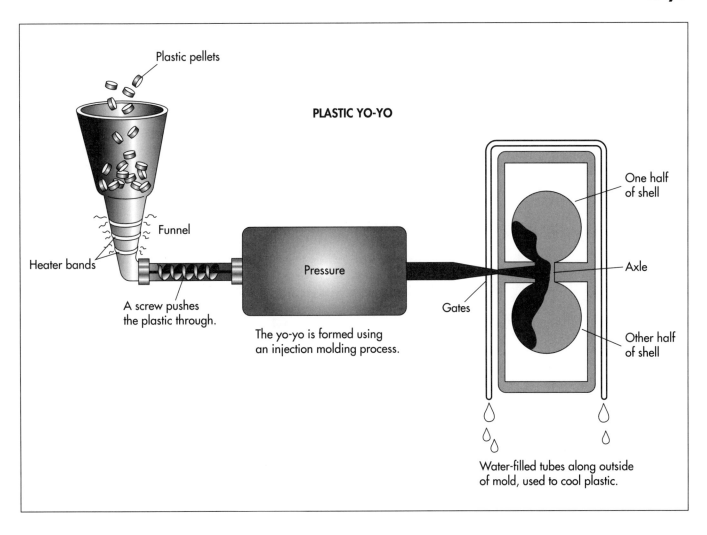

Plastic pellets

PLASTIC YO-YO

Funnel

Heater bands

A screw pushes
the plastic through.

Pressure

The yo-yo is formed using
an injection molding process.

Gates

One half
of shell

Axle

Other half
of shell

Water-filled tubes along outside
of mold, used to cool plastic.

the making of a wooden yo-yo because wood warps and shrinks as it dries, which is something that should happen before it is carved into the specific, balanced shape of a yo-yo. Once dried, the lumber was cut down to size. It was then put on a lathe and, using a master pattern to ensure the correct dimensions, it was shaved with a variety of chisels into a yo-yo. The yo-yo was either stained with wood stain or painted, and then it was finished with clear lacquer. Today, most manufacturers of wooden yo-yos still use this process except that very few make solid yo-yos anymore. Most drill into the wooden halves and connect them with a steel axle. However the majority of yo-yos made today are of plastic. That process has nine steps.

Shells and discs

1 Each half of a plastic yo-yo begins as two pieces: the flat, inner piece that will take the axle is called the disc, and the rounded outer piece is called the shell (the butterfly

type yo-yo is opposite, with a rounded disc and a flat shell. It is made via the same process but on a separate assembly line). The plastic used to make the discs and shells is fed in the form of solid pellets into a huge funnel with an equally huge screw inside. The outside of the funnel is wrapped with flexible strips called heater bands, which are like narrow electric blankets for the funnel. Each band gets progressively warmer as the funnel gets narrower, and as the screw pushes the pellets toward the bottom of the funnel, they become more and more liquid. This process ensures that the pellets will melt evenly and completely, which is critical to the next step. At the bottom of the funnel, the screw pushes the liquid plastic into a mold through a tube called a gate.

2 Inside the mold are the shapes of four discs and four shells, all connected by gates. The mold has a valve, which allows hot air to be pushed out ahead of the plastic and ensures that the mold is completely

Each half of a plastic yo-yos is composed of two parts, the outer shell and the inner disc. These two pieces are snapped together, and an axle joins two halves to form a yo-yo.

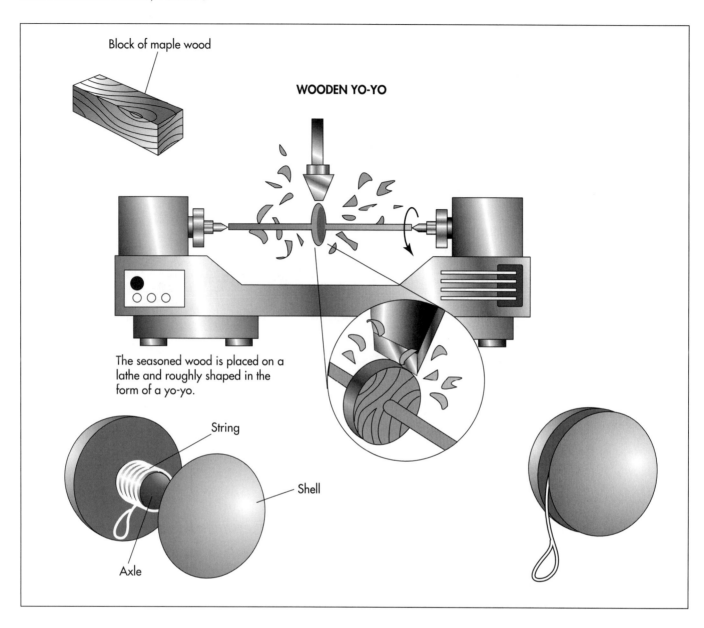

Block of maple wood

WOODEN YO-YO

The seasoned wood is placed on a lathe and roughly shaped in the form of a yo-yo.

String

Shell

Axle

Wooden yo-yos are made from lumber that has been dried to ensure the finished yo-yo will not change shape. Modern wooden models have two wooden halves that are connected by a metal axle.

filled. Because the mold is completely filled and because the plastic has been entirely and evenly melted, each shell and each disc will always be perfectly round and have perfectly even weight distribution. Once the mold is filled, it holds the plastic for 17 seconds and cools it with water-filled tubes running along the outside of the mold. Then the mold, which is actually two halves clamped tightly together, releases the shells and halves, all connected by the plastic that filled the gates, onto an assembly line to be fashioned into yo-yo halves.

Axles

3 While the molding line is making yo-yo halves, a separate machine is cutting the

axles to connect them. The machine is fed long pieces of round metal. The exact length of metal to make an axle is fed into a clamp where it is cut. The cut piece is then notched at each end so that it will grip the plastic.

Assembly

4 Once the mold releases the flat of shells and discs, a worker clips them apart, removes any extra plastic and snaps them together to make halves. Each shell and disc has a lip that fits into its complimentary piece. When the plastic is still slightly warm from the mold, these lips will snap together. Once fastened, they will not release.

5 The halves are sent down a belt to a machine called the hot stamper. The hot stamper has a circular tray with cup-like holders around its edge. Each holder takes a yo-yo half and rotates it to a heated press. The press feeds a roll of whatever logo the yo-yo will display onto the top of the yo-yo half. It then presses the logo onto the half and heats it, affixing the logo. Other designs use a round insert displaying the logo which fits into the disc and a clear shell so the insert can be seen.

6 The halves continue down the line to a worker who places an axle in one half and sets it with a mallet. The next worker places the other half on top and sets it with a mallet. Then the assembled yo-yo is placed in an air-driven press. The press has a tray in the shape of the yo-yo and a metal plate that fits between the yo-yo halves. The metal plate has a groove cut out to fit around the axle and is the exact thickness of the manufacturer's specified gap between the halves. When the press fires, it both permanently secures the two halves to the axle and sets the correct gap between the halves.

7 The finished yo-yo is now placed on a table with a spool of string and a cutter at one end and an upright, with several notches in it at the other. The distance between the cutter and the upright is exactly the length of the manufacturer's desired length of string for each yo-yo. The string is pulled down from the spool through the notch in the upright, where it is looped around the axle of the yo-yo. Then it is pulled back to the cutter, cut, and tied. The yo-yo is then allowed to hang from its string, held by the notch in the upright, to twist onto the axle.

8 The strung yo-yo moves down the line to a machine that resembles a miniature turntable. A worker places the yo-yo on the wheel, holds it down with one hand, holds tension in the string with the other hand, and with a foot-operated switch, spins the string onto the yo-yo.

Packaging

9 The most common packaging for a yo-yo is called a skin card. The skin card is the cardboard sheet with the product attached to it with what appears to be melted plastic wrap you see hanging on store displays. To package a yo-yo this way, a sheet of several cards is slid under a piece of wood with holes in it corresponding to the cards below. A yo-yo is dropped through each hole onto the card. This assures correct placement on the card. The sheet then moves on rollers to a machine that feeds plastic film over the entire sheet of cards and yo-yos. The film is made from a special plastic that conforms to any shape and shrinks when heated but does not melt. The machine heats the plastic, which tightly seals the yo-yos to the cards. The machine then cuts the sheet into individual cards and rolls them into a basket where they are collected and boxed for shipping.

The Future

The foremost modern advancement to yo-yo design was Pedro Flores' creation of the slip string. Some would argue that the advent of durable metal axles was an advancement. Others say that only wooden axles offer the proper friction to make a truly playable yo-yo. San Francisco yo-yo entrepreneur Tom Kuhn has invented a system that addresses both durability and spin. His yo-yos have a replaceable wooden sleeve that fits over a steel axle. His latest creation replaces the fixed axle altogether. It uses the same sealed bearing on which computer disk drives spin and is said to sleep 10 times longer than a conventional axle. But the basic design of yo-yos is the same as it has always been. And barring some radical change in the laws of physics, that design will never change.

Where to Learn More

Books

Cassidy, John. *The Klutz Yo-yo Book*. Klutz Press, 1987.

Malko, George. *The One and Only Yo-yo Book*. Avon Books, 1978.

Zeiger, Helane. *World on a String*. TK Yo-yos, Ltd., 1989.

Periodicals

Crump, Stuart, ed. *Yo-yo Times*. Creative Communications, Inc.

Kowalick, Vince. "Yo-yo Entrepreneur Had to Pull Some Strings." *Los Angeles Times,* May 23, 1994, pg. 5.

Other

"SoCool...Sonoma County ONLINE...Just Say Yo." http://www.socool.com/socool (1/29/97).

"The American Yo-yo Association Home Page." http://www.pd.net/yoyo (1/29/97).

—*Michael Cavette*

Index

Bold-faced terms indicate main entries in this volume. Entries from past volumes are listed with their volume and page number. For example, an entry on page 46 of the second volume appears as 2:46.